Intracellular
calcium regulation

Intracellular calcium regulation

edited by
H. Bader, K. Gietzen, J. Rosenthal,
R. Rüdel and H.U. Wolf

Proceedings of the International Symposium
'Intracellular Calcium Regulation'
Ulm/Neu-Ulm (West Germany), August 6–8, 1984

Manchester University Press

Copyright © Manchester University Press 1986.

Whilst copyright in the volume as a whole is vested in Manchester University Press, copyright in the individual chapters belongs to their respective authors, and no chapter may be reproduced in whole or in part without the express permission in writing of both author and publishers.

Published by
Manchester University Press
Oxford Road, Manchester M13 9PL, UK
27 South Main Street, Wolfeboro, NH 03894-2069, USA

British Library Cataloguing in Publication Data

Intracellular calcium regulation.
  1. Cell metabolism   2. Cellular control mechanisms   2. Calcium—Physiological effect
  I. Bader, H.
  574.87'61   QH634.5

Library of Congress cataloging in publication data applied for

ISBN 0-7190-1835-8 hardback

Printed and bound in Great Britain by
Biddles Ltd, Guildford and King's Lynn

CONTENTS

Preface — page ix

Part I Calcium detection

1 Determination of intracellular [Ca^{2+}] — 1
   John R. Blinks

Part II Calcium transport

2 Single calcium channels in isolated cardiac cells — 15
   Wolfgang Trautwein and Dieter Pelzer

3 Sodium–calcium exchange — 35
   P.F. Baker and T.J.A. Allen

4 The plasma membrane calcium pump — 47
   H.J. Schatzmann

5 Ca^{2+}-sequestration by the endoplasmic reticulum — 57
   Christian Petzelt

6 The regulation of mitochondrial calcium — 67
   M. Crompton, T.P. Goldstone and I. Al-Nasser

7 Calcium release from sarcoplasmic reticulum *in vitro* — 79
   N. Ikemoto, S. Danko, B. Antoniu, and D.H. Kim

Part III Calcium binding

8 Metal-chelating properties of organic compounds with biological significance — 95
   H.U. Wolf

9 Ca^{2+}-induced conformational changes of membranes — 105
   E. Sackmann

10 Calcium binding proteins — 121
   Dennis F. Michiel and Jerry H. Wang

Part IV Calcium activation

11 The calcium messenger system   139
Howard Rasmussen, Itaru Kojima, Kumiko Kojima,
Walter Zawalich, John Forder, Paula Barrett,
David Kreutter, William Apfeldorf, Suzanna Park,
Ann Caldwell, Dominique Delbeke and Priscilla Dannies

12 Intracellular mechanisms in the regulation of calcium metabolism by secretagogues in exocrine glands   151
I. Schulz, H. Streb, K. Imamura, E. Bayerdörffer, T. Kimura

13 Modulation of human platelet phospholipase A_2 activity   165
Leslie R. Ballou and Wai Yiu Cheung

14 Control of the initiation of chromosome replication by Ca^{2+}, Ca^{2+}·calmodulin and cAMP; an overview   179
J.F. Whitfield, M. Sikorska, D. Bossi, A.L. Boynton,
J.P. Durkin, L. Kleine, R.H. Rixon, and T. Youdale

15 Phosphoinositides and calcium mobilizing receptors   203
James W. Putney, Jr.

16 Calcium activation in skeletal muscle   213
G. Cecchi, P.J. Griffiths, J.R. Lopez,
S. Taylor and L.A. Wanek

17 Calcium and excitation-contraction coupling in smooth muscle   227
R. Casteels, G. Droogmans, F. Wuytack
and L. Raeymaekers

18 Contribution of calcium current to contraction in heart muscle   235
Martin Morad and Lars Cleemann

19 The role of calcium and calmodulin in higher plants   247
Dieter Marmé, Elisabeth Andrejauskas
and Rainer Hertel

Part V Calcium and pathophysiology

20 Calcium antagonists and atherosclerosis. Suppression of plaque formation and reversal of established lesions: an overview   253
Dieter M. Kramsch and Anita J. Aspen

21 Calcium - its presumptive role in hypertension   283
J. Rosenthal

22 Intracellular calcium and cardiac arrhythmias   291
U. Borchard, D. Hafner, B. Henning and C. Hirth

23 Ca^{2+}-related regulation of cells from bone marrow: overview including polycythaemia vera and other blood diseases   311
Ole Scharff and Birthe Foder

24 The state of calcium in sickle cell anaemia red cells 327
Robert M. Bookchin, Olga E. Ortiz, Austin Hockaday,
Maria Isabel Sepulveda and Virgilio L. Lew

25 Muscle calcium accumulation in muscular dystrophy 335
Genaro M.A. Palmieri, Tulio E. Bertorini,
Syamal K. Bhattacharya, David F. Nutting
and Abbie B. Hinton

26 Malignant hyperthermia 349
G.A. Gronert

27 Ca^{2+} and thyroid hormone action 355
Cornelis van Hardeveld and Torben Clausen

28 Calcium requirement for insulin secretion 367
V. Maier

Part VI Therapeutical implication

29 Drugs and muscle contraction 379
Marcus C. Schaub and Peter G. Waser

30 Calcium entry and calcium entry blockade:
therapeutic implications 393
T. Godfraind and R.C. Miller

31 Pharmacological regulation of the activity of calmodulin 405
Klaus Gietzen

32 Free intracellular magnesium in striated muscle 425
R. Weingart

33 Cardiac glycosides and intracellular calcium in mammalian cardiac
muscle 433
J. Daut

34 Mechanism of the cardiotonic and smooth muscle-relaxing effects of
methylxanthines 447
Hasso Scholz, Michael Böhm, Wilfried Meyer
and Wilhelm Schmitz

35 Molecular mechanism of antiarrhythmic action of diphenylhydantoin
(DPH), its effect on Ca^{2+}-ATPase activity 461
H.M. Rhee and S. Dutta

Index 467

PREFACE

In organizing the symposium 'Intracellular Calcium Regulation', held in Ulm/Neu-Ulm on 6–8 August 1984, it was our aim to obtain more insight into the immense amount of knowledge that has accumulated in recent years on the function and regulatory role of intracellular Ca^{2+}, and to discuss its bearing on human physiology and pathology. For this reason we invited experts from different fields to give us an overview on their specific subjects. We tried to cover as many aspects of intracellular Ca^{2+} regulation as possible in the time available, and the proceedings of this symposium are reported in this volume.

Calcium detection

The first chapter of this volume discusses the determination of intracellular free Ca^{2+}. Although the total amount of intracellular Ca^{2+}, expressed as the corresponding concentration, is somewhere near that of extracellular Ca^{2+}, the actual free intracellular $[Ca^{2+}]$ is about 10^{-7} M and therefore 10000 times lower than the extracellular $[Ca^{2+}]$. There are now several methods available to measure this low intracellular $[Ca^{2+}]$, and its steady state and dynamic changes. A knowledge of the real intracellular free Ca^{2+} is of fundamental importance for the understanding of the function of intracellular Ca^{2+} and, as we will see, for the function of the cell too.

Calcium transport

The next chapters cover the wide field of Ca^{2+} transport into and out of the cell. There are Ca^{2+} channels in cell membranes, by which Ca^{2+} can enter the cell. However, these channels are not just passive pores but are specialized instruments, which allow Ca^{2+} to cross the membrane at prescribed times to trigger events inside the cell. A second way to cross the cell membrane is by means of a Na^+–Ca^{2+} exchange mechanism, by which Ca^{2+} transport is linked to the transmembrane Na^+ gradient and the Na^+ pump. To maintain a 10000-fold Ca^{2+} gradient, the cell membrane contains a Ca^{2+} pump which continuously drives Ca^{2+} out of the cell. Similar Ca^{2+} transport mechanisms are located in the membranes of cell organelles such as the endoplasmic and sarcoplasmic reticula and the mitochondria. The mechanisms transport Ca^{2+} into these organelles for storage. From there, Ca^{2+} can be released on command to set forces in action, for instance muscle contraction.

Calcium binding

One of the main tasks of intracellular Ca^{2+} is its role as second messenger. Ca^{2+} is especially suited for this task because, among the divalent cations, it has specific binding properties with lipids and proteins. There are particular proteins in the cell, whose duty is to mediate many of the Ca^{2+} effects. Among these low molecular weight proteins are troponin C, the longest known, and calmodulin, the most ubiquitous. They have in common a homologous amino acid sequence, which forms the Ca^{2+}-binding domains. After complexing with Ca^{2+}, these proteins are able to activate many cell functions.

Calcium activation

A large part of this volume is devoted to Ca^{2+}-induced activation. Nearly every process in the cell is in one way or another regulated or controlled by Ca^{2+}, even Ca^{2+} transport itself. This becomes most evident in its nearly universal role as messenger in stimulus–response coupling. Some of the processes directly governed by Ca^{2+} are metabolism, exocytosis, cyclic nucleotide turnover, arachidonic acid cascade, chromosome replication, muscle contraction, and so on. The activation by Ca^{2+} is an important factor in all cells, whether they are of animal or plant origin. On the other hand, intracellular Ca^{2+} itself is regulated by specific mechanisms, for instance the phosphoinositide turnover.

Pathophysiology

Since Ca^{2+} plays a key role in the activity of the cell, it is not surprising that Ca^{2+} also participates in pathological conditions. Among these are not only such obvious Ca^{2+} disorders as diseases of the skeleton, but also many other ailments such as atherosclerosis, hypertension, cardiac arrhythmia, haematological problems, muscular diseases, hormonal disorders, mucoviscidosis, and many more. The role of Ca^{2+} in these diseases is just beginning to be understood.

Therapeutical implications

Some Ca^{2+}-induced pathological conditions have been found with the help of substances that interfere with Ca^{2+} movement or Ca^{2+} activation, namely the Ca^{2+}-entry blockers and the calmodulin antagonists. These substances are priceless tools for the investigation of the actions of Ca^{2+} on the one hand, and they can be used as therapeutical agents to counteract the role of Ca^{2+} in various diseases on the other hand. The latter is especially true for the Ca^{2+}-entry blockers. After the discovery of their mechanism of action, the blockade of the slow Ca^{2+} channel, they were introduced in the treatment of those diseases in which Ca^{2+} is involved. With the calmodulin antagonists it was just the other way round. Many of them were already established drugs before their calmodulin antagonistic properties were discovered. It will be the task of the future to determine how far their calmodulin antagonistic properties are involved in therapeutical actions or adverse reactions. There are still other substances which have been used as drugs for a long time, like the cardiac glycosides and the methyl xanthines. They have no direct effect on intracellular Ca^{2+}, but their actions are mediated by Ca^{2+}.

Conclusion

I am sure we are nowhere near the end of the Ca^{2+} story. However, I hope the contributions to the symposium have helped us, by enabling us to learn from basic and clinical research findings, to better understand the role of Ca^{2+} in health and disease. This understanding may provide ideas for new and better ways to prevent and cure ailments stemming from irregularities of intracellular Ca^{2+}.

Hermann Bader

DETERMINATION OF INTRACELLULAR [Ca^{2+}]

J. R. Blinks

Department of Pharmacology, Mayo Medical School, Rochester, Minnesota 55905, USA

Abstract

Five methods are currently available for the detection of changes of [Ca^{2+}] in living cells: 1) Ca^{2+}-selective microelectrodes, 2) metallochromic dyes, 3) fluorescent indicators, 4) Ca^{2+}-regulated photoproteins, and 5) nuclear magnetic resonance of ^{19}F-labelled chelators. Ca^{2+}-selective microelectrodes have the advantages that they can be calibrated directly and used to detect a very wide range of [Ca^{2+}]. Their overwhelming disadvantage is their slow speed of response, which precludes their use in the detection of many biological Ca^{2+} transients. Most metallochromic dyes respond rapidly, but their useful range is narrower, absolute calibration of signals is often difficult, and motion artifacts often present serious problems. Quin2 is a fluorescent indicator developed to measure cytoplasmic [Ca^{2+}]. It has the advantage of being easy to introduce into populations of some kinds of cells, but it saturates readily and tends to buffer changes of intracellular [Ca^{2+}]. Chlortetracycline gives information of a very different sort. It penetrates membranes readily, and its fluorescent Ca^{2+}-complex tends to associate with the surfaces of membranes bordering compartments of high [Ca^{2+}], such as the sarcoplasmic reticulum or mitochondria. The calcium-regulated photoproteins have the primary advantages of high sensitivity and relative freedom from motion artifacts. They can be used in conjunction with image intensification to localize Ca^{2+} within cells. However, their nonlinear responses to changes in [Ca^{2+}] present problems of interpretation, they are permanently discharged by exposure to high [Ca^{2+}], and they are sometimes difficult to introduce into small cells. The nuclear magnetic resonance technique has the primary advantages that the indicator is easy to introduce into cells, and that signals can be calibrated in absolute terms without damaging the cells. Its prime disadvantages are the expense of the equipment required, the long signal acquisition time, which essentially limits application of the technique to steady-state measurements, and the fact that the method can be used only with large populations of cells.

1.1. Introduction

Five general types of methods are currently available for the determination of ionised calcium concentrations in living cells: they are (1) Ca^{2+}-selective microelectrodes, (2) metallochromic dyes, (3) fluorescent indicators, (4) Ca^{2+}-activated photoproteins, and (5) nuclear magnetic resonance (NMR) of ^{19}F-labelled chelators. Each of these methods has its own set of advantages and disadvantages, which must be considered in choosing the most appropriate technique for a particular application. The properties of the first four groups of indicators have been discussed in great detail in a recent review from this laboratory (Blinks et al., 1982). The NMR method is very new, having been described in only a single short publication (Smith et al., 1983) at the time of writing. For these reasons, the various methods will be outlined here rather briefly, with emphasis on those features that are particularly pertinent to the determination of intracellular $[Ca^{2+}]$.

1.2. Ca^{2+}-selective microelectrodes

There seems now to be fairly general agreement that silanised glass micropipettes whose tips have been filled with a sensing mixture containing the neutral carrier (ETH 1001) developed in Simon's laboratory (Simon et al., 1978) seem to give the most satisfactory results. Debate continues as to whether the 'membrane' (actually a column of fluid) ought to be stabilised by adding polyvinylchloride (PVC), or whether the tips of the pipettes need to be bevelled. In any case, Ca^{2+} electrodes with tip diameters small enough to allow apparently non-damaging impalement of cells of ordinary size are now in use in a steadily increasing number of laboratories (for reviews see Blinks et al., 1982; Lee, 1981; Thomas, 1982; Tsien, 1983). Ca^{2+}-selective electrodes have as their main advantages the facts that they are sensitive to a wide range of $[Ca^{2+}]$ down to and including the concentrations encountered in most cells under resting conditions, and that they can be calibrated directly by dipping their tips into a series of solutions of known $[Ca^{2+}]$. Their principal disadvantage is that they usually respond far too slowly to permit their use in measuring the Ca^{2+} transients associated with the activity of excitable cells. Anecdotal reports of occasional electrodes with unusually rapid responses surface from time to time, so there seems reason to hope that methods can be developed to greatly improve the speed of the electrodes on a consistent basis. Whether Ca^{2+}-electrodes will ever be fast enough to track the Ca^{2+} transients of nerve and muscle is not clear, but they are generally far from it at present. Additional disadvantages of Ca^{2+}-selective microelectrodes lie in their instability (calibrations often change during the course of an experiment, and must be checked frequently), in the technical difficulty of fabricating and using them, and in the fact that accurate measurements are absolutely dependent on having clean, long-lasting impalements with two microelectrodes inserted into the cell within much less than a space constant of each other (the transmembrane potential must be subtracted from the potential recorded by the Ca^{2+}-electrode in order to obtain the Ca^{2+}-potential).

1.3. Metallochromic dyes

These substances change colour on binding Ca^{2+}, and therefore their use as intracellular Ca^{2+} indicators requires that the light absorption of the cell be recorded essentially as that of a cuvette full of solution would be in a spectrophotometer. Absorbance is usually measured at two or more wavelengths, including one at which Ca^{2+} binding has a major influence on the absorbance of the dye, and one at which it has very little influence. If the latter measurement is made at a wavelength where the absorbance of the dye is appreciable, it serves as a means of estimating the concentration of dye within the cell. If not, it allows the detection of non-specific changes in absorbance, which are particularly liable to result from motion of the cell within the beam of the microspectrophotometer. The problem of motion artifacts is obviously most serious in, but is by no means limited to studies on muscle cells. To some extent the Ca^{2+}-sensitive signal can be corrected (approximately) for changes in non-specific absorbance by recording it differentially with respect to the absorbance signal at the Ca^{2+}-insensitive wavelength. However, in contracting muscle the motion artifacts are usually too large to be corrected for, and most workers have found it necessary to prevent contraction by one means or another (for example, by stretching the muscle fibre beyond the point of filament overlap, by substituting D_2O for H_2O, or by applying hypertonic solutions) in order to study the Ca^{2+} transients of muscle cells. In only one study has a metallochromic dye been used successfully in a normally contracting muscle cell: Dubyak and Scarpa (1982) managed to immobilise segments of giant barnacle muscle fibres thoroughly enough to measure dye absorbance signals and isometric tension development simultaneously. This seems not to be possible in cardiac muscle or in skeletal muscle of ordinary size; at any rate no one has yet succeeded in doing it. The problem of motion artifacts is probably the most serious disadvantage in the use of the metallochromic dyes in muscle cells: if one is interested in studying the relation of intracellular Ca^{2+} to contraction, it is more than a little inhibiting not to be able to study muscles that can contract.

There are reports so far of attempts to use four different metallochromic dyes as cytoplasmic Ca^{2+} indicators in living cells: murexide, arsenazo III, antipyrylazo III, and dichlorophosphonazo III. Of these, only arsenazo III and antipyrylazo III can be regarded as having given satisfactory results. Murexide was used in the first report of the detection of Ca^{2+}-transients by any method (Jöbsis and O'Connor, 1966) but I have found only one other report of its use in living cells (Kometani and Sugi, 1978). Both reports were preliminary in nature and gave few details; neither has been followed up with additional results. Rumours abound of unsuccessful attempts to use murexide in various kinds of cells, and one has to conclude that the dye must be very difficult to use in practice. One reason may be that its affinity for Ca^{2+} is very much lower than that of the other metallochromic dyes. The affinity is so low ($K_D \sim 3 \times 10^{-3}$ M at pH 7; Ohnishi, 1978) that no more than about 1% of the dye could ever be complexed under physiological conditions, even at the height of a tetanic muscle contraction. In resting cells the percentage saturation would be two orders of magnitude lower.

In a painstaking comparison of the properties of arsenazo III, antipyrylazo III, and dichlorophosphonazo III, Baylor, Chandler and Marshall (1982a) came to the conclusion that dichlorophosphonazo III was not well suited for measuring Ca^{2+} transients in muscle because in most circumstances it was not possible to separate unambiguously the Ca^{2+} component of the absorbance change from those due to other factors (a dichroic signal and a late signal possibly due to changes in $[Mg^{2+}]$ or pH). Of the remaining two dyes, arsenazo III has seen the more widespread use, probably because it was introduced into biology before antipyrylazo III. However, it now seems clear that antipyrylazo III has a number of real advantages over arsenazo III. These are (1) that it responds more quickly to sudden changes in $[Ca^{2+}]$, (2) that the shape of its absorption spectrum is the same *in vitro* and *in vivo*, and (3) that its kinetic behaviour is consistent with only one Ca:dye stoichiometry (Baylor, Quinta-Ferreira and Hui, 1983). From a practical standpoint, the first of these differences is the most significant: arsenazo III has an 'off' rate constant of only about 85 s^{-1} at 20°C (Ogawa, Harafuji and Kurebayashi, 1980), which means that the dye cannot accurately track Ca^{2+}-transients as fast as those encountered in many types of excitable cells. (See review by Blinks et al., (1982), for a detailed discussion of the respective roles of the 'on' and 'off' rate constants in determining the ability of an indicator to track rapid Ca^{2+}-transients under various experimental conditions.) The 'on' and 'off' rate constants for antipyrylazo III are about 2.6×10^7 and 5.3×10^3 s^{-1} (Scarpa et al., 1978), which should not prove limiting in any foreseeable biological applications. In a head-to-head comparison of arsenazo III and antipyrylazo III in frog twitch muscle fibres, Baylor, Quinta-Ferreira and Hui (1983) showed that the difference in the reaction speeds of the two dyes was significant and plainly apparent.

Many of the metallochromic dyes have properties that may prove troublesome in certain applications. For example, arsenazo III has been shown to be reduced by rat liver microsomes to an azo-anion radical, the reoxidation of which is accompanied by the formation of active oxygen radicals (Docampo, Moreno and Mason, 1983). Murexide and antipyrylazo III undergo similar reactions, though to a lesser extent. Beeler, Schibeci and Martonosi (1980) have reported that for arsenazo III concentrations of $1-2 \times 10^{-3}$ M in skeletal muscle cells, 80–90% of the indicator is bound to cell proteins. The Ca^{2+}-affinity of the protein-bound dye was less than that of the free indicator. The binding of dyes to components of the myofibrils may well be responsible for the dichroic signals observed with arsenazo III and dichlorophosphonazo III (Baylor, Chandler and Marshall, 1982a and 1982b). Arsenazo III in concentrations as low as 5×10^{-5}M has been reported to inhibit the uptake of Ca^{2+} by the sarcoplasmic reticulum of frog skeletal muscle (Ogawa, Harafuji and Kurebayashi, 1980). The metallochromic dyes are all polar substances that do not cross cell membranes readily. (If they did, their utility as cytoplasmic Ca^{2+} indicators would be seriously compromised.) This means that for practical purposes they must be microinjected, iontophoresed, or otherwise assisted in their passage across the sarcolemma. Several reports suggest that murexide may penetrate cell membranes under some conditions. (Blayney et al., 1977; Jöbsis & O'Connor, 1966; Kometani & Sugi, 1978).

1.4. Fluorescent Ca^{2+}-indicators

Although the prospects for new and better indicators in this class seem promising, there is at present only one 'operational' fluorescent cytoplasmic Ca^{2+}-indicator: quin2. That compound was developed by R.Y. Tsien (Tsien, 1980, 1983; Tsien, Pozzan and Rink, 1982) in a systematic attempt to create a chelator that would (1) fluoresce, (2) change its fluorescence in a readily measurable way on binding Ca^{2+}, (3) be selective for Ca^{2+} over Mg^{2+}, and (4) be insensitive to changes in pH. In addition, Tsien wanted a compound that could be introduced into populations of cells and trapped in the cytoplasm by a technique that did not require microinjection or damage to the surface membrane. Quin2 came a long way toward meeting those specifications, though it is not ideal in all respects. Perhaps the most attractive feature of quin2 is the ease with which it can be introduced into many kinds of cells. To achieve this end, Tsien made use of the facts that the normally lipid-insoluble chelator can be made uncharged and lipid-soluble by esterifying its four carboxyl groups, and that many cells contain esterases capable of cleaving the ester bond (Tsien, 1981, Tsien, Pozzan and Rink, 1982). He found that the acetoxymethyl ester of quin2 readily penetrated cell membranes, and was hydrolysed by esterases found in the cytoplasm of many kinds of cells, so that these cells could be readily induced to accumulate millimolar concentrations of the original membrane-impermeant chelator in their cytoplasm. Unfortunately, millimolar concentrations of quin2 were needed in order to get useful fluorescence signals, and such concentrations of the dye have a number of undesirable effects. First, as might be predicted from the fact that quin2 has a high affinity for Ca^{2+} [$K_D \simeq 10^{-7}$M (Tsien, Pozzan and Rink, 1982)] and must be used in high concentrations, the indicator serves as a major Ca^{2+} sink and severely damps any efforts the cell might make toward mounting a Ca^{2+}-transient (Tsien, 1983; Tsien, Pozzan and Rink, 1982). Second, the enzymatic hydrolysis of the acetoxymethylester in the cytoplasm releases formaldehyde in potentially toxic amounts (Tsien, 1981, 1983). Third, high concentrations of quin2 have been found in some cells to produce effects resembling those of elevations of intracellular [Ca^{2+}] (Hesketh et al., 1983). Whether this reflects an alteration of cellular mechanisms for the regulation of intracellular [Ca^{2+}] or some other effect of the substance is not yet clear. In any case, it represents a worrying feature of the indicator. A fourth disadvantage of quin2 is that its Ca^{2+}-affinity is so high that it approaches saturation within the range of [Ca^{2+}] likely to be encountered during activity in some kinds of cells. Still another is that calibration of the fluorescence signals depends on determining the signals from the quin2 contained in the cells under conditions in which the dye is (1) free and (2) fully bound to Ca^{2+}. This requires that the cells be lysed or that ionophores be used to change the ionic content of the cell without loss of dye. For practical purposes the use of lysis is limited to well-stirred suspensions of cells in cuvettes, since the dye would diffuse away from the usual sort of superfused physiological preparations. Although frog skeletal muscle cells accumulate quin2 too slowly to make the acetoxymethylester loading technique useful in that tissue (Tsien, 1983) the method does seem to work satisfactorily with suspensions of rat ventricular myocytes (Murphy, Jacob and Lieberman, 1984; Sheu, Sharma and Banerjee, 1984). Furthermore, it has recently been demonstrated that it is possible to record Ca^{2+}-related fluorescence signals from individual thymocytes and basophils loaded with quin2 (Rogers et al., 1983). The signals were weak, and it is not yet clear whether quin2 will ever permit the localisation of Ca^{2+} within individual cells in the way that the photoproteins do. Many of the disadvantages of quin2 just discussed would be reduced if the extinction coefficient and fluorescence quantum yield of the compound could be increased, and if the emission spectrum could be red-shifted to a range of wavelengths where the intrinsic fluorescence of cells did not provide such a high background signal (Tsien, 1983). This is all the more desirable because the intensity of the background fluorescence is not constant, but varies with the redox state of pyridine nucleotides in the cell (see, for example, Chance et al., 1962; Charest et al., 1983; Murphy et al., 1980). Since calibration of quin2 signals requires either that the cell be lysed or that its function be destroyed by the application of an ionophore, a fluorescent Ca^{2+} indicator whose free and Ca^{2+}-bound forms fluoresced at different wavelengths would be a great improvement, because one could determine the relative proportions of the two forms in the cytoplasm without destroying the cell (Tsien, 1983). Tsien regards quin2 only as a step on the way towards the development of a fully satisfactory fluorescent Ca^{2+} indicator. It is to be hoped that more satisfactory analogues will be available before long.

A question which applies to all fluorescent Ca^{2+} indicators is the extent to which the Ca^{2+} signals will be subject to distortion by movement of the preparation. Ashley, Castell and Harvey (1983) have reported using quin2 in giant barnacle muscle fibres without evidence of motion artifacts. High concentrations of quin2 were used, however, and it is not clear that the fibres in question actually moved when stimulated. Furthermore, it should be kept in mind that the barnacle muscle is the only muscle in which the problem of motion artifacts with metallochromic dye signals has so far been overcome. It remains to be seen how serious the problem will be in cells of ordinary size. My guess, and it is only a guess, is that the problem will be intermediate in severity between that encountered with the photoproteins (minor) and that encountered with the metallochromic dyes (major).

In excitable tissues that can be driven electrically to give repetitive responses, signal averaging might make it possible to use quin2 at concentrations low enough so that the calcium-buffering effect of the dye would not be a serious problem. Under such circumstances a new question arises, however, and that concerns the response time of indicators such as quin2 which are chemically very closely related to EGTA. The rate constant for the dissociation of the Ca—EGTA complex is very low (approximately 0.8 s^{-1}) (Smith et al., 1977). I have not been able to find any published information on the kinetic properties of quin2, but from preliminary stopped-flow studies on the compound, Dr Ullrich Quast (personal communication) estimates the 'off' rate constant to be about 80 s^{-1} at $37\,°C$. (The 'on' rate constant was too fast to measure with the stopped-flow technique.)

The topic of fluorescent Ca^{2+} indicators cannot be left without brief mention of chlortetracycline. Although useful for some purposes, this compound cannot be regarded as comparable to the other indicators under discussion because it is not an indicator of cytoplasmic $[Ca^{2+}]$ (for further discussion see Blinks et al., 1982). Chlortetracycline binds Ca^{2+} (and Mg^{2+}), and increases its fluorescence intensity when it does so. The compound also penetrates cell membranes readily, and diffuses into (and out of) cells quite freely. The calcium—chlortetracycline complex tends to associate with membranes, and thus is found in association with membranes that border compartments of high $[Ca^{2+}]$, such as the sarcoplasmic reticulum and (under some circumstances) mitochondria. Though it does not provide a signal related in any simple way to cytoplasmic $[Ca^{2+}]$, chlortetracycline has proved useful in studies of the ways in which cells and subcellular organelles store and release calcium.

1.5. Ca^{2+}-activated photoproteins

These substances are self-contained bioluminescent systems extracted from a variety of luminescent marine organisms (mostly coelenterates) in which luminescence is controlled by changes in intracellular $[Ca^{2+}]$. In the photocytes of the parent organisms the Ca^{2+}-activated photoproteins serve as Ca^{2+}-regulated effector proteins whose function is to emit light. Since the range of concentrations over which cytoplasmic $[Ca^{2+}]$ is regulated in these cells seems to be similar to that in most other types of cells, the photoproteins can serve as indicators of ionised calcium in other types of cells into which they are transferred. They have now been used for this purpose in more than 50 different types of cells (see Blinks et al., 1982, for recent review), and are particularly well suited for the study of Ca^{2+}-transients in muscle because they respond reasonably quickly and are very much less prone to give motion artifacts than any other type of Ca^{2+} indicator.

Superficially, the photoproteins appear to differ radically from most other bioluminescent systems in that light emission results from the interaction of an inorganic ion (Ca^{2+}) with a single macromolecule (the photoprotein) in a reaction that requires no cofactors and is not influenced by the ambient oxygen concentration. This would appear to be a major departure from the classical pattern of bioluminescence in which an enzyme (the luciferase) catalyses the oxidation of a low molecular weight organic substrate (the luciferin) in a reaction requiring the presence of molecular oxygen. The departure is not as radical as might seem, however, since the photoprotein contains a low molecular weight chromophore that undergoes oxidative degradation in the course of the luminescent reaction. The chromophore is tightly, but not covalently bound to the apoprotein. The oxygen required for the oxidation is also tightly bound to the protein, though the nature and site of the binding are not yet known. Because the chromophore is degraded, the photoprotein does not 'turn over' as an enzyme does, and in most circumstances must be treated as if it were consumed in the course of the luminescent reaction.

The photoprotein, then, is a conveniently 'packaged' system containing all of the ingredients required for the bioluminescent reaction. That reaction does proceed slowly in the absence of any cofactors, but it is greatly accelerated in the presence of Ca^{2+} and certain other divalent and trivalent cations (notably the lanthanides). Since Ca^{2+} is the only one of these ions normally contained in physiological systems in concentrations sufficient to influence photoprotein luminescence, it serves as a relatively specific indicator of ionised calcium. The action of Ca^{2+} is inhibited competitively by Mg^{2+}, however, and this means that it is necessary to know the cytoplasmic Mg^{2+} in order to translate light measurements into Ca^{2+} concentrations. The luminescent reaction is also sensitive to the monovalent salt concentration, but operationally this poses much less serious uncertainties than does the Mg^{2+}-sensitivity. As one might assume from the physiological function of the photoproteins, they are sensitive to the range of Ca^{2+} concentrations likely to be found in the cytoplasm of living cells. They are not suitable for measurements of the Ca^{2+} concentrations normally present in the extracellular space because the luminescent reaction saturates at [Ca^{2+}] above about 10^{-4}M, and the photoprotein is consumed very rapidly at concentrations approaching that value.

Though the basic mechanism of luminescence appears to be the same in the Ca^{2+}-activated photoproteins isolated from the various organisms that have been studied in detail so far, significant differences in some of their properties have been observed. Only two photoproteins have been used as Ca^{2+} indicators to date. Aequorin, isolated from jellyfish of the genus *Aequorea*, has been used in the overwhelming majority of studies both on the biochemistry of photoprotein luminescence, and in which photoproteins have been used as intracellular Ca^{2+} indicators. Obelin, from hydroids of the genus *Obelia*, has seen very limited use as a Ca^{2+} indicator, and only a modest amount is known about its properties *in vitro*. Nevertheless, it is clear that obelin differs from aequorin in two significant respects: it is less sensitive to Ca^{2+} than aequorin, and it responds considerably faster to sudden changes in [Ca^{2+}] (Stephenson and Sutherland, 1981). Recently it has become clear that the properties of aequorin can be changed in significant ways by chemical modification of the protein. Shimomura and Shimomura (1982) first reported that the response to Ca^{2+} of certain acylated derivatives of aequorin differed from that of the native protein in significant ways. This general conclusion has been confirmed by experiments carried out in our laboratory, though our results differ from the earlier ones in several respects (see below) (Moore, Harrer and Blinks, 1984). While the photoproteins have the major advantage of being relatively immune to the motion artifacts that plague the metallochromic dyes, they have several disadvantages as well. Perhaps the most significant of these is the limited speed with which they respond to rapid changes in $[Ca^{2+}]$. Moore (1984) has recently found that the luminescent response to rapid changes of $[Ca^{2+}]$ is slower for aequorin that has been preincubated with physiological concentrations of magnesium ions than that for aequorin that has not. Since previous estimates of the distortions imposed by the reaction kinetics of aequorin on the light signals recorded during rapid Ca^{2+} transients were based on measurements made with aequorin that had not been preincubated with Mg^{2+} (Hastings et al., 1969), the whole question must be re-examined in detail. Moore is currently doing his PhD thesis on this topic. Obelin responds appreciably faster than aequorin (Stephenson and Sutherland, 1981) and for that reason may be preferable for applications in which it is necessary to track very rapid Ca^{2+} transients.

A property of the photoproteins that may be viewed either as an advantage or as a disadvantage is the steepness of the Ca^{2+} concentration–effect curve. The curves for both aequorin and obelin are sigmoid on log–log plots, and have maximum slopes of about 2.5 (Allen, Blinks and Prendergast, 1977; Stephenson and Sutherland, 1981). Over a large range of calcium concentrations the light signal increases in proportion to the $[Ca^{2+}]$ raised to a power of 2 or more. This means that the photoproteins are exceedingly sensitive to small changes in $[Ca^{2+}]$, and that when they are used in conjunction with microscopic image intensification to localise Ca^{2+} within cells there is high contrast between areas of differing $[Ca^{2+}]$. On the other hand, the non-linearity of the relation between $[Ca^{2+}]$ and light intensity complicates the quantitative interpretation of light signals recorded from preparations in which gradients of $[Ca^{2+}]$ may exist, because regions of high $[Ca^{2+}]$ contribute proportionally more to the overall signal than do regions of low $[Ca^{2+}]$. In practice, this means that whenever the possibility of $[Ca^{2+}]$ gradients cannot be ruled out, estimates of $[Ca^{2+}]$ derived from photoprotein signals must be regarded as upper limits on the average $[Ca^{2+}]$ rather than trustworthy figures. Shimomura and Shimomura (1982) first showed that acetylated aequorin had a Ca^{2+} concentration–effect curve that was less steep than that of native aequorin, and proposed that acetylated aequorin might be used in applications where a linear relation between $[Ca^{2+}]$ and luminescence would be desirable. We have confirmed their finding that acetylation reduces the slope of the Ca^{2+} concentration–effect curve for aequorin, but find that the slope of the mid-part of the log–log plot is still very much greater than 1.0 (Moore, Harrer and Blinks, 1984).

1.6. Nuclear magnetic resonance (NMR)

The most recently developed technique for measuring intracellular $[Ca^{2+}]$ involves the use of chelators which have been labelled with ^{19}F, one of the select group of nuclei with half-integral spin numbers that give rise to NMR signals. G. A. Smith and colleagues (1983) prepared a number of symmetrically substituted difluoro derivatives of the chelator BAPTA [1,2bis(o-aminophenoxy) ethane-N,N,N',N'-tetraacetic acid], which had earlier been developed by Tsien (1980) as one of a series of new Ca^{2+} chelators having selectivity against Mg^{2+} and protons. (The fluorescent Ca^{2+} indicator quin2 (see section 2.4 above) is a member of the same series.) Like quin2, BAPTA and its difluoro-derivatives can readily be introduced into many kinds of cells by the method of esterifying the four carboxyl groups that render the molecule charged at physiological pH levels (Tsien, 1981). As was described in section 2.4, the acetoxymethylesters of the BAPTA derivatives are hydrolysed by cytoplasmic esterases, leaving the charged forms of the chelators trapped in the cytoplasm.

Four different difluoro-derivatives (labelled symmetrically in each of the four previously unsubstituted positions of the benzene rings of BAPTA) were prepared and tested *in vitro*. All but one of the compounds (the 6F derivative) gave chemical shifts on binding Ca^{2+} that were clearly distinct from those for other ions likely to be found in living cells. Two of these compounds (the 3F and 5F derivatives) had very similar properties, the equilibrium dissociation constants of their Ca^{2+}-complexes being in the vicinity of pCa 6.2, and the exchange between the free and the Ca-bound forms being slow enough so that separate peaks representing the free and bound forms were discernible in the same NMR spectrum). For these two compounds, the determination of $[Ca^{2+}]$ would depend on estimating the relative areas under the peaks corresponding to the free and bound forms of the chelator. Their Ca^{2+}-affinities were such that they would be useful for estimating $[Ca^{2+}]$ in the range likely to be encountered in most kinds of cells under resting conditions. Physiologically plausible changes in pH or $[Mg^{2+}]$ did not interfere with the resolution of the resonance peaks corresponding to the free and bound forms.

The 4F derivative had a lower affinity for Ca^{2+} (-log $K_D \simeq 5.6$), and an exchange between free and bound forms that was fast enough so that only a single resonance peak was discernible, its position and breadth shifting with the relative amounts of the free and bound forms of the chelator present. The estimation of $[Ca^{2+}]$ with this compound would depend on the measurement of the chemical shift of the single resonance peak, which can be made with considerably greater precision than the estimation of relative areas. However the chelator's relatively low affinity for Ca^{2+} would limit its application to $[Ca^{2+}]$ between about 3×10^{-7} and 3×10^{-5} M, which is just above the range likely to be found in resting cells. The position of the resonance peak for the 4F derivative of BAPTA was virtually uninfluenced by changes in pH or $[Mg^{2+}]$ over a wide range.

Smith *et al.*, (1983) determined the ^{19}F NMR spectrum at 188.3 MHz in a suspension of mouse thymocytes loaded with 5FBAPTA, and on the basis of the areas of the resonance peaks corresponding to the free and Ca^{2+}-bound forms of the indicator, they estimated the cytoplasmic $[Ca^{2+}]$ to be approximately 2.5×10^{-7} M under resting conditions. (Experiments with quin2 in the same cells gave a value of 1.2×10^{-7} M.) Stimulation with concanavalin A (con A) or the administration of a Ca^{2+}-ionophore led to increases in cytoplasmic $[Ca^{2+}]$ as expected. NMR spectra recorded from 4FBAPTA in thymocytes indicated that under resting conditions virtually all of the indicator was in the free form, which is consistent with the other estimates, given the lower Ca^{2+}-affinity of this BAPTA derivative.

The ^{19}F NMR technique takes Tsien's elegant exercises in rational molecular design one step further, and is an intellectually very appealing development. It also seems to give reasonable answers. Drawbacks of the technique are the expense of the equipment required, the fact that relatively large volumes of cells and high concentrations of indicator must be used in order to get an

NMR signal, and the long accumulation time required to acquire the NMR spectrum. The last point makes it quite academic that the equilibration of the BAPTA derivatives with Ca^{2+} is intrinsically rather rapid (Smith et al., 1983). If it takes several minutes to accumulate a useful NMR spectrum, the method cannot be used to follow anything but exceedingly slow changes in cytoplasmic $[Ca^{2+}]$. In frog skeletal muscle, and perhaps other muscles as well, there is the additional complication that the acetoxymethyl ester method of getting tetracarboxylate chelators into the cells seems not to work (Tsien 1983).

1.7. Comparative merits of the various methods

The selection of the most appropriate intracellular Ca^{2+} indicator for a particular application almost always involves compromises between the desirable and the achievable at the present state of the art. Often, it may be helpful to use more than one method in approaching the answer to a particular question. If one is interested in measuring rapid Ca^{2+}-transients, such as those associated with the activity of excitable cells, one's choice is limited at present to the metallochromic dyes and the photoproteins. The metallochromic dyes, particularly antipyrylazo III, have the edge with respect to speed, and are currently the only available means of faithfully tracking the fastest biological Ca^{2+} transients. However, if the cell under examination is subject to motion, the photoproteins come to the fore. Their relative immunity to motion artifacts may more than offset their inferiority in response time, and make them the indicator of choice. The photoproteins usually provide the most satisfactory records of the Ca^{2+} transients in contracting muscle cells. They also lend themselves well to the localisation of Ca^{2+} through image intensification, and they are better able to provide absolute measurements of low $[Ca^{2+}]$, such as those encountered in resting cells, than are the metallochromic dyes in current use. On the other hand, when gradients of $[Ca^{2+}]$ exist in the preparation under examination, the quantitative interpretation of photoprotein signals becomes problematic.

The Ca^{2+}-selective microelectrodes have the advantages of responding to a very wide range of $[Ca^{2+}]$, and can be calibrated in a direct and straightforward manner. However, they are difficult to use, they respond very slowly, and they sometimes change their characteristics during an experiment. Of course they have the property, sometimes an advantage and sometimes not, of sensing $[Ca^{2+}]$ at a single potentially definable spot in the cytoplasm of a single cell, rather than providing some sort of spatial average, as most other indicators do.

Quin2 and the ^{19}F-labelled chelators used for NMR measurements have the principal advantage of being easy to introduce into cells, particularly large populations of small cells. Measurements with these indicators are technically easy, but at present the fluorescence and NMR methods are not capable of following anything but very slow changes in $[Ca^{2+}]$. Furthermore, the NMR method requires access to very expensive equipment that is not likely to be available to most physiologists. Both of these methods are best suited to carrying out measurements of static or slowly changing $[Ca^{2+}]$ in suspensions of cells in cuvettes. They do not interface well with mechanical, electrophysiological, or microscopical measurements, and are, in short, 'biochemists' Ca^{2+}-indicators.'

In several recent studies arsenazo III (Murphy et al., 1980; Murphy and Mandell, 1982; Williamson et al., 1983) and Ca^{2+}-selective electrodes (Simons,

1982) have been used in the extracellular fluid as 'null point indicators' of intracellular $[Ca^{2+}]$. In this approach aliquots of a suspension of cells are incubated in solutions of various Ca^{2+} concentrations, and then lysed with digitonin while the $[Ca^{2+}]$ in the medium is monitored. The solution in which there is no change in $[Ca^{2+}]$ on lysis is assumed to have the same $[Ca^{2+}]$ as the cytoplasm. Although use of this technique requires a number of major assumptions concerning the effects of digitonin on intracellular membranes and the effects of low extracellular $[Ca^{2+}]$ on cytoplasmic $[Ca^{2+}]$, the method does seem to have given plausible numbers where it has been applied so far. Its chief drawbacks would seem to be its inherently destructive nature, the fact that it can be applied only to suspensions of cells because of the need to have multiple identical aliquots of cells, and its unsuitability for the study of anything but the slowest changes in $[Ca^{2+}]$. There is also the potential problem that Ca^{2+}-buffers of sufficient capacity to reliably control the extracellular $[Ca^{2+}]$ may also control the cytoplasmic $[Ca^{2+}]$ when the cell membrane is lysed, thus giving a spurious 'null point' (Tsien,1983). So many better methods are available that for most purposes this approach seems hardly worth considering as a means of monitoring intracellular $[Ca^{2+}]$.

References

Allen, D.G., Blinks, J.R. and Prendergast, F.G. (1977). Aequorin luminescence: relation of light emission to calcium concentration. A calcium-independent component. *Science.* **195**, 996–8.

Ashley, C.C., Castell, L.M. and Harvey, D. (1983). Use of the fluorescent Ca^{2+}-indicator 'quin 2' in single barnacle muscle fibres. *J. Physiol. (Lond.).* **345**, 9P.

Baylor, S.M., Chandler, W.K. and Marshall, M.W. (1982a). Use of metallochromic dyes to measure changes in myoplasmic calcium during activity in frog skeletal muscle fibres. *J. Physiol. (Lond.).* **331**, 139–77.

Baylor, S.M., Chandler, W.K. and Marshall, M.W. (1982b). Dichroic components of Arsenazo III and Dichlorophosphonazo III signals in skeletal muscle fibres. *J. Physiol. (Lond.).* **331**, 179–210.

Baylor, S.M., Quinta-Ferreira, M.E. and Hui, C.S. (1983). Comparison of isotropic calcium signals from intact frog muscle fibers injected with arsenazo III or antipyrylazo III. *Biophys. J.* **44**, 107–12.

Beeler, T.J., Schibeci, A. and Martonosi, A. (1980). The binding of arsenazo III to cell components. *Biochim. Biophys. Acta.* **629**, 317–27.

Blayney, L., Thomas, H., Muir, J. and Henderson, A. (1977). Critical reevaluation of murexide technique in the measurement of calcium transport by cardiac sarcoplasmic reticulum. *Biochim. Biophys. Acta.* **470**, 128–33.

Blinks, J.R., Wier, W.G., Hess, P. and Prendergast, F.G. (1982). Measurement of Ca^{2+} concentrations in living cells. *Prog. Biophys. Molec. Biol.* **40**, 1–114.

Chance, B., Cohen, P., Jöbsis, F. and Schoener, B. (1962). Intracellular oxidation-reduction states *in vivo*. *Science.* **137**, 499–508.

Charest, R., Blackmore, P.F., Berthon, B. and Exton, J.H. (1983). Changes in free cytosolic Ca^{2+} in hepatocytes following α_1-adrenergic stimulation. Studies on quin-2-loaded hepatocytes. *J. Biol. Chem.* **258**, 8769–73.

Docampo, R., Moreno, S.N.J. and Mason, R.P. (1983). Generation of free radical metabolites and superoxide anion by the calcium indicators arsenazo III, antipyrylazo III, and murexide in rat liver microsomes. *J. Biol. Chem.* **258**, 14920–5.

Dubyak, G.R. and Scarpa, A. (1982). Sarcoplasmic Ca transients during the contractile cycle of single barnacle muscle fibres: measurements with arsenazo III-injected fibres. *J. Muscle Res. Cell Motility.* **3**, 87–112.

Hastings, J.W., Mitchell, G., Mattingly, P.H., Blinks, J.R. and van Leeuwen, M. (1969). Response of aequorin bioluminescence to rapid changes in calcium concentration. *Nature (Lond.).* **222**, 1047–50.

Hesketh, T.R., Smith, G.R., Moore, J.P., Taylor, M.V. and Metcalfe, J.C. (1983). Free cytoplasmic calcium concentration and the mitogenic stimulation of lymphocytes. *J. Biol. Chem.* **258**,. 4876–82.

Jöbsis, F.F. and O'Connor, M.J. (1966). Calcium release and reabsorption in the sartorius muscle of the toad. *Biochem. Biophys. Res. Commun.* **25**, 246–52.

Kometani, K. and Sugi, H. (1978). Calcium transients in a molluscan smooth muscle. *Experientia.* **34**, 1469–70.

Lee, C.O. (1981). Ionic activities in cardiac muscle cells and application of ion-selective microelectrodes. *Am. J. Physiol.* **241**, H459–78. (1981).

Moore, E.D.W. (1984). Effects of pre-equilibration with Mg^{2+} on the kinetics of the reaction of aequorin with Ca^{2+}. *J. Gen. Physiol.* **84**, 11a.

Moore, E.D.W., Harrer, G.C. and Blinks, J.R. (1984). Properties of acetylated aequorin relevant to its use as an intracellular Ca^{2+} indicator. *J. Gen. Physiol.* **84**, 11a.

Murphy. E., Coll, K., Rich, T.L. and Williamson, J.R. (1980). Hormonal effects on calcium homeostasis in isolated hepatocytes. *J. Biol. Chem.* **255**, 6600–8.

Murphy, E., Jacob, R., and Lieberman, M. (1984). Measurement of cytosolic free Ca (Ca_f) using quin2 in embryonic chick heart cells. *Fed. Proc.* **43**, 768.

Murphy E. and Mandel, L.J. (1982). Cytosolic free calcium levels in rabbit proximal kidney tubules. *Am. J. Physiol.* **242**, C124–8.

Ogawa, Y., Harafuji, H. and Kurebayashi, N. (1980). Comparison of the characteristics of four metallochromic dyes as potential calcium indicators for biological experiments. *J. Biochem.* **87**, 1293–303.

Ohnishi, S.T. (1978). Characterization of the murexide method: Dual-wavelength spectrophotometry of cations under physiological conditions. *Anal. Biochem.* **85**, 165–79.

Rogers, J., Hesketh, T.R., Smith, G.A., Beaven, M.A., Metcalfe, J.C., Johnson, P. and Garland, P.B. (1983). Intracellular pH and free calcium changes in single cells using quene 1 and quin 2 probes and fluorescence microscopy. *FEBS Lett.* **161**, 21–7.

Scarpa, A., Brinley, F.J., Tiffert, T. and Dubyak, G.R. (1978). Metallochromic indicators of ionized calcium. *Ann. N.Y. Acad. Sci.* **307**, 86–112.

Sheu, S.-S., Sharma, V.K. and Banerjee, S.P. (1984). Measurement of cytosolic free calcium concentration in isolated rat ventricular myocytes with quin2. *Circ. Res.* **55**, 830–4.

Shimomura, O. and Shimomura, A. (1982). EDTA-binding and acylation of the Ca^{2+}-sensitive photoprotein aequorin. *FEBS Lett.* **138**, 201–4.

Simon, W., Ammann, D., Oehme, M. and Morf, W.E. (1978). Calcium-selective electrodes. *Ann. N.Y. Acad. Sci.* **307**, 52–70.

Simons, T.J.B. (1982). A method for estimating free Ca within human red blood cells, with an application to the study of their Ca-dependent K permability. *J. Memb. Biol.* **66**, 235–47.

Smith, G.A., Hesketh, R.T., Metcalfe, J.C., Feeney, J. and Morris, P.G. (1983). Intracellular calcium measurements by ^{19}F NMR of fluorine-labeled chelators. *Proc. Natl. Acad. Sci. USA* **80**, 7178–82.

Smith, P.D., Berger, R.L., Podolsky, R.J. and Czerlinski, G. (1977). Stopped-flow study of the rate of calcium binding by EGTA. *Biophys. J.* **17**, 159a.

Stephenson, D.G. and Sutherland, P.J. (1981). Studies on the luminescent response of the Ca^{2+}-activated photoprotein, obelin. *Biochim. Biophys. Acta.* **378**, 65–75.

Thomas, M.V. (1982). *Techniques in Calcium Research.* Academic Press, New York.

Tsien, R.Y. (1980). New calcium indicators and buffers with high selectivity against magnesium and protons: design, synthesis, and properties of prototype structures. *Biochemistry.* **19**, 2396–404.

Tsien, R.Y. (1981). A non-disruptive technique for loading calcium buffers and indicators into cells. *Nature (Lond.).* **290**, 527–8.

Tsien, R.Y. (1983). Intracellular measurements of ion activities. *Ann. Rev. Biophys. Bioeng.* **12**, 91–116.

Tsien, R.Y., Pozzan, T. and Rink, T.J. (1982). Calcium homeostasis in intact lymphocytes: Cytoplasmic free calcium monitored with a new, intracellularly trapped fluorescent indicator. *J. Cell Biol.* **94**, 325–34.

Williamson, J.R., Williams, R.J., Coll, K.E. and Thomas, A.P. (1983). Cytosolic free Ca^{2+} concentration and intracellular calcium distribution of Ca^{2+}-tolerant isolated heart cells. *J. Biol. Chem.* **258**, 13411–4.

SINGLE CALCIUM CHANNELS IN ISOLATED CARDIAC CELLS

W. Trautwein and D. Pelzer

Department of Physiology 2, University of the Saarland, D-6650 Homburg/Saar, West Germany

Abstract

Currents through single Ca channels in the membrane of isolated guinea-pig ventricular cells were recorded by the patch-clamp technique in the cell-attached recording configuration using barium as charge carrier. Microscopic single Ca channel activity occurred as closely spaced bursts of brief unitary current pulses separated by wider periods of inactivity (millisecond (ms) time-scale). Bursts themselves were grouped into clusters of bursts (10s–100s of ms time-scale). The fast gating behaviour is related to Ca channel activation (that is, the rise of macroscopic current to its peak) and is explained by a sequential model in which two short-lasting shut states precede the open state. Clusters reflect the entrance of the Ca channel into an inactivated state being occupied during the long-lasting shut periods between clusters (10s–100s of ms time-scale). The mean cluster lifetime correlated well with macroscopic inactivation (that is, the decline of macroscopic current during maintained depolarisation). In a series of consecutive depolarisations, some clamp pulses did not elicit channel activity at all. These empty sweeps where the channel cannot open are shown to be related to the steady-state occupancy of the inactivated state. In the context of these data, the modulation of the Ca channel gating properties by D600 and β-adrenergic agents is discussed.

2.1. Introduction

An increase in the cytoplasmic free calcium ion concentration, $[Ca]_i$, within stimulated excitable cells underlies important biological processes such as muscle contraction, the synthesis and release of transmitters and hormones, the regulation of enzyme activities and the control of membrane permeabilities (cf. Reuter, 1983, 1984; Tsien, 1983; Trautwein and Pelzer, 1985a for references). The elevation of $[Ca]_i$ upon electrical or chemical stimulation is the result of (1) Ca influx across the cell membrane, (2) Ca release from intracellular stores, or (3) both mechanisms. This chapter intends to provide an overview of voltage-gated transmembrane Ca entry into cardiac cells upon electrical excitation (as a representative example of gated Ca transport across excitable biological membranes). The mechanism by which

$[Ca]_i$ is returned to that of the resting cardiac cell after stimulation by means of ATP-driven sarcolemmal Ca pumps, Na—Ca exchange or Ca uptake by mitochondria (the latter becoming particularly important when $[Na]_i$ is altered) (Chapman, 1983; Mullins, 1979), or Ca release (Chapman, 1983; Fabiato, 1983), will not be dealt with here. The interested reader is referred to the recent literature cited above.

The dominant mechanism for transmembrane Ca influx into cardiac cells during electrical stimulation (voltage-clamp pulse or action potential) is the Ca ion movement through a membrane channel down the electrochemical gradient for Ca ($[Ca]_o$ about 2mM, $[Ca]_i < 0.05$ μM). These Ca channels play a crucial role in coupling membrane excitation to cellular responses and are a major way of controlling $[Ca]_i$. The term 'Ca channel' is justified by the fact that this Ca-passing pore is highly selective for Ca ions, the respective permeability (P) of Ca over other ion species being P_{Ca}/P_{Na} and $P_{Ca}/P_K >$ 100 in ventricular muscle (Reuter and Scholz, 1977) and $P_{Ca}/P_{Na} \simeq 60-100$ in nodal tissue (Akiyama and Fozzard, 1979) although in the medium outside the channel, Ca ions are vastly outnumbered by other ions. A very recent report (Hess and Tsien, 1984) convincingly demonstrated for the first time that Ca channels are indeed multi-ion, single-file pores, rather than single-ion pores that function by simple mutual exclusion. However, ion-binding sites (energy wells) within the Ca channel seem to bind the preferred ion, that is, Ca, more strongly than other physiological ions, Ca being selected as much by deep-energy wells as by a relatively low-energy barrier ('selectivity filter'). In conclusion it was suggested (Hess and Tsien, 1984) that under physiological conditions, the channel is occupied almost continuously by one or more Ca ions which, by electrostatic repulsion, guard the channel against permeation by other ions.

As in other excitable membranes, cardiac Ca channels are controlled by voltage-dependent gating, that is, their opening and closing kinetics are the result of changes in membrane potential. Recent evidence suggests that a 'voltage sensor' within the membrane, for example a protein group with dipole properties that may be an integral part of an ion channel, reacts to the electric field. Any change in membrane potential will cause a reorientation of the charged 'sensor' within the field and thus a change in the ion flow through the channel (Coraboeuf, 1980; Reuter, 1983). Prior to 1980, information on Ca channels in the heart was obtained from voltage-clamp experiments on small multicellular preparations (cf. McDonald, 1982; Reuter, 1979 for review). Some of the difficulties in the interpretation of voltage-clamp data arose from the complex morphological and electrical features of cardiac tissue and from inadequate and unspecific tools for unequivocal separation of various overlapping ionic current components (Attwell and Cohen, 1977; McDonald, 1982; Reuter, 1979, 1983, 1984; Tsien, 1983; Trautwein and Pelzer, 1985a). Meanwhile, however, Ca-tolerant single cardiac cells from adult mammalian and frog hearts and embryonic heart tissue can be voltage-clamped and internally dialysed and single ion channels can be identified and analysed by the patch-clamp method (Reuter, 1983, 1984; Sakmann and Neher, 1983a; Trautwein and Pelzer, 1985 a and b; Tsien, 1983). Some of the previous results from intact cardiac tissue and some older concepts have been confirmed and important new insight has been gained by the new techniques (Reuter, 1983, 1984; Trautwein and Pelzer, 1985a; Tsien, 1983) which will be discussed below.

2.2. Materials and methods

Ca-tolerant single cells from adult mammalian hearts can be prepared by a number of methods (Trube, 1983) which follow a common sequence of steps: (1) washout of blood and Ca, (2) digestion by enzymes (mostly crude collagenase), (3) mechanical agitation, and (4) final cleaning and filtration. The solutions can be applied to the tissue by different means, but retrograde perfusion via the coronaries has emerged as the superior method both in terms of time and quantity of expensive enzyme (Trube, 1983). For the preparation of cells from large animal hearts, or where the perfusion method has not been chosen, the tissue is usually minced and the pieces then agitated in the enzyme medium (Isenberg and Klöckner, 1982a). Both methods can be combined (Powell, Terrar and Twist, 1980). Recent studies show that quiescent, rod-shaped adult mammalian ventricular myocytes in millimolar $[Ca]_o$ retain a normal energetic state (Dow, Harding and Powell, 1981; Piper et al., 1982) and that suction pipettes seem to be superior to conventional tapered microelectrodes for intracellular recording (Pelzer, Trube and Piper, 1984). Besides a low-ohm connective pathway between pipette and cell interior (see below, p.18), the probability of a shunt pathway for current flow due to membrane damage around the recording electrode is lower for suction pipettes with glass-membrane seal resistances > 10 gigaohms (GΩ) (see below, p.17) compared to about 500 megaohms (MΩ) for conventional microelectrodes (Marty and Neher, 1984). Most of the cells prepared from hearts of different adult mammalian species (guinea-pig, cat, rat, bull) respond to intracellular pulses with action potentials whose configurations resemble those recorded from multicellular preparations excised from the hearts of these animals (Trautwein and Pelzer, 1985a). In the present study, ventricular myocytes from hearts of adult guinea-pigs were allowed to attach at the culture dish bottom of the recording chamber mounted on the stage of an inverted microscope (Nikon, diaphot). Subsequently, the bath was continuously perfused (2-4 ml/min) with saline of the following composition (in mM): NaCl 131, KCl 10.8, $CaCl_2$ 3.6, $MgCl_2$ 1, glucose 10, HEPES 5. The pH was adjusted to 7.3–7.4 and the perfusate was prewarmed to 34–36 °C. Under these conditions, the cells yielded resting potentials of -65±8 mV on intracellular recording.

Recording pipettes were made from thick-walled, hard borosilicate glass (Pyrex, Jencons) on a vertical puller in a two-step process. The pipettes were then coated with insulating varnish (Sylgard, Dow Corning) and lightly fire-polished to give a final tip opening diameter of about 1 μm. Finer details on the construction, geometry and electrical properties of patch-recording pipettes can be found elsewhere (Corey and Stevens, 1983; Sakmann and Neher, 1983b). On mechanical contact of the pipette tip with the cell membrane, the pipette-membrane resistance is in the order of 50–100 MΩ. Upon slight suction (or application of negative pressure to the pipette interior), a small patch of membrane is drawn into the pipette tip, forming the cell-attached configuration for single-channel recording. The area of the membrane patches employed here was about 1.2–2.4 μm^2 (estimated from Fig. 2.8 in Sakmann and Neher, 1983b). This area is always larger than the tip opening area, consistent with the observation that membrane patches are pulled 2-3 μm into the pipette tip (Sakmann and Neher, 1983b). During the formation of the patch, the seal between membrane and pipette increases in resistance by two to three orders of magnitude. At this stage, the omega-shaped membrane protrusion can be destroyed by short pulses of suction or voltage applied to the pipette interior leading to the whole-cell configuration being suitable for the recording of action potentials and/or membrane currents (Hamill et al., 1981; Marty and Neher, 1983). Due to the large size of adult mammalian ventricular myocytes (about 100–140 μm in length, 15–30 μm in width), a second pipette is often used for homogeneous control of membrane potential in whole-cell voltage-clamp experiments (Trautwein and Pelzer, 1985 a and b). In the giga-seal cell-attached configuration, elementary current flow across the membrane patch can be recorded with high resolution under fast voltage-clamp control (Hamill et al., 1981; Sakmann and Neher, 1983a). With the pipette-filling solutions for single Ca channel recording (see below, p.18), however, it is not possible to measure directly the absolute membrane potentials in those cells from which elementary single channel currents are recorded because of the diffusional exchange between pipette and cell interior solutions after disruption of the membrane patch (see below, p.18). Thus, patch resting potentials are assumed to be similar to the resting potential measured in the whole-cell recording configuration in similar cells (see above, p.17). Since a zero potential in the pipette corresponds to the resting potential across the membrane patch, depolarisations are expressed as positive voltage displacements from the pipette zero potential. Voltage pulses and corresponding patch currents were collected on FM tape; subsequent automated analysis of the digitised (sampling rate 5–6.67 kHz, low-pass filtering at 1–3 kHz), leak and capacitance subtracted traces was carried out with procedures similar to those previously described (Cavalié et al., 1983; Sakmann and Neher, 1983a, Trautwein and Pelzer, 1985 a and b) and briefly summarized in the context with the appropriate figures.

For the whole-cell recording configuration of the action potentials and membrane currents (see above, p.17–18), pipettes were filled either with a solution containing (in mM): KCl 140, $MgCl_2$ 10, HEPES 5 (pH 7.15), or one containing glutamic acid 130, KOH 140, $MgCl_2$ 10, EGTA 5, HEPES 10 (pH 7.2). The composition of the pipette-filling solutions should be compatible with an intracellular environment since the solutes inside the pipette exchange rapidly with those of the cell interior (Marty and Neher, 1983). When immersed in the bath solution (see above, p.17), pipette resistances ranged between 3 and 5 MΩ. For elementary Ca channel current recordings in the cell-attached configuration (see above, p.17), pipettes were filled with solution containing (in mM) $BaCl_2$ 90, NaCl 2, KCl 4, TTX 0.02, HEPES 5, pH adjusted to 7.4, pipette resistances being between 5 and 10 MΩ. Ba, besides carrying charge through the Ca channel, abolishes currents through K channels (Sakmann and Trube, 1984); Na and Ba movements through Na channels are prevented by TTX. Thus, we are left with elementary current through a single ion channel species only, that is, Ca channels. In contrast to other ion channels, Ca channels do not function in isolated membrane patches (Cavalié et al., 1983). Thus, all data are derived from cell-attached recordings of single Ca channel activity.

2.3. Results

2.3.1. Time-course of Ca channel activity

Voltage-dependent Ca channels open transiently when activated by a sudden change in membrane potential by a voltage-clamp step (Fig.2.1). This behaviour results in the typical whole-cell Ca current wave form (I_{Ca}) (Fig.2.1Aa) which in the single ventricular myocyte represents the average opening (and closing) behaviour of about 10^4 to 10^5 individual Ca

Calcium channels in isolated cardiac cells

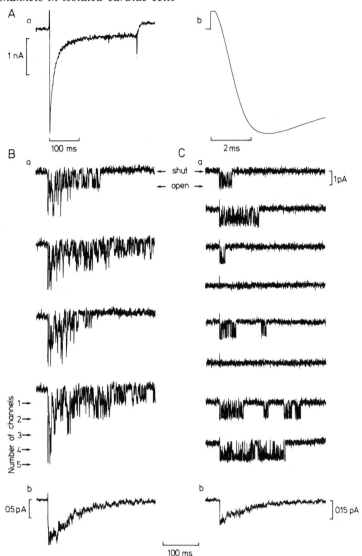

Fig.2.1. (A) Macroscopic net membrane current in response to a depolarising clamp-step (300 ms duration) from -50 (sodium current inactivated; see, for example, Fig.3A in Trautwein and Pelzer, 1985a) to 0 mV at two different time-scales. The current accompanying depolarisation started with a capacitive component which was outward directed (initial upward deflection) and chopped off (see especially part (A)b). The subsequent inward directed current (downward deflection) which peaks within about 3 ms (see especially part (A)b) is the Ca current, I_{Ca} (Trautwein and Pelzer, 1985a; Isenberg and Klöckner, 1982b; Josephson, Sanchez-Chapula and Brown, 1984; Mitchell et al., 1983). During maintained depolarisation, I_{Ca} declines again at a slower rate (see especially part (A)a) and at later times reverses its direction with respect to zero current (indicated by the upper horizontal bar of the current calibration) due to slow activation of the delayed potassium current system (I_K). (B) and (C) Elementary Ca channel currents (downward deflections) (parts (B)a and (C)a) and their ensemble averages (parts (B)b and (C)b) from membrane patches containing several (B) or a single (C) Ca channel. Depolarising voltage-clamp pulses (75 mV amplitude, 300 ms duration) were delivered from the pipette zero potential (= resting potential of the cell); leakage and capacitive current components resulting from steps of the command potential have been digitally subtracted from each record. The approximate number of active Ca channels in multichannel patches (D) was derived from superimposed steps of elementary Ca channel currents of unitary amplitude (bottom panel in (B)a; see Colquhoun and Hawkes, 1983). Single Ca channel traces were low-pass filtered at 2.5 kHz. Average currents were obtained from 150 averaged individual elementary current sweeps filtered at 500 Hz in (B) and (C), respectively. The temperature in parts (A), (B) and (C) was 36°C, and the pulsing rate 0.5 Hz.

channels (Bean, Nowycky and Tsien, 1983; Reuter, 1983, 1984). On single depolarisation of the whole-cell membrane, I_{Ca} rises rapidly to its peak (within about 2–4 ms at potentials close to 0 mV; Fig.2.1Ab). Subsequently, I_{Ca} decays at a 10–20 times slower rate (Fig.2.1Aa), the decline being attributed to an inactivation process as distinct from activation. Both processes are controlled by the membrane potential. Repetitive depolarisation of a small membrane patch resolves the opening and closing behaviour of several individual (Fig.2.1Ba) or even a single Ca channel (Fig.2.1Ca) as judged by the presence (Fig.2.1Ba) or absence (Fig.2.1Ca) of multiple conductance levels of N times the unitary single Ca channel current amplitude (i), where N represents the number of functional channels being present in the membrane patch (see, for example, Fig.2.1Ba). Channel openings and closings occur on a random basis 'as if each channel repeatedly flipped a coin to decide whether or not to open, to remain open, or to close' (Coraboeuf, 1980). According to this concept, a collection of single Ca channel current records during a series of identical voltage-clamp pulses forms a statistical ensemble (Aldrich and Yellen, 1983; Fig.2.1Ba and 2.1Ca show a few examples out of 150–300 single Ca channel current traces, respectively). In all ensembles, two types of current signals are regularly observed during depolarisation: (1) traces with closely spaced groups of brief channel openings and (2) traces with no detectable openings at all (Fig.2.1Ca). The impact of the respective modes of channel activity on our perceptions of Ca channel gating will be dealt with below (pp.23 and 25). Averaging channel behaviour over an entire ensemble (150–300 individual sweeps) describes the average time-course of the appearance and disappearance of single Ca channel activity in the patch, just as a record of whole-cell I_{Ca} reflects the average time-course of activity of the large number of Ca channels in the entire cell membrane (Aldrich and Yellen, 1983). These ensemble averages of individual Ca channel activity from multichannel (Fig.2.1Bb) and single-channel (Fig.2.1Cb) membrane patches share common properties with whole-cell I_{Ca} (Fig.2.1Aa), that is, fast activation to a peak followed by a slow decay. However, high-frequency components of the individual signals are eliminated by ensemble averaging as if the bandwidth of the filter system had been lowered (compare Fig.2.1Ba,b and 2.1Ca,b). Neither the elementary Ca channel current amplitude (see Fig.2.1Ca) nor the average lifetime and distribution of high-frequency brief channel opening and closing events (that is, burst activity, see below, pp.22 and 23) differ significantly at different times during the depolarising clamp pulse (Cavalié et al., 1983; Trautwein and Pelzer, 1985 a and b).

The unitary amplitude of elementary currents indicates that Ca channels have only two distinguishable conductance levels, open and closed (see Fig.2.1Ca). This allows to express the average Ca channel current (I) as

$$I = N \cdot i \cdot p, \qquad (1)$$

where N is the number of channels in the membrane patch, i the elementary current through the individual open channel and p the probability of an individual channel being open (Tsien, 1983); in other words, all open channels contribute to the same current and thus serve as counters, each of which adds a unit to the total whenever a channel is open. This simple expression is primarily substantiated by the powerful method of single-channel recording where one can directly observe the number of channels and the current through the individual open channel (Fig.2.1Ba,Ca). Furthermore, single-channel recordings give information about channel kinetics that is impossible to estimate from macroscopic whole-cell current measurements.

This advantage is most apparent with recordings from patches that contain only a single channel. Multiple channels in the membrane patch reduced the amount of information available on channel kinetics since it is impossible to determine the openings and closings of each individual channel in the record (Aldrich and Yellen, 1983; Fig.2.1Ba). In single channel patches, however, we know for certain the time of occurrence and the duration of every open and shut event that contributes to the total (Fig.2.1Ca). Thus, in the following sections, we will confine ourselves to data obtained from membrane patches with a single Ca channel only.

2.3.2. Kinetic properties of single Ca channel gating

Channel kinetics are usually described in terms of Markov processes. A lattice of states which presumably correspond to conformational states of the channel molecule is used to describe channel behaviour, and the channel may jump from one state to another at various transition rates. These rates seem most likely constants at a constant voltage (Trautwein and Pelzer, 1985 a and b); thus, the distribution of lifetimes of a single state is always exponential, with a characteristic lifetime determined by the reciprocal of the sum of the rate constants leaving that state. Thus, a histogram of the fraction of open lifetimes of a duration t' within an interval $t \pm \Delta t$ plotted against t gives information about the rate of closing. On the other hand, a histogram of the fraction of closed lifetimes of a duration t' within an interval $t \pm \Delta t$ plotted against t gives information about opening rates. In compiling both open and closed lifetime distributions from records during repetitive voltage pulses (Fig.2.1Ca), it is important to discard any event that is terminated by the end of the command pulse rather than by a subsequent gating transition, since it is impossible to know what the duration of the event would have been had the pulse not ended. However, this procedure can introduce a bias against observing long-lasting events (Aldrich and Yellen, 1983; Fukushima, 1981; Trautwein and Pelzer, 1985 a and b). Therefore, the distributions of open and closed times were first derived during long observation periods from steady-state records within the overlap region between the voltage ranges of activation and inactivation (Fig.2.2). For the computation of time histograms, a threshold level is installed halfway between the baseline and the average open channel current (see Fig.2.1Ca, upper panel). Shut—open transitions are detected as crossings of this threshold. Sojourn times are grouped in 0.15 ms bins and displayed as time histograms which are then fitted by a non-linear, unweighted, least-squares method. The distribution of the open times can consistently be fitted by a single exponential (Fig.2.2, left panel) whose time constant, τ_o, corresponds to the average lifetime of channel openings. By contrast, the shut-time distribution on a similar time-scale can only be described with the sum of two exponential terms whose time constants are well separated (Fig.2.2, right panel). This fit excludes all closed times longer than 10 ms which are condensed in the last bin of the shut-time distribution histogram (Fig.2.2, right panel) and which are far out of even the slow exponential term of the two-exponential best fitting density function. Their durations range from 10s—100s of ms (mean lifetime ~ 1300 ms) (see Fig.2.2 inset, lower panel) and thus their frequency of occurrence during step depolarisations of limited length would strongly depend on the duration of the command pulse. Especially during short voltage-clamp steps (<300 ms), the termination of the observation period by repolarisation would severely reduce the frequency of the long-lasting closed times in the entire shut-time histogram (see, for example, Figs.2.1Ca and 2.5) by disregarding them for reasons discussed above (p.21). Another reason will be outlined in Section 2.3.4

p.25. However, steady-state recordings of Ca channel activity are restricted to a very narrow potential range. At these positive steady potentials, the frequency to observe channel activity is low due to large inactiviation and a slow recovery (Cavalié et al., 1983; Trautwein and Pelzer, 1985 a and b; see Section 2.3.4 p.25). Thus, the distribution(s) and voltage dependence of these long-lasting closed events are not known exactly at present.

The inset in Figure 2.2 illustrates the occurrence of single Ca channel activity on different time-scales resulting from the existence of at least three distinct classes of gaps with well-separated lifetimes. Individual openings occur in groups, or bursts, and are separated from each other by short closures which correspond to the large and fast initial component of the shut-time histogram (τ_s) (Fig.2.2, inset, upper left panel). Each burst consists of n individual channel openings and (n-1) sojourns in the short-lasting shut state (Fig.2.2, inset, upper left panel). Thus, the average burst length ($\bar{t}_b$) is

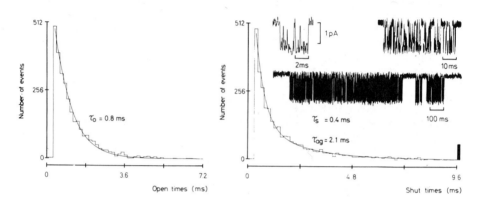

Fig.2.2. Kinetic properties of single Ca channel currents at RP + 65 mV as derived from open (left panel) and shut time (right panel) distributions. For time analysis, single Ca channel current traces were digitised at 6.67 kHz; linear components of leakage and capacitance were digitally subtracted. For further explanation, see text. The inset illustrates the grouping of single Ca channel activity at different time-scales resulting from the multiexponential shape of the shut-time distribution. The upper two panels were low-pass filtered at 3 kHz; the lower panel at 1.5 kHz. The temperature was 34–35 °C. The original sweeps show one group (or burst) of individual single Ca channel openings separated by brief closures (upper left panel), bursts of openings (upper right panel) and clusters of bursts (lower panel), respectively. This classification of single Ca channel activity is based on the distribution of all shut periods which can be expressed as a mixture of the distributions of (1) the shut periods within bursts, (2) the shut periods between bursts, and (3) the fraction of very long-lasting shut periods which are out of even the slow exponential term of the two-exponential best fitting density function, that is, the shut periods between clusters of bursts. The probability that a shut period of length t separates two bursts rather than being within a burst was determined by classifying shut periods as short/fast or long/slow depending on whether the fast (τ_s) or slow (τ_{ag}) exponential term of the best fitting density function is larger for the observed lifetime of a shut interval. This classification can be performed by noting that for a time, t', defined by the relation

$$\frac{N_s \exp(-t'/\tau_s)}{\tau_s} = \frac{N_{ag} \exp(-t'/\tau_{ag})}{\tau_{ag}} \quad (6)$$

the probability densities of the short and long lifetimes are equal. Thus, if a shut period has a lifetime shorter than t', it has a higher probability of being a member of the fast component, that is, a gap within a burst, whereas if a shut interval has a duration longer than t', it is more likely to belong to the slow component, that is, to gaps separating bursts. Although the distribution of shut intervals separating clusters is not known, the number of gaps incorrectly classified as gaps between clusters can be calculated from $N_{ag} \exp(-t''/\tau_{ag})$. Minimising this quantity by variation of t'' renders possible a reliable classification of the long-lasting gaps.

Calcium channels in isolated cardiac cells

$$\bar{t}_b = n \cdot \tau_o + (n-1) \cdot \tau_s \tag{2}$$

Bursts are separated from each other by shut periods of intermediate length corresponding to the small and slow secondary component of the shut-time histogram (τ_{ag}) (Fig.2.2, inset, upper right panel). Furthermore, bursts of openings themselves are grouped together into clusters of bursts (Colquhoun and Hawkes, 1982; Sakmann and Neher, 1983a; Trautwein and Pelzer, 1985 a and b) which are separated by the longest closed times (that is, the last bin of the shut-time histogram) whose lifetimes are in the order of 10s–100s of ms (Fig.2.2, inset, lower panel). Cluster durations likewise range on a 10s–100s of ms time-scale (Fig.2.2, inset, lower panel; see also Fig.2.5 and Trautwein and Pelzer, 1985 a and b). All the foregoing findings imply that the cardiac Ca channel has only one open conducting conformation but at least three shut non-conducting states (see p.21). Referring back to the kinetic properties of macroscopic whole-cell I_{Ca} and average Ca channel current (see Fig.2.1 and Section 2.3.1, p.18) we tend to suggest that the bursting behaviour with its fast ms gating transitions reflects Ca channel activation (see Fig.2.1Ab, Bb and Cb) whereas the clustering behaviour results from the existence of a long-lasting inactivated state being occupied during the shut periods between clusters, entrance into which would then be responsible for the slow decay of macroscopic whole-cell I_{Ca} and average Ca channel current during maintained depolarisation, in other words, inactivation (see Fig.2.1Aa, Bb and Cb). This hypothesis will be examined in detail in the next sections.

2.3.3. Ca channel activation kinetics and conductance properties

As illustrated above (Fig.2.2), single Ca channel activity on a ms time-scale appears as bursts of closely spaced brief unitary current pulses separated by wider periods of channel inactivity. This gating behaviour is inconsistent with a simple first-order closed-open reaction and, if reflecting channel activation, implies that there are at least two shut channel states related to the activation sequence (Cavalié *et al.*, 1983; Colquhoun and Hawkes, 1981, 1982, 1983; Colquhoun and Sakmann, 1983; Reuter, 1984; Trautwein and Pelzer, 1985 a and b; Tsien, 1983). If C_1 and C_2 denote two closed states, one of very short average lifetime, and O denotes the open state, then two obvious alternative general schemes of Ca channel activation are

$$C_1 \rightleftharpoons C_2 \rightleftharpoons O \tag{3}$$

and

$$C_1 \rightleftharpoons O \rightleftharpoons C_2. \tag{4}$$

These three states are connected by four independent, voltage-dependent rate constants (k_i). A possibility of discriminating between the two schemes during voltage-pulse experiments is provided by measuring the time between the onset of the voltage step and the first channel opening; this time is called the first latency or latency-to-first-opening (see Aldrich and Yellen, 1983). First latencies are treated in the theory of stochastic processes dealing with first passage times (Cox and Miller, 1965; Feller, 1950, 1966). The first passage time is the time at which a state is first entered, given an initial probability distribution among the states and a set of transition probabilities (rate constants). Upon depolarisation, the Ca channel opens with various latencies (Fig.2.3A). The distribution of the first latencies (Fig.2.3B, lower panel) reaches a maximum at a time later than zero, thus indicating a distinct rising phase with subsequent slower decay. This is the expected result for a process in which multiple closed states precede the open state. The arrangement of the open state, O, between the two closed states C_1 and C_2, would result in a distribution with a simple exponential fall. For finding the time-course of the probability to enter the open state, the open state is considered as an absorbing state (that is, a state with no exit). Integration of the latency-to-first-opening histogram then yields a cumulative probability distribution function of the first latencies which estimates the time-course of the first entrance of the Ca channel into the open state. This function correlates fairly well with the activation phase of the average Ca channel current (Fig.2.3B, upper panel) as can be seen by the coincidence of the latency distribution histogram with the rate of activation of the average Ca channel current both being displayed at the same fast time-scale (Fig.2.3B). With decreasing depolarisation, both the rising phase and the decay of the

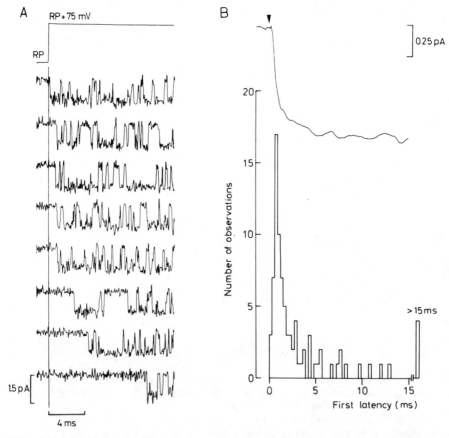

Fig.2.3. Correlation between the first latencies and the rise of the ensemble average Ca channel current. The latencies-to-first-openings were measured from records such as those shown in (A) during potential steps (300 ms total length) of 75 mV amplitude delivered from the pipette zero potential at 0.5 Hz. Leakage and capacitive currents have been digitally subtracted from each record. The vertical line indicates the onset of the command pulse (A, uppermost panel). The distribution of the latencies from 85 records is shown in the lower panel of (B). Data are grouped in 0.3 ms bins. The corresponding average Ca channel current obtained from these records is shown in the upper panel of (B) at the same time-scale. The sampling rate in this experiment was 5 kHz and the records were low-pass filtered at 2.5 kHz. The temperature was 34°C.

latency distribution become slower, and on the other hand, bursts of Ca channel openings accumulate earlier on depolarisation at more positive membrane potentials (Cavalié et al., 1983). This finding is compatible with increasing rates of whole-cell I_{Ca} activation with increasing membrane potentials in multicellular preparations (Reuter, 1979; Reuter and Scholz, 1977) and single cells (Lee and Tsien, 1982; Trautwein and Pelzer, 1985b).

Since the theoretical first latency is a complex function of the transition rates between the open and closed conformations of the Ca channel in the lattice of states describing channel activation (Cavalié et al., 1983), the voltage dependence of the first latencies suggests voltage-dependent opening and closing of the Ca channel. In the experiment summarised in figure 2.4, the membrane patch was depolarised from the pipette zero potential by increasing amplitudes (ΔV). The single-channel current amplitude (i) decreases continuously with positive membrane potential (Fig.2.4A), the i-ΔV relation being approximately linear over a wide potential range with a slope conductance of about 20 pS (Fig.2.4B; range 15–25 pS) (Cavalié et al., 1983; Reuter, 1983, 1984; Reuter et al., 1982; Trautwein and Pelzer, 1985a). Slope conductances at lower divalent cation concentrations and/or different cation species can be found elsewhere (see Trautwein and Pelzer, 1985b for numbers and respective references). It has been suggested that, in analogy to neurones (Hagiwara and Byerly, 1981), cardiac Ca channels are rectifying and can pass Ca ions only into the cell, since (1) clear elementary Ca or Ba outward currents through the Ca channel have been never observed even during strong depolarisations to about +135 mV absolute membrane potential (Cavalié et al., 1983; Trautwein and Pelzer, 1985 a and b), and (2) the time-dependent whole-cell outward current at very positive membrane potentials has been identified as K ion movement through Ca channels (Lee and Tsien, 1982). However, the direct experimental proof is still lacking in cardiac cells under the present single Ca channel recording conditions (see Section 2.2, p.17).

Referring back to figure 2.4A, there seems to exist a striking visual relation between single Ca channel gating and membrane potential. Quantitative kinetic analysis by means of time histograms yields a steep increase of the open channel lifetime (Fig.2.4C) and a decrease of both short-lasting shut-time components (Fig.2.4D) with increasing depolarisation. Since in single-channel patches, the average Ca channel current (I) is equal to the product $p \cdot i$ (see equation (1)), the data on the voltage dependence of the fast gating transitions and the single-channel current amplitude provide a suitable explanation for the bell-shaped voltage dependence of the peak average Ca channel current (Cavalié et al., 1983) and whole-cell I_{Ca} in multicellular preparations (McDonald, 1982; Reuter, 1979) and single cells (see Trautwein and Pelzer, 1985 a and b for references). The probability of the Ca channel being open (p) increases with voltage (Cavalié et al., 1983; Trautwein and Pelzer, 1985a) due to the prolongation of τ_o and the simultaneous shortening of both τ_s and τ_{ag}. Thus, the product $p \cdot i$ increases to a maximum although i declines linearly with increasing depolarisation. At positive membrane potentials, however, the decrease of i outweighs the increase of p and the product $p \cdot i$ and thus, I and whole-cell I_{Ca} decline again.

2.3.4 Ca channel inactivation kinetics

The grouping of bursts of Ca channel openings into clusters of bursts with lifetimes of 10s–100s of ms separated from each other by long-lasting shut periods (also 10s–100s of ms in duration) was hypothetically attributed to Ca channel inactivation (see Section 2.3.2, p.21). If so, clusters should reflect the time to the entrance of the Ca channel into an inactivated state; they might

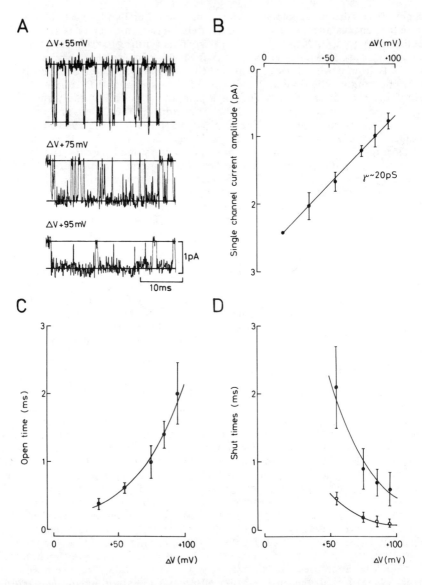

Fig.2.4. Voltage dependence of the single Ca channel current and the fast components of single Ca channel gating. (A) Cell-attached recordings of elementary current through a single Ca channel during 300 ms depolarisations of various amplitudes (ΔV) from the pipette zero potential delivered at 0.5 Hz. In each record, linear components of leakage and capacitance have been subtracted; the baseline and average elementary current amplitude are indicated by the solid lines. The records were low-pass filtered at 3 kHz. (B) Current–voltage relation of the elementary current amplitude. Each value represents the mean ± standard deviation from four experiments as shown in (A). In each experiment the individual single-channel current amplitude was derived from amplitude histograms and/or visually. The straight line is drawn by eye and represents the single-channel (chord) conductance (γ). (C) Open and (D) fast components of closed times (n = 4, mean ± standard deviation) derived from time histograms at different patch membrane potentials. The solid lines represent monoexponential approximations of the mean values. The temperature was 34–35°C.

be considered as 'apparent open times' taken at very low time resolution and bandwidth where all the high-frequency components are eliminated (see Section 2.3.1). According to this hypothesis, the relation between cluster duration and the time-course of average Ca channel current decay during depolarisation (that is, inactivation; see Fig.2.1) was examined during long-lasting voltage-clamp pulses (Fig.2.5). Figure 2.5A illustrates the appearance of clusters during step depolarisation. A considerable fraction of pulses did not elicit detectable Ca channel activity at all (topmost panel; see also Fig.2.1). If, however, Ca channel opening was triggered by the clamp step, clusters are most frequently observed only once per depolarisation. The chance to observe two, three, four and more clusters per single pulse in this experiment was about 0.12, 0.04, 0.01 and 0, respectively, being similar in other experiments. On average, 0.83 clusters occur per step depolarisation with 75 mV command pulses of 300 ms to 1s in duration. Thus, the long-lasting

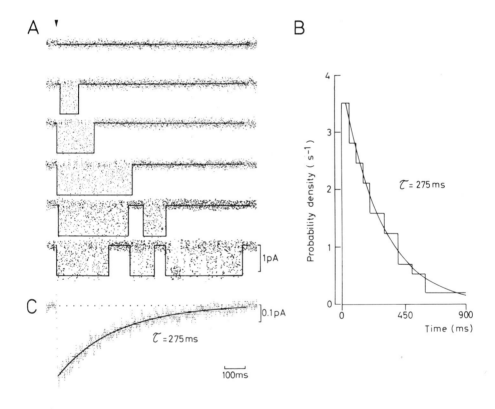

Fig.2.5. Correlation between cluster lifetime and the decay of the ensemble average Ca channel current. (A) Examples of the appearance of clusters during 900 ms voltage-clamp pulses of 75 mV amplitude delivered from the pipette zero potential at a rate of 0.2 Hz. Capacitive transients and leakage currents have been digitally subtracted from each record; the signals were low-pass filtered at 1 kHz. The temperature was 35°C. Clusters in the displayed traces were idealised (solid lines) using a critical gap of 25 ms (see legend of Fig.2.2). (B) Distribution of cluster lifetimes. The histogram was compiled by measuring the open intervals in the idealised records. Cluster durations up to 200 ms were grouped in 50 ms bins whereas for longer cluster lifetimes, 100 ms bins have been used. The data were fitted by a single exponential with a time constant corresponding to the mean cluster lifetime. (C) Corresponding ensemble average Ca channel current. The declining phase was fitted by a regression method with a monoexponential function using the time constant of the cluster lifetimes from (D). For this fit, the shut intervals between clusters can be neglected since the average number of clusters per voltage-clamp step was 0.83.

shut periods between clusters of bursts might be easily overlooked since they appear rarely even during long-lasting depolarisations, and mostly last until the termination of the observation period by the end of the command pulse, especially if short voltage-clamp pulses are used. Thus, a major fraction of them is discarded for reasons outlined above (see Section 2.3.2) and the remaining one contributes only a small number of events to the histogram of all closed times. The distribution of all cluster lifetimes is well fitted by a single exponential with a time constant corresponding to the mean cluster lifetime (Fig.2.5B). By using this time constant, the time-course of inactivation of the corresponding average Ca channel current can well be approximated (Fig.2.5C) indicating that the time constant of the average Ca channel current decay approximately mirrors cluster lifetime distribution and/or average duration. In the light of these data, it seems not unlikely that the empty sweeps without apparent gating transitions called blanks (Fig.2.5A, topmost panel) reflect the entrance of the Ca channel into the inactivated state before/without channel activation. Then, the complete reaction scheme of Ca channel gating must be generally written in a cyclic form (Colquhoun and Hawkes, 1983):

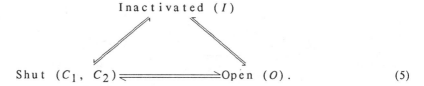

$$\text{Shut } (C_1, C_2) \rightleftharpoons \text{Open } (O). \tag{5}$$

with Inactivated (I) at the apex.

For the sake of simplicity, the two short-lasting closed states related to channel activation are lumped together.

Blanks refer to a condition or conditions where the Ca channel is unavailable for opening. Referring back to the last paragraph of this section (p.28), voltage-dependent inactivation might be one factor that stabilises the channel in this condition (Cavalié et al., 1983). This hypothesis is tested by adopting a voltage-clamp protocol used to study steady-state inactivation in multicellular preparations and single cells (Fig.2.6); that is, variation of the holding or conditioning potential (ΔV) which precedes clamp pulses to a constant test potential (see in particular Fig.2.6B, inset, upper panel). Depolarisation of the conditioning potential results in a decrease of the frequency of records with Ca channel activity within a series of depolarisations due to an increased fraction of blanks per ensemble of single Ca channel current traces (Fig.2.6A). The elementary Ca channel current amplitude and the distributions and average lifetimes of remaining Ca channel openings and closings, the mean number of interruptions per burst, and the mean burst length remain unaffected (Trautwein and Pelzer, 1985a). The probability that the Ca channel is available to open at the constant test potential has a sigmoidal dependence on the potential of the conditioning depolarisation (Fig.2.6B, filled circles), the relation being described by a Boltzmann distribution (Fig.2.6B, solid line). The decreased availability of the Ca channel to open upon depolarisation from depolarised conditioning membrane potentials results in smaller amplitudes of the ensemble averages of single Ca channel activity (Fig.2.6B, inset, lower panel). The dependence of the amplitudes of the average Ca channel currents normalised with respect to maximal current on conditioning membrane potential closely matches the probability- conditioning potential relation (Fig.2.6B) which therefore strongly resembles the relation between voltage and the steady-state inactivation variable for whole-cell I_{Ca} in single cells (Isenberg and Klöckner, 1982b; Josephson, Sanchez-Chapula and Brown, 1984; Mitchell *et al.*, 1983) and multicellular tissue (McDonald, 1982; Reuter, 1979). Thus, the fraction of blanks per ensemble of single Ca channel current records (or the corresponding probability of the Ca channel being able to open on depolarisation during a series of clamp pulses) seems to provide a measure of the steady-state occupancy of the inactivated state(s) of the Ca channel.

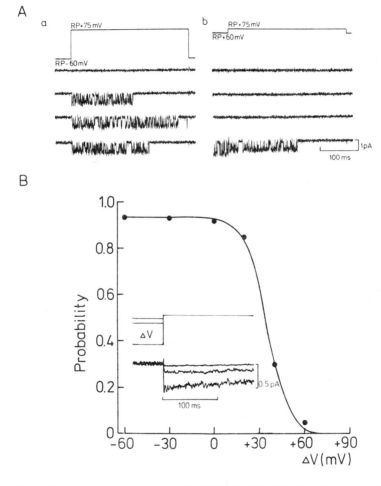

Fig.2.6. Dependence of the fraction of blanks per ensemble of elementary Ca channel current records on the conditioning prepotential (= holding potential). (A) Test pulses (300 ms duration) to a constant potential (RP + 75 mV) were imposed from conditioning potentials either 60 mV negative ((A)a) or 60 mV positive ((A)b) from the pipette zero potential. The traces were low-pass filtered at 2.5 kHz (sampling rate 5 kHz). (B) The probability that a depolarising clamp pulse elicits at least one burst of Ca channel openings is plotted against the conditioning potential, ΔV. The solid line is a Boltzmann distribution,

$$p = 0.93 / \{1 + \exp [(V - V_{0.5}) / k]\}, \tag{7}$$

with parameters chosen for the best visual fit: $V_{0.5}$ + 35 mV, k 7 mV. The inset shows ensemble average Ca channel currents obtained from 150 elementary current sweeps, respectively (conditioning potentials ΔV from the bottom toward the top: 60, +40 and +60 mV). Average currents were superimposed with respect to the zero current level (current at the pipette zero potential). As in (A), leakage and capacitive currents have been digitally subtracted. The temperature was 34°C.

2.4 Concluding remarks

Calcium channels in cardiac membranes to a large extent regulate the functional state of the cell by controlling Ca influx and thereby [Ca]$_i$ by means of voltage-dependent gating. The present experiments have in detail explored the gating properties of the Ca channel on the microscopic single-channel level free from some of the limitations of macroscopic whole-cell current recording (see Section 2.1; Aldrich and Yellen, 1983; Trautwein and Pelzer, 1985a). The information gleaned from isolated single cardiac cells and single-channel recordings has, no doubt, a major impact on our perceptions of Ca channel gating and function on a molecular level. Besides analysis of Ca channel conductance and kinetics, this preparation and technique seem to meet precisely the requirements for quantitative studies of chemical modulation of Ca channels by drugs and Ca channel phosphorylation by β-adrenergic agents and the intracellular intermediates of the β-adrenergic pathway on the single-channel level. Very recent studies (Cavalié, Pelzer and Trautwein, 1985; Pelzer, Cavalié and Trautwein, 1984 a and b; Trautwein and Pelzer, 1985a) show that for example D600, the methoxyderivative of verapamil, decreases the probability of the Ca channel being open (p) in traces with channel openings by a shortening of the open channel lifetime, a reduction of the frequency of openings per burst, and a prolongation of the closed times between bursts of openings. The brief closures within bursts, the first latencies, and the single-channel current amplitude are not affected by the drug. Beside the drug-channel interactions on a ms time-scale, D600 distinctly increases the frequency of blanks; these D600-induced extra empty sweeps tend to cluster in groups of consecutive records suggesting an additional modulation in channel gating properties by the drug over a time-scale of several seconds. Due to both fast and slow alterations of the gating properties, the Ca channel is stabilised in the non-conducting state. Consequently, the ensemble average Ca channel current is reduced; its time-course, however, is not affected by the drug. The 1,4-dihydropyridines nifedipine, nitrendipine and especially nimodipine in Ca-antagonistic concentrations (at low concentrations these compounds have mixed agonistic and antagonistic effects) exclusively seem to increase the fequency of blanks without any clear evidence for changes in open or closed-time distributions (Hess, Lansman and Tsien, 1984). However, the decay of the reduced average Ca channel current is speeded by these agents being interpreted as acceleration of the transition to that channel state(s) being responsible for the occurrence of blanks (Hess, Lansman and Tsien, 1984). In our terminology, this would correspond to a shortening of the cluster duration being equivalent to the finding of Hess, Lansman and Tsien (1984) that 'channel openings show a tendency to occur toward the beginning of the pulse and not toward the end'.

In contrast, adrenaline increases the ensemble average Ca channel current (Brum, Osterrieder and Trautwein, 1984; Reuter, 1983; Trautwein and Pelzer, 1985a; Trautwein, Brum and Osterrieder, 1985). This increase is mediated by an increased p due to a slight prolongation of the open channel lifetime, a marked shortening of the shut periods between bursts and a reduction of the number of blanks per ensemble of single Ca channel current records (Brum, Osterrieder and Trautwein, 1984; Reuter, 1983; Trautwein and Pelzer, 1985a; Trautwein, Brum and Osterrieder, 1985). The adrenaline effect can be quantitatively mimicked by the application of 8-bromocyclic AMP (Cachelin et al., 1983) and the intracellular injection of cAMP (Brum, Osterrieder and Trautwein, 1984) and more than likely the catalytic subunit of the cAMP-dependent protein kinase (Brum, Osterrieder and Trautwein, 1984; Brum et al., 1983; Osterrieder et al., 1982; Trautwein, Brum and Osterrieder, 1985). In the light of these data, the fact that Ca channels do not function in isolated membrane patches (Cavalié et al., 1983) leads to the notion that cAMP-dependent phosphorylation and/or ATP regeneration might turn out to be a prerequisite for Ca channel gating and function. The conversion of blanks to sweeps with channel activity by β-adrenergic agents (see above, p.30) indicates, in addition to voltage-dependent inactivation, the existence of some sort of phosphorylation-dependent 'chemical inactivation'. In this context, it is very interesting that the dihydropyridine derivative BAY K 8644 which has been shown to be a very potent Ca agonist (Hess, Lansman and Tsien, 1984; Kokubun and Reuter, 1984; Ochi, Hino and Nimii, 1984) seems not to affect (that is, decrease) the frequency of blanks at least in two out of the three studies (Kokubun and Reuter, 1984; Ochi, Hino and Nimii, 1984). In contrast to β-adrenoreceptor agonists, Kokubun and Reuter (1984) also found no elevation of cAMP levels by BAY K 8644 in their cell system. Future work on chemical modulation and modifaction of Ca channels will no doubt provide further insight into structures involved in Ca channel gating.

Acknowledgements

This chapter incorporates data obtained in collaboration with Dr A. Cavalié. His experiments, suggestions and discussions formed a substantial contribution to the chapter. We are grateful to H. Ehrler for skilful electronics support, D. Steimer and S. Bastuck for technical assistance, and B. Bresser for excellent computer programs. The work was supported by the DFG, SFB 38 (Membranforschung), project G1.

References

Akiyama, T. and Fozzard, H.A. (1979). Ca and Na selectivity of the active membrane of rabbit AV nodal cells. *Am. J. Physiol.* 236 (1), C1–C8.
Aldrich, R.W. and Yellen, G. (1983). Analysis of nonstationary channel kinetics. In B. Sakmann and E. Neher (eds.) *Single-channel Recording*, pp. 287–99. Plenum Press, New York and London.
Attwell, D. and Cohen, I. (1977). The voltage clamp of multicellular preparation. *Prog. Biophys. Mol. Biol.* 31, 201–45.
Bean, B.P., Nowycky, M.C. and Tsien, R.W. (1983). Electrical estimates of Ca channel density in heart cell membranes. *Biophys. J.* 41, 295a.
Brum, G., Flockerzi, V., Hofmann, F., Osterrieder, W. and Trautwein, W. (1983). Injection of catalytic subunit of cAMP-dependent protein kinase into isolated cardiac myocytes. *Pflügers Arch.* 398, 147-54.
Brum, G., Osterrieder, W. and Trautwein, W. (1984). β-adrenergic increase in the calcium conductance of cardiac myocytes studied with the patch clamp. *Pflügers Arch.* 401, 111–8.
Cachelin, A.B., dePeyer, J.E., Kokubun, S. and Reuter, H. (1983). Ca^{2+} channel modulation by 8-bromocyclic AMP in cultured heart cells. *Nature* 304, 462–4.
Cavalié, A., Ochi, R., Pelzer, D. and Trautwein, W. (1983). Elementary currents through Ca^{2+} channels in guinea pig myocytes. *Pflügers Arch.* 398, 284–97.
Cavalié, A., Pelzer, D. and Trautwein, W. (1985). Modulation of the gating properties of single calcium channels by D600 in guinea-pig ventricular myocytes. *J. Physiol. (Lond.)* 358, 59P.

Chapman, R.A. (1983). Control of cardiac contractility at the cellular level. *Am. J. Physiol.* **245** (*Heart Circ. Physiol.* 14), H535–H552.

Colquhoun, D. and Hawkes, A.G. (1981). On the stochastic properties of single ion channels. *Proc. R. Soc. Lond. B.* **211**, 205–35.

Colquhoun, D. and Hawkes, A.G. (1982). On the stochastic properties of bursts of single ion channel openings and of clusters of bursts. *Phil. Trans. R. Soc. Lond. B.* **300**, 1–59.

Colquhoun, D. and Hawkes, A.G. (1983). The principles of the stochastic interpretation of ion-channel mechanisms. In B. Sakmann and E. Neher (eds.) *Single-channel Recording*, pp. 135–75. Plenum Press, New York and London.

Colquhoun, D. and Sakmann, B. (1983). Bursts of openings in transmitter-activated ion channels. In B. Sakmann and E. Neher (eds.) *Single-channel Recording*, pp. 345–64. Plenum Press, New York and London.

Coraboeuf, E. (1980). Voltage clamp studies of the slow inward current. In D.P. Zipes, J.C. Bailey and V. Elharrar (eds.) *The Slow Inward Current and Cardiac Arrhythmias*, pp. 25–95. Martinus Nijhoff, The Hague, Boston, London.

Corey, D.P. and Stevens, C.F. (1983). Science and technology of patch-recording electrodes. In B. Sakmann and E. Neher (eds.) *Single-channel Recording*, pp. 53–68. Plenum Press, New York and London.

Cox, D.R. and Miller, H.D. (1965). *The Theory of Stochastic Processes*. Chapman and Hall, New York.

Dow, J.W., Harding, N.G.L. and Powell, T. (1981). Isolated cardiac myocytes. I. Preparation of adult myocytes and their homology with the intact tissue. *Cardiovasc. Res.* **15**, 483–514.

Fabiato, A. (1983). Calcium-induced release of calcium from the cardiac sarcoplasmic reticulum. *Am. J. Physiol.* **245** (*Cell Physiol.* 14), C1–C14.

Feller, W. (1950, 1966). *An introduction to Probability Theory and its Applications*, Vols. I and II. John Wiley and Sons, New York.

Fukushima, Y. (1981). Identification and kinetic properties of the current through a single Na channel. *Proc. Natl. Acad. Sci. USA* **78**, 1274–7.

Hagiwara, S. and Byerly, L. (1981). Calcium channel. *Rev. Neurosci.* **4**, 69–125.

Hamill, O.P., Marty, A., Neher, E., Sakmann, B. and Sigworth, F.J. (1981). Improved patch-clamp techniques for high-resolution current recording from cells and cell-free membrane patches. *Pflügers Arch.* **391**, 85–100.

Hess, P. and Tsien, R.W. (1984). Mechanism of ion permeation through calcium channels. *Nature* **309**, 453–6.

Hess, P., Lansman, J.B. and Tsien, R.W. (1984). Different modes of calcium channel gating favored by calcium agonists and antagonists. *Nature* **311**, 538–44.

Isenberg, G. and Klöckner, U. (1982a). Ca-tolerant ventricular myocytes prepared by preincubation in a 'KB medium'. *Pflügers Arch.* **395**, 6–18.

Isenberg, G. and Klöckner, U. (1982b). Calcium currents of isolated bovine ventricular myocytes are fast and of large amplitude. *Pflügers Arch.* **395**, 30–41.

Josephson, I.R., Sanchez-Chapula, J. and Brown, A.M. (1984). A comparison of calcium currents in rat and guinea-pig single ventricular cells. *Circ. Res.* **54**, 144–56.

Kokubun, S. and Reuter, H. (1984). Dihydropyridine derivatives prolong the open state of Ca channels in cultured cardiac cells. *Proc. Natl. Acad. Sci. USA.* **81**, 4824–7.

Lee, K.S. and Tsien, R.W. (1982). Reversal of current through calcium channels in dialysed heart cells. *Nature* **297**, 498-501.

Marty, A. and Neher, E. (1983). Tight-seal whole-cell recording. In B. Sakmann

and E. Neher (eds.) *Single-channel Recording*, pp. 107–22. Plenum Press, New York and London.
McDonald, T.F. (1982). The slow inward calcium current in the heart. *Ann. Rev. Physiol.* **44**, 425–34.
Mitchell, M.R., Powell, T., Terrar, D.A. and Twist, V.W. (1983). Characteristics of the second inward current in cells isolated from rat ventricular muscle. *Proc. R. Soc. Lond. B.* **219**, 447–69.
Mullins, J.L. (1979). The generation of electric currents in cardiac fibers by Na/Ca exchange. *Am. J. Physiol.* **236** (3), C103–C110.
Ochi, R., Hino, N. and Nimii, Y. (1984). Prolongation of calcium channel open time by the dihydropyridine derivative BAY K 8644 in cardiac myocytes. *Proc. Japan Acad. B.* **60**, 153–6.
Osterrieder, W., Brum, G., Hescheler, J., Trautwein, W., Hofmann, F. and Flockerzi, V. (1982). Injection of subunits of cyclic AMP-dependent protein kinase into cardiac myocytes modulates Ca^{2+} current. *Nature* **298**, 576–8.
Pelzer, D., Cavalié, A. and Trautwein, W. (1984a). Modulation of single calcium channel gating by D600 in cardiac myocytes. *Pflügers Arch.* **400**, R 29.
Pelzer, D., Cavalié, A. and Trautwein, W. (1984b). Modulation of the gating properties of single calcium channels in cardiac cell membranes by D600. In J. Mironneau (ed.) *Calcium Regulations in Smooth Muscle: Biochemical and Physiological Aspects.* pp. 415–24. INSERM Symposia Series, Vol. 124.
Pelzer, D., Trube, G. and Piper, H.M. (1984). Low resting potentials in single isolated heart cells due to membrane damage by the recording microelectrode. *Pflügers Arch.* **400**, 197–9.
Piper, H.M., Probst, I., Schwartz, P., Hütter, F.J. and Spieckermann, P.G. (1982). Culturing of calcium stable adult cardiac myocytes. *J. Mol. Cell. Cardiol.* **14**, 397-412.
Powell, T., Terrar, D.A. and Twist, V.W. (1980). Electrical properties of individual cells isolated from adult rat ventricular myocardium. *J. Physiol. (Lond.)* **302**, 131–53.
Reuter, H. (1979). Properties of two inward membrane currents in heart. *Ann. J. Physiol.* **41**, 413–24.
Reuter, H. (1983). Calcium channel modulation by neurotransmitters, enzymes and drugs. *Nature* **301**, 569–74.
Reuter, H. (1984). Ion channels in cardiac cell membranes. *Ann. Rev. Physiol.* **46**, 473–84.
Reuter, H. and Scholz, H. (1977). A study of the ion selectivity and the kinetic properties of the calcium-dependent slow inward current in mammalian cardiac muscle. *J. Physiol. (Lond.)* **264**, 17–47.
Reuter, H., Stevens, C.F., Tsien, R.W. and Yellen, G. (1982). Properties of single calcium channels in cardiac cell culture. *Nature* **297**, 501–4.
Sakmann, B. and Neher, E. (1983a). *Single-channel Recording.* Plenum Press, New York and London.
Sakmann, B. and Neher, E. (1983b). Geometric parameters of pipettes and membrane patches. In B. Sakmann and E. Neher (eds.) *Single-channel Recording*, pp. 37–51. Plenum Press, New York and London.
Sakmann, B. and Trube, G. (1984). Voltage-dependent inactivation of inward-rectifying single-channel currents in the guinea-pig heart cell membrane. *J. Physiol. (Lond.)* **347**, 659–83.
Trautwein, W. and Pelzer, D. (1985a). Voltage-dependent gating of single calcium channels in the cardiac cell membrane and its modulation by drugs. In D. Marmé (ed.) *Calcium and Cell Physiology*, pp. 53–93. Springer, Berlin, Heidelberg, New York, Tokyo.

Trautwein, W. and Pelzer, D. (1985b). Gating of single calcium channels in the membrane of enzymatically isolated ventricular myocytes from adult mammalian hearts. In D.P. Zipes and J. Jalife (eds.) *Cardiac Electrophysiology and Arrhythmias*, pp. 31–42. Grune and Stratton, Orlando and London.

Trautwein, W., Brum, G. and Osterrieder, W. (1985). Effects of cAMP or catalytic subunit of protein kinase on cardiac calcium channels. In *Proceedings of the 16th Meeting of the Federation of European Biochemical Societies*; VNU Science Press BV, Utrecht.

Trube, G. (1983). Enzymatic dispersion of heart and other tissue. In B. Sakmann and E. Neher (eds.). *Single-channel Recording*, pp. 69–76. Plenum Press, New York and London.

Tsien, R.W. (1983). Calcium channels in excitable cell membranes. *Ann. Rev. Physiol.* **45**, 341–58.

3

SODIUM—CALCIUM (Na—Ca) EXCHANGE

P.F. Baker and T.J.A. Allen

Department of Physiology, King's College London, Strand, London WC2R 2LS, UK.

Abstract

The partial reactions of the sodium—calcium exchanger have been well characterised in only a few large cells. The main reactions are Na_o-dependent Ca efflux and Na_i-dependent Ca influx. Although these share a number of common properties such as activation by intracellular alkalinity, they also display a number of features that suggest the exchanger is far from simple. Most notable are (1) the exchanger is asymmetric — mM Ca activating externally and μM internally; (2) Ca_o–Na_i exchange seems to require ATP, (3) Ca_o–Na_i exchange is activated by intracellular Ca. This constitutes a positive feedback system the physiological implications of which are only just beginning to be explored. Plasma membrane vesicles from many excitable cells also exhibit Na—Ca exchange but it appears not to display the full range of partial reactions seen in intact cells. In both cells and vesicles, at least three Na ions seem to exchange for one Ca and fluxes through the exchange are sensitive to membrane potential, depolarisation favouring Ca uptake into cells.

3.1. Introduction

Multiple pathways exist for transporting calcium across membranes. These include calcium channels, ATP-dependent calcium pumps and sodium—calcium exchange. This last system is particularly interesting as it can move Ca in either direction depending on the prevailing conditions. Since its discovery in invertebrate nerve (Baker et al., 1967a,b, 1969; Baker and Blaustein, 1968) and cardiac muscle (Reuter and Seitz, 1968) sodium—calcium exchange has been shown to be widespread, though by no means universal, in the plasma membranes of excitable cells and epithelia (see Baker, 1972; Blaustein, 1974). In addition, a rather similar exchanger has been described in certain intracellular organelles including mitochondria (Crompton, Capano and Carafoli, 1976; Nicholls and Akerman, 1982) and secretory vesicles (Saermark, Thorn and Gratzl, 1983).

It has always proved difficult to obtain really clean data on calcium fluxes and information about the detailed properties of the exchanger rely rather heavily on studies on a few large cells, especially the squid axon (Baker, 1970, 1972, 1978; Baker and Dipolo, 1984). This situation has, however, changed in the last few years with the demonstration of sodium—calcium exchange in plasma membrane vesicles (Bers, Philipson and Nishimoto, 1980; Pitts 1979; Reeves and Sutko, 1979). As will become apparent the exchanger of intact cells exhibits a number of unusual features, and it is by no means clear whether all these persist in membrane vesicles. For these reasons this chapter will provide a detailed account of sodium–calcium exchange in one particular cell, the squid axon; but wherever possible will compare this preparation with other cell types and also membrane vesicles

3.2. Partial reactions

3.2.1. General

The sodium–calcium exchanger exhibits at least three, possibly four, modes of operation. These are illustrated in Fig.3.1. Some doubt exists whether the Na–Na exchange mode is operative in intact axons. This immediately highlights one of the problems in studying sodium–calcium exchange: the lack of a really specific inhibitor with which to distinguish fluxes through the exchanger from other transmembrane fluxes of Na and Ca. From time to time various substances have been suggested as inhibitors but none has so far proved to be specific for Na–Ca exchange in the sense that the cardiac glycosides are for the (Na + K) exchange pump. Thus Na–Ca exchange in squid can be inhibited by tetracaine and by a variety of divalent and trivalent cations, including Mn, Co, Ni, Cd and La; but these substances also block Ca channels (Baker, Meves and Ridgway, 1973; Hagiwara and Byerly, 1981; Kostyuk, 1981). On the other hand, the Ca channel antagonists D-600 (10^{-4} M) and nifedipine (10^{-4} M) have rather little effect on the exchanger (Allen and Baker, unpublished data). Doxorubicin (Adriamycin) which has been reported to inhibit the exchanger in isolated vesicles (Caroni, Villani and Carafoli, 1981), has no detectable effect on intact cells at 10^{-4} M. Amiloride which does not seem to affect Na–Ca exchange in axons at 10^{-4} M, inhibits the exchanger in vesicles at somewhat higher concentrations (Luciani, 1984, Schellenberg and Swanson, 1982b). Very recently Kaczorowski, Barros, Dethmers, Trumble and Cragoe (1985) have described inhibition in vesicles by a group of amiloride derivatives that are effective at 10^{-5} M. These seem to be specific for the Na binding site of the Na–Ca exchanger, but their effects on the full range of fluxes seen in intact cells have yet to be examined.

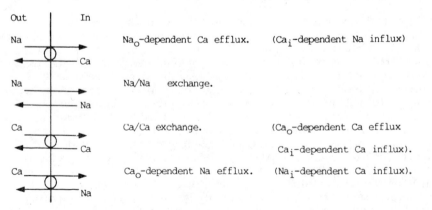

Fig.3.1. The partial reactions of the exchanger.

In the absence of really specific inhibitors, fluxes through the Na—Ca exchanger have to be defined operationally as indicated in Fig.3.1. Some of the more important features of these partial reactions are summarized below.

3.2.2. Na_o-dependent Ca efflux

About 20% of the resting Ca efflux from unpoisoned axons is dependent on external sodium, the rest apparently occurring via the ATP-dependent Ca pump. As Ca_i is raised, the Na_o-dependent Ca efflux increases dramatically until it greatly exceeds the pumped flux. It appears that Na_o-dependent Ca efflux is far from saturation at the level of Ca found in resting cells (~100 nM): but raising Ca_i into the μM range (Fig.3.2.) reveals a transport capacity through the exchanger that greatly exceeds that of the ATP-dependent pump. At these high levels of Ca_i it is possible to detect a Ca_i-dependent Na influx (Blaustein and Russell, 1975).

Provided that Ca_i is kept constant, Na_o-dependent Ca efflux is not markedly affected by the presence of the Ca buffer EGTA inside the axon and persists in the absence of ATP, although under these conditions the apparent affinity for external Na is reduced (Baker and Glitsch, 1973; Baker and McNaughton, 1976; and Fig.3.3). In these latter two properties, Na_o-dependent Ca efflux differs markedly from Na_i-dependent Ca influx which seems to require ATP and is inhibited by intracellular EGTA.

3.2.3. Na_i-dependent Ca influx

This flux can be monitored either as the Ca_o-dependent Na efflux (Baker *et al.*, 1969) or as the Na_i-dependent Ca influx (Baker *et al.*, 1969; Dipolo, 1979). The results obtained by both approaches agree closely. Na_i-dependent Ca inflow has an absolute requirement for intracellular ATP and is activated by intracellular Ca (Fig.3.4.). It follows that the flux disappears in metabolically poisoned axons (Baker *et al.*, 1969) and can be inhibited by injection of EGTA to buffer intracellular Ca at a low level (Baker, 1970). This activation of Na_i-dependent Ca influx by internal Ca constitutes a positive feedback loop,

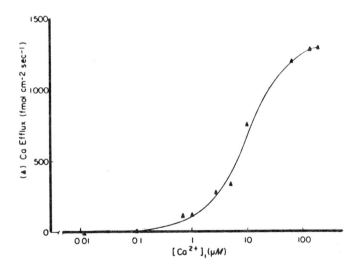

Fig.3.2. Dependence of the Na_o-dependent Ca efflux on internal Ca. Data from axons dialysed with ATP_i (2 mM) and Na_i (45 mM). (Data of Dipolo and Beaugé, from Baker and Dipolo, 1984.)

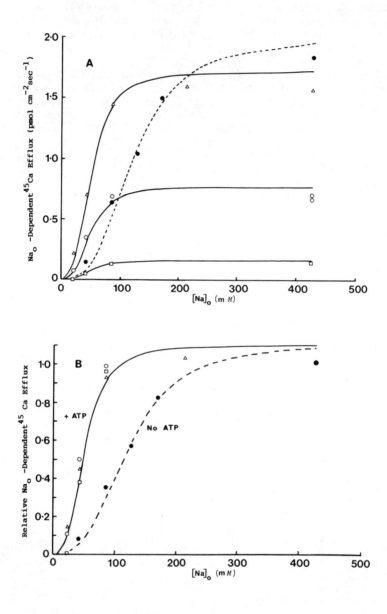

Fig.3.3. Dependence on Na_o of Na_o-dependent Ca efflux in an axon of *Loligo pealei* dialysed with different concentrations of free Ca. (A) Raw data; (B) normalised data. Open symbols: in the presence of ATP (4 mM); filled symbols: in the nominal absence of ATP. Internal free Ca 2.5 μM (Δ), 0.5 μM (O) 0.3 μM (□), and 100 μM (●). The dialysis fluids also contained FCCP (10 μM) and oligomycin (31 μg/ml). Temperature 15°C. The curves have been drawn on the assumption of an exchange of 3Na for 1Ca. The observation in (A) that the apparent affinity for Na_o is unaffected by alterations in Ca_i is consistent with some sort of simultaneous exchange of Na_o for Ca_i (from Blaustein, 1977).

the physiological significance of which is only just beginning to be explored.

In the presence of both ATP and intracellular Ca, the flux is activated by external Ca and this activation is subject to competitive inhibition by external Na; but the actual affinity for Ca is determined by the external cation used to substitute for Na. Thus the apparent K_m is quite high (many mM) in the presence of Na or choline but is decreased to about 2 mM when these are replaced by Li or K (Fig.3.5.). These findings have been interpreted in terms of competition between Na and Ca for the calcium binding site, together with an external monovalent cation activation site which influences the affinity of the exchanger for calcium (Baker et al., 1969). Na_o-dependent Ca efflux seems not to have a comparable site at the external face of the axon membrane.

3.2.4. Ca–Ca exchange

This is most marked under conditions where Ca_i is elevated. It appears not to require ATP, nor is it affected - at constant Ca_i- by intracellular EGTA. It is, however, activated by monovalent cations in a manner that closely resembles the activation of Na_i-dependent Ca inflow.

3.2.5. General Conclusions

The striking feature of work on intact axons is that the exchange seems not to be a simple symmetrical process. Not only does the apparent affinity for Ca appear to be different on the two sides of the membrane; but the requirement for ATP of the Ca inflow mode suggests an altogether more complex system. It might even be argued that Na_o-dependent Ca efflux and Na_i-dependent Ca influx represent different systems but their similar response to changes in intracellular pH argues against this; both are strongly inhibited by intracellular acidity and activated by intracellular alkalinisation (Baker and McNaughton, 1977; Baker and Honerjager, 1978; Baker and

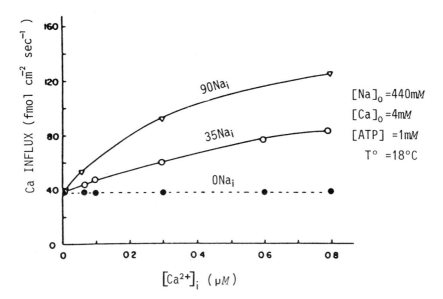

Fig.3.4. Calcium influx as a function of $[Ca^{2+}]_i$ at different Na_i in dialysed squid axons. In all cases the free EGTA was greater than 200 μM. (Data were calculated from Dipolo, 1979.)

Dipolo, 1984).

3.3. Stoichiometry of the exchange

It is a matter of considerable importance to determine whether or not the Na—Ca exchanger is electroneutral. If the exchanger is electrogenic, the exchange fluxes will be sensitive to membrane potential. As such it could contribute to both current flow and voltage-sensitive Ca movements in excitable and other cells. The early experiments on Ca_o-dependent Na efflux in squid axon showed it to be activated by K depolarisation and to have a stoichiometry of 1 Ca : 3—5 Na (Baker et al., 1969). Subsequent work has shown that K depolarisation inhibits Na_o-dependent Ca efflux consistent with the operation of an electrogenic process (Baker and McNaughton, 1976; Blaustein, Russell and De Weer 1974; Mullins and Brinley, 1975).

The use of K is, however, open to the objection that at least some of its actions may be chemical rather than through changes in membrane potential. This possibility has recently been examined by measuring fluxes under conditions where the membrane potential is altered electrically by means of a voltage clamp (Allen and Baker, 1983). The conclusions seem quite clear. K_o and electrical depolarisation exert equivalent effects on the Na_o-dependent Ca efflux, but not on the Na_i-dependent Ca influx where K_o can exert both

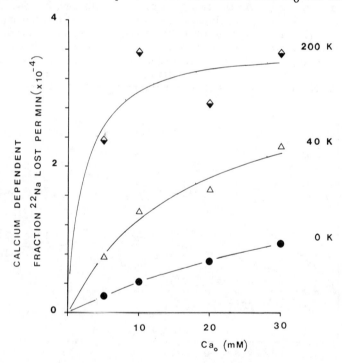

Fig.3.5. Potassium increases the apparent affinity of Na_i/Ca_o exchange for external calcium. Data obtained in choline sea water in the presence of 10^{-5}M ouabain; axon diameter 714 μm. Temperature 17°C. The curves drawn are sections of rectangular hyperbolae with respective K_m and V_{max} x 10^{-4} in 0 K_o, 40 K_o, and 200 K_o of: 53.2 mM, 2.6; 17.0 mM, 3.4; and 2.1 mM, 3.7.

chemical and electrical effects. This seems dependent on the degree of saturation at the Ca binding site. Illustrated in Fig.3.6. is an experiment which shows that in choline-substituted media only part of the stimulation of Ca_o-dependent Na efflux brought about by K can be removed by electrical repolarisation to the resting potential. Further, electrical depolarisation fails to mimic the effects of an equivalent potassium depolarisation. This chemical effect of K appears directed at the monovalent cation activating site, controlling the affinity of the Na_i-dependent Ca influx for external Ca.

Figure 3.7 summarises the voltage sensitivity of the two exchange fluxes determined in the absence of external K. It leaves little doubt that the exchanger is electrogenic with net Ca entry increasing with depolarisation up to at least +40 mV. Ca_o-dependent Na efflux increases and Na_o-dependent Ca efflux decreases with depolarisation in an apparently symmetrical fashion consistent with the energy barrier to Ca movement being located close to the middle of the membrane. Their net slope suggests an exchange of 1 Ca for 3 Na. Figure 3.8. shows a very simple model that would behave electrically in the way observed.

3.4. Studies on isolated plasma membrane vesicles

The demonstration by Reeves and Sutko (1979), Pitts (1979) and Bers, Philipson and Nishimoto (1980) of Na—Ca exchange in plasma membrane

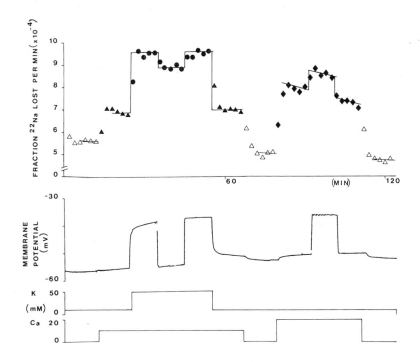

Fig.3.6. Comparison of potassium and electrically evoked depolarisations on Ca_o-dependent Na efflux in choline. Axon prepared by preinjection of HEPES, MOPS, and TEA (pH 7.3) to control internal pH and block the voltage-sensitive K conductance. Superfusate contained 0.6 μM TTX and 10^{-5}M ouabain, to block the Na conductance and Na pump fluxes respectively; axon diameter 1057 μm. Temperature 18 °C.

vesicles provided a major step forward in the biochemical analysis of the exchanger which has opened the way to the reconstruction of the exchanger in liposomes (Luciani, 1984; Miyamoto and Racker, 1980; Wakabayashi and Goshima, 1982). Essentially similar results have been obtained with axonal (Peterson, Matsumura and McGroarty, 1984) and synaptosomal vesicle preparations rich in plasmalemma (Gill, 1982; Gill, Grollman and Kohn, 1981; Schellenberg and Swanson, 1981, 1982a).

Work on vesicles and reconstituted preparations should, however, always be related to fluxes found in intact cells. Comparison between sarcolemmal and synaptosomal vesicles with squid axon fluxes leaves the overwhelming impression that the vesicle preparations currently in use may not exhibit the full range of features that have been described in intact cells. Thus the vesicle exchanger appears to be symmetrical, displaying a high affinity for Ca on both sides of the membrane (apparent K_m 10–50 μM), and there is no evidence for anything resembling the ATP_i-dependent Ca inflow mode of intact cells with its low apparent affinity (mM) for Ca_o. Most workers have found very little evidence for an effect of ATP on the vesicular exchanger, although Caroni and Carafoli (1983) have reported that the exchange fluxes can be modified by ATP, apparently via a calcium plus calmodulin-dependent protein kinase.

These differences between vesicles and intact cells deserve close examination. A symmetrical exchange such as that described for vesicles seems not to exist in intact axons where a high affinity calcium-binding site is exposed only at the internal face of the membrane and the external face displays a low affinity. It is perhaps worth speculating that this alteration in affinity may reflect a component that is missing in vesicles - either because

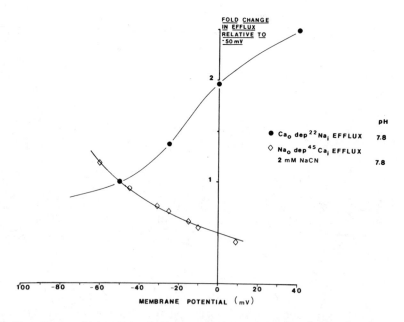

Fig.3.7. The voltage sensitivity of the Na_i/Ca_o and Ca_i/Na_o exchange fluxes under voltage clamp conditions. Points are from two experiments. They are the fold changes in the Na_o-dependent Ca efflux and Ca_o-dependent Na efflux normalised to the fluxes at -50 mV, following electrical manipulation of V_m. Determined in Na/choline superfusate containing 0.6 μM TTX and 10^{-5}M ouabain. Axons preinjected with TEA, MOPS and/or HEPES (pH 7.3) as in Fig.3.6.

it is soluble and lost during vesicle purification or because it is modified during vesicle isolation. One possibility might be proteolytic modification during vesicle preparation. Exposure of the external face of the squid axon to pronase has no obvious effect on Na—Ca exchange (Baker and McNaughton, 1978); but proteolysis of vesicles enhances exchange activity (Philipson and Nishimoto, 1982). Treatment of vesicles separately with phospholipases C and D is also known to stimulate Na—Ca exchange activity, and phospholipase D increases the affinity for Ca (Philipson and Nishimoto, 1984; Philipson, Frank and Nishimoto, 1983).

Bearing these *caveats* in mind, there is a wealth of data on the properties of the vesicular Na—Ca exchanger. It appears not to require Mg (Gill, Grollman and Kohn, 1981) and is inhibited by a range of divalent and trivalent cations, including Cd, Sr, Mn, Ba, Mg, and La (Ledvora and Hegyvary, 1983; Trosper and Philipson, 1983). It is not clear whether these cations are also transported - although work on the squid axon has revealed transport of Sr, Mg and Mn in a Na-dependent fashion (Baker and Crawford, 1972; Baker and Singh, 1982). The vesicular exchanger resembles the squid axon in being activated by alkaline pH (Philipson, Bersohn and Nishimoto, 1982; Wakabayashi and Goshima, 1982), and seems to be electrogenic (Caroni, Reinlib and Carafoli, 1980; Ledvora and Hegyvary, 1983; Philipson and Nishimoto, 1980, Reeves and Sutko, 1980). It appears therefore to share many features in common with the exchanger of intact cells and elucidation of its molecular properties are eagerly awaited.

3.5. Physiological role

The roles played by the exchanger are still controversial. Under resting conditions it may represent an important pathway for Ca entry into cells. Additionally, Ca entry via this route will be enhanced by electrical depolarisation, by a rise in intracellular sodium and by the positive feedback

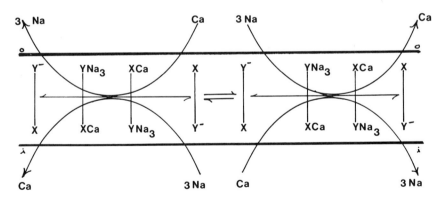

Fig.3.8. One possible model based on a dipole exchanger bearing a single negative charge. Exchange is shown as simultaneous. This model would be symmetrical and needs some other reaction to accommodate the experimental finding that a high affinity Ca-binding site is not normally exposed at the external face of the membrane.

process described earlier. Ca efflux via the exchanger seems only likely to be important under conditions where cytosolic Ca is elevated especially under conditions where the membrane potential is hyperpolarised; but this may include many physiologically important conditions. A clear answer on the role of Na–Ca exchange in a particular tissue must wait the development of specific inhibitors of the exchanger; but it is of considerable interest that the amiloride derivatives mentioned earlier seem to reverse the positive inotropic action of the cardiac glycosides (Siegl et al., 1984) which have long been suspected to exert their action via increased Na_i-dependent Ca influx (Baker et al., 1969).

References

Allen, T.J.A. and Baker, P.F. (1983). Comparison of the effects of potassium and electrical depolarization on Na/Ca exchange in squid axons. *J. Physiol.* 345, 80P.

Baker, P.F. (1970). Sodium-calcium exchange across nerve cell membranes. *Calcium and cellular function.* p. 96–107. A.W. Cuthbert (Ed.), Macmillan, London.

Baker, P.F. (1972). Transport and metabolism of calcium ions in nerve. *Prog. Biophy. Mol. Biol.*, 24, 177–223b.

Baker, P.F. (1978). The regulation of intracellular calcium in giant axons of *Loligo* and *Myxicola*. *Ann. NY Acad. Sci.* 307, 250–68.

Baker, P.F. and Blaustein, M.P. (1968). Sodium-dependent uptake of calcium by crab nerve. *Biochim. Biophys. Acta.* 150, 167–70.

Baker, P.F., Blaustein, M.P., Hodgkin, A.L. and Steinhardt, R.A. (1967a). The effect of sodium concentration on calcium movements in giant axons of *Loligo forbesi*. *J. Physiol.* 192, 43P–4P.

Baker, P.F., Blaustein, M.P., Hodgkin, A.L. and Steinhardt, R.A. (1969). The influence of calcium on sodium efflux in squid axons. *J. Physiol.* 200, 431–58.

Baker, P.F., Blaustein, M.P., Manil, J. and Steinhardt, R.A. (1967b) Ouabain-insensitive, calcium-sensitive sodium efflux from giant axons of *Logigo*. *J. Physiol.* 191, 100P–2P.

Baker, P.F. and Crawford, A.C. (1972). Mobility and transport of magnesium in squid giant axons. *J. Physiol.* 227, 855–74.

Baker, P.F. and Dipolo, R. (1984). Axonal calcium and magnesium homeostasis. *Curr. Top. Memb. Transp.* 22, 195–248.

Baker, P.F. and Glitsch, H.G. (1973). Does metabolic energy participate directly in the Na-dependent extrusion of Ca ions from squid axons? *J. Physiol.* 233, 44P.

Baker, P.F. and Honerjager, P. (1978). Influence of carbon dioxide on level of ionised calcium in squid axons. *Nature (Lond.)* 273, 160–1.

Baker, P.F., Meves, H. and Ridgway, E.B. (1973). The effects of manganese and other agents on the calcium uptakes that follows depolarization of squid axons. *J. Physiol.* 231, 511-26.

Baker, P.F. and McNaughton, P.A. (1976). Kinetics and energetics of calcium efflux from intact squid axons. *J. Physiol.* 259, 103–44.

Baker, P.F. and McNaughton, P.A. (1977). Selective inhibition of the calcium-dependent sodium efflux from intact squid axons by a fall in intracellular pH. *J. Physiol.* 269, 78P.

Baker, P.F. and McNaughton, P.A. (1978). The influence of extracellular binding on the calcium efflux from squid axons. *J. Physiol.* 276, 127–50.

Baker, P.F. and Singh, R. (1982). Metabolism and transport of strontium in giant axons of *Loligo*. *J. Physiol.* 330, 373–92.

Bers, D.M., Philipson, K.D. and Nishimopo, A.Y. (1980). Sodium–calcium exchange and sidedness of isolated cardiac sarcolemmal vesicles. *Biochim. Biophys. Acta.* 601, 358–71.

Blaustein, M.P. (1974). The interrelationship between sodium and calcium fluxes across cell membranes. *Rev. Physiol. Bioch. Exp. Pharmacol.* 70, 33–82.

Blaustein, M.P. (1977). Effects of internal and external cations and of ATP on sodium–calcium and calcium–calcium exchange in squid axons. *Biophys. J.* 20, 79–'0.

Blaustein, M.P. and Russell, J.M. (1975). Sodium–calcium exchange and calcium–calcium exchange in internally dialysed squid giant axons. *J. Memb. Biol.* 22, 285–312.

Blaustein, M.P., Russell, J.M. and De Weer, P. (1974). Calcium efflux from internally dialysed squid axons: the influence of external and internal cations. *J. Supramol. Struct.* 2, 558–81.

Caroni, P. and Carafoli, E. (1983). The regulation of the Na^+-Ca^{2+} exchanger of heart sarcolemma. *Eur. J. Biochem.* 132, 451–60.

Caroni, P., Reinlib, L. and Carafoli, E. (1980). Charge movements during the Na^+-Ca^{2+} exchange in heart. *Proc. Natl. Acad. Sci.* 77, 6354–8.

Caroni, P., Villani, F. and Carafoli, E. (1981). The cardiotoxic antibiotic doxorubicin inhibits the Na^+/Ca^{2+} exchange of dog heart sarcolemmal vesicles. *FEBS Lett.* 130(2), 184–6.

Crompton, M., Capano, M. and Carafoli, E. (1976). The sodium-induced efflux of calcium from heart mitochodria. *Eur. J. Biochem.* 69, 453–62.

Dipolo, R. (1979). Calcium influx in internally dialysed squid giant axons. *J. Gen. Physiol.* 73, 91–113.

Gill, D.L. (1982). Sodium channel, sodium pump, and sodium–calcium exchange activities in synaptosomal plasma membrane vesicles. *J. Biol. Chem.* 257, 10986–90.

Gill, D.L., Grollman, E.F. and Kohn, L.D. (1981). Calcium transport mechanisms in membrane vesicles from guinea-pig brain synaptosomes. *J. Biol. Chem.* 256, 184–92.

Hagiwara, S. and Byerly, L. (1981). Calcium channel. *Ann. Rev. Neuosci.* 4, 69–125.

Kaczorowski, G.J., Barros, F., Dethmers, J.K., Trumble, M.J. and Cragoe, E.J. (1985). Inhibition of Na^+/Ca^{2+} exchange in pituitary plasma membrane vesicles by analogues of amiloride. *Biochemistry* 24, 1394–403.

Kostyuk, P.G. (1981). Calcium channels in the neuronal membrane. *Biochim. Biophys. Acta.* 650, 128–50.

Ledvora, R.F. and Hegyvary, C. (1983). Dependence of Na^+-Ca^{2+} exchange and Ca^{2+}-Ca^{2+} exchange on monovalent cations. *Biochim. Biophys. Acta.* 729, 123–36.

Luciani, S. (1984). Reconstitution of the sodium–calcium exchanger from cardiac sarcolemmal vesicles. *Biochim. Biophys. Acta.* 772, 127–34.

Miyamoto, H. and Racker, E. (1980). Solubilization and partial purification of the Na^+/Ca^{2+} antiporter from the plasma membrane of bovine heart. *J. Biol. Chem.* 255, 2656–8.

Mullins, L.J. and Brinley, F.J. (1975). The sensitivity of calcium efflux from squid axons to changes in membrane potential. *J. Gen. Physiol.* 50, 2333–55.

Nicholls, D. and Akerman, K. (1982). Mitochondrial calcium transport. *Biochim. Biophys. Acta.* 683, 57–88.

Peterson, A.A., Matsumura, F. and McGroarty, E.J. (1984). Na^+–Ca^{2+} exchange in the axolemma-rich membrane vesicle preparation from the walking-leg nerve of the American lobster. *Biochim. Biophys. Acta.* **771**, 53–8.

Philipson, K.D., Bersohn, M.M., Nishimoto, A.Y. (1982). Effects of pH on Na^+-Ca^{2+} exchange in canine cardiac sarcolemmal vesicles. *Circ. Res.* **50**, 287–93.

Philipson, K.D., Frank, J.S. and Nishimoto, A.Y. (1983). Effects of phospholipase C on the Na^+-Ca^{2+} exchange and Ca^{2+} permeability of cardiac sarcolemmal vesicles. *J. Biol. Chem.* **258**, 5905–10.

Philipson, K.D. and Nishimoto, A.Y. (1980). Na^+-Ca^{2+} exchange is affected by membrane potential in cardiac sarcolemmal vesicles. *J. Biol. Chem.* **255**, 6880–2.

Philipson, K.D. and Nishimoto, A.Y. (1982). Stimulation of Na^+-Ca^{2+} exchange in cardiac sarcolemmal vesicles by proteinase pretreatment. *Am. J. Physiol.* **243**, C191–5.

Philipson, K.D. and Nishimoto, A.Y. (1984). Stimulation of Na^+-Ca^{2+} exchange in cardiac sarcolemmal vesicles by phospholipase D. *J. Biol. Chem.* **259**, 16–9.

Pitts, B.R.J. (1979). Stoichiometry of Na-Ca exchange in cardiac sarcolemmal vesicles. *J. Biol. Chem.* **254**. 6232–5.

Reeves, J.P. and Sutko, J.L. (1979). Sodium–calcium exchange in cardiac membrane vesicles. *Proc. Natl. Acad. Sci.* **76**, 590–4.

Reeves, J.P. and Sutko, J.L. (1980). Sodium-calcium exchange activity generates a current in cardiac membrane vesicles. *Science.* **208**, 1461–3.

Reuter, H. and Seitz, N. (1968). The dependence of calcium efflux from cardiac muscle on temperature and external ion composition. *J. Physiol.* **195**, 451–70.

Saermark, T., Thorn, N.A. and Gratzl, M. (1983). Calcium/sodium exchange in purified secretory vesicles from Bovine neurohypophyses. *Cell Calcium* **4**, 151–70.

Schellenberg, G.D. and Swanson, P.D. (1981). Sodium-dependent and calcium-dependent calcium transport by rat brain microsomes. *Biochim. Biophys. Acta.* **648**, 13–27.

Schellenberg, G.D. and Swanson, P.D. (1982a). Solubilisation and reconstitution of membrane containing the Na^+-Ca^{2+} exchange carrier from rat brain. *Biochim. Biophys. Acta.* **690**, 133–44.

Schellenberg, G.D. and Swanson, P.D. (1982b). Properties of the Na^+-Ca^{2+} exchange transport system from rat brain: inhibition by amiloride. *Fed. Proc.* **41**, 673.

Siegl, P.K.S., Cragoe, E.J.Jr., Trumble, M.J. and Kaczorowski, G.J. (1984). Inhibition of Na^+-Ca^{2+} exchange in membrane vesicles and papillary muscle preparations from guinea-pig heart by analogs of amiloride. *Proc. Natl. Acad. Sci.* **81**, 3238–42.

Trosper, T.L. and Philipson, K.D. (1983). Effects of divalent and trivalent cations on Na^+-Ca^{2+} exchange in cardiac sarcolemmal vesicles. *Biochim. Biophys. Acta* **731**, 63–8.

Wakabayashi, S. and Goshima, K. (1982). Partial purfication of Na^+-Ca^{2+} antiporter from plasma membrane of chick heart. *Biochim. Biophys. Acta.* **693**, 125–33.

# THE PLASMA MEMBRANE CALCIUM PUMP

H.J. Schatzmann

Department of Veterinary Pharmacology, University of Bern, Switzerland

Abstract

Ca is removed from the cytosol by ATP-fuelled Ca-pumps located in the endoplasmic reticulum (notably in skeletal muscle amd myocardium) and in the plasma membrane. The two pumps perform similarly but differ in molecular weight, antigenicity, substrate specificity, calmodulin sensitivity and a few more details. They are able to operate against a gradient of 10 000 and keep the $[Ca^{2+}]_i$ at $<10^{-7}$M. It is likely that Ca^{2+} movement is achieved by exchange for H^+. ATP hydrolysis proceeds under transient phosphorylation of the protein and two conformational states of the P-protein can be discerned that obviously correspond to forms with the Ca^{2+} binding site in the in and out position, respectively. The calmodulin mechanism in the plasma membrane pump introduces hysteretic behaviour leading to a prolongation of high $[Ca^{2+}]_i$ after a shortlived increase in Ca^{2+} permeability. The plasma membrane pump has been found in animal, plant and bacterial cells. In animal cells the Na—Ca exchange system establishes a link between the Na pump and the Ca pump such that the concentration of either cation depends on the activity of both the Na pump and the Ca pump. The biological meaning of the sarcoplasmic pump is to mop up the Ca^{2+} released in twitch activity while the plasma membrane pump keeps the overall cellular Ca concentration low indefinitely.

4.1. Introduction

It seems to be a general principle for living cells to keep the intracellular free Ca^{2+} concentration very far below that existing in the external medium. In many instances reliable estimates exist indicating that $[Ca^{2+}]_i$ is < 10^{-7} M in a medium of $[Ca]_o \sim 1$ mM (mammalian extracellular fluid) or even 10 mM (seawater). One way of maintaining a ratio $[Ca]_o:[Ca]_i$ of 10 000 across the plasma membrane (PM) which is by no means tight to Ca^{2+} is the Na—Ca exchange mechanism described for many animal cells (Baker, 1972; Blaustein and Nelson, 1982; Reuter, 1982; Schatzmann, 1984). This system exploits the inward Na gradient created by the Na—K pump to energise the uphill Ca^{2+} extrusion from the cell. Another way of ridding the cytosol of Ca^{2+} is the plasma membrane Ca pump driven directly by ATP (Penniston, 1983; Sarkadi, 1980; Schatzmann 1975, 1982, 1985).

In addition to these mechanisms that not only maintain the cytosolic Ca^{2+} concentration at its low value but are able to keep the overall Ca content of the cell low, there are internal Ca storage organelles that accumulate Ca^{2+} against a gradient. They are membrane-enclosed spaces such as the endoplasmic reticulum in all cells, the sarcoplasmic reticulum (SR) in muscle and the dense tubular system in platelets. The SR Ca pump is very similar to, though not identical with that of the plasma membrane. Its study started with the discovery by Hasselbach and Makinose (1961) that the 'relaxing factor' of Marsh is not a protein but is made up of Ca-accumulating vesicles. It is by now the best known type of ATP-driven Ca pumps.

Mitochondria accumulate Ca^{2+} by moving a (charged) Ca–carrier complex across the inner membrane along the potential gradient created by the proton extrusion mechanism. This Ca^{2+} transport becomes prominent only if the Ca^{2+} exporting contrivances at the plasma membrane fail (Somlyo, 1984), but it might be important for keeping mitochondrial enzymes at the proper Ca^{2+} concentration in the intact cells.

The ATP-driven Ca pump of the plasma membrane (PM) has been demonstrated in very many cells (Martonosi, 1983; Schatzmann, 1982, 1985; Sulakhe and St Louis, 1980) and seems to be a universal attribute of all cells, whereas the SR Ca pump of which we have a more detailed knowledge is a highly specialised arrangement adapted to muscle.

Basically the two Ca pumps perform very similarly (Fig.4.1.), accepting the terminal phosphate group (P) from ATP and forming an acyl-P-bond in one conformation (E_1) of the protein if Ca^{2+} is attached to the high affinity site

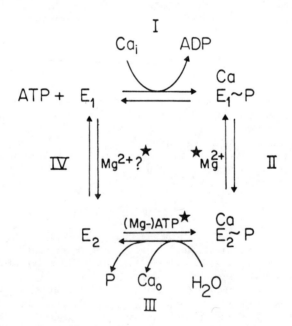

Fig.4.1. Reaction cycle of the Ca pump. E_1 and E_2 are two conformations of the protein. The cycle starts in the left upper corner by phosphorylation of the protein (I) when Ca^{2+} is attached to the high affinity site (internally in red cells) and proceeds clockwise. E_1~P undergoes a conformational change (II) translocating Ca^{2+} across the membrane (and lowering the affinity of the Ca^{2+}-binding site). Step II requires Mg^{2+} ions. Ca^{2+} is released (outside in red cells) and E_2~P is hydrolysed (III). Step III requires high concentration of ATP, possibly as MgATP.E_2 returns to E_1 with the Ca^{2+} site probably in the protonated form (IV).

of the system, whereupon a conformational change of the $E_1 \sim P$ to the $E_2 \sim P$ occurs. The $E_1 \sim P \to E_2 \sim P$ conformational change probably achieves the translocation of the Ca^{2+} ion across the membrane and the drop of the affinity for Ca^{2+} such that first Ca^{2+} release and then hydrolysis of the $E_2 \sim P$-bond ensue. Finally E_2 returns to the form E_1 and the cycle starts again. However, there are differences between the two types of Ca pumps:

(1) The monomeric molecular weight is ~ 110 kD in SR and ~ 140 kD in the PM.
(2) The peptide pattern obtained after proteolysis is different.
(3) The SR pump is not directly affected by calmodulin, whereas the PM pump is stimulated by it.
(4) The SR pump moves Ca^{2+} at the expense of acylphosphates or paranitrophenylphosphate instead of ATP, a behaviour not seen in the PM pump.
(5) Antibodies against one do not react with the other.

For further differences see the review by Schatzmann (1985).

The degree of coupling between the ATP hydrolysis reaction and the Ca^{2+} transport is unknown, because the Ca^{2+} leak and the presence of other ATPases in the available membrane preparations in which $[Ca]_i$, $[Ca]_o$, $[ATP]$, $[ADP]$ and $[P]_i$ can be set at will (resealed red cells) have so far prevented precise measurements of efficiency.

The following deals with the PM Ca pump as studied in the human red cell, which does not possess a Na—Ca exchange system and in which 95% of the Ca is confined to a single compartment (Simonsen, Gomme and Lew, 1982). This makes it a seemingly simple object to study.

4.2. The red cell Ca pump

4.2.1. *Functional aspects*

Estimates of the free Ca^{2+} concentration inside human red cells have been decreasing continuously. The latest figure is $\leq 3 \cdot 10^{-8}$ M (Lew et al., 1982) and the passive (inward) leak under physiological conditions is ~ 50 μmol·(1 cell)$^{-1}$· h^{-1} (Lew et al., 1982). These two facts can only be reconciled by assuming an active uphill extrusion mechanism whose existence was first demonstrated in 1966 by Schatzmann. This Ca pump extrudes some 10 mmol·(1 cell)$^{-1}$· h^{-1} at saturating $[Ca]_i$.

The Ca^{2+} affinity at the internal side rises with increasing pH (Luterbacher, unpublished observations), the K_{Ca} being slightly less than 10^{-6} M at pH 7. Interestingly, the Hill coefficient for Ca^{2+} binding rises with increasing pH (Luterbacher, unpublished observations). This is similar to what is observed in the SR (Hill and Inesi, 1982) and indicates that protons act competitively with Ca^{2+} at the high affinity transport site but that the interaction is complex. Complicated effects might be due to proton—Ca^{2+} interaction at calmodulin and interference of protons at the dephosphorylation of $E_2 \sim P$. There is also preliminary evidence for H^+-Ca^{2+} competition at the low affinity Ca^{2+} site (Schatzmann, unpublished observations). The competitive behaviour between Ca^{2+} and H^+ at the internal side (high affinity site) is compatible with the claim that the pump exchanges Ca^{2+} for protons (Niggli, Sigel and Carafoli, 1982; Smallwood et al., 1983), whereas an exchange of Ca^{2+} for K^+, Na^+ or Mg^{2+} or a transport of an ion pair (Ca^{2+} with 2 Cl$^-$) can be ruled out (Rossi and Schatzmann, 1982;

Schatzmann, 1975). Nevertheless, K^+ and Na+ have a stimulatory effect (some 30–50%) (Kratje, Garrahan and Rega, 1983; Romero and Romero, 1984), which is also true for the SR Ca pump.

The system is a Ca^{2+}-dependent ATPase which appears only in the presence of free Mg^{2+} ions. The Mg^{2+} requirement is crucial, is displayed by the SR Ca pump as well and was found in all PM Ca pumps. It is almost certain that Ca ATPases not requiring Mg^{2+} (or being inhibited by Mg^{2+}) are not Ca pumps. There is good evidence that the Mg^{2+} is necessary at the conformational change $E_1 \sim P \rightarrow E_2 \sim P$ (Garrahan and Rega, 1978; Luterbacher and Schatzmann, 1983).

For ATP there are two sites of different affinity (Muallem and Karlish, 1979, 1980, 1981; Richards, Rega and Garrahan, 1978). The enzymic site proper has a K_m of 1–3 μM, and in addition there is a 'regulatory' site with a K_{ATP} of ~100 μM or higher. Since Garrahan and Rega (1978) have shown that high concentration of ATP ($\geq$100 μM) is required for dephosphorylation, the regulatory site must be attributed to the E_2 conformation. Interpreting this behaviour as two distinct sites for ATP may be erroneous. It might be that the same site appears with two different affinities in the two conformations E_1 and E_2.

ATP is by far the best substrate, other nucleoside triphosphates sustaining only minor activity (Sarkadi, 1980). There is no definitive answer to the question whether Mg–ATP or free ATP (or even Ca–ATP) is required at the enzymic site. The fact that formation of the phosphoprotein ($E_1 \sim ^{32}P$) is possible in the absence of added Mg^{2+} (but has a strict Ca^{2+} requirement) argues against Mg–ATP as the unique substrate. Thus far Sr^{2+} was the only other cation transported by the Ca pump. Very recently Simons (1984) discovered that red cells also transport Pb^{2+} actively. The Pb^{2+} transport is antagonised by Ca^{2+} and might be mediated by the Ca pump.

The stoichiometry of the process (g-atoms Ca^{2+} transported per mol of ATP hydrolysed) is not settled (Sarkadi, 1980; Sarkadi et al., 1977). Whereas in the SR it seems to be 2, in the red cell several independent measurements gave values nearer to 1 than 2 (Clark and Carafoli, 1983; Ferreira and Lew 1976; Muallem and Karlish, 1979; Schatzmann and Roelofsen, 1977). If the figure for $Ca_i \leq 3 \cdot 10^{-8}$ M of Lew et al. (1982) is correct, the stoichiometry cannot be 2 for thermodynamic reasons.

The SR protein can be phosphorylated from inorganic phosphate (^{32}P) if a Ca^{2+} gradient is set up in the direction opposite to that of the transport movement. If the pH is made low, $[Mg^{2+}]$ high and $[Ca^{2+}]$ = O there is even formation of $E_2 \sim P$ from phosphate without a Ca^{2+} gradient at elevated temperature (De Meis, Martins and Alves, 1980; Kolassa et al., 1979; Punzengruber et al., 1978). This would suggest that $E_2 \sim P$ is a 'low energy' phosphate compound. The same experiments have not been successful so far in red cells (Luterbacher, 1982). However, a recent communication by Chiesi, Zurini and Carafoli (1985) has shown that if the water activity is reduced (by adding 40% DMSO, a technique established for the SR Ca pump; De Meis, Martins and Alves, 1980), some phosphate can be incorporated even in the absence of a Ca^{2+} gradient. It seems, therefore, that the $E_2 \sim P$ hydrolysis is reversible, but that the equilibrium is more strongly poised to the left in the red cell than in the SR. Upon restoring 100% water activity and adding Ca^{2+} the ^{32}P-protein (which is presumably $E_2 \sim P$ after backward phosphorylation) phosphorylated ADP to $[^{32}P]$-ATP. This observation is compatible with the assumption that the change in Ca^{2+} affinity occurs at reaction II, that is, before dephosphorylation in the forward mode of the pump.

As in the SR Ca pump the reversibility of the whole cycle under a Ca^{2+}

gradient opposite to the Ca pumping is ascertained by the fact that the system can generate ATP on account of PO_4^{3-} and ADP (Ferreira and Lew, 1975; Rossi, Garrahan and Rega, 1978; Wüthrich, Schatzmann and Romero, 1979). The latter authors found an ATP production of ~15 pmol· min^{-1}· (mg protein)$^{-1}$ and an ADP-dependent Ca^{2+} efflux of ~240 pmol· min^{-1}· (mg protein)$^{-1}$ in inside-out red cell vesicles with ~2×10^{-2} M Ca^{2+} inside and $\leqslant$ 10^{-8} M outside during 15 min at 37°C in a medium containing 2 mM ADP, 10 mM PO_4^{3-} and ~1 µM ATP. On average the maximal rate of the forward reaction in red cells is ~30 nmol· min^{-1}· (mg protein)$^{-1}$. Thus the rate of the backward reaction is, under the conditions mentioned, somewhere between 1/2000 to 1/100 of that of the forward reaction, which may be due to reaction III ($E_2 \sim P \rightarrow E_2 + P$) being strongly exergonic. This in turn means that in the red cell $E_2 \sim P$ is not a 'low energy' phosphate compound.

4.2.2. Stimulation by calmodulin
The performance of the pump and the ($Ca^{2+} + Mg^{2+}$)-ATPase so far described holds when the solutions for preparing the membranes are not made rigorously Ca^{2+}-free. Treatment of the membranes with Ca chelators leads to removal of calmodulin. The discovery (Gopinath and Vincenzi, 1977; Jarrett and Penniston, 1978; Larsen and Vincenzi, 1979) that Ca calmodulin binds to the PM Ca pump and thereby increases the Ca^{2+} affinity of the transport site by a factor of 30 and the v_{max} by some 30% came rather as a surprise because the physiological significance of this is not at first sight obvious. The human red cells contain some 3–5 µmol·(l cells)$^{-1}$ calmodulin. Scharff and Foder (1978, 1982) have shown quite convincingly that at physiological [Ca]$_i$ the pump calmodulin complex dissociates and that beyond [Ca^{2+}]$_i$ = 10^{-6}M it associates and that the process is rather slow. This may be the reason for the observation made by Ferreira and Lew (1976) that stepping up the Ca^{2+} permeability of pumping cells by adding traces of the ionophore for divalent cations A 23187 does not lead immediately to a new elevated steady state [Ca^{2+}]$_i$ but that [Ca^{2+}]$_i$ first passes through a considerably higher maximum (Scharff, Foder and Skibsted, 1983). In excitable cells this slow activation of the Ca pump by incoming Ca^{2+} might have the consequence that a shortlived increase in Ca^{2+} permeability (for example, in an action potential) is followed by a prolonged state of elevated [Ca^{2+}]$_i$ outlasting the short burst of Ca^{2+} influx, whereas a Ca pump having high Ca^{2+} affinity at all Ca^{2+} concentrations might effectively 'exclude' Ca^{2+} from the cell interior. This may be meaningful for cellular processes triggered by Ca^{2+} that are slow compared to the duration of an action potential.

4.2.3. Lipid requirement
The protein spans the membrane and thus must have a hydrophobic domain. It is not surprising, therefore, that its function depends on the phospholipid environment. All glycerophospholipids are able to maintain function (Roelofsen, 1981) (and protect against irreversible deterioration). What seems to matter is their presence in a position corresponding to the inner leaflet of the bilayer. Interestingly acidic lipids such as phosphatidylserine are in addition able to mimic the action of calmodulin (Niggli, Adunyah and Carafoli, 1981; Niggli et al., 1981; Stieger and Luterbacher, 1981).

In the natural surroundings the enzyme is stimulatable by calmodulin, behaving as if it were surrounded by lecithin. Enyedi et al. (1982) have pointed out that phosphatidylinositol phosphates, another class of acidic lipids known to stimulate the pump, have a rapid turnover in red cell membranes and might, therefore, play a role in regulating the Ca pump.

4.2.4. Biochemical approach

The protein has been isolated in more than 95% pure form on a calmodulin affinity column (Gietzen, Tejcka and Wolf, 1980; Niggli, Penniston and Carafoli, 1979). Unfortunately a large part of the purified protein is not in a functional state. However, it was possible to incorporate the protein into phospholipid vesicles and to demonstrate transport and $(Ca^{2+} + Mg^{2+})$-ATPase in such vesicles (Niggli et al., 1981; Stieger, 1982).

Controlled proteolytic attack has provided interesting new insight. It was shown that when the molecular weight is reduced by about 30 kD, the requirement for calmodulin vanishes and calmodulin no longer binds to the pump protein (Niggli, Adunyah and Carafoli 1981; Stieger and Schatzmann, 1981; Sarkadi, Enyedi and Gardos, 1980). Zurini et al. (1984) have proposed that a 10 kD fragment appearing in the hydrolysate is responsible for calmodulin binding. Further, the fragment carrying the ATP-binding site and the hydrophobic segment have been identified. With these achievements the (long) way is paved which will eventually lead to the elucidation of the primary structure of the molecule. The low abundance of the protein in red cell membranes makes chemical work difficult. It amounts to about 0.1% of the membrane proteins or a few 1000 copies per cell, whereas in the SR the pump protein amounts to some 70% of all membrane proteins.

An important question is whether the protein aggregates within the membrane. What is interesting is not the purely morphological question of whether the protein forms dimers (or high aggregates) in the membrane, but rather the question whether a dimerisation is necessary for function. The technique of radiation inactivation seems suitable for answering this latter question. Two studies (Cavieres, 1984; Minocherhomjee et al., 1983) were carried out with this elegant technique and both are much in favour of the assumption that it is the dimeric form which is actively transporting Ca^{2+}. The work by Cavieres (1984) in addition suggests the interesting possibility that higher aggregates are inert and that calmodulin prevents their formation within the membrane.

4.3. General aspects

A Ca pump of the red cell type has been found in nerve, synaptosomes, cardiac muscle, pancreas islets, intestine and kidney epithelia, the liver cell, adipocytes, macrophages, lymphocytes, neutrophil granulocytes, Ehrlich ascites tumour cells, fibroblasts, bone cells, corpus luteum cells and spermatozoa. It seems to be present also in bacteria and green plants (Martonosi, 1983; Schatzmann, 1985).

Unlike the human red cell, many of these cells of higher organisation also posses the Ca–Na exchange system in the plasma membrane. In at least some of them the ATP-driven pump has a higher Ca^{2+} affinity and a lower capacity than the exchange mechanism (Di Polo, 1978; Di Polo and Beaugé, 1979, 1980). If the leak for Ca^{2+} is negligible, the exchange system will bring the $[Na^+]_i$ and the $[Ca^{2+}]_i$ to an equilibrium value. Any action of the pump superimposed on this state will necessarily lead to Ca^{2+} entry and Na^+ extrusion through the exchange system. In other words, both the Na–K pump and the Ca pump can serve to remove Na^+ or Ca^{2+} according to circumstances, owing to the coupling across the Ca–Na exchange system. A natural experiment putting this notion to test is the dog red cell, which lacks a Na–K pump but possesses a Ca pump and a Ca–Na exchange system. It is believed that Na^+ is prevented from swamping the cell (which would ultimately lead to the cell bursting) by the combination of the Ca pump with the Ca–Na exchange mechanism.

In the general case, the $[Ca^{2+}]_i$ may never fall to the value dictated by the equilibrium across the Na–Ca exchange system owing to considerable Ca^{2+} leakage such that the pump and the Ca–Na exchange may have their share in Ca^{2+} extrusion. At high $[Ca^{2+}]_i$ the exchange system will account for most of the Ca^{2+} efflux while the pump will become more and more prominent as the $[Ca^{2+}]_i$ decreases.

References

Baker, P. (1972). Transport and metabolism of calcium ions in nerve. *Prog. Biophys. Molec. Biol.* **24**, 177–223.
Blaustein, M.P. and Nelson, M.T. (1982). Sodium–calcium exchange: its role in the regulation of cell calcium. In *Membrane Transport of Calcium*, E. Carafoli (ed.), pp. 217–36. Academic Press London, New York.
Cavieres, J.D. (1984). Calmodulin and the target size of the $(Ca^{2+} + Mg^{2+})$-ATPase of human red cell ghosts. *Biochim. Biophys. Acta.* **771**, 241–4.
Chiesi, M., Zurini, M. and Carafoli, E. (1984). ATP synthesis catalyzed by the purified erythrocyte Ca-ATPase in the absence of Ca-gradients. *Biochemistry* **23**, 2595–9.
Clark, A. and Carafoli, E. (1983). The stoichiometry of the Ca^{2+} pumping ATPase of erythrocytes. *Cell Calcium.* **4**, 83–8.
De Meis, L., Martins, O.B. and Alves, E.W. (1980). Role of water, hydrogen ion and temperature on the synthesis of adenosine triphosphate by the sarcoplasmic reticulum adenosinetriphosphatase in the absence of a calcium gradient. *Biochemistry.* **19**, 4252–61.
Di Polo, R. (1978). Ca^{2+}-pump driven by ATP in squid axons. *Nature.* **274**, 390–2.
Di Polo, R. and Beaugé, L. (1979). Physiological role of ATP driven calcium pump in squid axon. *Nature.* **278**, 271–3.
Di Polo, R. and Beaugé, L. (1980). Mechanisms of calcium transport in the giant axon of the squid and their physiological role. *Cell Calcium.* **1**, 147–69.
Enyedi, A., Sarkadi, B., Nyers, A. and Gardos, G. (1982). Effects of divalent metal ions on the calcium pump and membrane phosphorylation in human red cells. *Biochim. Biophys. Acta.* **690**, 41–9.
Ferreira, H.G. and Lew, V.L. (1975). Ca transport and Ca-pump reversal in human red blood cells. *J. Physiol. (Lond.).* **252**, 86P.
Ferreira, H.G. and Lew, V.L. (1976). Use of ionophore A 23187 to measure cytoplasmic Ca-buffering and activation of the Ca-pump by internal calcium. *Nature.* **259**, 47–9.
Garrahan, P.J. and Rega, A.F. (1978). Activation of partial reactions by the Ca^{2+}-ATPase from human red cells by Mg^{2+} and ATP. *Biochim. Biophys. Acta.* **513**, 59–65.
Gietzen, K., Tejcka, M. and Wolf, U. (1980). Calmodulin affinity chromatography yields a functional purified $(Ca^{2+} + Mg^{2+})$-dependent adenosine triphosphatase. *Biochem. J.* **188**, 47–54.
Gopinath, R.M. and Vincenzi, F.F. (1977). Phosphodiesterase protein activator mimics red blood cell cytoplasmic activator of $(Ca^{2+} + Mg^{2+})$-ATPase. *Biochem. Biophys. Res. Comm.* **77**, 1203–9.
Hasselbach, W. and Makinose, M. (1961). Die Calciumpumpe der 'Erschlaffungsgrana' des Muskels und ihre Abhängigkeit von der

ATP-Spaltung. *Biochem. Z.* **333**, 518–28.
Hill,T.L. and Inesi, G. (1982). Equilibrium cooperative binding of calcium and protons by sarcoplasmic reticulum ATPase. *Proc. Natl. Acad. Sci. USA.* **79**, 3978–82.
Jarrett, H.W. and Penniston, J.T. (1977). Partial purification of the Ca^{2+} + Mg^{2+} ATPase activator from human erythrocytes. Its similarity to the activator of 3':5'-cyclic nucleotide phosphodiesterase. *Biochem. Biophys. Res. Comm.* **77**, 1210–6.
Kolassa, N., Punzengruber, C., Suko, J. and Makinose, M. (1979). Mechanism of calcium-independent phosphorylation of sarcoplasmic reticulum ATPase by orthophosphate. *FEBS Lett.* **108**, 495–500.
Kratje, R.B., Garrahan, P.J. and Rega, A.F. (1983). The effects of alkali metal ions on active Ca^{2+} transport in reconstituted ghosts from human red cells. *Biochim. Biophys. Acta.* **731**, 40–6.
Larsen, V.L. and Vincenzi, F.F. (1979). Calcium transport across the plasma membrane: stimulation by calmodulin. *Science.* **204**, 306–9.
Lew, V.L., Tsien, R.Y., Miner, C. and Bookchin, R.M. (1982). Physiological (Ca^{2+}) level and pump-leak in intact red cells measured using an incorporated Ca-chelator. *Nature.* **298**, 478–81.
Luterbacher, S. (1982). *Die Teilreaktionen der ATP-Spaltung durch das isolierte Protein der Ca^{2+}-Pumpe aus Erythrocytenmembranen.* Thesis, University of Bern.
Luterbacher, S. and Schatzmann, H.J. (1983). The site of action of La^{3+} in the reaction cycle of the human red cell membrane Ca^{+}-pump ATPase. *Experientia.* **39**, 311–2.
Martonosi. A.N. (1983). The regulation of cytoplasmic Ca^{2+} concentration in muscle and non-muscle cells. In *Muscle and Non-muscle Motility Vol. 1*, pp. 233–357. Academic Press, London, New York.
Minocherhomjee, A.M., Beauregard, G., Potier, M. and Roufogalis, B.D. (1983). The molecular weight of the calcium transport ATPase of the human red blood cell determined by radiation inactivation. *Biochem. Biophys. Res. Comm.* **116**, 895–900.
Muallem. S. and Karlish, S.J.D. (1979). Is the red cell calcium pump regulated by ATP? *Nature.* **277**, 238–40.
Muallem, S. and Karlish, S.J.D. (1980). Regulatory interaction between calmodulin and ATP on the red blood cell Ca^{2+}-pump. *Biochim. Biophys. Acta.* **597**, 631–6.
Muallem. S. and Karlish, S.J.D. (1981). Studies on the mechanism of regulation of the red cell Ca^{2+}-pump by calmodulin and ATP. *Biochim. Biophys. Acta.* **647**, 73–86.
Niggli, V., Adunyah, E.S. and Carafoli, E. (1981). Acid phospholipids, unsaturated fatty acids and limited proteolysis mimic the effect of calmodulin on the purified erythrocyte Ca^{2+}-ATPase. *J. Biol. Chem.* **256**, 8588–92.
Niggli, V., Adunyah, E.S., Penniston, J.T. and Carafoli, E. (1981). Purified (Ca^{2+} + Mg^{2+})-ATPase of the erythrocyte membrane. Reconstitution and effect of calmodulin and phospholipids. *J. Biol. Chem.* **256**, 395–401.
Niggli, V., Penniston, J.T. and Carafoli, E. (1979). Purification of the (Ca^{2+} + Mg^{2+})-ATPase from human erythrocyte membranes using a calmodulin affinity column. *J. Biol. Chem.* **254**, 9955–8.
Niggli, V., Sigel, E. and Carafoli, E. (1982). The purified Ca^{2+}-pump of human erythrocyte membranes catalyzes an electroneutral Ca^{2+}-H^{+} exchange in reconstituted liposomal systems. *J. Biol. Chem.* **257**, 2350–6.
Penniston, J.T. (1983). Plasma membrane Ca^{2+}-ATPase as active Ca^{2+}-pumps.

In *Calcium and Cell Function Vol. IV*, pp. 100–49. Academic Press, New York, London.

Punzengruber, C., Prager, R., Kolassa, N., Winkler, F. and Suko, J. (1978). Calcium gradient-dependent and calcium gradient-independent phosphorylation of sarcoplasmic reticulum by orthophosphate. *Eur. J. Biochem.* **92**, 349–59.

Reuter, H. (1982). Na–Ca countertransport. In *Membranes and Transport*, A.M. Martonosi (ed.), pp. 623–31. Plenum Press, New York, London.

Richards, D.E., Rega, A.F. and Garrahan, P.J. (1978). Two classes of site for ATP in the Ca^{2+}-ATPase from human red cell membranes. *Biochim. Biophys. Acta.* **511**, 194–201.

Roelofsen, B. (1981). The (non)-specificity in the lipid requirement of calcium-and (sodium plus potassium)-transporting ATPases. *Life Sci.* **29**, 2235–47.

Romero, P.J. and Romero, E. (1984). The modulation of the calcium pump of human red cells by Na^+ and K^+. *Biochim. Biophys. Acta.* **778**, 245–52.

Rossi, J.P.F.C., Garrahan, P.J., Rega, A.F. (1978). Reversal of the calcium pump in human red cells. *J. Memb. Biol.* **44**, 37–46.

Rossi, J.P.F.C. and Schatzmann, H.J. (1982). Is the red cell calcium pump electrogenic? *J. Physiol.* **327**, 1–15.

Sarkadi, B. (1980). Active calcium transport in human red cells. *Biochim. Biophys. Acta.* **604**, 159–90.

Sarkadi, B. Enyedi, A. and Gardos, G. (1980). Molecular properties of the red cell calcium pump I/II. *Cell Calcium.* **1**, 287–310.

Sarkadi, B., Szasz, I., Gerloczi, A. and Gardos, G. (1977). Transport parameters and stoichiometry of active calcium ion extrusion in intact human red cells. *Biochim. Biophys. Acta.* **464**, 93–107.

Scharff, O. and Foder, B. (1978). Reversible shift between two states of Ca^{2+}-ATPase in human erythrocytes mediated by Ca^{2+} and a membrane bound activator. *Biochim. Biophys. Acta.* **509**, 67–77.

Scharff, O. and Foder, B. (1982). Rate constants for calmodulin binding to Ca^{2+}-ATPase in erythrocyte membranes. *Biochim. Biophys. Acta.* **691**, 133–43.

Scharff, O., Foder, B. and Skibsted, U. (1983). Hysteretic activation of the Ca^{2+}-pump revealed by calcium transients in human red cells. *Biochim. Biophys., Acta.* **730**, 295–305.

Schatzmann, H.J. (1966). ATP-dependent Ca-extrusion from human red cells. *Experientia* **22**, 364–8.

Schatzmann, H.J. (1975). Active calcium transport and Ca^{2+} activated ATPase in human red cells. *Curr. Top. Memb. Trans.* **6**, 125–68.

Schatzmann, H.J. (1982). Active calcium transport in human red blood cells. In *Membranes and Transport, Vol. 1*, A.M. Martonosi (ed.), pp. 601–5. Plenum Press, New York, London.

Schatzmann, H.J. (1985). Calcium extrusion across the plasma membrane by the Ca^{2+}-pump and the Ca^{2+}–Na^+ exchange system. In *Calcium and Cell Physiology*, D. Marmé (ed.). pp. 18–52. Springer, Berlin, New York.

Schatzmann, H.J. and Roelofsen, B. (1977). Characteristics of the Ca-pump in human red blood cells. In *Biochemistry of Membrane Transport*, Semenza and Carafoli (eds.) pp. 389–400. Springer, Berlin, New York.

Simons, T.J.B. (1984). Active transport of lead by human red blood cells. *FEBS letters* **172**, 250–4.

Simonsen, L.O., Gomme, J. and Lew, V.L. (1982). Uniform ionophore A 23187 distribution and cytoplasmic calcium buffering in intact human red cells. *Biochim. Biophys. Acta,* **692**, 431–40.

Smallwood, J.I., Waisman, D.M., Lafreniere, D. and Rasmussen H. (1983).

Evidence that the erythrocyte calcium pump catalyzes a Ca^{2+} : n H^+ exchange. *J. Biol. Chem.* **258**, 11092—7.

Somlyo, A.P. (1984). Cellular site of calcium regulation. *Nature.* **309**, 516—7.

Stieger, J. (1982). Charakterisierung und Rekonstitution der isolierten Ca^{2+}-Pumpe aus der Erythrocytenmembran. Thesis, University of Bern.

Stieger, J. and Luterbacher, S. (1981). Some properties of the purified (Ca^{2+} + Mg^{2+})-ATPase from human red cells. *Biochim. Biophys. Acta.* **641**, 270—5.

Stieger, J. and Schatzmann, H.J. (1981). Metal requirement of the isolated red cell Ca^{2+}-pump ATPase after elimination of calmodulin dependence by trypsin attack. *Cell Calcium* **2**, 601—16.

Sulakhe,P.V. and St Louis, P.J. (1980). Passive and active calcium flux across plasma membranes. *Prog. Biophys. Molec. Biol.* **35**, 135—95.

Wüthrich, A., Schatzmann, H.J. and Romero, P. (1979). Net ATP-synthesis by running the red cell calcium pump backwards. *Experientia* **35**, 1589—90.

Zurini, M., Krebs, J., Penniston, J.T. and Carafoli. E. (1984). Controlled proteolysis of the purified Ca^{2+}-ATPase of the erythrocyte membrane. A correlation between structure and function of the enzyme. *J. Biol. Chem.* **259**, 618—27.

CA^{2+} SEQUESTRATION BY THE ENDOPLASMIC RETICULUM

Christian Petzelt

Institute of Cell and Tumour Biology, German Cancer Research Centre, D-69 Heidelberg, West Germany.

Abstract

The intracellular Ca^{2+} regulation is performed to a great extent by one or more Ca^{2+}-sequestering systems derived from the endoplasmic reticulum. Biochemical and physiological facts on the intracellular Ca^{2+}-control are reviewed and as one example the Ca^{2+}-regulation at mitosis is described in more detail. It appears very probable now that not the whole endoplasmic reticulum is involved in the intracellular Ca^{2+} regulation but only certain definable membranous parts. First attempts are described to analyse its components and understand its control.

In order to function as one of the 'second messengers', the intracellular concentration of free Ca^{2+} has to be maintained not only at a very low level but internal systems are required inside the cell to transduce the small Ca^{2+}-changes used for transferring the 'message'. The huge difference in free intracellular and extracellular Ca^{2+}, which is in the order of 10^4, is mainly maintained by the Ca^{2+}-control systems of the plasma membrane: the Ca^{2+}-channels regulate the influx of Ca^{2+} whereas the Na$^+$–Ca^{2+} exchanger is mainly responsible for the efflux control of Ca^{2+}. Finally, the Ca^{2+}-transport-ATPase system of the plasma membrane is used to exert the fine adjustments of the overall low Ca^{2+} concentration inside the cell.

The slow Ca^{2+} channels have been described in cardiac and smooth muscle where they are an essential part of the excitation–concentration coupling (Bolton, 1979; Reuter, 1983) and in somatic membranes of nerve cells both from invertebrates and vertebrates (Kostyuk and Doroshenko, 1983). A fair concept on the mechanism of the Ca^{2+} permeation has been developed recently (Hess and Tsien, 1984). The channels show a typical response to the so-called Ca^{2+} antagonists (Fleckenstein, 1977) which are now widely used to study their properties in other cell types too (Belleman *et al*, 1972; Lee and Tsien, 1983). Recently it was shown that small modifications of the nifedipine molecule result in dihydropyridines which show an opposite effect - that is, are positively inotropic (Schramm *et al.*, 1983 a and b; Towart and Schramm, 1984). These findings open up an exciting field for studying the part of the Ca^{2+} control by Ca^{2+} channels of the plasma membrane and possibly also of internal membranes.

Many parameters of the Na$^+$–Ca^{2+}-exchange system are known today (Caroni and Carafoli, 1983; Caroni *et al.*, 1980; Philipson and Nishimoto, 1980;

Reeves and Sutko, 1979, 1980), and they show this system as having a high capacity for the Ca^{2+}-transport but a rather low affinity. It represents most probably the main part of the mechanism which maintains the high intracellular and extracellular Ca^{2+} gradient. Unfortunately, many of its properties are still unknown, thus making it difficult to establish fully its physiological role for the cell.

The Ca^{2+} transport-ATPase of the plasma membrane, on the other hand, has been fully described (Niggli, Penniston and Carafoli (1979)) and its system reconstituted (Niggli et al., 1981). It is either activated by calmodulin, limited proteolysis or by acid phospholipids (Niggli, Adungah and Carafoli, 1981). Because of its rather low capacity but high affinity it is most probably used to regulate the internal Ca^{2+} concentration in very small increments. For a recent extensive review on its properties see Penniston (1983).

This low internal Ca^{2+} concentration is necessary for Ca^{2+} to transfer 'messages' from extracellular to intracellular targets. The compartmentalisation in the cell allows processes with different Ca^{2+} dependencies to occur at the same time separated either physically by membranes or by different modes of regulation.

The physical separation of the various Ca^{2+} domains is accomplished by membranous systems thought to belong to the endoplasmic reticulum. Mitochondria, as the other intracellular organelles capable of regulating Ca^{2+}, are probably not involved in the local control of Ca^{2+}, but because of their high capacity and low affinity for this ion may serve as the internal reservoir capable of removing or providing large amounts of Ca^{2+} when needed (Fiskum and Lehninger, 1982).

The use of different modes of regulation for perhaps identical Ca^{2+}-sequestering systems is a rather new concept, and the few facts available will be described at the end of this paper as well as its synergism and interdependence with the membrane-surrounded domains.

That intracellular Ca^{2+}-regulating systems must exist has been indicated by results obtained as early as 1937 when Mazia showed an increase in free Ca^{2+} after sea urchin eggs are fertilised. With eggs, in particular, these results have been extended using a variety of methods and species (Jaffe, 1980). The activation potential in frog eggs is initiated by Ca^{2+} (Cross, 1981) and exocytosis of the cortical granules induced (Hollinger, Dumont and Wallace, 1979). The activating effects of hormones on amphibian oocytes are associated with an increase in free Ca^{2+} (Moreau, Doree, Guerrier, 1976; Wassermann et al., 1980), and this increase has been shown to transmit the initiation of meiosis. In the Medaka fish egg a transient rise in free Ca^{2+} at fertilisation was demonstrated using the Ca^{2+} photoprotein aequorin (Gilkey et al., 1978). Steinhardt, Zucker and Schatten (1977) used similar methods to demonstrate that in the sea urchin egg the increase in free Ca^{2+} at fertilisation is brought about by the release of Ca^{2+} from internal stores rather than by an influx from the outside. Additionally, they showed that this internal Ca^{2+} increase causes the exocytosis of the cortical granules. Using Vacquier's (1975) method of isolating intact granules this approach provided informations on the intracellular concentration of free Ca^{2+} with yet another method. This relationship between Ca^{2+} and the exocytosis of the cortical granules in sea urchin eggs is completely described in Baker, Knight and Whitaker (1980).

Besides fertilisation, a cell cycle typical event is mitosis which appears to be correlated with an increase in free Ca^{2+}. If aequorin is injected into *Xenopus* eggs, a transient rise of free Ca^{2+} is observed (Baker and Warner, 1972). Electron microprobe analysis of mitotic sea urchin eggs reveal an increase of calcium above the site of the mitotic apparatus (Timourian, Jotz and Clothier,

1974). This mechanism responsible for the process of mitosis, the mitotic apparatus, has been recognised as a definite entity since its first isolation by Mazia and Dan (1952). One of its main properties is its calcium lability, and isolation procedures preserving this sensitivity have been developed (Salmon and Segall, 1980).

That this Ca^{2+} lability is indeed functionally related to the process of mitosis was demonstrated by the elegant experiments of Kiehart (1981). By microinjecting sea urchin eggs in mitosis with millimolar Ca^{2+} concentrations on various sites of the cell he was able to show that an injection of Ca^{2+} directly onto the mitotic spindle immediately reduced its birefringence, and the chromosomes ceased to move. A similar injection into the cytoplasm had no effect. However, this result was not so dramatic *per se*, since its cogent relation to the *in vivo* mechanism with regard to the actual Ca^{2+} effect is thereby not proven. Only a demonstration of a specific system capable of the regulation and sequestration of Ca^{2+} could have been taken as such a decisive support for the importance of Ca^{2+} at mitosis. Kiehart's further observations of the microinjected cell provided just this insight. Within minutes after the treatment the cell was able to reduce the locally very destructive Ca^{2+} concentration, restore the spindle birefringence and resume the movement of chromosomes. Mitosis was completed normally.

The existence of such intracellular Ca^{2+}-sequestering system has been shown by morphological and biochemical methods.

By the 1960s membranous intracellular elements were described which were thought to be involved in the intracellular Ca^{2+} regulation. Porter and Machado (1960) reported the distribution of the endoplasmic reticulum in mitotic plant cells (onion root tips), and Harris (1961, 1962) found an elaborate system of the endoplasmic reticulum (ER) around the mitotic apparatus of sea urchin eggs. Further work by Harrison on the ultrastructural distribution of the ER membranes in mitotic sea urchin eggs gave not only a complete overview of their arrangement (Harris, 1975), but led her also to the hypothesis that this membrane system may be the Ca^{2+}-sequestering entity controlling mitosis (Harris, 1978).

For mitotic plant cells a similar characterisation of ER membranes was carried out by Hepler (1971, 1980) and his group. They, however, extended these studies considerably by showing that some but not all of these membranes around and in the spindle contained calcium (Wick and Hepler, 1980; Wolniak, Hepler and Jackson, 1981). Using the fluorescence of a calcium—chlorotetracycline complex, this group confirmed the electron microscopical results described, namely that calcium-containing stores occur in and around the spindle (Wolniak, Hepler and Jackson, 1980). In isolated mitotic apparatus of sea urchin eggs a typical vesicular distribution of the calcium—chlorotetracycline complex was also found by Schatten, Schatten and Simerly (1982). Using, however, the isolated mitotic apparatus taken out of the cell, it is not possible to be sure if the vesicles described occur also *in vivo* as vesicles, or are merely disrupted membranes having joined on to vesicles after the isolation.

One of the few cases where investigations are extended beyond the mere ultrastructural level is found in the elegant work of Walz (1982 a and b). He describes a network of smooth endoplasmic reticulum cisternae in the photoreceptor cells of the blowfly *Calliphora erythrocephala*. He then shows that these membranes of the smooth endoplasmic reticulum are capable of accumulating Ca^{2+} in an ATP-dependent fashion. Very similar results were obtained by Ungar, Piscopo and Holtzmann (1981) and Ungar *et al.* (1984) for vertebrate terminal rods and for presynaptic nerve terminals from rat brain by McGraw, Somlyo and Blaustein (1980).

Besides this morphological evidence for an membranous intracellular Ca^{2+}-sequestering system, there are many biochemical observations pointing to a 'microsomal' Ca^{2+} pump. Selinger, Naim and Lasser (1970) described such a microsomal system from rat parotid and submaxillary glands. They used ouabain to discriminate between mitochondrial and microsomal transport; but they did not separate into plasma and intracellular membranes. Moore et al. (1974) isolated microsomal membranes of rat kidney, and by using sucrose density gradient centrifugation they could show the existence of an ER-derived membrane system which could accumulate Ca^{2+}. A similar system was shown to occur in brain microsomes (Trotta and de Meis, 1975) and in rat liver (Moore et al. (1975)). The microsomal membranes were insensitive to the mitochondrial inhibitors azide and oligomycine and, using density gradient centrifugation, evidence was given that the majority of these Ca^{2+}-pumping membranes belonged to the smooth ER. Support for these results was obtained when the Ca^{2+}-transport characteristics of these microsomes were determined (Bygrave, 1978). Their intracellular distribution and the ineffectiveness of the mitochondrial inhibitor ruthenium red confirmed that they were ER-derived (Bygrave and Trauter, 1978). These authors documented also the dependence of the Ca^{2+}-pump activity on the hormonal status of the liver. A Ca^{2+} uptake of the endoplasmic reticulum of adipocytes has been shown to be dependent in its activity on the Ca^{2+}-concentration previous to its isolation suggesting a feedback control mechanism for this ion (Black, Jarett and McDonald, 1980, 1981).

Such Ca^{2+}-transporting microsomal systems also exist in plants (Dieter and Marmé, 1980 a and b; Gross and Marmé, 1978; Marmé and Dieter, 1983). However, except in a paper by Gross (1982), no proper distinction between ER and plasma membrane-derived systems has been made. All the systems described were positively affected by calmodulin.

In contrast to the large amount of information on the microsomal Ca^{2+}-uptake system we are only beginning to understand its molecular composition. From centrifugation studies it was concluded that it consists of membranes of the smooth ER (Heilmann, Spanner and Gerok, 1983; Immelmann and Sölling, 1983). These results confirmed the morphological findings described earlier (Walz, 1982 a and b). Heilmann, Spanner and Gerok (1983) describe as one of its constituents a $Ca^{2+}-Mg^{2+}$-ATPase with an apparent molecular weight of 118 kilodaltons (kD) which can be phosphorylated by ATP. They stress its resemblance to the Ca^{2+}-ATPase of the sarcoplasmic reticulum. Silver, Cole and Cande (1980) reported the isolation of Ca^{2+}-sequestering vesicles of isolated mitotic apparatus of sea urchin eggs, and preliminary evidence was given that antibodies to such a vesicular fraction crossreact with the Ca^{2+}-ATPase of the sarcoplasmic reticulum (Silver, 1983).

Using a different approach we isolated first a membrane-bound Ca^{2+}-ATPase from isolated mitotic apparatus of sea urchin eggs (Petzelt, 1979). This protein is different from the Ca^{2+}-ATPase of the sarcoplasmic reticulum (Petzelt and Auel, 1978) and has been characterised extensively (Petzelt, 1979). During the cell cycle it shows characteristic fluctuations in its activity with an activity maximum at mitosis, and it was named, therefore, the 'mitotic' Ca^{2+}-ATPase (Petzelt and Auel, 1977). Such a protein was found in all eucaryotic cells studied so far. By means of immunocytochemical techniques it was localised in the cytoplasm, tentatively as part of the endoplasmic reticulum (Petzelt, 1984). On the basis of its biochemical properties, its intracellular location and its physiological behaviour during the

Ca^{2+}-sequestration by the e.r.

cell cycle we had postulated that it is an essential part of the intracellular Ca^{2+}-regulating system. However, one of the main arguments against this role was its low affinity (>10^{-5}M) to Ca^{2+} (Baker, 1976). However, this property was typical for the detergent-solubilised enzyme. Recently, we succeeded in isolating a vesicular Ca^{2+}-sequestering system where the 'mitotic' Ca^{2+}-ATPase is an essential part of it (Petzelt and Wülfroth, 1985). This system now has a Ca^{2+}-affinity of 5 <10^{-6}M. It has been found in sea urchin eggs as well as in mammalian cells and many of its properties have been described. There is, therefore, now sufficient evidence to be confident that the 'mitotic' Ca^{2+}-ATPase is an universal and necessary constituent of the intracellular Ca^{2+}-regulating system.

If such Ca^{2+}-sequestering membranes are isolated from a synchronous cell population at different points of the cell cycle, alterations in the Ca^{2+}-uptake capacity very similar to the cyclic behaviour of the solubilized enzyme can be observed. At least during the cell cycle of fertilised sea urchin eggs the Ca^{2+}-transport activity increases during mitosis in every cell cycle thus confirming the Ca^{2+}-regulation of the mitotic process described earlier (Petzelt and Wülfroth, 1985). Recently, similar results have been described by Suprynowicz, Poenie and Mazia (1983).

What do we know about the regulatory mechanisms of such intracellular Ca^{2+}-sequestering systems? Almost nothing. Calmodulin may be involved in some of the systems described (Marmé and Dieter, 1983) but at least the Ca^{2+}-sequestering system described by us neither contains calmodulin nor can it be activated by it. Although the Ca^{2+}-uptake is sensitive to halogenated phenothiazines (Petzelt and Wülfroth, 1985) we can show that this effect is caused by a strong binding of the Ca^{2+}-ATPase itself onto the phenothiazines (Wülfroth and Petzelt, unpublished observations). We must, therefore, assume that other regulatory components exist for the control of the intracellular Ca^{2+}-sequestering activity.

One additional difficulty is that the requirement for such mechanisms is such that they are not only able to exert their function in such a way that the Ca^{2+} concentration varies only in the range between 10^{-7} and 10^{-6}M. The compartmentalisation of the cell often requires that different Ca^{2+} concentrations for different processes occur at the same time. This can be brought about not only be varying the regulator for the Ca^{2+}-sequestering system but also by modulating its receptor. Conformational changes of the system may allow only a limited access of the regulatory components. First indications for the existence of such a mode were obtained using monoclonal antibodies to the 'mitotic' Ca^{2+}-ATPase in order to localise the enzyme in the isolated mitotic apparatus of sea urchin eggs (Petzelt and Wülfroth, 1984). We could show that different monoclonal antibodies stain differently the membranes in the mitotic apparatus varying with the progress through mitosis, and indicating therefore, the unequal exposure and conformation of such a system.

Summary and conclusions

All the evidence presented confirms the existence of an intracellular membrane system capable of sequestering Ca^{2+} ions. It can be clearly distinguished from the mitochondrial transport. It derives from the endoplasmic reticulum, but there are increasing indications that the morphological term 'reticulum' is becoming less and less useful for the positioning of such a system in the cell. Probably only some differentiated part of these membranes functions as a Ca^{2+}-pump. Since it seems to be possible now to analyse its components on a molecular level, it should become clear not only of what necessary parts it consists, but also which parts of the 'endoplasmic reticulum' form the Ca^{2+}-transporting apparatus. The formerly little known mechanism of the regulation of such a system in the cell may also become better understood. *In vitro* studies on the reaction of the Ca^{2+}-transport activity to a variety of calcium antagonists in its broadest sense should yield some information on its regulation. Finally, using some of the cell systems where transformation is expressed in an altered calcium metabolism, it should be possible to approach a causal analysis of this typical response to such a genome alteration.

Acknowledgements

I thank Petra Wülfroth for many helpful discussions. Part of this work was supported by the Deutsche Forschungsmeinschaft.

References

Baker, P.F. (1976). In *Mitosis, Facts and Questions.* M. Little, N. Paweletz, C. Petzelt, H. Ponstingl, D. Schroeter, H.P. Zimmermann (eds), p.48. Springer-Verlag, Berlin, Heidelberg, New York.

Baker, P.F. and Warner, A.E. (1972). Intracellular calcium and cell cleavage in early embryos of *Xenopus laevis. J. Cell Biol.* **53**, 579–81.

Baker, P.F., Knight, D.E. and Whitaker, M.J. (1980). The relation between ionized calcium and cortical granule exocytosis in eggs of the sea urchin *Echinus esculentus. Proc. R. Soc. Lond.,* **B207**, 149–61.

Bellemann, P., Ferry, D., Lübbecke, F. and Glossmann, H. (1982). ^{3}H-Nimodipine and ^{3}H-Nitredipine as tools to direct identify the sites of action of 1,4-dihydropyridine calcium antagonists in guinea-pig tissues. *Arzneim.-Forsch./Drug Res.* **32**, 361–3.

Black, B.L., Jarett, L. and McDonald, J.M. (1980). Relationship between calcium ion transport and (Ca^{2+}–Mg^{2+})-ATPase activity in adipocyte endoplasmic reticulum. *Biochim. Biophys. Acta.* **596**, 359–71.

Black, B.L., Jarett, L. and McDonald, J.M. (1981). The regulation of endoplasmic reticulum calcium uptake of adipocytes by cytoplasmic calcium. *J. Biol. Chem.* **256**, 322–9.

Bolton, T.B. (1979). Mechanism of action of neurotransmitters and other substances on smooth muscle. *Physiol. Rev.* **59**, 607–718.

Bygrave, F. (1978). Properties of energy-dependent calcium transport by rat liver microsomal fraction as revealed by initial-rate measurements. *Biochem. J.* **170**, 87–91.

Bygrave, F. and Trauter, C. (1978). The subcellular location, maturation and response to increased plasma glucagon of ruthenium red-insensitive calcium ion transport in rat liver. *Biochem. J.* **174**, 1021–30.

Caroni, P. and Carafoli, E. (1983). The regulation of the Na^+–Ca^{2+} Exchanger of heart sarcolemma. *Eur. J. Biochem.* **132**, 451–60.

Caroni, P., Reinlib, L. and Carafoli, E. (1980). Charge movements during the Na^+–Ca^{2+} exchange in heart sarcolemmal vesicles. *Proc. Natl. Acad. Sci. (USA).* **77**, 6354–8.

Cross, N.L. (1981). Initiation of the activation potential by an increase in intracellular calcium in the eggs of the frog *Rana pipiens. Devel. Biol* **85**, 380–4.

Dieter, P. and Marmé, D. (1980a). Calmodulin-activated plant microsomal Ca^{2+}-uptake and purification of plant NAD kinase and other proteins by calmodulin-sepharose chromatography. *Ann. NY Acad. Sci.* **356**, 371–3.

Dieter, P. and Marmé, D. (1980b). Calmodulin activation of plant microsomal Ca^{2+}-uptake. *Proc. Natl. Acad. Sci. USA.* **77**, 7311–4.

Fiskum, G. and Lehninger, A.L. (1982). Mitochondrial regulation of intracellular calcium. In *Calcium and Cell Function*, W.Y. Cheung, Vol. 2 (ed.), pp. 39–80, Academic Press, New York.

Fleckenstein, A. (1977). Specific pharmacology of calcium in myocardium, cardiac pacemakers and vascular smooth muscle. *Ann. Rev. Pharmacol. Toxicol.* **17**, 149–66.

Gilkey, J.C., Jaffe, L.F., Ridgway, E.B. and Reynolds, G.T. (1978). A free calcium wave traverses the activating egg of the medaka *Oryzias latipes*. *J. Cell Biol.* **76**, 448–66.

Gross. J. (1982). Oxalate-enhanced active calcium uptake in membrane fractions from zucchini squash. In: *Plasmalemma and Tonoplast, Their Functions in the Plant Cell.* D. Marmé, E. Marre and R. Hertel (eds.), pp. 369–76. Elsevier, Amsterdam.

Gross, J. and Marmé, D. (1978). ATP-dependent Ca^{2+}-uptake into plant membrane vesicles. *Proc. Natl. Acad. Sci. USA.* **75**, 1232–6.

Harris, P. (1961). Electron microscope study of mitosis in sea urchin blastomeres. *J. Biochem. Biophys. Cytol.* **11**, 419–31.

Harris, P. (1962). Some structural and functional aspects of the mitotic apparatus in sea urchin embryos. *J. Cell Biol.* **14**, 475–89.

Harris, P. (1975). The role of membranes in the organization of the mitotic apparatus. *Exp. Cell Res.* **94**, 409–25.

Harris, P. (1978). Triggers, trigger waves, and mitosis: A new model. In *Monographs on Cell Biology.* J.R. Jeter, I.L. Cameron, G.M. Padilla and A.M. Zimmerman, (eds.) pp. 75–104, Academic Press, New York.

Heilmann, C., Spanner, C. and Gerok, W. (1983). The phosphoprotein intermediate of a Ca^{2+}-transport ATPase in rat liver endoplasmic reticulum. *Biochem. Biophys. Res. Comm.* **114**, 584–92.

Hepler, P.K. (1977). Membranes in the spindle apparatus: their possible role in the control of microtubule assembly. In *Mechanisms and Control of Cell Division.* T.L. Rost and E.M. Gifford (eds.), pp. 212–22, Dowden, Hutchison and Ross, Stroudsburg, PA.

Hepler, P.K. (1980). Membranes in the mitotic apparatus of barley cells. *J. Cell Biol.* **86**, 490–9.

Hess, P. and Tsien, R.W. (1984). Mechanism of ion permeation through calcium channels. *Nature.* **309**, 453–6.

Hollinger, T.G., Dumont, J.N. and Wallace, R.A.: Calcium-induced dehiscence of cortical granules in Xenopus laevis oocytes. *J. Exp. Zool.* **210**, 107–16.

Immelmann, A. and Sölling, H.-D. (1983). ATP-dependent calcium sequestration and calcium/ATP stoichometry in isolated microsomes from guinea-pig parotid glands. *FEBS Lett.* **162**, 406–10.

Jaffe, L. (1980). Calcium explosions as triggers of development. *Ann. NY. Acad. Sci.* **339**, 86–101.

Kiehart, D.E. (1981). Studies on the *in vivo* sensitivity of spindle microtubules to calcium ions and evidence for a vesicular calcium-sequestering system. *J. Cell Biol.* **88**, 604–17.

Kostyuk, P. and Doroshenko, P. (1983). Calcium binding and calcium channels. In: *Calcium-binding Proteins.* de Bernard *et al.*, eds, pp. 331–8, Elsevier Science Publishers.

Lee, K.S. and Tsien, R.W. (1983). Mechanism of calcium channel blockade by verapamil, D 600, diltiazem and nitredipine in single dialysed heart cells. *Nature.* 302, 790–4.

Marmé, D. and Dieter, P. (1983). Role of Ca^{2+} and calmodulin in plants. In: *Calcium and Cell Function, Vol. IV.* W.Y. Cheung (ed.), pp. 263–311. Academic Press, New York.

Mazia, D. (1937). The release of calcium in *Arbacia* eggs on fertilization. *J. Cell. Comp. Physiol.* 10, 291–304.

Mazia, D. and Dan, K. (1952). The isolation and biochemical characterization of the mitotic apparatus of dividing cells. *Proc. Natl. Acad. Sci. USA.* 38, 826–8.

McGraw, C.F., Somlyo, A.V. and Blaustein, M.P. (1980). Localization of calcium in presynaptic nerve terminals. An ultrastructural and electron microprobe analysis. *J. Cell Biol.* 85, 228–41.

Moore, L., Fitzpatrick, D.F., Chen, T. and London, E.J. (1974). Calcium pump activity of the renal plasma membrane and renal microsomes. *Biochim. Biophys. Acta.* 345, 405–18.

Moore, L., Chen, T., Knapp, H. and London, E. (1975). Energy-dependent calcium sequestration activity in rat liver microsomes. *J. Biol. Chem.* 250, 4562–8.

Moreau, M., Doree, M. and Guerrier, P. (1976). Electrophoretic induction of calcium ions into the cortex of *Xenopus laevis* oocytes triggers meiosis reinitiation. *J. Exp. Zool.* 197, 443–9.

Niggli, V., Adungah, E.S. and Carafoli, E., (1981). Acidic phospholipids, unsaturated fatty acids, and limited proteolysis mimic the effect of calmodulin on the purified erythrocyte Ca^{2+}-ATPase. *J. Biol. Chem.* 256, 8588–92.

Niggli, V., Penniston, J.T. and Carafoli, E. (1979). Purification of the Ca^{2+}-Mg^{2+}-ATPase from human erythrocyte membranes using a calmodulin affinity column. *J. Biol. Chem.* 254, 9955–8.

Niggli, V., Adungah, E.S., Penniston, J.T. and Carafoli, E. (1981). Purified Ca^{2+}-Mg^{2+}-ATPase of the erythrocyte membrane; reconstitution and effect of calmodulin and phospholipids. *J. Biol. Chem.* 256, 395–401.

Penniston, J.T. (1983). Plasma membrane ATPases as active Ca^{2+}-pumps. In *Calcium and Cell Function, Vol. 4.* W.Y. Cheung, (ed.) pp. 99–149, Academic Press, New York.

Petzelt, C. (1979). Biochemistry of the mitotic spindle. *Int. Rev. Cytol.* 60, 53–92.

Petzelt, C. (1984). Localization of an intracellular membrane-bound Ca^{2+}-ATPase in PtK-cells using immunofluorescence techniques. *Eur. J. Cell Biol.* 33, 55–9.

Petzelt, C. and Auel, D. (1978). Purification and some properties of the mitotic Ca^{2+}-ATPase. In *Cell Reproduction - Honoring Daniel Mazia.* D. Prescott and E.R. Dirksen (eds.), pp. 487–94.

Petzelt, C. and Auel, D. (1977). Synthesis and activation of the mitotic Ca^{2+}-ATPase during the cell cycle of mouse mastocytoma cells. *Proc. Natl. Acad. Sci. USA.* 74, 1610–3.

Petzelt, C. and Wülfroth, P. (1985). Cell cycle specific variations in transport capacity of an isolated Ca^{2+}-transport system. *Cell Biol. Int. Rep.* 8, 823–40.

Petzelt, C. and Wülfroth, P. (1984). On the dissociability of Ca^{2+}-dependent mitotic events in sea urchin eggs. *Adv. Invert. Repr. Vol. 3.* W. Engels and A. Fischer (eds), pp. 107–14.

Philipson, K.D. and Nishimoto, A.Y. (1980). Na^+-Ca^{2+} exchange is affected by membrane potential in cardiac sarcolemmal vesicles. *J. Biol. Chem.* 255, 6880–2.

Porter, K.R. and Machado, R.D. (1960). Studies of the endoplasmic reticulum. IV. Its form and distribution during mitosis in cells of onion root tips. *J. Biochem. Biophys. Cytol.* **7**, 167–80.

Reeves, I.P. and Sutko, J.L. (1979). Sodium–calcium ion exchange in cardiac membrane vesicles. *Proc. Natl. Acad. Sci (USA).* **76**, 590–4.

Reeves, I.P. and Sutko, J.L. (1980). Sodium–calcium exchange activity generates a current in cardiac membrane vesicles. *Science.* **208**, 1461–4.

Reuter, H. (1983). Calcium channel modulation by neurotransmitters, enzymes and drugs, *Nature,* **301**, 569–74.

Salmon, E. and Segall, R.R. (1980). Calcium-labile mitotic spindles isolated from sea urchin eggs (*Lytechinus variegatus*). *J. Cell Biol.* **86**, 355–65.

Schatten, G., Schatten, H. and Simerly, C. (1982). Detection of sequestered calcium during mitosis in mammalian cell cultures and in mitotic apparatus isolated from sea urchin zygotes. *Cell Biol. Int. Rep.* **6**, 717–24.

Schramm, M., Thomas, G., Towart, R. and Franckowiak, F. (1983a). Activation of calcium channels by novel 1,4-dihydropyridine. A new mechanism for positive inotropics or smooth muscle stimulants. *Arzneim. Forsch./Drug Res.* **33**, 1268–72.

Schramm, M., Thomas, G., Towart, R. and Franckowiak, G. (1983b). Novel dihydropyridines with positive inotropic action through activation of Ca^{2+}-channels. *Nature.* **303**, 535–7.

Selinger, Z., Naim, E. and Lasser, M. (1970). ATP-dependent calcium uptake by microsomal preparations from rat parotid and submaxillary glands. *Biochim. Biophys. Acta.* **203**, 326–34.

Silver, R.B. (1983). Co-localization of calcium sequestered in vivo and the calcium transport enzyme in isolated sea urchin mitotic apparatus. *J. Cell Biol.* **98**, 41a.

Silver, R.B., Cole, R.D. and Cande, W.Z. (1980). Isolation of mitotic apparatus containing vesicles with calcium sequestration activity. *Cell* **19**, 505–16.

Steinhardt, R., Zucker, R. and Schatten, G. (1977). Intracellular calcium release at fertilization in the sea urchin egg. *Devel. Biol.* **58**, 185–96.

Suprynowicz, F.A., Poenie, M. and Mazia, D. (1983). Calcium sequestering system of the sea urchin embryo. *J. Cell Biol.* **97**, 30a.

Timourian, H., Jotz, M.M. and Clothier, G.E,. (1974). Intracellular distribution of calcium and phosphorus during the first cell division of the sea urchin egg. *Exp. Cell Res.* **83**, 380–6.

Towart, R. and Schramm, M. (1984). Recent advances in the pharmacology of the calcium channel. *Trends Pharm. Sci.(TIBS)*, **5** 111–3.

Trotta, E. and de Meis, L. (1975). ATP-dependent calcium accummulation in brain microsomes, enhancement by phosphate and oxalate. *Biochim. Biophys. Acta.* **394**, 239–47.

Ungar, F., Piscopo, I. and Holtzman, E. (1981). Calcium accumulation in intracellular compartments of frog retina rod photoreceptors. *Brain Res.* **205**, 200–6.

Ungar, F., Piscopo, I., Letizia, J. and Holtzman, E. (1984). Uptake of calcium by the endoplasmic reticulum of the frog photoreceptor. *J. Cell Biol.* **98**, 1645–55.

Vacquier, V.D. (1975). The isolation of intact cortical granules from sea urchin eggs: calcium ions trigger granule discharge. *Devel. Biol.* **43**, 62–74.

Walz, B. (1982a). Ca^{2+}-sequestering smooth endoplasmic reticulum in an invertebrate photoreceptor. I. Intracellular topography as revealed by OsFeCN staining and *in situ* calcium accumulation. II. Its properties as revealed by morphometric measurements. *J. Cell Biol.* **93**, 839–59.

Walz, B. (1982b). Calcium-sequestering smooth endoplasmic reticulum in retinula cells of the blowfly. *J.Ultrastruct. Res.* **81**, 240—8.

Wassermann, W.J. and Masui, Y. (1975). Initiation of meiotic maturation in *Xenopus laevis* oocytes by the combination of divalent cations and ionophore A 23187. *J. Exp. Zool.* **193**, 369—75.

Wassermann, W.J., Pinto, L.H., O'Conner, C.M. and Smith, L.D. (1980). Progesterone induces a rapid increase in $(Ca^{2+})_{in}$ of *Xenopus laevis* oocytes. *Proc. Nat. Acad. Sci. USA.* **77**, 1534—6.

Wick, S.M. and Hepler, P.K. (1980). Localization of Ca^{2+}-containing antimonate precipitates during mitosis. *J. Cell Biol.* **86**, 500—13.

Wolniak, S.M., Hepler, P.K. and Jackson, W.T. (1980). Detection of the membrane—calcium distribution during mitosis in *Haemanthus* endosperm with chlorotetracycline. *J. Cell Biol.* **87**, 23—32.

Wolniak, S.M., Hepler, P.K. and Jackson, W.T. (1981). The coincident distribution of calcium-rich membranes and kinetochore fibers at metaphase in living endosperm cells of *Haemanthus*. *Eur. J. Cell Biol.* **25**, 171—4.

6

THE REGULATION OF MITOCHONDRIAL CALCIUM

M. Crompton, T.P. Goldstone and I. Al-Nasser

Department of Biochemistry, University College, London. WC1E 6BT, UK

Abstract

Mitochondrial Ca^{2+} in mammalian tissues is established by steady state recycling mediated by the Ca^{2+} uniporter (Ca^{2+} influx) and one (Na^+-Ca^{2+} carrier) or more (Na^+-independent) systems for Ca^{2+} efflux. In principle the cycle may control intramitochondrial-free Ca^{2+} or provide Ca^{2+} for cytosolic processes, although both functions cannot operate simultaneously. In its simplest form, the cycle allows changes in the free $[Ca^{2+}]$ of the mitochondrial matrix to occur in parallel with those of the cytosol. Thus, in heart, increase in cytosolic Ca^{2+} (time-averaged) may be transmitted to the mitochondrial matrix and allow control of oxidative metabolism. This basic relay in heart is characterised by amplification in which an increase in cytosolic Ca^{2+} produces a proportionally larger increase in matrix Ca^{2+}. Several factors contribute to this amplification including the sigmoidal kinetic properties of the uniporter, inhibition of the Na^+-Ca^{2+} carrier by external Ca^{2+} and, possibly, α-adrenergic activation of the uniporter. In contrast, matrix free Ca^{2+} in liver may undergo different changes during β-adrenergic and glucagon action. A key factor in these cases may be the cAMP-mediated activation of the Na^+—Ca^{2+} carrier by these agonists, a feature not evident in heart mitochondria.

6.1. Introduction

It is only quite recently that the means and importance of regulating intramitochondrial Ca^{2+} have become appreciated. For many years, ideas of why mitochondria transport Ca^{2+} were dominated by a general property of mitochondria *in vitro*, namely the capacity to accumulate massive amounts of Ca^{2+}, but with relatively low affinity. It was widely thought that the significance of mitochondrial Ca^{2+} transport must somehow reflect this property. Thus mitochondria were often considered to be potential sinks for intracellular Ca^{2+}, to be used perhaps as an emergency measure when cytosolic Ca^{2+} exceeded normal limits. Indeed, this concept still persists, although in a more plausible form, taking into account developments in our understanding of the mitochondrial transport system (Nicholls, Snelling and Rial, 1984). Perhaps the most significant advances, however, concern the role of mitochondrial Ca^{2+} transport during normal cell function. A fairly compelling case can now be made that intramitochondrial-free Ca^{2+} in some tissues at least is normally maintained at levels not grossly different from those of the cytosol, and that, at such levels, intramitochondrial-free Ca^{2+} may control the activities of certain dehydrogenases (Denton and McCormack, 1980).

The realisation that intramitochondrial Ca^{2+} is not biochemically inert and, in consequence, may need to be controlled very precisely according to physiological circumstances, is seemingly at odds with the sink concept by which excess Ca^{2+} may simply be shunted to the intramitochondrial compartment with impunity. In attempting to resolve this and related issues recourse must be made to the transport systems themselves. The conceptual advances mentioned above are based firmly on the recognition that Ca^{2+} enters and leaves mitochondria by separate transport processes, thereby establishing a steady state Ca^{2+} distribution across the inner membrane (Crompton, Capano and Carafoli, 1976; Puskin et al., 1976). This cycle, depicted in Fig.6.1, as involving distinct Na^+-dependent and

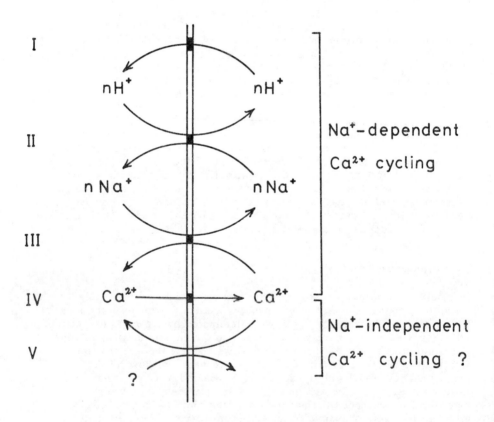

Fig.6.1. Ca^{2+} cycling across the inner mitochondrial membrane. Symbols: I, respiratory chain; II, Na^+–H^+ antiporter; III, Na^+–Ca^{2+} carrier; IV, Ca^{2+} uniporter; V, Na^+-independent Ca^{2+} release. The stoichiometric coefficient n is undetermined.

Na^+-independent processes of Ca^{2+} efflux has remained unchanged since its inception (Crompton, Capano and Carafoli. 1976; Crompton et al., 1978). Above all else, the scheme implies that the gradient of free Ca^{2+} across the inner membrane will be established kinetically. That is, the capacity of the cycle to establish and to change the Ca^{2+} distribution will be governed by the relative kinetic properties of the transport systems and how these properties are modified. Moreover, these properties must be analysed within the context of steady state Ca^{2+} cycling, the operational whole through which these properties are expressed. This brief overview is not intended to cover all aspects of mitochondrial Ca^{2+} transport. The purpose is simply to examine the above concept and to stress those unresolved and often controversial issues that have a direct bearing on it.

6.2. Na^+-dependent and Na^+-independent Ca^{2+} cycling

The Ca^{2+} cycle is conventionally represented as in Fig.6.1 as a single mode of Ca^{2+} entry opposed by two modes of Ca^{2+} exit, and ancillary reactions. It is worthwhile examining briefly the extent to which this simple representation is actually indicated explicitly by experimental data.

Na^+-dependent Ca^{2+} cycling, involving the uniporter, Na^+-Ca^{2+} carrier, Na^+–H^+ antiporter and the respiratory chain (Crompton and Heid, 1978; Crompton, Capano and Carafoli, 1976; Crompton, Kunzi and Carafoli, 1977; Crompton et al., 1978) is now supported by a diversity of evidence, which has been recently reviewed in some detail (Crompton, 1985). There seems to be little doubt that the separate transport systems as indicated do exist. In outline, the component reactions of the Na^+-dependent Ca^{2+} cycle may be resolved by their differing sensitivities to inhibitors (ruthenium red, lanthanides, antiglycoprotein antibodies, Ca^{2+} antagonist drugs, Mg^{2+}), activators (such as K^+) and substrates (Ca^{2+}, Na^+). Na^+–Ca^{2+} exchange has not been demonstrated directly since means for the selective suppression of the competing Na^+–H^+ exchange are not available. Nevertheless, the Na^+–Ca^{2+} carrier mediates other exchanges characteristic of such a mechanism, such as Ca^{2+}–Ca^{2+} exchange which is inhibited by Na^+ (as would be predicted) and which exhibits the same sensivity to inhibitors (such as La^{3+}) and activators (such as K^+) as does Na^+-induced efflux of Ca^{2+}. In addition the steady state gradient of Ca^{2+} across the inner membrane may be manipulated in the manner predicted by changing $[Na^+]$. Continuous steady state cycling requires that the net flux of each ionic intermediate of the cycle is zero. Since the electrophoretic influx of Ca^{2+} is supported by active H^+ extrusion by the respiratory chain, Ca^{2+} efflux must be accompanied by the re-entry of H^+. This is achieved via the Na^+-H^+ exchange.

In contrast, exactly how the steady state flux requirements may be met in the case of Na^+-independent Ca^{2+} cycling is not clear. One proposal, that Ca^{2+} efflux may be coupled directly to H^+ influx (Akerman, 1978; Fiskum and Lehninger, 1979) is not corroborated by recent studies which revealed highly variable H^+–Ca^{2+} stoichiometries and a contrary dependence on intramitochondrial pH (Gunter et al., 1983). In our opinion there is as yet no firm evidence for the widespread assumption that the phenomenon of Na^+-independent Ca^{2+} release reflects the existence of a distinct carrier. Indeed, the process is essentially insensitive even to lanthanides (Crompton et al., 1979), agents which typically substitute for Ca^{2+} with high affinity at Ca^{2+}-selective sites. It may be timely therefore to consider alternative possible explanations for the net losses of mitochondrial Ca^{2+} that occur in the absence of Na^+ when the uniporter is inhibited by ruthenium red.

In some mitochondria, such as heart, such losses occur extremely slowly, whereas in others, notably liver, the process is considerably more rapid (Crompton et al., 1978). Mitochondria of heart are also considerably more resistant to Ca^{2+} load than those of liver (Palmer and Pfeiffer, 1981). Is this correlation coincidental, or not? Intolerance to Ca^{2+} load is characterised by an accumulation of lysophospholipids and free fatty acids, and a non-specific increase in membrane permeability to low molecular weight solutes of the mitochondrial matrix, including Ca^{2+} (Beatrice, Stiers and Pfeiffer, 1982, and refs. therein). Attention to this phenomenon was first drawn by the work of Fiskum and Lehninger (1979), which showed that it may be reversible, depending on mitochondrial redox state. Conceivably the levels of lysophospholipids and free fatty acids may be determined by cyclical deacylation/acylation reactions (Beatrice, Stiers and Pfeiffer, 1982). Although mitochondrial destabilisation is affected by a bewildering array of factors (including redox state, P_i, adenine nucleotides), the indispensable condition appears to be the level of matrix Ca^{2+}. Indeed, Lötscher et al. (1980) reported that the deleterious effects of excess Ca^{2+} load on $\Delta\psi$ (and, presumably, on membrane permeability) were fully reversed by EGTA. This observation has been confirmed in this laboratory (T.P. Goldstone and M. Crompton, unpublished work); moreover, if sucrose, a classical non-penetrant, is added to the mitochondria in the phase of low $\Delta\psi$, and then the mitochondria are repolarised by addition of EGTA, significant quantities of sucrose become entrapped, presumably in the matrix space. If EGTA is not added, then very little sucrose is entrapped and retained by the mitochondria. Considered as a whole, the observations outlined strongly suggest that mitochondria may become freely permeable to low molecular weight solutes generally when matrix Ca^{2+} exceeds certain limits but, and most importantly, that the selectively permeable properties of the inner membrane are restored when matrix free Ca^{2+} is decreased. In fact quite analogous conclusions were drawn some years ago by Hunter, Howarth and Southard (1976), but their possible significance has not been fully appreciated, due perhaps to the use of heart mitochondria and the consequent need for very high Ca^{2+} loads for Ca^{2+}-induced permeability changes.

If this is the case, it seems not unreasonable to suggest that mitochondria *in vitro* may generally undergo such Ca^{2+}-induced transitions between 'intact' and 'permeable' forms as shown schematically in Fig.6.2. Under non-deleterious *in vitro* conditions, the permeable fraction may be extremely small indeed to account for the maintainence of matrix constituents generally. Increased Ca^{2+} load, however, may displace the steady state towards the permeable form to an extent dependent on incubation conditions (redox state, etc.) and mitochondrial type. Inhibition of Ca^{2+} uptake would allow net Na^+-independent Ca^{2+} release at a rate equal to that of the intact-permeable transition and would in no way reflect the activity of a distinct carrier.

This model is no more than an attempt to rationalise various observations concerning Na^+-independent interactions between Ca^{2+} and mitochondria. If nothing else, it serves to underline the fact that Na^+-independent Ca^{2+} cycling cannot be assumed to be analogous to Na^+-dependent Ca^{2+} cycling. If Na^+-independent Ca^{2+} cycling genuinely involves Ca^{2+}-induced permeabilisation then its significance is surely quite different from that of Na^+-dependent recycling. Indeed, Ca^{2+} cycling would be a misnomer, the essential event being the recycling of mitochondrial forms.

It must be stressed that the rate of Na^+-independent release of Ca^{2+} is quite insignificant in mitochondria of heart and many other tissues at physiological levels of matrix Ca^{2+}, for example, $<1\%$ of Na^+–Ca^{2+} carrier activity in heart (Hayat and Crompton, 1982). Thus the nature of Na^+-independent cycling is irrelevant to the control of matrix-free Ca^{2+} in heart and many other tissues in which the steady state Ca^{2+} distribution may be analysed in terms of Na^+-dependent Ca^{2+} cycling alone.

6.3. The control of intramitochondrial Ca^{2+} by Na^+-dependent Ca^{2+} cycling

The discovery that certain intramitochondrial enzymes of oxidative metabolism are acutely sensitive to Ca^{2+} represents one of the most important developments in our perception of mitochondrial transport. These enzymes, α-oxoglutarate dehydrogenase, NAD-linked isocitrate dehydrogenase, and pyruvate dehydrogenase phosphatase, are all greatly activated by increase in free Ca^{2+} over the range $0.1-10$ μM. Apart from the implication that oxidative metabolism may be subject to matrix free Ca^{2+} (Denton amd McCormack, 1980), the findings raised a number of important issues concerning the operation of the Na^+-dependent Ca^{2+} cycle. These issues are most fruitfully considered with respect to heart mitochondria, in which the relevant kinetic properties of the enzymes and transport systems have been studied.

The immediate question is whether the Na^+-dependent Ca^{2+} cycle is capable of setting matrix-free Ca^{2+} at the low levels required for dehydrogenase control and, if so, whether this is reconcilable with the older sink concept. When isolated heart mitochondria are incubated in saline approximating in ionic composition to the cytosol the dehydrogenases respond to changes in

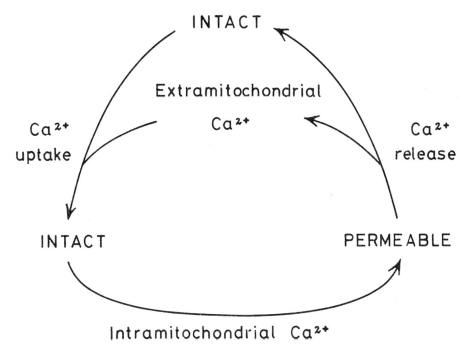

Fig.6.2. A hypothesis for Na^+-independent release of mitochondrial Ca^{2+}.

extramitochondrial free Ca^{2+} over the range 0.1–1 μM. Moreover, the activity/external [Ca^{2+}] relations are modulated in a coherent manner by changes in uniporter and Na^+–Ca^{2+} carrier activity. (Denton, McCormack and Edgell, 1980; Hansford and Castro, 1980; Kessar and Crompton, 1981). It appears then that the Na^+-dependent Ca^{2+} cycle can set matrix-free Ca^{2+} at the appropriate levels, at least *in vitro*. This conclusion is supported by estimates of matrix-free Ca^{2+} under the same conditions with a technique employing the ionophore A23187 (Coll *et al.*, 1981; Hansford and Castro, 1982). Such estimates indicate that as little as 0.1% of total matrix Ca^{2+} may be free. If this fraction is applied, albeit with reservations, to determinations of total heart mitochondrial Ca^{2+} *in vivo* (approximately 2–4 mM total Ca^{2+}; Crompton, Kessar and Al-Nasser, 1983), then again one arrives at a suggestive value for matrix-free Ca^{2+} in the μM range.

The maintenance of low matrix-free Ca^{2+} then does not appear to exclude the presence of very much larger amounts of total Ca^{2+} in mitochondria. Recent estimates suggest that about 13% of total cellular Ca^{2+} may be mitochondrial in unstimulated heart cells (Kessar and Crompton, 1983; Williamson *et al.*, 1983). Conflict with the idea that mitochondria may provide a source or sink for cytosolic Ca^{2+} arises on other grounds, namely that this cannot easily be envisaged without simultaneously perturbing oxidative metabolism. In view of the amplification properties of the Na^+-dependent Ca^{2+} cycle (see below), even small adjustments of cytosolic Ca^{2+} would lead to significant changes in matrix-free Ca^{2+}. According to a recent sink model, in which a proportion of matrix Ca^{2+} is complexed with P_i, matrix-free Ca^{2+} would be effectively buffered at a precise value determined by the solubility product of the complex (Nicholls, Snelling and Rial, 1984). While this model would allow Ca^{2+} accumulation without metabolic consequences, it leads to predictions quite different from those advanced in this article, namely that matrix-free Ca^{2+} would be a function of matrix free [P_i] and pH, rather than the kinetic properties of the component carriers of the cycle. The weight of evidence strongly indicates that any such complexation in heart mitochondria occurs only when matrix Ca^{2+} exceeds the range required for dehydrogenase control (see Crompton, 1985).

The behaviour of the Na^+-dependent Ca^{2+} cycle may be appropriately referred to as a relay. When extramitochondrial Ca^{2+} is increased, for example, the cycle will readjust from its former steady state to another in which external Ca^{2+}, internal Ca^{2+} and cycling rate are all increased (Crompton, 1985; Nicholls and Crompton, 1980). In this sense the cycle may be regarded as relaying an increase in extramitochondrial-free Ca^{2+} into an increase in intramitochondrial-free Ca^{2+}. The possible significance of such a relay is immediately apparent in the case of heart, since it might allow oxidative metabolism to adjust to the demands imposed by increased contractility, when cytosolic Ca^{2+} increases (Denton and McCormack, 1980). A critical aspect of this concept concerns the behaviour of the cycle when faced with the extreme amplitude of the changes in cytosolic Ca^{2+} from beat to beat. There is no obvious way in which this question can be experimentally investigated directly. A theoretical model (Fig.6.3) may be constructed however from the measured kinetic constants of the uniporter and Na^+–Ca^{2+} carrier (see Legend, Fig.6.3) and the cytosolic Ca^{2+} transients (see curve (a), Fig 6.3). While the data emanating from this model cannot be taken to be precise, being subject to any inaccuracies in the constants and cytosolic [Ca^{2+}] used, the picture provided is instructive in indicating what the cycle in principle could achieve. With peak cytosolic Ca^{2+} of 2.1 μM (see curve (a), Fig.6.3), the computed matrix free [Ca^{2+}] oscillates around 1 μM (see curve (b), Fig.6.3.); it is thus reassuring to find that calculations based solely on the measured kinetic constants fall within the range required for dehydrogenase control. The predicted beat-to-beat variation in matrix Ca^{2+} is very small, amounting to about 3% of total. In other words it seems that the cycle would not oppose the rapid changes in cytosolic Ca^{2+} (upon which heart

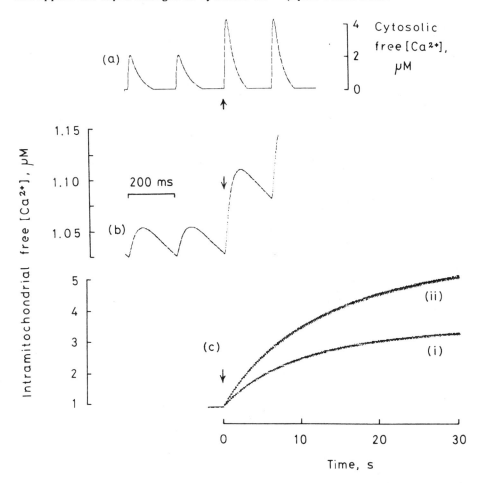

Fig.6.3. Simulated matrix free Ca^{2+} concentrations in rat heart mitochondria. The cytosolic Ca^{2+} transients (curve (a)) are adapted from Allen and Kurihara (1980). Net Ca^{2+} flux into mitochondria (v) was calculated as the difference between the activities of the uniporter (v_{in}) and the Na^+–Ca^{2+} carrier (v_{eff}) at the free Ca^{2+} concentrations of the cytosol ($[Ca]_c$) and the mitochondrial matrix ($[Ca]_m$), that is, $v = v_{in} - v_{eff}$. The following relations were used: $v_{in} = V_{in} [Ca]_c^{1.5}/300 + [Ca]_c^{1.5}$ as in Crompton et al. (1976), where V_{in} = 10 nmol Ca^{2+}/mg protein·sec in the presence of 5 mM Mg^{2+} (Scarpa and Graziotti, 1973). $v_{eff} = V_{eff} [Ca]_m/K + [Ca]_m$ where K = 6 μM as in Coll et al. (1981). V_{eff} was calculated taking into account the effect of cytosolic Ca^{2+} as follows: $V_{eff} = V_{eff}$ -0.7 V_{eff} ($[Ca]^2_c/0.6 + [Ca]^2_c$) with V_{eff} = 0.17 nmol Ca^{2+}/mg protein·sec at 6 mM Na^+ (Hayat and Crompton, 1982). All kinetic constants refer to 25°C. The free $[Ca^{2+}]$ in the mitochondrial matrix was assumed to change by 1 μM/nmol total Ca^{2+}·mg protein as reported in Coll et al. (1981) and Hansford and Castro (1982). A BASIC program was used to simulate $[Ca]_m$ according to the above equations with a BBC microcomputer interfaced with a dot matrix printer. Curve (b) shows the short-term changes in matrix free Ca^{2+} accompanying the changes in cytosolic Ca^{2+} of curve (a). Curve (c,i) shows the longer term changes in matrix free Ca^{2+} when the amplitude of the cytosolic Ca^{2+} transient is increased at the arrow by 2-fold (as in curve (a)). Curve (c,ii) refers to the same conditions as in curve (c,i) except that, in addition, uniporter activity was increased by 35%.

performance depends) to any significant degree, but rather relay these into a very largely damped-out ripple of matrix Ca^{2+}. If the cytosolic Ca^{2+} transient is increased, as indicated by the arrows, net Ca^{2+} accumulation would be predicted during each contraction–relaxation cycle (see curve (b), Fig.6.3) until a new quasisteady state is attained after about 30 s (see curve (c,i), Fig.6.3).

A further feature of relay behaviour that emerges is that a change in cytosolic Ca^{2+} yields a proportionally larger change in matrix Ca^{2+}. In the example considered here, a 2-fold increase in peak cytosolic Ca^{2+} gives rise to an approximate 3-fold increase in matrix Ca^{2+} (from about 1 μM to 3 μM; see curve (c,i), Fig.6.3). This amplification in the relay response is caused by two factors. Firstly, the uniporter activity displays a sigmoidal dependence on extramitochondrial $[Ca^{2+}]$ (Crompton et al., 1976). It may be remarked that the relatively low affinity of the uniporter for extramitochondrial Ca^{2+}, which has often been taken superficially to imply a secondary role of mitochondria in Ca^{2+} metabolism, now emerges as highly significant in allowing matrix-free Ca^{2+} to increase disproportionately with cytosolic Ca^{2+}. Secondly, the Na^+–Ca^{2+} carrier is inhibited by extramitochondrial Ca^{2+} in the physiological range. In essence, extramitochondrial Ca^{2+} acts as a partial, non-competitive (with Na^+) inhibitor with half maximal inhibition at about 0.6 μM Ca^{2+} (Hayat and Crompton, 1982). Maximal inhibition decreases V_{max} by about 70%. These characteristics suggest that the Na^+–Ca^{2+} carrier may contain regulatory sites for Ca^{2+}, distinct from the substrate (Ca^{2+}) binding sites. The two factors outlined may well have evolved in order to introduce amplification in the relay and permit matrix-free Ca^{2+} to respond in a sensitive manner to changes in cytosolic Ca^{2+}. As discussed previously (Crompton, 1985), the degree of amplification is not constant, but increases as cytosolic Ca^{2+} increases.

6.4. External control of the Na^+-dependent Ca^{2+} cycle

The above description of the Na^+-dependent Ca^{2+} cycle amounts to a purely passive relay in which matrix-free Ca^{2+} is fixed by cytosolic Ca^{2+}. In other words, although the means for Ca^{2+}-control of oxidative metabolism resides in the matrix space, the essential determinant would be cytosolic Ca^{2+}. The implications of this need to be carefully considered. For instance, one element in β-adrenergic stimulation of heart is decreased sensitivity of the myofibrils to Ca^{2+} (Katz, 1979). On the other hand, very recent evidence suggests that the Ca^{2+}-sensitivity of the myofibrils may be increased during α-adrenergic stimulation (Endoh and Blinks, 1984). If the Na^+ dependent Ca^{2+} cycle really does provide one means of coordinating contractility and oxidative metabolism, then one might expect that a changed contractile response to Ca^{2+} would be paralleled by changes affecting the oxidative response as well, either in the relay behaviour (amplification) of the Na^+-dependent Ca^{2+} cycle or the Ca^{2+}-sensitivity of the dehydrogenases themselves.

In this regard, there is evidence that the Ca^{2+} uniporter of heart mitochondria may be activated by an α-adrenergic mechanism (Kessar and Crompton, 1981). The weight of evidence indicates that the activation is caused by a changed uniporter conductance, rather than reflecting an increased driving force ($\Delta\psi$) for Ca^{2+} influx (Crompton, Kessar, Al-Nasser, 1983). It is important to discriminate between these two possible causes. Whereas increased uniporter conductance would presumably be expressed in vivo, and permit selective control of matrix Ca^{2+}, one cannot assume the same for $\Delta\psi$-mediated effects as discussed previously (Crompton, 1985). The measured increase in uniporter conductance is small (30–40%), but the effect can only be assessed in terms of the consequent change in Ca^{2+} distribution. To illustrate this, if the activation is applied to the theoretical model (Fig.6.3) matrix-free Ca^{2+} is increased from about 3 μM (curve (c,i, Fig.6.3.), to about 5 μM (curve (c) (ii), Fig.6.3.). Again one cannot take these values too precisely; but they do illustrate the inbuilt amplification in the system. In fact, α-adrenergic stimulation of rat ventricle leads to an increase of about 50% in mitochondrial endogenous Ca^{2+} and a maximal increase of about 100% in α-oxoglutarate dehydrogenase activity during in vitro Ca^{2+} cycling (Crompton, Kessar and Al-Nasser, 1983; Kessar and Crompton, 1981). Conceivably then, α-adrenergic activation of the uniporter may potentiate amplification in the relay and allow matrix Ca^{2+} to rise more in response to a small change in the Ca^{2+} transient. In this way, α-adrenergic changes in contractility, which may involve very small changes in cytosolic Ca^{2+} (Endoh and Blinks, 1984), might be adequately compensated by oxidative metabolism.

If these lines of thought are extended to β-adrenergic stimulation, one might anticipate deamplification in the Ca^{2+} relay, either inhibition of Ca^{2+} influx or promotion of Ca^{2+} efflux. Beta-adrenergic stimulation leads to an approximate 2-fold increase in endogenous Ca^{2+} of rat venticular mitochondria, but whether or not this is the expected consequence of the change in amplitude and frequency of the Ca^{2+} transient cannot be answered (Crompton, Kessar and Al-Nasser, 1983; McCormack and Denton, 1984). No effects of β-adrenergic agonists on either the uniporter or the Na^+–Ca^{2+} carrier in heart mitochondria have been detected (Kessar and Crompton, 1981). However β-adrenergic agonists do stimulate Na^+–Ca^{2+} carrier activity in liver mitochondria (see below) specifically under conditions of low Ca^{2+} load, so that a reinvestigation of this question under varied conditions with heart mitochondria may be advisable before firm conclusions are drawn.

The control of the Na^+-dependent Ca^{2+} cycle in liver mitochondria may have to be viewed somewhat differently from that in heart. Whereas the periodicity of the changes in cytosolic Ca^{2+} in heart is too short to allow mitochondria to interfere with the physiological response of the tissue, this does not appear to be the case in liver. For example, during glycogenolysis induced by α-adrenergic agonists and vasopressin, the rise in cytosolic Ca^{2+} in liver is maintained for at least 1 min (Charest et al., 1983), sufficient time in principle for mitochondria to accumulate Ca^{2+} and oppose the physiological response. The obvious question is why this does not occur. Indeed, many authors have proposed that mitochondria may actually release Ca^{2+} (Williamson, Cooper and Hock, 1981), although the immediate hormone action may well involve Ca^{2+} release from non-mitochondrial stores (Joseph et al., 1984). No means have yet been recognised whereby these agents promote net Ca^{2+} efflux from mitochondria, or restrict net accumulation in the face of raised cytosolic Ca^{2+}. On the other hand, β-adrenergic agonists and glucagon activate the liver mitochondrial Na^+–Ca^{2+} carrier quite markedly (approximately 3-fold; Goldstone and Crompton, 1982; Goldstone, Duddridge and Crompton, 1983). This could account for the losses of mitochondrial Ca^{2+} (Baddams, Chang and Barritt, 1983; Taylor et al., 1980) and the rise of cytosolic Ca^{2+} (Charest et al., 1983) induced by glucagon although the significance of such changes would be unclear. The actions of glucagon and β-adrenergic agonists are generally considered to be independent of changes in cytosolic Ca^{2+}, suggesting that the significance may be to limit mitochondrial uptake of Ca^{2+}. However, there are no reported effects of glucagon on Ca^{2+}-sensitive reactions of the mitochondrial matrix, for example, on pyruvate dehydrogenase (Wieland, 1983), to support this idea.

In conclusion it appears that the Na^+-dependent Ca^{2+} cycle may provide one locus for the action of certain hormones on mitochondria. The molecular nature of the modifications responsible have yet to be resolved. Although very nearly all mitochondria examined transport Ca^{2+}, it would be premature to assume that this property has evolved strictly with relevance to oxidative metabolism. The possibility that, in other tissues, different matrix reactions are Ca^{2+}-sensitive, or that mitochondria provide a source of cytosolic Ca^{2+}, cannot be excluded.

Abbreviations

$\Delta\psi$, the mitochondrial inner membrane potential; ΔpH, the pH gradient across the mitochondrial inner membrane.

Acknowledgment

The authors' work was supported by grants from the SERC and the Wellcome Trust.

References

Akerman, K.E.O. (1978). Effect of pH and Ca^{2+} on the retention of Ca^{2+} by rat liver mitochondria. *Arch. Biochem. Biophys.* **189**, 256–62.

Allen, D.G. and Kurihara, S. (1980). Calcium transients in mammalian ventricular muscle. *Eur. Heart J. 1 (suppl. A).* 5–15.

Baddams, H.M., Chang, L.B.F. and Barritt, G.F. (1983). Evidence that glucagon acts on liver to decrease mitochondrial Ca^{2+} stores. *Biochem. J.* **210**, 73–7.

Beatrice, M.C., Stiers, D.L. and Pfeiffer, D.R. (1982). Increased permeability of mitochondria during Ca^{2+} release induced by t-butylhydroperoxide or oxalacetate. *J. Biol. Chem.* **257**, 7161–71.

Charest, R., Blackmore, P.F., Berthon, B. and Exton, J.H. (1983). Changes in free cytosolic Ca^{2+} in hepatocytes following α-adrenergic stimulation. Studies on quin-2 loaded hepatocytes. *J. Biol. Chem.* **258**, 8766–733.

Coll, K.E., Joseph, S.K., Corkey, B.E. and Williamson, J.R. (1981). Determination of the matrix free Ca^{2+} concentration and kinetics of Ca^{2+} efflux in liver and heart mitochondria. *J. Biol. Chem.* **257**, 8696–704.

Crompton, M. (1985). The calcium carriers of mitochondria: In *The Enzymes of Biological Membranes, Vol. 3*, A.N. Martonosi (ed.), Plenum Press.

Crompton, M. and Heid, I. (1978). The cycling of calcium, sodium and protons across the inner membrane of cardiac mitochondria. *Eur. J. Biochem.* **91**, 599–608.

Crompton, M., Capano, M. and Carafoli, E. (1976). The sodium induced efflux of calcium from heart mitochondria. A possible mechanism for the regulation of mitochondrial calcium. *Eur. J. Biochem.* **69**, 453–62.

Crompton, M., Kessar, P. and Al–Nasser, I. (1983). The α-adrenergic mediated activation of the cardiac mitochondrial Ca^{2+} uniporter and its role in the control of intramitochondrial Ca^{2+} *in vivo*. *Biochem. J.* **216**, 333–42.

Crompton, M., Kunzi, M. and Carafoli, E. (1977). The calcium-induced and sodium-induced effluxes of calcium from heart mitochondria. Evidence for a sodium–calcium carrier. *Eur. J. Biochem.* **79**, 549–58.

Crompton, M., Sigel, E., Salzman, M. and Carafoli, E. (1976). A kinetic study of the energy linked influx of Ca^{2+} into heart mitochondria. *Eur. J. Biochem.* **69**, 429–34.

Crompton, M., Moser, R., Ludi, H. and Carafoli, E. (1978). The interrelations between the transport of sodium and calcium in mitochondria from various mammalian tissues. *Eur. J. Biochem.* **82**, 25–31.

Crompton, M., Heid, I., Baschera, C. and Carafoli, E. (1979). The resolution of the calcium fluxes in heart and liver mitochondria using the lanthanide series. *FEBS Lett.* **104**, 352–4.

Denton, R.M. and McCormack, J.G. (1980). On the role of the calcium transport cycle in heart and other mitochondria. *FEBS Lett.* **119**, 1–8.

Denton, R.M., McCormack, J.G. and Edgell, N.J. (1980). The role of calcium ions in the regulation of intramitochondrial metabolism. Effects of Na^+, Mg^{2+}, and ruthenium red on the Ca^{2+}-stimulated oxidation of oxoglutarate and pyruvate dehydrogenase activity in intact rat heart mitochondria. *Biochem. J.* **190**, 107–17.

Endoh, M. and Blinks, J.R. (1984). Regulation of the intracellular Ca^{2+} transient and Ca^{2+} sensitivity of myofibrils in rabbit papillary muscle. *Cell Calcium* **5**, 301.

Fiskum, G. and Lehninger, A.L. (1979). Regulated release of Ca^{2+} from respiring mitochondria by $Ca^{2+}/2H^+$ antiport. *J. Biol. Chem.* **254**, 6236–9.

Goldstone, T.P. and Crompton, M. (1982). Evidence for β-adrenergic activation of Na^+-dependent efflux of Ca^{2+} from isolated liver mitochondria. *Biochem. J.* **204**, 369–71.

Goldstone, T.P., Duddridge, R.J. and Crompton, M. (1983). The activation of Na^+-dependent Ca^{2+} efflux from liver mitochondria by glucagon and β-adrenergic agonists. *Biochem. J.* **210**, 463–72.

Gunter, T.E., Chace, J.H., Puskin, J.S. and Gunter, K.K. (1983). Mechanism of sodium-independent calcium efflux from rat liver mitochondria. *Biochemistry* **22**, 6341–51.

Hansford, R.G. and Castro, F. (1981). Effects of micromolar concentrations of free calcium ions on pyruvate dehydrogenase interconversions in intact rat heart mitochondria. *Biochem. J.* **194**, 721–32.

Hansford, R.G. and Castro, F. (1982). Intramitochondrial and extramitochondrial–free Ca^{2+} ion concentrations of suspensions of heart mitochondria with low, plausibly physiological contents of total calcium *J. Bioenerg. Biomemb.* **14**, 361–76.

Hayat, L.H. and Crompton, M. (1982). Evidence for the existence of regulatory sites for Ca^{2+} on the Na^+/Ca^{2+} carrier of cardiac mitochondria. *Biochem. J.* **202**, 509–18.

Hunter, D.R., Haworth, R.A. and Southard, J.H. (1976). Relationship between configuration, function and permeability in calcium treated mitochondria. *Eur. J. Biochem.* **251**, 5069–77.

Joseph, S.K., Thomas, A.P., Williams, R.J., Irvine, R.F. and Williamson, J.R. (1984). Myoinositol 1,4,5-trisphosphate. A second messenger for the hormonal mobilization of intracellular Ca^{2+} in liver. *J. Biol. Chem.* **259**, 3077–81.

Katz, A.M. (1979). The role of the contractile proteins and sarcoplasmic reticulum in the response of the heart to catecholamines. In *Advances in Cyclic Nucl. Research Vol. 11*, Greengard and Robison, pp. 303–43.

Kessar, P. and Crompton, M. (1981). The α-adrenergic mediated activation of Ca^{2+} influx into cardiac mitochondria. *Biochem. J.* **200**, 379–88.

Kessar, P. and Crompton, M. (1983). The sequestration of Ca^{2+} by mitochondria in rat heart cells. *Cell Calcium.* **4**, 295–305.

Lötscher, H.R., Winterhalter, K.H., Carafoli, E. and Richter, C. (1980). The energy state of mitochondria during the transport of Ca^{2+}. *Eur. J. Biochem.*

110, 211–16.

McCormack, J.G. and Denton, R.M. (1984). Role of Ca^{2+} in the regulation of intramitochondrial metabolism in rat heart. Evidence from studies with isolated mitochondria that adrenaline activates the pyruvate and 2-oxoglutarate dehydrogenase complexes by increasing the intramitochondrial concentration of Ca^{2+}. *Biochem. J.* **218**, 235–47.

Nicholls. D.G. and Crompton, M. (1980). Mitochondrial calcium transport. *FEBS Lett.* **111**, 261–8.

Nicholls, D.G., Snelling, R. and Rial, E. (1984). Proton and calcium circuits across the mitochondrial inner membrane. *Biochem. Soc. Trans.* **12**, 388–90.

Palmer, J.W. and Pfeiffer, D.R. (1981). The control of Ca^{2+} release from heart mitochondria. *J. Biol. Chem.* **256**, 6742–50.

Puskin, J.S., Gunter, T.E., Gunter, K.K. and Russell, P.R. (1976). Evidence for more than one Ca^{2+} transport mechanism in mitochondria. *Biochemistry.* **15**, 3834–42.

Scarpa, A. and Graziotti, P. (1973). Mechanisms for intracellular calcium regulation in heart. *J. Gen. Physiol.* **62**, 756–72.

Taylor, W.M., Prpic. V., Exton, J.H. and Bygrave, F.L. (1980). Stable changes to calcium fluxes in mitochondria isolated from rat livers perfused with α-adrenergic agonists and glucagon. *Biochem. J.* **188**, 433–50.

Wieland, O.H. (1983). The mammalian pyruvate dehydrogenase complex. Structure and regulation. *Rev Physiol. Biochem. Pharmacol.* **96**, 123–70.

Williamson, J.R., Williams, R.J., Coll, K.E. and Thomas, A.P. (1983). Cytosolic-free Ca^{2+} concentration and intracellular calcium distribution of calcium tolerant isolated heart cells. *J. Biol. Chem.* **258**, 13411–4.

Williamson, J.R., Cooper, R.H. and Hoek, J.B. (1981). Role of calcium in the hormonal regulation of liver metabolism. *Biochim. Biophys. Acta.* **639**, 243–95.

CALCIUM RELEASE FROM SARCOPLASMIC RETICULUM *IN VITRO*

N. Ikemoto, S. Danko, B. Antoniu, and D.H. Kim

*Department of Muscle Research, Boston Biomedical Research Institute;
Department of Neurology, Harvard Medical School, Boston, Massachusetts, USA*

Abstract

One of the outstanding questions in muscle physiology is how depolarization of the transverse tubule (T-tubule) membrane triggers rapid Ca^{2+} release from sarcoplasmic reticulum (SR). We have investigated various key processes involved in Ca^{2+} release *in vitro*. Simultaneous replacement of cations and anions in the reaction solution produces changes in the T-tubule membrane potential, as determined by stopped flow fluorometry using membrane-bound potential-sensitive dyes. The T-tubule membrane depolarization leads to rapid Ca^{2+} release from SR, $t_{\frac{1}{2}}$ = 5–9 ms. The time constants are of the same order of magnitude as those expected for the Ca^{2+} release in the intact muscle. Dissociation of the T-tubule from SR blocks the rapid Ca^{2+} release without affecting the T-tubule depolarization. Re-association of the T-tubule leads to almost complete restoration of Ca^{2+} release, indicating that the T-tubule/SR linkage is essential for the triggering mechanism. However, Ca^{2+} and drug-induced Ca^{2+} release, whose half time (0.7–1.4 s) is much longer than that of depolarization-induced Ca^{2+} release, is not affected by dissociation of the T-tubule. A number of Ca^{2+} release inhibitors (such as ruthenium red, tetracaine, dantrolene Na) have been investigated in order to identify the molecular components involved in Ca^{2+} release. Our recent data suggest that an intrinsic SR protein, Mr = 32k, is one of the prime candidates for the Ca^{2+} release channel protein.

7.1. Introduction

According to the generally accepted view, excitation initiated at the surface of the muscle cell is propagated to the inside of the cell through the transverse tubule membrane system (T-tubule). The excitation is communicated to the sarcoplasmic reticulum (SR), which in turn releases the Ca^{2+} from the SR lumen. Binding of the released Ca^{2+} to the Ca^{2+} binding subunit of troponin located in the thin filaments (Ebashi, 1976; Ebashi, Endo and Ohtsuki, 1969; Sandow, 1965; Weber, 1966), releases the inhibition of the interaction between actin and myosin, which is manifested in muscle contraction. The reaccumulation of cytoplasmic Ca^{2+} into the SR by the ATP-dependent Ca^{2+} pump reverses the above process and muscle relaxation ensues.

The Ca^{2+} uptake process has been extensively investigated, while little is known about the mechanism of Ca^{2+} release. A number of methods have been reported in the literature which permit Ca^{2+} release from sarcoplasmic reticulum (SR) *in vitro*: for example (a) an increase of the extravesicular $[Ca^{2+}]$ to the order of several μM (Ca^{2+}-induced Ca^{2+} release) (Endo, 1977; Endo *et al.*, 1983; Fabiato and Fabiato, 1977; Ford and Podolsky, 1972; Inesi and Malan, 1976; Kim, Ohnishi and Ikemoto, 1983; Kirino and Shimizu, 1982; Nagasaki and Kasai, 1981; Ohnishi, 1979; Ohnishi, Taylor and Gronert, 1983); (b) the addition of drugs such as caffeine (Endo, 1975, 1977; Katz *et al.*, 1977; Kim, Ohnishi and Ikemoto, 1983; Miyamoto and Racker, 1982; Ohnishi, 1979b, 1981), quercetin (Kim and Ikemoto, 1983; Kim, Ohnishi and Ikemoto, 1983; Kirino and Shimizu, 1982; Watras *et al.*, 1983), and halothane alone or with Ca^{2+} (Kim *et al.*, 1984, Ohnishi, 1979b, 1981; Ohnishi, Taylor and Gronert, 1983); (c) substitution of permeable anions for impermeable ones or substitution of impermeable cations for permeable ones, or both (depolarization-induced Ca^{2+} release: (Campbell and Shamoo, 1980; Caswell and Brandt, 1981; Endo, 1977; Endo and Nakajima, 1973; Ikemoto, Antoniu and Kim, 1984; Kasai and Miyamoto, 1976; Kim, Ohnishi and Ikemoto, 1983; Kim *et al.*, 1985; Ohnishi, 1979a); (d) the addition of proton ionophores (Shoshan, MacLennan and Wood, 1981) and organic anions (MacLennan, Shoshan and Wood, 1983; Shoshan MacLennan and Wood, 1983) and (e) chemical modification of -SH groups (Abramson *et al.*, 1983). One of the important tasks imposed on the studies of SR *in vitro* is to establish the relation of these *in vitro* types of Ca^{2+} release to the mechanism operating *in situ*.

An outstanding question in muscle physiology is the mechanism by which transient changes in the membrane potential of the T-tubule lead to triggering of rapid Ca^{2+} release from SR (Ebashi, 1976; Ebashi, Endo and Ohtsuki, 1969). An abrupt change of the T-tubule membrane potential in the intact muscle fibre is followed by a non-linear charge movement, which appears to be one of the key processes in the functional coupling between the T-tubule and SR membranes (Adrian and Peres, 1979; Chandler, Rakowski and Schneider, 1976; Horowicz and Schneider, 1981; Schneider and Chandler 1973; Vergara and Caputo, 1983). T-tubules and SR are linked by electron-microscopically discernible structures called feet or pillars (Cadwell and Caswell, 1982; Eisenberg and Gilai, 1979; Eisenberg and Eisenberg, 1982; Franzini-Armstrong, 1975, 1980; Mitchell *et al.*, 1983; Somlyo, 1979) consisting of several substructures (Cadwell and Caswell, 1982; Franzini-Armstrong and Nunzi, 1983; Mitchell *et al.*, 1983). The pillar may work as a plunger whose voltage-dependent movement leads to the opening of the SR Ca^{2+} channel (Chandler, Rakowski and Schneider, 1976; Schneider and Chandler, 1973), or there might be an ionic flow within the feet to trigger Ca^{2+} release. In view of the crucial role of the T-tubule in Ca^{2+} release *in vivo*, it is especially important to investigate whether any of the above types of Ca^{2+} release *in vitro* is triggered via the T-tubule. In the first part of this chapter we will focus our discussion on this question. As demonstrated here, dissociation of the T-tubule/SR complex by treatment with a French press (Caswell *et al.*, 1979; Ikemoto, Antoniu and Kim, 1984) inhibits rapid Ca^{2+} release induced by ion substitution ('depolarization'-induced Ca^{2+} release). Furthermore, reformation of the T-tubule/SR complex by incubation of the purified T-tubules with SR vesicles in cacodylate (Ikemoto, Antoniu and Kim, 1984) leads to restoration of depolarization-induced rapid Ca^{2+} release, suggesting that the ion replacement produces membrane potential changes in the

T-tubule moiety leading to rapid Ca^{2+} release from SR. On the other hand, dissociation of the T-tubule affects neither Ca^{2+} nor drug-induced Ca^{2+} release, suggesting that these types of Ca^{2+} release are produced by a direct stimulation of the SR membrane.

The Ca^{2+} release function is much higher in heavy SR fractions than light SR fractions with almost all types of Ca^{2+} release described above (Campbell and Shamoo, 1980; Caswell and Brandt, 1981; Ikemoto, Antoniu and Kim, 1984; Kim, Ohnishi and Ikemoto, 1983; Kim et al; 1985; Miyamoto and Racker, 1982; Nagasaki and Kasai, 1981; Ohnishi, 1979 a and b, 1981; Ohnishi, Taylor and Gronert,. 1983). Since the heavy and light fractions are derived from terminal cisternae and tubular portions of SR, respectively (Meissner, 1975), it appears that the molecular mechanisms responsible for these different types of Ca^{2+} release are localized in the same region, that is, the terminal cisternae region of the SR membrane. We have proposed a hypothesis that a similar or identical set of proteins would serve as the Ca^{2+} channel for different types of Ca^{2+} release (Kim and Ikemoto, 1983, 1984; Kim, Ohnishi and Ikemoto, 1983; Kim et al., 1984). Several facts described here are consistent with this hypothesis. Furthermore, the results described here also permit insight into the candidate proteins for the Ca^{2+} release channel.

The use of appropriate drugs in kinetic studies of Ca^{2+} release may eventually permit further resolution of the molecular mechanism involved in Ca^{2+} release. Careful comparison of Ca^{2+}-release functions in normal and diseased systems will also reveal crucial components or steps involved in the Ca^{2+}-release mechanism. This chapter contains a brief discussion of our recent studies of pharmacological and pathophysiological aspects of Ca^{2+} release.

7.2. Results and discussion

7.2.1. Depolarization-induced Ca^{2+} release

7.2.1.1. Appropriate methods to produce depolarization-induced Ca^{2+} release. There has been an extensive dispute in the literature regarding the involvement of artifacts in Ca^{2+} release produced by ionic replacement, which is generally called 'depolarization'-induced Ca^{2+} release. It has been recognised that anion replacement alone, which was used in earlier years to induce membrane depolarization (Endo and Nakajima, 1973; Kasai and Miyamoto, 1976), leads to a swelling of SR vesicles, and hence the observed Ca^{2+} release may be an artifactual result of membrane lysis (Meissner and McKinley, 1976). On the other hand, cation replacement alone (for example, choline.Cl replacement of K.Cl), which we previously used to induce membrane depolarization (Kim, Ohnishi and Ikemoto, 1983), produces a shrinkage of vesicles as judged from our recent light scattering measurements (Ikemoto, Antoniu and Kim, 1984). Therefore, cation replacement alone is also likely to produce artifactual effects on Ca^{2+} release. The vesicles that had been incubated in 0.15 M KCl were incubated with complexes of K^+ with several 'impermeable' anions such as gluconate, methanesulfonate, propionate, or thiosulfonate and replaced with choline.Cl (that is, replacement of both cation and anion). Under these conditions, light scattering changes were much smaller than with cation replacement or anion replacement alone. In particular, when K.gluconate was replaced with choline.Cl, there were virtually no light scattering changes (see Ikemoto, Antoniu and Kim, 1984; and Fig.7.1B).

7.2.1.2. *Time course of Ca^{2+} release.* The Ca^{2+} release induced by replacement of K·gluconate with choline·Cl occurs in three phases (Fig.1A, see also Fig.7.2B). In the first phase (0–50 ms), there is virtually no Ca^{2+} release. In the second phase (50–65 ms) about 15 nmol Ca^{2+}/mg protein is released with $t_{\frac{1}{2}} \approx 10$ ms (Fig.7.1A). In the third phase (65 ms–3 s), a larger amount of Ca^{2+} (for example, 56 nmol/mg) is released at a much slower rate. In the control experiment, in which K·gluconate vesicles were mixed with K·gluconate (that is, no ionic replacement), there was no Ca^{2+} release.

7.2.1.3. *Role of the T-tubule in triggering of 'depolarization'-induced Ca^{2+} release.* Caswell et al. (1979) previously reported that treatment of the T-tubule–SR complexes with a French press leads to dissociation of the T-tubule for SR, and incubation of the French-press treated membranes at higher concentrations of K·cacodylate (for example, 0.42 M) leads to reassociation. We have investigated whether or not T-tubules are involved in the induction of rapid Ca^{2+} release using these methods. As shown in Table 7.1 upon dissociation of T-tubules the amount of Ca^{2+} released in the rapid phase (A) decreased considerably; and upon reassociation the original amount of rapid Ca^{2+} release was restored. The rate constant of Ca^{2+} release (k) is virtually unaffected by the T-tubule dissociation and reassociation.

We purified SR and T-tubule vesicles from the French press-treated preparation by means of sucrose density gradient fractionation. The SR fraction was incubated in 0.4 M cacodylate without (Fig.7.2A) or with (Fig.7.2B) addition of the isolated T-tubules. Fig.7.2A illustrates that SR alone is incapable of producing Ca^{2+} release upon choline·Cl replacement of

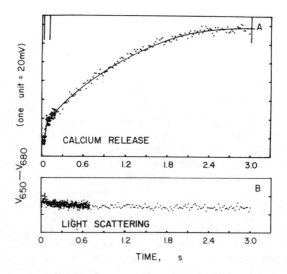

Fig.7.1. Time-course of Ca^{2+} release induced by replacement of K·gluconate with choline·Cl (A) and control experiment showing lack of light-scattering changes in the chemical depolarisation reaction when K.gluconate is replaced with choline·Cl in the absence of arsenazo III (B). Active Ca^{2+} loading of K·gluconate vesicles was done in solution A containing 1.6 mg/ml protein, 200 μM $CaCl_2$, 0.15 M K·gluconate, 0.5 mM Mg.ATP, 2.5 mM phosphoenol pyruvate, 10 units of pyruvate kinase per ml, 9 μM (A) or 0 μM (B) arsenazo III, and 20 mM MES (pH 6.8). Solution A was mixed with solution B containing 0.15 M choline·Cl, 9 μM (A) or 0 μM (B) arsenazo III, and 20 mM MES (pH 6.8) to induce Ca^{2+} release. Changes in $V_{650}-V_{680}$ were recorded with a dual-beam stopped-flow spectrophotometer as described elsewhere (Ikemoto, 1976). The upward excursion represents an increase of extravesicular [Ca^{2+}]. $V_{650}-V_{680}$ = 10 mV corresponds to 5.6 nmol Ca^{2+} released per mg protein.

Table 7.1. Effect of dissociation and reassociation of the T-tubule–SR complexes on depolarization-induced Ca^{2+} release.

Treatment	A nmol Ca^{2+}/mg	k s^{-1}
None	6.8	93.7
French press	2.2	72.9
Cacodylate	7.3	77.0

K·gluconate even after incubation in K·cacodylate. However, incubation of SR with the T-tubule fraction in K·cacodylate leads to reconstitution of rapid Ca^{2+} release. The results shown in Table 7.1 and Fig.7.2. indicate that the attachment of T-tubules to SR is an essential requirement for induction of rapid Ca^{2+} release.

7.2.1.4. Action potential of the isolated T-tubule. Several membrane-impermeable potential-sensitive dyes bound to the external membranes of the muscle fibre (such as WW375, NK2367, and merocyanine 540) have proven to be useful for optical measurements of the action potential (Heiny and Vergara, 1982; Morad

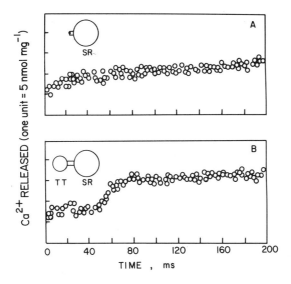

Fig.7.2. Reconstitution of the T-tubule/SR complex capable of producing rapid Ca^{2+} release. The T-tubule and SR fractions were isolated by a density gradient fractionation of the French press-treated vesicles, and homogenised in 0.15 M K gluconate. (A) The isolated K·gluconate SR (20 mg/ml) was mixed with the isolated T-tubules (1 mg/ml), and incubated as above. SR vesicles ((A) and (B)) were actively loaded with Ca^{2+}, and mixed with the choline·Cl solution in a stopped-flow apparatus as described in the legend to Fig.7.1.

and Salama, 1979; Nakajima and Gilai, 1980; Vergara and Bezanilla, 1976). Muscles were homogenized in a solution containing merocyanine 540, and the merocyanine 540-stained T-tubules were isolated by means of sucrose density gradient fractionation. Upon choline·Cl replacement of K·gluconate, the fluorescence intensity of the T-tubule membrane-bound merocyanine 540 decreased with $t_{\frac{1}{2}} \approx 10$ ms (Ikemoto, Antoniu and Kim, 1984). The rapid decrease of the dye fluorescence in the stopped-flow experiments represents the production of a negative-inside potential of the isolated T-tubule (Ikemoto, Antoniu and Kim, 1984). However, whether or not this process represents depolarization of the T-tubules depends upon the resting potential of the membrane at the time of ionic replacement, which remains to be investigated. The most important aspect of the kinetics of the membrane potential change described here is that Ca^{2+} release does not occur until the potential change is completed. This indicates that the T-tubule 'depolarization' is prerequisite for Ca^{2+} release. The results also suggest that the lag phase which precedes rapid Ca^{2+} release (see Figs.7.1A and 7.2B) probably consists of the time required for a relatively slow membrane potential change and that required for communication of the T-tubule with SR.

7.2.2. Ca^{2+} and drug-induced Ca^{2+} release

7.2.2.1. Kinetic characteristics

The Ca^{2+}-induced Ca^{2+} release and drug (such as caffeine, halothane)-induced Ca^{2+} release share a number of common properties; for example (a) the rate constant of Ca^{2+} release is slow in both cases (k = 0.2–1.1 s^{-1}; Ikemoto, Antoniu and Kim, 1984; Kim, Ohnishi and Ikemoto, 1983); (b) $[Ca^{2+}_o]$-dependence of drug-induced Ca^{2+} release is similar to that of Ca^{2+}-induced Ca^{2+} release (Kim, Ohnishi and Ikemoto, 1983); and (c) Ca^{2+}_o and drug have an additive effect on Ca^{2+} release (Kim, Ohnishi and Ikemoto, 1983; Kim et al., 1984). Several papers (Kim amd Ikemoto, 1983; Kim, Ohnishi

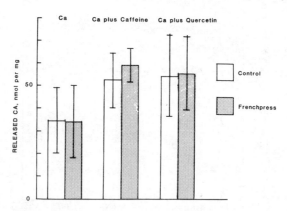

Fig.7.3. Dissociation of the T-tubule/SR complex has little effect on the Ca^{2+} and (Ca^{2+} + drug)-induced Ca^{2+} release. Control and French press-treated preparations (KCl vesicles) were loaded actively with Ca^{2+} in solution A. One part of solution A was mixed with one part of solution B containing 100 μM Ca^{2+} for Ca^{2+}-induced Ca^{2+} release. For (Ca^{2+} + drug)-induced Ca^{2+} release, solution B contained both 100 μM Ca^{2+} and drug (either 4 mM caffeine or 100 μM quercetin). The amounts of Ca^{2+} released were calculated by iterative computer fitting (Kim, Ohnishi, and Ikemoto, 1983) after signal averaging of total 70–95 traces for each type of experiment originated from three to five preparations.

and Ikemoto, 1983; Kirino and Shimizu, 1982; Watras *et al*, 1983) suggest that quercetin has a potentiation effect of Ca^{2+} release as caffeine does. As previously described (Kim, Ohnishi and Ikemoto, 1983), the V_{max} value (20–25 nmol/mg/s) obtained from $1/V$ versus $1/[drug]$ (V = rate of Ca^{2+} release) in the presence of a relatively low ATP concentration is about the same for both caffeine and quercetin, though K_m is much lower for quercetin than caffeine. This suggests that caffeine and quercetin react with the same receptor responsible for the induction of Ca^{2+} release (Kim, Oshnishi and Ikemoto, 1983).

Figure 7.3 illustrates that treatment of vesicles with a French press has no effect on any of these types of Ca^{2+} release, indicating that the T-tubule plays no appreciable role in the Ca^{2+} release (Ikemoto, Antoniu and Kim, 1984). These results also suggest that the mechanism involved in triggering these types of Ca^{2+} release, that is, receptors for Ca^{2+} and drugs, are localized in the SR membrane.

An improved time resolution in the stopped-flow measurements has revealed some new features in the kinetics of Ca^{2+} and drug-induced Ca^{2+} release. As

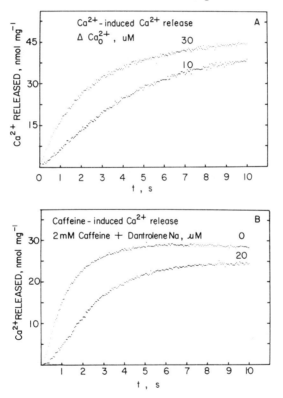

Fig.7.4. Stopped-flow spectrometric traces of Ca^{2+}-induced Ca^{2+} release (A) and caffeine-induced Ca^{2+} release (B). The SR vesicles were loaded with Ca^{2+} release (B). The SR vesicles were loaded with Ca^{2+} by ATP-dependent Ca^{2+} accumulation in a reaction mixture containing 1.6 mg protein/ml, 75 μM $CaCl_2$, 0.15 M KCl, 0.5 mM Mg.ATP, 2.5 mM phosphoenol pyruvate, 10 units per ml pyruvate kinase, 9 μM arsenazo III, and 20 mM MES, pH 6.8 (solution A). The reaction mixture was then loaded in syringe A of a stopped-flow apparatus. When the extravesicular Ca^{2+} concentration became virtually zero, the content of syringe A was mixed with an equal volume of solution B to induce Ca^{2+} release. In case A, solution B contained various concentrations of $CaCl_2$ to produce the final Δ Ca_o^{2+} as indicated. In case B, solution B contained 4 mM caffeine ([caffeine] after mixing = 2 mM) and 0 or 20 μM dantrolene sodium.

shown in Fig.7.4A, there is an appreciable lag phase before Ca^{2+} release, especially when submaximal $[Ca_o^{2+}]$ is used for induction of Ca^{2+} release. Figure 7.4B illustrates that in the case of caffeine-induced Ca^{2+} release also there is a lag phase, which becomes more pronounced if dantrolene sodium is present during the induction of Ca^{2+} release. Thus, a single exponential model used previously (Kim, Ohnishi and Ikemoto, 1983) is not suitable for the fitting of these traces.

$$C_1 \xrightarrow{k_1} C_2 \xrightarrow{k_2} C^*_3 \qquad \text{Model 1}$$

Model 1 gave a reasonably good fit to these traces. This model assumes that opening of the putative Ca^{2+} channel takes place in two steps. The first step ($C_1 \longrightarrow C_2$, the rate constant k_1) is a preparative step for the second step ($C_2 \longrightarrow C^*_3$), in which the channel opens with k_2. C and C^* represent closed and opened states of the channel, respectively.

7.2.3. Molecular components involved in Ca^{2+} release
(a) Depolarization-induced Ca^{2+} release, and (b) Ca^{2+} and drug-induced Ca^{2+} release are governed by different triggering mechanisms as described above. However, it appears that the sites of Ca^{2+} release are localized in SR in both types of Ca^{2+} release. The following data suggest that a 32 kD protein

Table 7.2. Effect of N-(7-dimethylamino-4-methyl-3-coumarinyl) maleimide (DACM) on various types of Ca^{2+} release. SR was actively loaded with Ca^{2+} as described in the legends to Figs.7.1 and 7.4. After the maximal loading was accomplished (at 3–4 min after addition of ATP), 1.6 mg/ml SR protein reacted with 2 μM DACM for 3 min, and the reaction mixture was placed in syringe A of a stopped-flow apparatus (solution A). One part of solution A was mixed with an equal volume of solution B to induce Ca^{2+} release. The compositions of solution B (in syringe B) for depolarization-induced Ca^{2+} release and for (Ca^{2+} and drug)-induced Ca^{2+} release were as described in the legends to Fig.7.1 and Fig.7.4, respectively.

Triggering method	Control		DACM	
	A	k	A	k
Depolarisation (Ca_i = 75 nmol/mg)	3.0	74.5	6.5	86.5
Increase of $[Ca_o^{2+}]$ (Ca_i = 31 nmol/mg, $[Ca_o^{2+}]$ = 10 mM)	24.6	0.8	43.0	0.8
100 μM quercetin (Ca_i = 47 nmol/mg)	26.0	0.4	36.0	0.4

A, nmol Ca^{2+} released per mg protein; k, s^{-1}

component of SR may serve as a Ca^{2+} channel protein for both types of Ca^{2+} release (a) and (b).

It was found that reaction of SR with a limited amount of fluorescent derivative of maleimide, N-(7-dimethylamino-4-methyl-3-coumarinyl) maleimide (DACM, for example 1.25 nmol/mg SR protein) exerts acceleration effects on Ca^{2+} release as shown in Table 7.2. If triggering is done in suboptimal conditions (that is, lower concentrations of Ca_o^{2+} and drugs) the amount of Ca^{2+} released is nearly doubled by the DACM modification in all types of Ca^{2+} release investigated. If triggering is done under optimal conditions, however, DACM modification has little effect on any kind of Ca^{2+} release, suggesting that the DACM modification of the channel protein decreases the threshold of the opening response to the triggering signal without changing the maximal size of the response. It is important to note that at the DACM:SR ratio used for these experiments almost all of the added DACM molecules were incorporated into the 32 kD protein of SR as shown in Fig.7.5. These results suggest that the 32 kD protein is one of the constituents of the Ca^{2+} channel.

The 32 kD protein is practically insoluble in the non-ionic detergent $C_{12}E_8$, but soluble in zwitterionic detergents, such as Zwittergent 3-14. Thus, successive treatment of SR with the two different types of detergent led to a partial purification of the 32 kD protein (Kim and Ikemoto, 1984). The specific binding activity of ruthenium red, which appears to be a blocker of the Ca^{2+} channel (see section 7.2.4), increased significantly upon partial

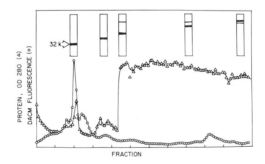

Fig.7.5. DACM is almost exclusively incorporated into the 32 kD protein under the conditions which produce acceleration effects on various types of Ca^{2+} release. DACM reacted with SR as described in the legend to Table 7.1; 5% beta-mercaptoethanol and 1% SDS were added to the DACM reaction mixture, boiled for 3 min, and 1.5–2.0 mg protein was placed on the top of a 1.0 x 10.5 cm column of a gel consisting layered polyacrylamide (7.5%, 3.5%). Electrophoresis was carried out at 3 mA for 1 hour and at 6 mA for 16 hours at 22°C. Fractions (1.5 ml each) eluted through a preparative gel were collected with a fraction collector. Protein was determined by OD_{280}, and fluorescence intensity of each fraction (excitation = 395 nm; emission = 465 nm) was determined in a Perkin-Elmer fluorometer.

purification of the 32 kD protein. This is consistent with the notion that this protein is one of the prime candidates for the Ca^{2+} channel component.

According to the recent report of Campbell, Lipschutz and Denney (1984), reaction of the plasma membrane Ca^{2+} channel blocker [^{3}H]-nitrendipine with cardiac muscle SR membrane led to a selective labelling of the 32 kD component. It is possible that the 32 kD protein is a common component of different types of Ca^{2+} channels of different origins.

7.2.4. Inhibitors

It would be expected that inhibitors reacting with the 'common' channel affect all types of Ca^{2+} release, whereas inhibitors reacting with different receptors exert selective effects depending on the types of Ca^{2+} release. Table 7.3 depicts the results of our recent studies of several reagents. It was found that ruthenium red (RR) and tetracaine, which are known inhibitors of Ca^{2+} and caffeine-induced Ca^{2+} release (Kim and Ikemoto, 1983, 1984; Miyamoto and Racker, 1982; Ohnishi, 1979b, 1981; Volpe et al., 1983), also inhibit the T-tubule-mediated rapid Ca^{2+} release. It should be noted that inhibition of different types of Ca^{2+} release occurs at about the same I_{50} (drug concentration that produces half maximal inhibition) in each case of RR and tetracaine. Thus, it appears that both RR and tetracaine work as a blocker of the 'common' channel. Dantrolene sodium, on the other hand, produces appreciable inhibition of caffeine-induced Ca^{2+} release (see Table 7.3; also halothane-induced Ca^{2+} release; compare Danko and Ikemoto, 1984; Kim and Ikemoto, 1983; Ohnishi, Taylor and Gronert, 1983); whereas there is no appreciable effect on Ca^{2+}-induced Ca^{2+} release (Table 7.3, compare Kim and Ikemoto, 1983; and for opposite results, Van Winkle, 1976). The rate of depolarization-induced Ca^{2+} release was reduced by dantrolene sodium in some but not all experiments. Thus, dantrolene sodium is regarded as an example of a receptor inhibitor. According to our recent dantrolene sodium-binding studies (Danko and Ikemoto, 1984), SR has at least two classes of dantrolene sodium-binding sites: one is available for caffeine binding, but the other is not. Presumably, the former class represents the sites involved in the drug-induced Ca^{2+} release.

7.2.5. Ca^{2+} release functions in normal and diseased muscles

We have investigated the time-course of Ca^{2+} release for SR isolated from

Table 7.3. Inhibitors of several types of Ca^{2+} release.

Method of triggering	Ruthenium red	Reagent Tetracaine	Dantrolene Na
Ca^{2+} jump (50 µM)	I (0.20 µM)	I (0.10 mM)	N
Caffeine (2 mM)	I (0.15 µM)	I (0.08 mM)	I (3.0 µM)
Ionic replacement	I (0.20 µM)	I (0.11 mM)	I (2.0 µM), or N

I, inhibition; N, no effect; I or N, occasional inhibition. The number in the parenthesis indicates I_{50} (drug concentration that produces half maximal inhibition).

Table 7.4. The ratios of the kinetic parameters (A, the amount of Ca^{2+} released; k, the rate constant; $A \cdot k$, the initial rate) of malignant hyperthermia SR / those of normal SR. The ratios were calculated from the data originating from four different preparations of each normal SR and malignant hyperthermia SR.

Malignant hyperthermia/normal	Ca^{2+}	Halothane	Ca^{2+} + Halothane	Depolarisation
A	1.2	1.7	1.2	1.4
k	1.6	1.7	1.5	2.2
$A \cdot k$	1.7	2.6	1.8	1.6

The concentration of reagents added: Ca^{2+}, 1 µM; halothane, 0.2 mM; depolarisation, 0.15 M K-gluconate → 0.15 M choline-Cl.

muscles of normal (N) pigs and those of pigs susceptible to malignant hyperthermia (MH) using several different triggering methods: (a) addition of halothane (for example, 0.2 mM); (b) an increase of extravesicular Ca^{2+} concentration; (c) combination of (a) and (b); and (d) replacement of ions (K·gluconate with choline·Cl) to produce depolarization of the T-tubule membrane (Kim et al., 1984). As shown in Table 7.4, all types of Ca^{2+} release investigated ((a), (b), (c), and (d)) have higher rates in malignant hyperthermia SR than normal SR. This confirms and further extends the previous findings that SR has higher Ca^{2+}-induced Ca^{2+} release (Endo et al., 1983; Ohnishi, Taylor and Gronert, 1983), caffeine-induced Ca^{2+} release (Nelson, 1983) and halothane-induced Ca^{2+} release (Ohnishi, Taylor and Gronert, 1983) than normal SR. The most likely explanation for these findings is that the Ca^{2+} channels (probably common for different types of Ca^{2+} release) are altered in malignant hyperthermia SR. As a result of such modification, more Ca^{2+} is released per unit of trigger.

7.3. Concluding remarks

Recent methodological advances in kinetic studies of Ca^{2+} release have permitted to induce and monitor rapid Ca^{2+} release from SR *in vitro*. There are at least two clearly distinguishable routes for triggering of Ca^{2+} release: one involves transmission of the signal elicited by the T-tubule membrane depolarization from T-tubules to SR; and the other, a direct stimulation of SR receptors by Ca^{2+} and drugs. However, a single class of Ca^{2+} release channel is sufficient to account for various types of Ca^{2+} release at the moment. The fact that some reagents such as ruthenium red and tetracaine inhibit different types of Ca^{2+} release suggest that they work as blockers of the Ca^{2+} channel. Our data described here suggest that the 32 kD component of SR is one of the prime candidates for the channel protein. Some new experiments described in this article, such as production of rapid Ca^{2+} release by reconstituting the triadic complex from the purified T-tubule and SR fractions, have opened new possibilities for studies of the molecular mechanisms of E-C coupling.

Acknowledgement

This work was supported by Grant AM-16922 from National Institutes of Health and a grant from the Muscular Dystrophy Association of America. D.H. Kim is a postdoctoral fellow of National Institute of Health.

References

Abramson, J.J., Trimm, J.L., Weden, L. and Salama, G. (1983). Heavy metals induce rapid calcium release from sarcoplasmic reticulum vesicles isolated from skeletal muscle. *Proc. Natl. Acad. Sci.* **80**, 1524–30.

Adrian, R.H. and Peres, A.R. (1979). Charge movement and membrane capacity in frog muscle. *J. Physiol (Lond.)*, **289**, 83–97.

Cadwell, J.J.S. and Caswell, A.H. (1982). Identification of a constituent of the junctional feet linking terminal cisternae to transverse tubules in skeletal muscle. *J. Cell Biol.* **93**, 543–50.

Campbell, K.P. and Shamoo, A.E. (1980). Chloride-induced release of actively loaded calcium from light and heavy sarcoplasmic reticulum vesicles. *J.*

Memb. Biol. **54**, 73–80.

Campbell, K.P., Lipshutz, G.M. and Denney, G.H. (1984). Direct photoaffinity labeling of the high affinity nitrendipine-binding site in subcellular membrane fractions isolated from canine myocardium. *J. Biol. Chem.* **259**, 5384–7.

Caswell, A.H., Lau, Y.H., Garcia, M. and Brunschwig, J-P. (1979). Recognition and junction formation by isolated transverse tubules and terminal cisternae of skeletal muscle. *J. Biol. Chem.* **254**, 202–8.

Caswell A.H. and Brandt, N.R. (1981). Ion-induced release of calcium from isolated sarcoplasmic reticulum. *J. Memb. Biol.* **58**, 21–33.

Chandler, W.K., Rakowski, R.F. and Schneider, M.F. (1976). Effects of glycerol treatment and maintained depolarization of charge movement in skeletal muscle. *J. Physiol. (Lond.).* **254**, 245–86.

Danko, S. and Ikemoto, N. (1984). Dantrolene Na binding sites involved in inhibition of calcium release from SR *in vitro*. *Biophys. J.* **45**, 319a.

Ebashi, S. (1976). Excitation-contraction coupling. *Ann. Rev. Physiol.* **38**, 293–313.

Ebashi, S., Endo, M. and Ohtsuki, I. (1969). Control of muscle contraction. *Q. Rev. Biophys.* **2**, 351–84.

Eisenberg, B.R. and Gilai, A. (1979). Structural changes in single muscle fibers after stimulation at a low frequency. *J. Gen. Physiol.* **74**, 1–16.

Eisenberg, B.R, and Eisenberg, R.S. (1982). The T-SR junction in contracting single skeletal muscle fibers. *J. Gen. Physiol.* **79**, 1-19.

Endo, M. (1975). Mechanism of action of caffeine on the sarcoplasmic reticulum of skeletal muscle. *Proc. Jap. Acad.* **51**, 479–84.

Endo, M. (1977). Calcium release from the sarcoplasmic reticulum. *Physiol. Rev.* **57**, 71–108.

Endo, M. and Nakajima, Y. (1973). Release of calcium induced by depolarization of the sarcoplasmic reticulum membrane. *Nature New Biol.* **246**, 216–8.

Endo, M., Yagi, S., Ishizuka, T., Horiuti, K., Koga, Y. and Amaha, K. (1983). Changes in the Ca^{2+}-induced Ca^{2+} release mechanism in the sarcoplasmic reticulum of the muscle from a patient with malignant hyperthermia. *Biomed. Res.* **4**, 83–92.

Fabiato, A. and Fabiato, F. (1977). Calcium release from the sarcoplasmic reticulum. *Circ. Res.* **40**, 119–29.

Ford, L.E. and Podolsky, R.J. (1972). Intracellular calcium movements in skinned muscle fibers. *J. Physiol (Lond.).* **223**, 21–33.

Franzini-Armstrong, C. (1975). Membrane particles and transmission at the triad. *Fed. Proc.* **34**, 1382–9.

Franzini-Armstrong, C. (1980). Structure of sarcoplasmic reticulum. *Fed. Proc.* **39**, 2403–9.

Franzini-Armstrong, C. and Nunzi, G. (1983). Junctional feet and particles in the triads of a fast-twitch muscle fibre. *J. Muscle Res. Cell. Motility.* **4**, 233–52.

Heiny, J.A. and Vergara, J. (1982). Optical signals from surface and T system membranes in skeletal muscle fibers. *J. Gen. Physiol.* **80**, 203–30.

Horowicz, P. and Schneider, M.F. (1981). Membrane charge movement in contracting and non-contracting skeletal muscle fibres. *J. Physiol. (Lond.).* **314**, 565–93.

Ikemoto, N. (1976). Behaviour of the Ca^{2+} transport sites linked with the phosphorylation reaction of ATPase purified from the sarcoplasmic reticulum. *J. Biol. Chem.* **251**, 7275–7.

Ikemoto, N., Antoniu, B. and Kim. D.H. (1984). Rapid calcium release from the isolated sarcoplasmic reticulum is triggered *via* the attached transverse tubular system. *J. Biol. Chem.* 259, 13151–8.

Inesi, G. and Malan, N. (1976). Mechanisms of calcium release in sarcoplasmic reticulum. *Life Sci.* 18, 773–9.

Kasai, M. and Miyamoto, H. (1976). Depolarization-induced calcium release from sarcoplasmic fragments. I. Release of calcium taken up upon using ATP. II. Release of calcium incorporated without ATP. *J. Biochem.* 79, 1053–76.

Katz, A.M., Repke, E.I., Dunnett, J. and Hasselbach, W. (1977). Dependence of ionophore and caffeine-induced calcium release from sarcoplasmic reticulum vesicles on external and internal calcium ion concentrations. *J. Biol. Chem.* 252, 1938–49.

Kim, D.H. and Ikemoto, N. (1983). Kinetic resolution of the components involved in calcium release from sarcoplasmic reticulum. *Biophys. J.* 41, 232a.

Kim, D.H. and Ikemoto, N. (1984). Identification of candidate components for the calcium release channel of sarcoplasmic reticulum using ruthenium red as a probe. *Biophys. J.* 45, 399a.

Kim, D.H., Ohnishi, S.T. and Ikemoto, N. (1983). Kinetic studies of calcium release from sarcoplasmic reticulum. *J. Biol. Chem.* 258, 9662–8.

Kim, D.H., Sreter, F.A., Ohnishi, S.T., Ryan, J.F., Roberts, J., Allen, P.D., Meszaros, L.G., Antoniu, B. and Ikemoto, N. (1984). Kinetic studies of Ca^{2+} release from sarcoplasmic reticulum of normal and malignant hyperthermia susceptible pig muscles. *Biochim. Biophys. Acta.* 775, 320–7.

Kirino, Y. and Shimizu, H. (1982). Ca^{2+}-induced Ca^{2+} release from sarcoplasmic reticulum. *J. Biochem.* 92, 1287–96.

MacLennan, D.H., Shoshan, V. and Wood, D.S. (1983). Studies of Ca^{2+} release from sarcoplasmic reticulum. *Ann. NY. Acad. Sci.* 402, 470–6.

Mathias, R., Levis, R.A. and Eisenberg, R.S. (1980). Electrical models of excitation-contraction coupling and charge movement in skeletal muscle. *J. Gen. Physiol.* 76, 1–31.

Meissner, G. (1975). Isolation and characterization of two types of sarcoplasmic vesicles. *Biochim. Biophys. Acta.* 389, 51–68.

Meissner, G. and McKinley, D. (1976). Permeability of sarcoplasmic reticulum membrane. The effect of changed ionic environments on Ca^{2+} release. *J. Memb. Biol.* 58, 21–33.

Mitchell, R.D., Saito, A., Palade, P. and Fleischer, S. (1983). Morphology of isolated triads. *J. Cell. Biol.* 96, 1017–29.

Miyamoto, H. and Racker, E. (1982). Mechanism of calcium release from skeletal sarcoplasmic reticulum. *J. Memb. Biol.* 66, 193–201.

Morad, M. and Salama, G. (1979). Optical probes of membrane potential in heart muscle. *J. Physiol.* 292, 267–95.

Nagasaki, K. and Kasai, M. (1981). Ca^{2+}-induced calcium release from sarcoplasmic reticulum. *J. Biochem.* 90, 749–55.

Nakajima, S. and Gilai, A. (1980). Action potentials of isolated single muscle fibers recorded by potential-sensitive dyes. *J. Gen. Physiol.* 76, 729–50.

Nelson, T.E. (1983). Abnormality in calcium release from skeletal sarcoplasmic reticulum of pigs susceptible to malignant hyperthermia. *J. Clin. Invest.* 72, 862–70.

Ohnishi, S.T. (1979a). A method for studying the depolarization-induced calcium ion release from fragmented sarcoplasmic reticulum. *Biochim. Biophys. Acta.* 587, 217–30.

Ohnishi, S.T. (1979b). Calcium-induced calcium release from fragmented sarcoplasmic reticulum. *J. Biochem.* 86, 1147–50.

Ohnishi, S.T. (1981). Calcium-induced calcium release as a gated calcium transport. In *The Mechanism of Gated Calcium Transport across Biological Membranes.* S.T. Ohnishi and M. Endo (eds.), pp. 275–93. Academic Press, New York.

Ohnishi, S.T., Taylor, S. and Gronert, G.A. (1983). Calcium-induced Ca^{2+} release from sarcoplasmic reticulum of pigs susceptible to malignant hyperthermia. *FEBS Lett.* **161**, 103–7.

Sandow, A. (1965). Excitation-contraction coupling in skeletal muscle. *Pharmacol. Rev.* **17**, 265–320.

Schneider, M.G. and Chandler, W.K. (1973). Voltage dependent charge movement in skeletal muscle: a possible step in excitation-contraction coupling. *Nature (Lond.)* **242**, 244–6.

Shoshan, V., MacLennan, D.H. and Wood, D.S. (1981). A proton gradient controls a Ca^{2+} release channel in sarcoplasmic reticulum. *Proc. Natl. Acad. Sci.* **78**, 4828–32.

Shoshan, V., MacLennan, D.H. and Wood, D.S. (1983). Tetraphenylboron causes Ca^{2+} release in isolated sarcoplasmic reticulum. *J. Biol. Chem.* **258**, 2837–42.

Somlyo, A.V. (1972). Bridging structures spanning the junctional gap at the triad of skeletal muscle. *J. Cell. Biol.* **80**, 743–50.

Van Winkle, W.B. (1976). Calcium release from skeletal muscle sarcoplasmic reticulum: site of action of dantrolene sodium. *Science.* **193**, 1130–31.

Vergara, J. and Bezanilla, F. (1976). Fluorescence changes during electrical activity in frog muscle stained with merocyanine. *Nature (Lond.).* **259**, 684–6.

Vergara, J. and Caputo, C. (1983). Effects of tetracaine on charge movements and calcium signals in frog skeletal muscle fibers. *Proc. Natl. Acad. Sci.* **80**, 1477–81.

Volpe, P., Palade, P., Costello, B., Mitchell, R.D. and Fleischer, S. (1983). Spontaneous calcium release from sarcoplasmic reticulum. *J. Biol. Chem.* **258**, 12434–42.

Watras, J., Glezen, S., Seifert, C. and Katz, A. (1983). Quercetin stimulation of calcium release from rabbit skeletal muscle sarcoplasmic reticulum. *Life Sci.* **32**, 213–19.

Weber, A. (1966). Energized calcium transport and relaxing factors. In *Current Topics in Bioenergetics.* D.R. Sanadi (ed.), pp. 203–54. Academic Press, New York.

METAL-CHELATING PROPERTIES OF ORGANIC COMPOUNDS WITH BIOLOGICAL SIGNIFICANCE: AN OVERVIEW

H.U. Wolf

Abteilung Pharmakologie und Toxikologie der Universität Ulm, Oberer Eselsberg N 26, D-7900 Ulm/Donau, West Germany

Abstract

Almost all metal ions of the periodic table can be bound by low molecular weight chelators or by biological macromolecules. Binding occurs via oxygen, nitrogen or sulphur atoms as electron donor atoms either by replacement of hydrogen ions or by coordination. Selectivity in binding of special metal ions by proteins and other biological macromolecules generally is based on (1) the total concentrations of the metal ions in the biological system; (2) the design of the binding site, that is, the presence of oxygen, nitrogen or sulphur as electron donor atoms; and (3) the total concentration of binding sites with oxygen donor atoms and those with nitrogen or sulphur atoms. In particular the selectivity of Ca^{2+} over Mg^{2+} binding occurs in those biological macromolecules (1) which are rather flexible than rigid; (2) in which global conformational changes can occur concomitantly with the binding of Ca^{2+}; and (3) in which three or more acidic carboxyl groups can be involved into the formation of the chelate, that is, in acidic protein molecules.

It is well known today that there is a considerable number of proteins which can bind divalent metal ions tightly and in some cases with rather high selectivity. On the other hand, there is a series of different biological aims and consequences for binding a metal ion in biological systems. This binding occurs for example with the aim of increasing the catalytic activity of enzymes (Zn^{2+} in phosphatases), of being transported across a membrane (binding of Ca^{2+} to Ca^{2+}-transport ATPases), of designing a specific part of a biological molecule to bind certain ligands (Fe^{2+} in haemoglobin), or simply of stabilising a protein. Furthermore, it is also well known that Ca^{2+} plays a unique biological role among all divalent cations considering the ubiquitous occurrence of this metal ion, its role as a second messenger, its distribution to different sides of biological membranes resulting from a very efficient transport, the selectivity of its binding to modular proteins, and so on.

In order to explain at least in part the outstanding role of calcium within the biological processes mentioned, one should try to answer two questions, which will be discussed in this chapter. These are:

(1) Which functional groups present in low and high molecular weight chelating compounds are able to bind metal ions tightly?
(2) Are there significant differences in the chelating behaviour of Ca^{2+} on

one side, and other divalent metal ions, especially Mg^{2+}, on the other side, which can explain the selectivity and tightness of binding to biological macromolecules?

The first of the two questions may be answered as follows: according to Mellor (1964) there are two prerequisites which have to be fulfilled in order that a certain compound may function as a metal ion chelator:

The first of these prerequisites is that the molecule under consideration has at least two functional groups with donor atoms, which are able to donate a pair of electons to the metal ion to be bound or chelated. This pair of electrons may origin from acidic groups which have lost a proton such as the dissociated carboxyl group -COO$^-$ or from basic coordinating groups such as the -NH$_2$ group.

It is known that considerable number of functional groups can donate a pair of electrons to form such a complex. As shown in Table 8.1, there is a category of functional groups which can react with metal ions by proton exchange. These are the following:

Table 8.1. Functional groups with metal ion binding properties

---

1. Groups interacting with metal ions after dissociation of a proton

    1.1 Oxygen as donor atom
    -COOH
    -OH(enolic or phenolic)
    -SO$_3$H
    -PO$_3$H$_2$

    1.2 Nitrogen as donor atom
    -NH$_2$
    -NH-R

    1.3 Sulphur as donor atom
    -SH

2. Groups interacting with metal ions by coordination

    2.1 Oxygen as donor atom
    =O
    -OH
    -O-

    2.2 Nitrogen as donor atom
    -NH$_2$
    -NH-R
    -N=

    2.3 Sulphur as donor atom
    -S-

---

(1) The carboxyl group which is present in a large number of metal ion chelating model compounds and in biological macromolecules such as proteins.
(2) The enolic or phenolic hydroxyl group which is present in the amino acid tyrosine.
(3) The sulphur-containing group with oxygen as donor atom is the sulphonic acid group present in heparin.
(4) The phosphorus-containing group with oxygen as donor atom of particular interest is the phosphate present for example in ATP, which readily binds Mg^{2+} and Ca^{2+}, in phospholipids, and in phosphorylated proteins such as phospholamban.
(5) In addition, there are also some functional groups containing a heteroatom, such as nitrogen and sulphur.
(6) Functional groups containing nitrogen are the $-NH_2$ group, and the -NH-R group of secondary amines, in both of which one proton can be replaced by metal ions. The incorporation of Fe^{2+} into the haem moiety is one of the best-known examples for the latter case.
(7) A functional group containing sulphur is the -SH group present for example in cysteine with a very high affinity to heavy metal ions such as Zn^{2+}, Cd^{2+} and Hg^{2+}.
(8) The category of coordinating functional groups contain also examples with oxygen, nitrogen, and sulphur atoms (see Table 8.1).
(9) Oxygen donor atoms are the oxygen atom of the carbonyl function, present in organic acids, amino acids and proteins, the alcoholic -OH function, for example, that of the amino acid threonin, and the ether -O- atom.
(10) Groups with nitrogen atoms are primarily the NH_2 group of amino acids and of proteins containing diaminomonocarboxylic acids like lysine. The =NH group of secondary amines and the -N= group of tertiary amines occur in the imidazole moiety of histidine.
(11) A group with sulphur as donor heteroatom occurs as a thioether group in methionine.

The second prerequisite for an effective chelation of metal ions is the fact that these groups have to be positioned in a molecule not too far from each other so that a ring system can be formed, the metal ion being the closing member. These two conditions are necessary for an effective chelating of metals, but they are not always sufficient (Mellor, 1964). One of the additional conditions which have to be met is for example the maintenance of a sufficiently high pH value, since many metal ion-chelator complexes become unstable under the influence of high proton concentrations resulting in a partial replacement of metal ions by protons, which eventually leads to a dissociation of the complex.

If we try to classify the different types of chelates, it might be of advantage to use low molecular weight compounds as model substances for this purpose. A distinction between different types of chelates is possible with

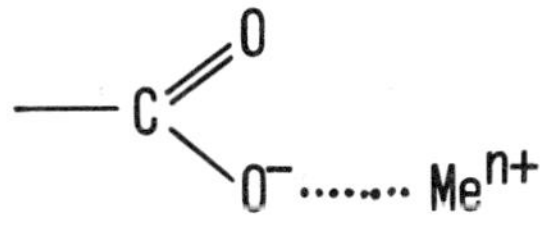

Fig.8.1. The carboxylic group of organic acids as an example of a unidentate chelating group.

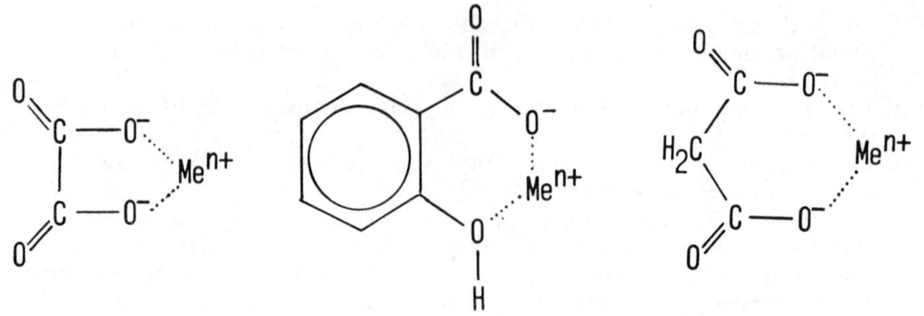

Fig.8.2. Oxalic acid, salicylic acid, and malonic acid as examples of bidentate chelating agents.

respect to:

(1) the number of donor atoms interacting with the metal ion; and
(2) the number of atoms forming a ring system necessary for the stability of the chelate.

Concerning the first method of distinction, the most simple case of interaction between a metal ion and a chelator is where only one donor atom is involved in the interaction. This case, for example, between a metal ion and one oxygen atom of a carboxyl group, as shown in Fig.8.1, is called unidentate. However, these complexes are rather unstable, since two donor atoms are necessary to form a stable complex as mentioned above. Particularly in the case of monovalent cations ($n = 1$), these complexes are very weak. In contrast, bidentate complexes are much more stable, and indeed, there is a great variety of compounds, which meet the prerequisites of a chelating compound mentioned above. Examples for this group of chelators are given in Fig.8.2 - oxalic acid and its homologue malonic acid, with two carbolic groups, and salicylic acid with one carboxylic group and one phenolic OH function. Penicillamine (Fig.8.3) is an example of a tridentate chelator since it contains a carboxylic acid group, an amino group and an SH group. Since the stability of the metal ion—chelator complexes increases with the number of donor atoms interacting with the metal ion, the chelates with quadridentate, quinquedentate, and especially with sexadentate chelators, are generally more stable than the bidentate and tridentate chelator complexes. The best-known compound forming sexadentate complexes with a series of metal

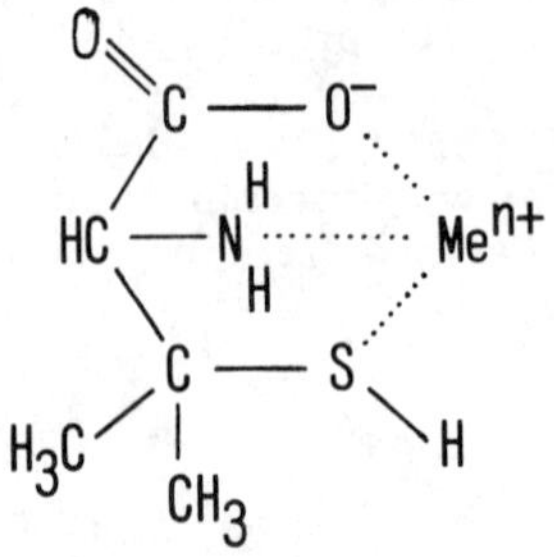

Fig.8.3. Penicillamine as an example of a tridentate chelating agent.

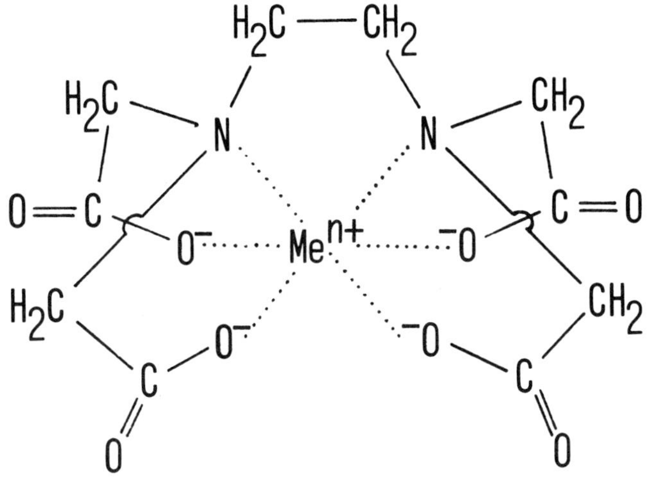

Fig.8.4. EDTA as an example of a sexadentate chelating agent.

ions including calcium, is EDTA (Fig.8.4). In this case, four negatively charged carboxylic groups and two neutral nitrogen donor atoms are involved in the chelate formed from bivalent or trivalent metal ions and EDTA.

As mentioned above, in order to form a stable metal ion–chelator complex, two donor atoms of the chelator have to interact with the metal ion. As a consequence of this, a ring system is formed containing among others the two donor atoms and the metal ion as the closing member. Generally these ring systems must exceed a minimum size; in other words they have to contain a minimum number of five atoms including the two donor atoms and the metal ion. Since the formation of four–membered rings involves a considerable strain in the molecule, they are rather rare and fairly unstable. Thus, the anionic and the neutral oxygen atoms of a carboxylic group do not form a stable ring system with a metal ion. Ring systems of biological significance are five-membered like the ring system of oxalic acid, and six-membered like that of malonic acid, and salicylic acid (see Fig.8.2). Penicillamine (Fig.8.3) contains two five-membered ring-systems (N-C-C-O-Me^{n+} and N-C-C-S-Me^{n+}), but a six-membered ring system (S-C-C-C-O-Me^{n+}) is also involved. In EDTA (see Fig.8.4) there are four five-membered ring-systems containing N-C-C-O-Me^{n+}, and one five-membered ring with N-C-C-N-Me^{n+}. Ring systems with a higher number of atoms than six become increasingly less stable.

Since the compounds discussed here contain the same functional groups also found in proteins, it is not surprising that the binding behaviour of these two categories of compounds is rather similar. This does not only concern the binding of metal ions itself, but also the relative strength of the binding. In fact, this is the reason why the low molecular weight chelators are very effective antidotes in cases of acute and chronic intoxications by metal ions, since they are able to compete for metal ion binding and thus dissociate the complexes which have been formed from structural and functional body proteins and metal ions.

In addition, the stability of the metal ion complexes of both the low molecular weight model substances and the proteins are similarly dependent on the hydrogen ion concentration; in other words, the stability of the

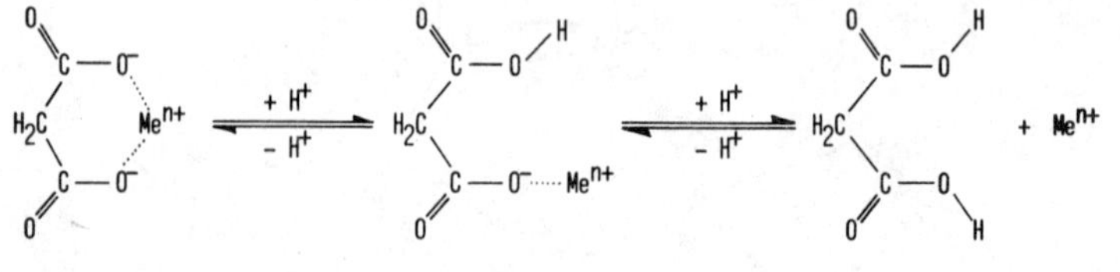

Fig.8.5. Stepwise protonisation of a malonic acid–Me^{n+} complex and subsequent dissociation of the complex into free malonic acid and Me^{n+}.

complexes decreases with increasing proton concentration on account of the competition of metal ions and protons for the same acidic groups. As shown in Fig.8.5, binding of a proton considerably decreases the stability of a bidentate complex, and binding of a second proton leads to the complete degradation of the complex.

The control of the numerous biological and biochemical processes in which Ca^{2+} is involved is based on the interaction of this ion with specific proteins by a binding-dissociation equilibrium. As we know, Ca^{2+} exhibits a special function, which renders it different from all other metal ions including monovalent alkali cations, Mg^{2+}, and the transition metal ions, Zn^{2+}, Cd^{2+}, and Pb^{2+}. One of the numerous examples of the special physiological function of Ca^{2+} is the activation of the Ca^{2+}-transport–ATPase activity. But as we could show some years ago, this activation can also be done under certain conditions by a series of different metal ions as this is shown in Fig.8.6. Here the relative activation potency of different metal ions was plotted as a function of their non-solvated ionic radius (Pfleger and Wolf, 1975). We obtained an almost symmetrical bell-shaped curve, which starts with Mg^{2+} as a very poor activator of this enzyme (around 20% of the activity shown by Ca^{2+}). The increase of the ionic radius leads to an increase of the activation potency in the case of Co^{2+}, Ni^{2+}, Mn^{2+}, Cu^{2+}, and Zn^{2+}, reaching its optimum with Ca^{2+} and Sr^{2+}. However, the further increase then leads to a decrease of the activation potency, which is shown in the case of Pb^{2+} and Ba^{2+}. With the exceptions of Cd^{2+} and Hg^{2+}, the activation potency of any ion is thus rather a function of their ionic radius than of the nature of the activating ion. In particular, it is not dependent on the dissociation constant of the metal ion–ATPase complex.

As can be seen there is a considerably high number of metal ions which could compete more or less effectively for the binding site of the Ca^{2+}-transport ATPase as an activating ion. Why does this competition occur under physiological conditions to a much lesser extent than one could presume on account of the data presented here? In other words, what is the reason for the high selectivity of Ca^{2+} binding relative to that of transition metals including Ca^{2+} and Pb^{2+} on one side, and relative to Na^+, K^+ and especially Mg^{2+} on the other side?

Starting with the group of heavy metal ions, we see that according to Williams (1977) there are three different factors which determine the ability of these ions in competing against Ca^+.

(1) First the total concentration of heavy metal ions present in biological systems ions is much lower than that of Ca^{2+}. The experiments presented before were done in the presence of a metal ion buffer system, in which total metal ion concentration of 20–380 μM resulted in free metal ion concentrations of 10^{-20}–10^{-13}M. In contrast, total metal ion concentrations for heavy metal ions are generally much lower in

Metal-chelating by organic compounds

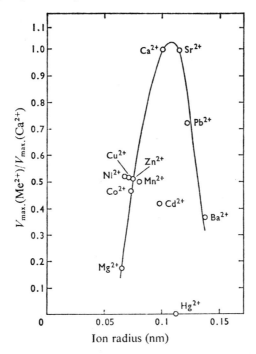

Fig.8.6. Relative potency of various bivalent metal ions in activating the ATP-hydrolysing activity of the Ca^{2+}-transport ATPase of human erythrocyte membranes (potency of $Ca^{2+} = 1.0$). Reprinted by permission from the Biochemical Journal, vol. 147, pp. 359–61 copyright © 1975, The Biochemical Society, London.

biological systems. Consequently, the free concentrations of these metal ions are also much lower so that they cannot generally compete against Ca^{2+}. An example of these very low concentrations of metal ions is the Cd^{2+} level in blood plasma; even in the case of acute poisoning the total Cd^{2+} level is below 0.2 μM.
(2) The heavy metal ions are bound very tightly by binding sites containing nitrogen and sulphur donor atoms, to which Ca^{2+} has a much lower binding affinity. Consequently, heavy metal ions are more likely bound by these binding sites with comparatively high affinity.
(3) In the biological systems the concentration of these high-affinity nitrogen and sulphur-containing binding sites is considerably higher than the concentration of heavy metal ions. The consequence is a relatively low free concentration of these ions.

In summary, on account of these three factors, a competition between Ca^{2+} and heavy metal ions such as Zn^{2+}, Cu^{2+}, Cd^{2+}, and Pb^{2+} for Ca^{2+}-binding sites does not in most cases play a practical role in biological systems.

The second group of metal ions, which can mainly compete with Ca^{2+} for Ca^{2+}-binding sites involves Na^+, K^+ and Mg^{2+}. These cations are present in biological systems at concentrations which, in contrast to the group discussed before, exceed the concentration of Ca^{2+} considerably. Why do they not compete against Ca^{2+} for its specific binding sites, or only to a low extent. The answer is easy in the case of Na^+ and K^+. These ions are monovalent, and as mentioned above, monovalent cations form very weak complexes with donor atoms like oxygen, the log of the binding constant being in the range of about 2 or even much lower, whereas the correspondent value for Ca^{2+} in the case of the Ca^{2+}-transport ATPase exceeds 6, and in the case of EDTA it is more than 10 (Williams, 1980). So, these ions also cannot compete effectively against Ca^{2+} for the Ca^{2+}-binding site in spite of their high physiological concentrations.

To answer the question of the reason for the selectivity in the competition between Ca^{2+} and Mg^{2+}, I would like to explain some fundamental facts for the binding behaviour of Ca^{2+} and Mg^{2+}. At present Ca^{2+} is known to bind preferably via oxygen as donor atoms - that is, both the anionic groups such as carboxyl or phosphate groups, and the neutral oxygen atoms of the carbonyl, ether or alcohol type. These groups are present in the main polypeptide chain and in the amino acid side-chains of proteins as well. Although nitrogen atoms in the absence of oxygen are weak donor atoms for most bivalent metal ions, as can be seen in the case of ethylenediamine, these nitrogen atoms are involved as donor atoms simultaneously into bivalent metal ion chelates (see Fig.8.4). This is also true to a certain extent in the case of Mg^{2+}, but apparently to a much lesser extent, if at all, in the case of Ca^{2+}.

Another specific fact is that by binding Ca^{2+} the chelating molecules, especially proteins, undergo a considerable conformational change, which is not local, but global; in other words, these protein molecules have to possess a considerable flexibility. In fact, Ca^{2+} can be considered as a crosslink between different domains of a protein molecule.

In contrast to the behaviour of Ca^{2+}, Mg^{2+} demands a specific environment at its binding site, which is characterised by a high order rather than by the type of the donor atoms, which have to be prearranged to form an octahedral symmetry. One very well-known example for this prearranged environment, the so-called cryptate, is found in the molecule of chlorophyll which binds Mg^{2+} much better than Ca^{2+}. In other words, Mg^{2+} is much more sensitive than Ca^{2+} to changes in the distance of the donor atoms involved in its binding, as can be seen in the case of the Ca^{2+}-ATPase mentioned above, and for example, in the case of EDTA and EGTA. Both chelators which are considerably flexible, bind to Ca^{2+} with almost the same affinity, the log of the binding constant being 10.7 (Williams, 1980). Binding of Mg^{2+} by EDTA occurs with a log K of 8.9, whereas the log of the binding constant is 5.4 in the case of EGTA, where the distance between the two nitrogen donor atoms has been increased considerably in comparison to the distance in EDTA. Furthermore, the molecule has become more flexible with a loss of the preformed octahedral conformation of its donor atoms.

So in general one can observe that the selectivity for Mg^{2+} over Ca^{2+} binding occurs in molecules,

(1) which are predominantly rigid (rather than flexible);
(2) which have a preformed cryptate, the size of which is suitable for incorporating a Mg^{2+} atom without developing much constraint of the molecule; and
(3) in which there is an octahedral symmetry after binding of magnesium.

These prerequisites are fulfilled in the molecules of phosphomutases.

On the other side, the selectivity for Ca^{2+} binding over Mg^{2+} occurs in those molecules,

(1) which are flexible rather than rigid;

(2) in which global conformational changes can occur concomitantly with the binding of Ca^{2+}, this ion acting as a crosslink between different domains of the proteins;
(3) in which three or more acidic carboxyl groups can be involved in the formed chelate, that is, in acidic protein molecules.

Indeed, many Ca^{2+}-binding proteins such as parvalbumins, troponin, etc., and especially calmodulin have a considerably high number of acidic aspartic and glutamic residues, and Ca^{2+} is bound by these proteins, mostly via 3 or 4 carboxylates/Ca^{2+}. In some cases other oxygen functions are additionally involved like the alcoholic OH function of the amino acid threonine (Vanaman, 1980).

In summary, it seems to be rather obvious that there are some specific properties of the Ca^{2+}-binding behaviour, which indeed can explain the selectivity of binding of this bivalent cation over that of all other monovalent and bivalent cations of biological significance, such as Mg^{2+}, Na^+, K^+, and most of the bivalent heavy metal ions.

References

Mellor, D.P. (1964). Historical background and fundamental concepts. In *Chelating Agents and Metal Chelates*. F.P. Dwyer and D.P. Mellor (eds), pp.1–50. Academic Press, New York and London.

Pfleger, H. and Wolf, H.U. (1975). Activation of membrane-bound high-affinity calcium ion-sensitive adenosine triphosphatase of human erythrocytes by bivalent metal ions. *Biochem. J.* 147, 359–61.

Vanaman, T.C. (1980). Structure, function, and evolution of calmodulin. In *Calcium and Cell Function, Vol. 1, Calmodulin*. W.Y. Cheung (ed.), pp. 41–58. Academic Press, New York, London, Toronto, Sydney, San Francisco.

Williams, R.J.P. (1977). Calcium chemistry and its relation to protein binding. In *Calcium Binding Proteins and Calcium Function, Proceedings of the International Symposium on Calcium-Binding Proteins and Calcium Function in Health and Disease, Cornell University, 5–9 June 1977*. R.H. Wasserman, R.A. Corradino, E. Carafoli, R.H. Kretsinger, D.H. MacLennan and F.L. Siegel (eds), pp. 3–12. Elsevier North-Holland Biomedical Press, New York, Amsterdam.

Williams, R.J.P. (1980). A general introduction to the special properties of the calcium ion and their deployment in biology. In *Calcium-Binding Proteins: Structure and Function, Proceedings of an International Symposium on Calcium-Binding Proteins and Calcium Function in Health and Disease, Madison, Wisconsin, 8–12 June 1980*, pp. 3–10. F.L. Siegel, E. Carafoli, R.H. Kretsinger, D.H. MacLennan, and R.H. Wasserman (eds). Elsevier North-Holland, New York, Amsterdam, Oxford.

CA^{2+}-INDUCED CONFORMATIONAL CHANGES OF MEMBRANES

E. Sackmann

Physik Department E22 (Biophysics Group), Technical University of Munich, D-8046 Garching, West Germany

Abstract

This chapter presents first an overview of Ca^{2+}-induced phase transitions, lateral phase separation processes and vesicle—vesicle fusion triggered by Ca^{2+} in mixed bilayers of charged and uncharged lipids. Secondly, possible effects of Ca^{2+} on the structure of biological membranes and the head group conformation of integral membrane proteins are considered.

9.1. Introduction

Lipids play a two-fold role in biological membrane processes. First, they function as structural elements and may thus play an active role for the organisation of enzymes into functional units. Secondly, they may function as biochemical intermediates. It appears that a very fascinating confirmation concerning the first role is the rapid lipid turnover associated with signal transmission processes such as the action of hormones.

I shall review some aspects of the action of Ca^{2+} on the structure of membranes. Concerning natural membranes it is, however, hard to distinguish direct effects of Ca^{2+} on the membrane structure from indirect ones in which Ca^{2+} plays a role as a modulator of biochemical processes which then alter the microstructure of the membranes. Thus the addition of Ca^{2+} to erythrocytes causes crenation of the cells. This shape change occurs, however, with a time delay of about an hour and is therefore certainly an indirect Ca^{2+}-effect.

For a direct effect of Ca^{2+} both the charged lipids and the integral proteins with head groups carrying excess negative charges may play a role. Concerning the lipids it is important to realise that phosphatidylserine (PS) is often present in rather high concentrations. The inner monolayer of erythrocytes for instance contains about 30% of PS. At such a high concentration, Ca^{2+}-induced lateral phase separation processes may come into play.

Most of our knowledge on possible Ca^{2+}-effects on membranes comes from studies of model membranes. For that reason I shall review mainly experimental studies of these systems. In the last, more speculative part, I shall discuss which Ca^{2+}-induced conformational changes of natural membranes we can expect from our knowledge on model systems.

9.2. Charge induced lipid phase transitions

9.2.1. *Theoretical considerations*

Three types of lipid conformations may play a role in the microscopic organisation of biological membranes: (1) the liquid crystalline or fluid bilayer phase, (2) the solid bilayer phase, and (3) the micellar structure. The transition temperatures between these phases depend in a complex way on the length of the hydrocarbon chains, on the lateral packing density of the lipid molecules, and on the intrinsic lateral pressure of the monolayers. The latter two conformations are functions of the state of hydration of the lipid—water interface or of the degree of saturation of the chains.

For charged lipids the packing density is also sensitively dependent on the charging state of the lipid head groups. For that reason transitions between the different states of the lipids may be triggered by variations in the local pH as well as by the adsorption of charged proteins or ions. The most drastic effects are evoked by bivalent ions such as Ca^{2+}.

A much debated question is whether lipid phase transitions play a regulatory role in biological membrane processes. Although there is no direct evidence for this, it is well established that biological membranes exhibit local liquid crystalline-to-solid phase transitions leading to local reorganisations of the membrane proteins. Thermally induced lateral phase transitions have been observed in erythrocyte plasma membranes at 34°C, 18°C and 6°C. At the middle transition the membrane elasticity changes abruptly. Below the lowest transition one observes a reorganisation of the proteins due to localised lateral phase separation processes. This transition is accompanied by a strong increase in the haemolytic fragility of the cells.

The phase behaviour of charged lipid layers can be generally described in terms of the additional lateral pressure, $\Delta\Pi_{el}$, which arises due to the Coulomb replusion between the head groups. This electrostatic pressure is related to the surface potential Ψ_0 and the concentration, c, of counter ions (compare Sackmann, 1983; Träuble and Eibl, 1975) according to:

$$\Delta\Pi_{el} = 6.1 \sqrt{c} [\cosh(e\Psi_0/2kT) - 1] \qquad (1)$$

The surface potential, Ψ_0, however, depends on the degree of dissociation, α, of the lipid head groups and on the area per lipid molecule, A, as follows

$$\Psi_0 = (2kT/e) \operatorname{arcsinh}(134\alpha/A\sqrt{c}) \qquad (2)$$

where the numerical factor of 134 holds for a dielectric constant of $\epsilon = 80$ and a temperature of $T = 300$ K. The transition temperature T_{tr} is a function of the state of the head group charging. This follows from the Clausius—Clapeyron equation according to which the electrostatic pressure leads to a shift in the transition temperature of

$$\Delta T_m = (\Delta A/\Delta S) \cdot \Delta\Pi_{el} \qquad (3)$$

where ΔA and ΔS are the changes in the area and the entropy, respectively, of the given phase transition.

9.2.2. Ca^{2+}-induced shifts of the chain-melting transitions

The effect of Ca^{2+} on the chain-melting transition temperature of charged lipids depends on the charge and probably also on the structure of the head group. Thus addition of Ca^{2+} to vesicles of negatively charged dilaurylophosphatidylglycerol leads to a sharp increase in the transition temperature T_m from about 0 to 23°C at increasing the Ca^{2+}-to-lipid ratio from 0 to about 1:10. Then T_m remains constant up to a ratio of 1:2 and jumps again to a value of 83°C upon approaching this limit. Simultaneously, the vesicles condense into tightly packed stacks of bilayers (Verkleij et al., 1974). As will be discussed below this second sharp rise in T_m is caused by the dehydration of the lipid head groups which is also responsible for the condensation.

The Ca^{2+}-effect on the chain-melting transition of membranes composed of lipids carrying two negative charges such as phosphatidylserine (PS) or phosphatidic acid (PA^{2-}) is much stronger. The shift in the transition temperature is of the order of some 50°C. Again the binding leads to the condensation of the vesicles into multilamellar structures due to the dehydration of the lipid head groups.

This dehydration is most probably due to a specific binding of the Ca^{2+} to the lipid, L, according to

$$nL^- + Ca^{2+} \rightleftharpoons CaL_n$$

The dissociation constant, K, and thus the degree of dissociation depends on the surface potential because

$$K = K_0 \exp[e\Psi_0/kT] \qquad (4)$$

where K_0 is the dissociation constant of isolated lipid molecules. The equilibrium is shifted to the left side if the negative surface charge of the membrane is screened by other counterions such as Na^+. Therefore the limiting Ca^{2+}-concentration where condensation sets in depends on the ionic strength of the aqueous phase. The effect of Ca^{2+} can thus be modulated by the addition of salt. It should be emphasised that this shift of the chain-melting transition temperature of negatively charged lipids is also observed if basic polypeptides such as polylysine or cytochrome c adsorb to the bilayer (Sackmann 1983; Sackmann, Kotulla and Heiszler, 1984).

9.2.3. Binding-site of Ca^{2+} ions

The binding of Ca^{2+} to the lipid head groups has been studies by NMR and infrared spectroscopy (Dluhly et al., 1983). According to these spectroscopic studies the Ca^{2+} binds to the phosphate group of the PS and causes dehydration of the phosphate ester. These studies confirm also the crystallisation of the lipid bilayer caused by the binding.

9.3. Ca^{2+}-induced lateral phase separation

Since biological membranes are multicomponent lipid systems, the Ca^{2+}-induced lateral phase separation in mixtures containing charged (PS or PA^{2-}) and uncharged (PC) lipids which was discovered about 10 years ago (Galla and Sackmann, 1975; Ito and Ohnishi, 1974) is of uttermost importance.

In these experiments the uncharged component was marked by spin probes and the Ca^{2+}-induced lateral phase separation could directly be followed by analysing the modification of the ESR-spectra caused by the spin–spin exchange interaction. This technique permitted detection of subtle changes in the concentration of the labelled lipid caused by lateral phase separation.

A first important result of these studies is that lateral phase separation sets in above a certain threshold concentration which depends on the content of the charged lipid component. For twofold negative PA, the threshold concentration increases from $C = 5 \cdot 10^{-5}$M for pure PA^{2-} to $4 \cdot 10^{-4}$M for a 1:1 mixture of this lipid with phosphatidylcholine (Fig.9.1). The threshold concentration does not depend sensitively on the total lipid content. This follows also from most intriguing studies of the carrier-mediated ion translocation as well as the transport of hydrophobic ions through planar lipid membranes composed of mixtures of phosphatidylcholine and two-fold negative PA (Schmidt, Eibl and Knoll, 1982, Miller et al, to be published). In these studies it was shown that the addition of Ca^{2+} to the aqueous phase causes lateral phase separation into a solid phase of $(PA^{2-}-Ca^{2+})$-complexes and fluid PC-rich phase. The conductivity of the solid domains is practically abolished. By analysing the voltage–jump–current relaxation times as a

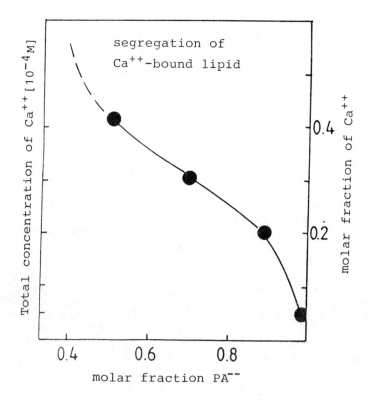

Fig.9.1. Dependence of threshold Ca^{2+}-concentration at which lateral phase separation sets in for mixture of two-fold negative phosphatidic acid (PA^{2-}) with phosphatidylcholine (PC) as function of molar fraction of Ca^{2+} with respect to PA^{2-} (right ordinate). The left ordinate gives the absolute Ca^{2+}-concentration.

function of the lipid composition and the Ca^{2+} concentration these authors could determine a composition—Ca^{2+} phase diagram of the PC/PA mixture which is shown in Fig.9.2. For a 1:1 mixture the threshold concentration where phase separation sets in is of the same order of magnitude as that found for vesicle suspension (see Fig.9.1). Ca^{2+}-induced lateral phase separation has also been observed in PS/PC mixtures although a composition—Ca^+ phase diagram has not been determined yet. A composition—temperature phase diagram of a bovine brain PS–DPPC mixture has, however, been determined by Hui *et al.* (1983), and is discussed in section 9.4.

A second important feature is the following: in spin label and freeze fracture studies it was shown that both the thermally and the charge-induced lateral phase separation leads to a domain-like lateral organisation of the components. An example of such a domain structure is shown in Fig.9.3 for the case of the Ca^{2+}-induced phase separation in a 1:1 PA^{2-}/PC mixture. The domain structure is a consequence of the different local curvatures of the patches of different composition. As illustrated in Fig.9.3b this leads to an abrupt change in the local lipid orientation at the boundaries of the domains. Since the formation of such orientational defects is associated with an elastic energy, a restoring force arises which prevents the growth of the precipitates

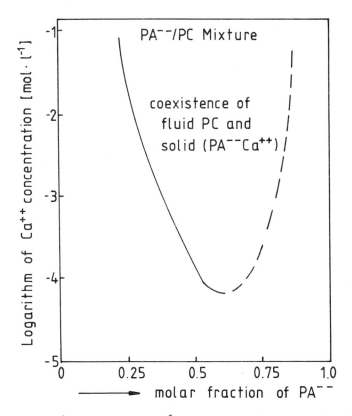

Fig.9.2. Composition — Ca^{2+} phase diagram of PA^{2+}/PC mixture as obtained from voltage jump current relaxation studies of the ion translocation in black lipid membranes (Schmidt, Eibl and Knoll, 1982).

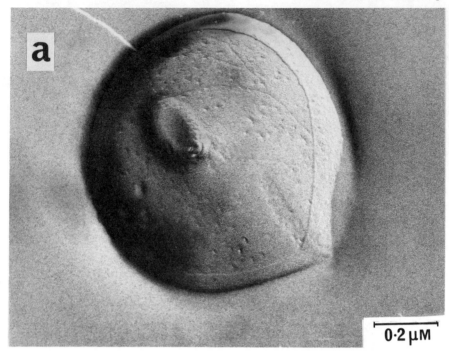

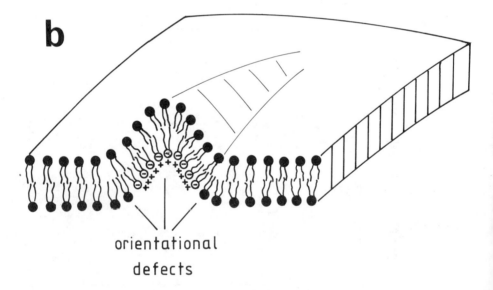

Fig.9.3. (a) Freeze fracture micrograph of domains formed in membrane of dioleyl-phosphatidic acid (PA^{2-}) and dioleylphosphatidylcholine (PC) (1:1 mixture). (b) Schematic representation of formation of domains which protrude sharply into the third dimension in order to minimise the elastic energy at the boundary between phases (Sackmann, 1983).

(Gebhardt, Gruler and Sackmann,1977). As discussed below, the vesicles may become unstable if the size of the domains exceeds a certain limit. This type of instability may play an important role for the Ca^{2+}-induced membrane fusion.

A third important characteristic feature of the charge induced lateral phase separation, its pronounced hysteresis, is discussed below.

9.4. Ca^{2+}-induced fusion in model membranes: threshold behaviour

9.4.1. *Some important results*
The Ca^{2+}-induced fusion of phosphatidylserine bilayers is an often studied effect. The important results of these studies are:

(1) Fusion is only observed above a threshold concentration of the bivalent ion (Schuber *et al.*, 1983). For pure PS vesicle this threshold concentration is about $c_{th} \approx 2\ 10^{-3}$M at a ten-fold PS concentration.
(2) Above the threshold the fusion consists of a fast process characterised by a rate constant (the so-called initial rate of fusion) of the order of $f_1 \approx 1$ sec^{-1} and a slow process, with a time constant of the order of hours (Nir, Düzgünes and Bentz, 1983).
(3) The initial rate of fusion, f_1, is strongly influenced by the presence of monovalent counterions such as Na$^+$. Thus Fig.9.4 shows that f_1 decreases exponentially with increasing NaCl concentrations (Nir, Düzgünes and Bentz, 1983).

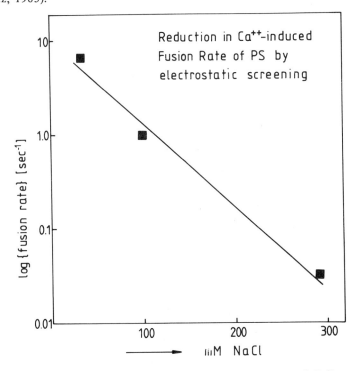

Fig.9.4. Initial rate of fusion, f_1, of phosphatidylserine vesicles as a function of NaCl concentration of the aqueous phase according to data by Nir, Düzgünes and Bentz (1983).

(4) The fusion can be described in terms of two kinetically linked steps (1) the association of two vesicles (V_1 and V_2) according to equation (5) (association rate constant a_1) and (2) the fusion (rate constant f_1) of the two partners caused by a destabilisation of the dimer V_1V_2

$$V_1 + V_2 \underset{a_{-1}}{\overset{a_1}{\rightleftharpoons}} (V_1\ V_2) \overset{f_1}{\to} V_{12} \qquad (5)$$

As shown below, the first process must be accompanied by a dehydration at the vesicle–vesicle contact area in order to lead to a successful fusion.
(5) In mixtures of charged and uncharged lipids the fusion depends critically on the relative concentration of the charged component. Again threshold behaviour is observed. Hui et al. (1983) showed that at 0–30 mol % of PS the Ca^{2+}-binding does not affect the vesicle stability but leads to lateral phase separation within the mixed bilayers. Between 30 and 60% PS one observes simultaneously the formation of a condensed bilayer fraction

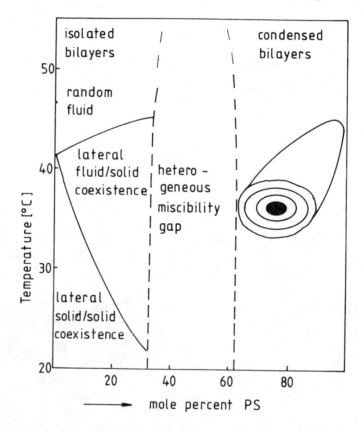

Fig.9.5. Temperature-composition phase diagram of mixture of bovine brain PS and dipalmitoylphosphatidylcholine (DPPC) at excess Ca^{2+} (20 mM Ca^{2+} and 2 mM of total lipid) as obtained from calorimetric measurements of Hui et al. (1983). Note that fusion leading to condensed bilayers is only observed above a threshold PS concentration. Within the miscibility gap the suspension decomposes into a vesicle fraction exhibiting lateral phase separation and aggregated bilayers forming jellyroll structures.

exhibiting a jellyroll-like structures (Fig.9.5). An important lesson to be learned from Fig.9.5 is that one must carefully distinguish between (homogenous) phase separation within one bilayer and a (heterogeneous) decomposition of the components into different aggregates.
(6) A very interesting feature of the fusion of PS vesicles with planar bilayer membranes of the Muller—Montral type was reported by Akabas, Cohen and Finkelstein (1984). These authors showed that the first step - the vesicle-membrane attachment - requires Ca^{2+} while the actual fusion process is accelerated by osmotic swelling. A striking finding is that the Ca^{2+} is only necessary for the contact formation and that the contact is not disrupted if the Ca^{2+} is removed again from the solution after vesicle attachment. Below, this behaviour is explained in terms of hysteresis effects.

9.4.2. *Ca^{2+} fulfils several purposes in the fusion process (lateral phase separation and lipid dehydration)*
It appears that two conditions must be fulfilled for the formation of the prefusion state (step 1 in equation (5)). First, the area of contact between the vesicles must be large enough. Second, water must be removed from this area in order to overcome the very large repulsive dehydration forces (Cowley *et al.*, 1978). Ca^{2+} can serve both purposes.

(1) Ca^{2+}-binding causes solidification of the charged lipid which precipitates into two-dimensional crystalline patches. Due to the very high curvature-elastic constant of the solid phase these patches become flat and thus form large contact areas to which other vesicles can become strongly attached (Fig.9.6b). Simultaneously the vesicle-vesicle contact area is dehydrated. Evidence for the necessity of a large contact area comes from experiments by Kwok and Evans (1981). These authors showed by the micropipette method that mixed PC:PS vesicles can adhere only if they can form a large area of contact. This could be achieved for example by osmotic shrinking of the initially spherical vesicles.
(2) The vesicle—vesicle interaction is governed by three types of forces (Cowley *et al.*, 1978): (1) by the attractive Van der Waals interaction, (2) by the electrostatic repulsion, and (3) by the strongly repulsive dehydration forces. The latter decay exponentially with the distance (Marcelja and Radic, 1976).

If the distance between two adjacent bilayers is larger than about 25Å, the interaction is determined by the balance of the first two forces. At distances below 10Å the dehydration force becomes dominant which is by at least one order of magnitude larger than the Coulomb interaction. This is, of course, a consequence of the fact that at such short distances the solvation shells of the two monolayers start to interpenetrate which would have to be removed for a closer approach of the bilayers. The second purpose of Ca^{2+} is obviously the elimination of water from the lipid head groups. Moreover it is very likely that Ca^{2+} enforces the attachment of the bilayers by the formation of salt-bridges.

9.4.3. *Destabilisation of attached bilayers*
The next question concerns the actual fusion process where the diaphragm separating the inner compartments of the fusing systems becomes unstable and breaks. This could be explained by the strong change in curvature of the vesicles at the boundary between the flat areas of contact (formed by

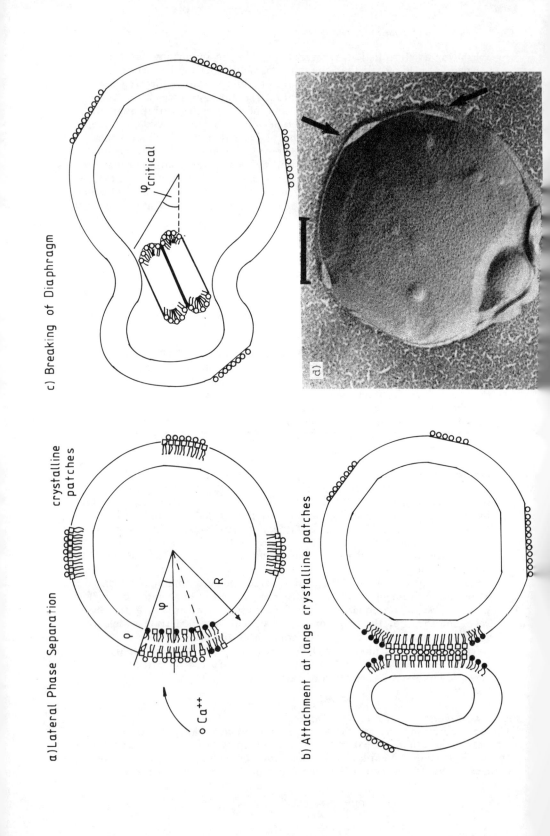

crystalline bilayers) and the non- diaphragmatic region of the vesicles. As is well known, even single vesicles can break if the patches formed by lateral phase separation become too large (Gebhardt, Gruler and Sackmann, 1977). The elastic energy associated with the change in the average lipid orientation at the boundary clearly increases with the diameter, ρ, of the contact area (see Fig.9.6a). Following the theory of membrane-binding elasticity (Helfrich 1974; Sackmann 1984) this elastic energy may be expressed as

$$F_{el} = \tfrac{1}{2} K \, (\sin\varphi/\Delta)^2 \, 2\Pi\rho \cdot \Delta \tag{6}$$

where K is the curvature elastic constant and where Δ is the width of the boundary. If the diaphragm breaks, a small disc of bilayer has to be formed. This formation is, however, associated with an elastic line energy

$$W_{edge} = 2\Pi\rho \cdot \gamma \tag{7}$$

where γ is the line energy per unit length of the edge of the diaphragm detached (see Fig.9.6c).

The diaphragm is expected to break if the two energies are about equal. According to equations (6) and (7) this limiting size ρ_c of ρ is given by ($\sin\varphi = \rho/R$)

$$\rho_c = \sqrt{2\gamma\Delta/K} \cdot R$$

With $K = 10^{-12}$ erg, $\gamma = 10^{-6}$ erg/cm (Helfrich, 1974) $R = 5 \cdot 10^{-4}$cm and a value of $\Delta \approx 10^{-7}$ (which is suggested by the freeze fracture results) it follows $\delta \approx 10^{-4}$ cm which is in agreement with the experiments.

The above consideration strongly suggests that the tendency for an instability would be greatly enhanced by small amphipathic solutes (such as fatty acids) which could reduce the edge energy. Experimental evidence for the above model comes from freeze fracture studies of the early event of the fusion process. As shown in Fig.9.6d small vesicles adhere to a large one and the area of contact is flat.

9.4.4. *Hysteresis effects may stabilise a structural change caused by a transient avalanche of Ca^{2+}*

Charge-induced structural changes exhibit very often pronounced hysteresis effects. This arises if the charge-binding is accompanied by a solidification of the lipid. It is due to the dependence of the dissociation constant K on the surface potential (see equation (4)). Due to the higher lipid packing density, the negative surface potential is higher in the solid phase than in the fluid one and therefore the dissociation constant is smaller in the first phase. Experimentally this hysteresis effect was established for the pH-induced

Fig.9.6. Three-step model of vesicle–vesicle fusion. (a) Lateral phase separation caused by binding of Ca^{2+} to charged lipid which solidifies and therefore becomes flat. The binding may already be accompanied by dehydration of the lipid head group region. (b) Attachment of small vesicles to larger patches of Ca^{2+}-bound lipid formed by merging of smaller domains (see Fig.9.3). Note that the area is flat. (c) Breaking of diaphragm due to the bilayer instability which arises as a consequence of the elastic energy associated with the abrupt change in the average lipid orientation at the transition from the flat diaphragm to the free bilayer. (d) Freeze fracture electron micrograph demonstration of vesicle–vesicle attachment (arrows). Vesicles are composed of 1:1 mixture of two-fold negative dioleyl-phosphatidic acid (PA^{2-}) and dimyristoyl phosphatidylcholine (bar 1 μm).

phase separation in an equimolar mixture of phosphatidic acid (DPPA) and DPPC. At high H^+ concentrations the system decomposes laterally into a fluid phase rich in DPPC and solid domains which consists mainly of DPPAH$^-$). At low H^+-contents the latter domains melt due to the dissociation of the phosphatidic acid. According to Fig.9.7 the H^+ concentrations where the transition occurs at increasing and decreasing H^+-content differ by two orders of magnitude. A similar type of hysteresis effect was observed for the Ca^{2+}-induced transition (E. Sackmann, unpublished results).

It is most likely that the attachment of the PS-vesicles to planar bilayer membranes after removal of the Ca^{2+} from the bulk aqueous phase reported by Akabas, Cohen and Finkelstein (1984) is due to such a hysteresis effect. In biological membranes such hysteresis effects could become important because a burst of Ca^{2+} flowing into a cell is rapidly sequestered. However, if some of the Ca^{2+} causes the attachment of vesicles to the plasma membrane before being sequestered it would no longer be removed from the site of attachment owing to such hysteresis effects. A possible example is the fusion of synaptic vesicles with the presynaptic membrane.

9.4.5 Do lipid particles play a role for fusion?

In a large number of papers it was shown by Cullis, de Kruyff, Verkleij and their group (Cullis and de Kruyff, 1979; Verkleij, 1984) that addition of Ca^{2+} to lipid mixtures containing charged lipids leads to the formation of particles of 200 Å diameter which have been called lipidic particles and which are presumably made up of inverted micelles. It was also postulated that these particles are a necessary intermediate for fusion. One can indeed easily imagine that the accumulation of such micelles at the edge of the diaphragm would cause vesicle instability.

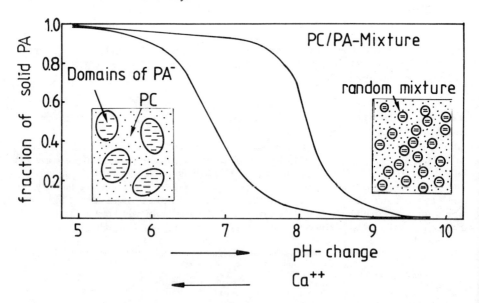

Fig.9.7. Hysteresis of charge induced lateral phase separation in mixed bilayers of dipalmitoylphosphatidylcholine (DPPC) and dipalmitoylphosphatidic acid (DPPA). The numbers on the abscissa correspond to the case of a pH-change (according to Sackmann, 1983). Temperature 20°C.

Ca^{2+}-induced conformational changes

However, it appears that the lipid particles are generally observed at high lipid concentrations or they are preferentially intercalated between the condensed bilayers. The small protrusions observed in Fig.9.6d are certainly not lipid particles.

It should, however, be noted that according to current models of membrane bending elasticity fusion could also occur at pit-like protrusions as discussed previously (Sackmann, 1984).

9.5 Possible Ca^{2+}-effects in biological membranes

9.5.1. *Ca^{2+}-induced fusion*

No direct evidence for a Ca^{2+}-induced effect on the lateral organisation of lipids in biological membranes is known yet. Nevertheless Ca^{2+} affect biological membranes in many ways:

(1) It stabilises electrically excitable membranes of nerve and muscles due to a shift of the conductance-voltage curves along the voltage axis. The most natural explanation for this effect is that the bivalent ions change the membrane surface potential (McLaughlin, 1977).
(2) Very striking is the role of Ca^{2+} for the synaptic signal transmission. A gated Ca^{2+} influx into the synapses of the axons leads to an increase in the Ca^{2+}-concentration by one to two orders of magnitude and triggers the release of neurotransmitter. It is very likely that the Ca^{2+}-influx induces the fusion of the synaptic vesicles with the presynaptic membrane. The Ca^{2+} increase is short-lived because the ion is rapidly sequestered by proteins. However, such a transient Ca^{2+} increase would be sufficient to cause the attachment of the vesicles to the membrane while the above mentioned hysteresis effect would stabilise this prefusion state. Another important point is that the cytoplasmic side of the presynaptic membrane is covered by a protein-layer. This would have to be removed for the fusion process. On the other side it could exhibit a synergic effect as observed by Schuber et al. (1983).

A more indirect effect of Ca^{2+} on the membrane microstructure may be caused by a structural change of that part of the cytoskeleton which is directly coupled to the membrane such as the spectrin—actin network of erythrocytes. It is well known that Ca^{2+} may modulate the effect of factors which control the gel-to-sol transition of the cytoplasmic networks (see Weeds, 1982).

9.5.2. *Possible effect of Ca^{2+} on lipid protein interaction (on the possibility of a Ca^{2+}-induced collapse of the protein head groups*

A large fraction of integral membrane proteins is sticking out of the lipid bilayer. The most prominent examples are the glycoproteins such as the insulin receptor, the glycophorin of the erythrocyte plasma membrane or the enzymes of the mitochondrial electron transfer chain. These hydrophilic head groups are polyelectrolytes and carry a substantial number of negatively charged amino acids or acidic sugars (such as glycophorin contains about 30 sialic acid groups). In particular the latter can provide binding places for Ca^{2+}.

In recent model membrane studies evidence was provided that the head groups of glycoproteins such as glycophorin or the insulin receptor tend to spread completely at the lipid—water interface forming a quasi-two-dimensional macromolecular layer. Thus it was shown (Heiszler, 1984; Rüppel et al., 1983) that one glycophorin (or insulin-receptor) molecule interacts with about 1000 lipid molecules. Due to this strong tendency for spreading the protein head groups may form a network of entangled chains (as shown in Fig.9.8). This spreading tendency is strongly enhanced by the electrostatic repulsion of the charged groups.

As is well known from studies of ordinary solutions of polyelectrolytes the conformation of the head group can go from an expanded to a completely collapsed state by small variations in the counterion concentration, particularly if these can bind specifically to the charged side groups.

The behaviour is described in terms of the Flory interaction parameter X whcih is a measure for the gain in energy if the chain configuration goes from a completely explanded into a collapsed state. It is determined (1) by a contribution, X_v, which accounts for the difference in the Van der Waals interaction between the monomers of the macromolecule and between these and the solvent (water or lipid), respectively, and (2) by a contribution X_{el} which accounts for the electrostatic repulsion of the sidegroups. The macromolecule goes from an expanded to a collapsed state if X changes from a value $X < \frac{1}{2}$ to a value $X > \frac{1}{2}$. Now X is given by

$$X = X_v - \text{const } \alpha^2$$

where α is the fraction of charged side groups (or degree of dissociation). If X_v is near a value of $\frac{1}{2}$ a transition between the two states of the macromolecule can be induced by the binding or dissociation of a few (such as two) charges. Clearly such a type of phase transition can provide an enormous factor of amplification.

The second important point is that the transition from the expanded to the

Fig.9.8. Model of Ca^{2+}-induced aggregation of integral membrane proteins triggered by a transition of the head group from a swollen to a collapsed state. The aggregation could be enhanced by lateral segregation of the charged lipid component caused by Ca^{2+}-binding.

collapsed state is accompanied by an aggregation of the macromolecules. For that reason the Ca^{2+}-binding to membrane proteins could well trigger a lateral reorganisation of the membrane proteins as shown in Fig.9.8. Such a process could, for instance, trigger the binding of synaptic vesicles to the membrane.

Although there is no experimental proof yet for a Ca^{2+}-induced conformational change of membrane proteins there exists evidence that the binding of insulin to the insulin receptor induces a conformational change of the latter type in such a way that the number of lipids interacting with the receptor is drastically reduced (about by a factor of 10) which points to a head group conformational change of the above-mentioned type (Heiszler, 1984).

Ca^{2+} plays an important role for the uncoupling of cells which are connected by gap junctions. This is achieved by a large increase in the cellular free Ca^{2+}. This decoupling could be achieved by a conformational change of the macromolecules forming the channels, or by a decoupling of the protomers forming the connections. This could be caused by a Ca^{2+}-induced phase change of the lipid bilayer moiety.

References

Akabas, M.H., Cohen, F.S. and Finkelstein, A. (1984). Separation of the osmotically driven fusion event from vesicle-planar membrane attachment in a model system for exocytosis. *J. Cell Biol.* **98**, 1003–71.

Cowley, A.C., Fuller, N.L., Rand, R.P. and Parsegian, V.A. (1978). Measurement of repulsive forces between charged phospholipid bilayers. *Biochemistry.* **17**, 3163–8.

Cullis, P.R. and De Kruyff (1979). Lipid polymorphism and the functional roles of lipids in biological membranes. *Biochim. Biophys. Acta.* **559**, 399–420.

Dluhly, R.A., Cameron, D.G., Mantsch, H.H. and Mendelsohn, R. (1983). Fourier transform infrared spectroscopic studies of the effect of calcium ions on phosphatidylserine. *Biochemistry.* **22**, 6318–25.

Galla, H.J. and Sackmann, E. (1975). Chemically induced lipid phase separation in model membranes containing charged lipids: a spin label study. *Biochim. Biophys. Acta.* **401**, 509–29.

Gebhardt, C., Gruler, H. and Sackmann, E. (1977). On domain structure and local curvature in lipid bilayers and biological membranes. *Z. Naturforsch.* **32C**, 581–96.

Giese, G., Fromme, I. and Wunderlich, F. (1979). Suppression of thermotropic lipid clustering in tetrahymena nuclear membranes upon Ca^{++}/Mg^{++}-induced membrane contraction. *Eur. J. Biochem.* **95**, 275–85.

Harlos, K. and Eibl, H. (1981). Hexagonal phases in phospholipids with saturated chains: phosphatidylethanolamines and phosphatidic acids. *Biochemistry (USA)* **20**, 2888–92.

Hartmann, W., Galla, H.J. and Sackmann, E. (1977). Direct evidence of charge-induced lipid domain structure in model membranes. *FEBS Lett.* **78**, 169–72.

Heiszler, F.J. (1984). *Rekonstruktion des Insulin Rezeptors in Modellmembranen.* Doctoral thesis, Munich.

Helfrich, W. (1974). The size of bilayer vesicles generated by sonication. *Phy. Lett.* **50A**, 115–6.

Hui, S.W., Boni, L.T., Stewart, T.P. and Isac, T. (1983). Identification of phosphatidylserine and phosphatidylcholine in calcium induced phase

separation domains. *Biochemistry.* 22, 3511–6.
Ito, T. and Ohnishi, S. (1974). Ca^{2+}-induced lateral phase sepatations in phosphatidic acid–phosphatidylcholine membranes. *Biochem. Biophys. Acta.* 352, 29–37.
Jürgens, E., Hoehne, G. and Sackmann, E. (1983). Calorimetric study of the dipalmitoylphosphatidylcholine/water phase diagram. *Ber. Bunsenges. Phys. Chem.* 87, 95–101.
Kapitza, H.J. and Sackmann, E. (1980). Local measurement of lateral motion in erythrocyte membranes by photobleaching technique. *Biochim. Biophys. Acta.* 595, 56–9.
Kwok, R. and Evans, E. (1981). Thermoelasticity of large lecithin lipid bilayer vesicles. *Biophys. J.* 35, 637–52.
Marcelja, S. and Radic, N. (1976). Repulsion of interfaces due to boundary water. *Chem. Phys. Lett.* 42, 129–31.
McLaughlin, S. (1977). Electrostatic potentials at membrane–solution Interfaces. In *Current Topics in Membranes and Transport, Vol. 9.* pp. 71–144. F. Bonner and A. Kleinzeller (eds.).
Nir, S., Düzgünes, N. and Bentz, J. (1983). Binding of monovalent cations to phosphatidylserine and modulation of Ca^{2+} and Mg^{2+}-induced vesicle fusion. *Biochim. Biophys. Acta.* 735, 160–72.
Rüppel, D., Kapitza, H.J., Galla, H.J., Sixl, F. and Sackmann, E. (1983). On the microstructure and phase diagram of dimyristoylphosphatidylcholine glycophorin bilayers. *Biochim. Biophys. Acta.* 692, 1–17.
Sackmann, E., (1984). Physical basis of trigger processes and membrane structure. In *Biological Membranes, Vol. 5.* D. Chapman (ed.). Academic Press, London, New York.
Sackmann, E., Kotulla, R. and Heiszler, F.J. (1984). On the role of lipid bilayer elasticity for the lipid-protein interaction and the indirect protein–protein coupling. *Can. J. Biochem.* 62, 778–88.
Sackmann, E. (1983). Physical foundations of the molecular organization and dynamics of membranes. In *Biophysics.* W. Hoppe, W. Lohmann, H. Marki and H. Ziegler (eds.). Springer–Verlag, Berlin, Heidelberg.
Schmidt, G., Eibl, H. and Knoll, W. (1982). Carrier–mediated ion transport through black membranes of lipid mixtures and its coupling to Ca^{2+}-induced phase separation. *J. Membrane Biol.* 70, 147–55.
Schuber, F., Hong, K., Düzgünes, N. and Papahadjopoulos (1983). Polyamines as modulators of membrane fusion: aggregation and fusion of liposomes. *Biochemistry.* 22, 6134–40.
Träuble, H. and Eibl, H. (1975). Electrostatic effects on lipid phase transition: membrane structure and ionic environment. *Proc. Natl. Acad. Sci. USA.* 71, 214–9.
Verkleij, A.J. (1984). Lipidic intramembranous particles. *Biochim. Biophys. Acta.* 779, 43–63.
Verkleij, A.J., De Kruyff, B., Ververgaert, P.H., Tocanne, J.F. and Van Deenen, L.L.M. (1974). The Influence of pH, Ca^{2+} and proteins on the thermotropic behaviour of the negatively charged phospholipid phosphatidylglycerol. *Biochim. Biophys. Acta.* 339, 432–7.
Weeds. A. (1982). Actin-binding proteins - regulators of cell architecture and motility. *Nature.* 296, 881–6.

CALCIUM BINDING PROTEINS

Dennis F. Michiel and Jerry H. Wang

Department of Medical Biochemistry, University of Calgary, Calgary, Alberta, Canada

Abstract

Many of the physiological effects of calcium are mediated by a family of proteins that contain a homologous amino acid sequence that forms the Ca^{2+}-binding domains. Some of these proteins have specialized roles. The ancestor of these proteins probably resembled calmodulin which is more general in its distribution and function. Calmodulin mediates Ca^{2+} effects by interacting with several target proteins to regulate smooth muscle contraction, calcium efflux, membrane phosphorylation, microtubule assembly—disassembly, secretion, glycogen metabolism and others. The mechanisms involved in calcium activation of calmodulin and its interaction with the target proteins are beginning to be understood. The calcium—calmodulin system is closely linked to the cAMP second messenger system through its effects on common regulatory proteins and through regulation of cyclic nucleotide metabolism. Some recent studies which further elucidate the complexity of calmodulin regulation of cyclic nucleotide metabolism and implicate calmodulin in other regulatory systems will be discussed.

10.1. Introduction

The importance of intracellular calcium ions as second messengers in bioregulation was first emphasized by Rasmussen (1970). Normally Ca^{2+} levels are extremely low intracellularly (10^{-8}–10^{-7} M) and high extracellularly (10^{-3} M). Several 'activators' of the cell such as hormones, mitogens, depolarizing currents, drugs and toxins cause a transient increase in cytosolic Ca^{2+} which is rapidly removed by membrane-bound calcium pumps. The calcium-binding proteins upon which our interest has centred are those capable of transducing this transient Ca^{2+} signal into physiological effects. Some other proteins also bind calcium but they do not appear to be involved in modulating the intracellular effects of calcium, proteins such as concanavalin A (Becker et al., 1975, 1976), prothrombin (Nelsestuen, 1975), phospholipase A_2 (Van den Bosch, 1980), Ca^{2+}-ATPase (MacLennan and Holland, 1975), calsequestrin (Han and Benson, 1970; Ostwald and MacLennan, 1974) and protein kinase C (Takai et al., 1979). For example, protein kinase C is a Ca^{2+}-activated, phospholipid-dependent protein kinase, however, the concentration of Ca^{2+} needed for half-maximal activity is in the 10^{-7} M range (Kaibuchi, Takai and Nishizuka, 1981) so small amounts of diacylglycerol will cause activation of the enzyme in the absence of any increase in intracellular calcium (Kishimoto et al., 1980). Several reviews (Cheung, 1980; Klee and Vanaman, 1982; Klee, Crouch and Richman, 1980; Levine and Dalgarno, 1983; Means and Dedman, 1980; Van Eldik et al., 1982; Wang and Waisman, 1979) and books (Cheung, 1980) have been written on calcium-binding proteins and calmodulin, so many areas of interest have to be left out of a review of this size.

10.2. Proteins containing the E-F hand

Since even during the activation phase the Ca^{2+} levels reach only the μM range in the presence of approximately 1 mM Mg^{2+}, the intracellular calcium receptors must have an affinity for calcium in the μM range and a selectivity for Ca^{2+} over Mg^{2+} by a factor of 10^3. Table 10.1 shows a partial list of the intracellular proteins with dissociation constants in this range. They are generally low molecular weight proteins with an acidic isoelectric point. Among these proteins a great deal of homology has been noted especially in the calcium-binding structure termed the E-F hand, named from the calcium-binding loop found in parvalbumin (Kretsinger, 1976, 1980; Moews and Kretsinger, 1975). This structure is formed by two α-helices, the E-helix and the F-helix of parvalbumin, connected by a peptide loop that chelates the calcium ion. The homology is probably indicative of structural elements underlying their mechanism of response to calcium. This group of proteins is sometimes referred to as the troponin C superfamily.

Table 10.2 shows the sequence of three of these calcium-binding proteins, calmodulin (Grand, Shenolikar and Cohen, 1981), troponin C (Collins, 1976) and calcineurin B (Aitken et al., 1983). Each of these calcium-binding proteins has four homologous domains that can bind one calcium ion each. Goodman et al. (1979) and Weeds and McLachlan (1974) have suggested that this internal homology arose through a pair of tandem duplications of a primordial one-domain polypeptide.

With the widespread physiological effects of calcium, it is surprising that the function of many of these calcium-binding proteins is unknown. Only two Ca^{2+}-binding proteins, calmodulin and troponin C, appear to have unequivocally established functions. Troponin C functions specifically as the subunit of a protein complex which binds calcium and confers calcium sensitivity to the contraction process in striated muscle. The complex also contains troponin I and troponin T. Calmodulin can substitute for troponin C in reconstitution experiments, albeit without the same efficiency (Amphlett, Vanaman and Perry, 1976). Calmodulin appears to have a general Ca^{2+} mediating role. It is ubiquitous in eukaryotes and its structure has been highly conserved throughout evolution. Calmodulin has been implicated in the regulation of many proteins (Waisman, Stevens and Wang, 1975) and in many different physiological processes. Since calmodulin is involved in many Ca^{2+}-dependent processes and its mechanism of activation has been studied in some detail, we will use it as a general model for the activation process of calcium-binding proteins. Calmodulin does not mediate all of the Ca^{2+}-dependent processes; some of these may be regulated wholly or in part by other calcium-binding proteins. These proteins may be viewed as having a structure similar to calmodulin but which have evolved for a more

Table 10.1. Troponin C superfamily.

Protein	Ca^{2+}/mol	Molecular weight	pI	Distribution	Function	Reference
Calmodulin	4	16 700	3.9	ubiquitous	several	Dedman and Kaetzel (1983); Klee and Vanaman (1982)
Troponin C	4	17 800	4.1	striated muscle	contraction	Ebashi (1974); Hartshorne and Dreizen (1972)
Calcineurin B	4	19 200	4.8	nerve, muscle	phosphatase	Aitken et al. (1983); Klee and Krinks (1978); Klee and Krinks (1979)
Vitamin D dependent Ca^{2+} binding proteins avian intestine	4	28 000	4.2	chicken gut, bovine and rat brain and kidney	Ca^{2+} metabolism	Klee and Vanaman (1982)
mammalian intestine	2	9 700	4.7	mammalian intestine	Ca^{2+} metabolism	Klee and Vanaman (1982); Szebenji, Obendorf and Moffat (1981)
Parvalbumin	2	12 000	4.5	skeletal muscle	Ca^{2+} buffer?	Haiech et al. (1979)
Oncomodulin	2	11 500	3.9	tumour cells, tissue	?	MacManus (1980); MacManus, Watson and Yaguchi (1983)
Myosin light chain	1	20 000	—	with myosin	contraction	Alexis and Gratzer (1978)
S100a	2?	10 400	acidic	brain	?	Isobe and Okuyama (1981); Isobe, Nakajima and Okuyama (1977)
S100b	2?	10 500	acidic	brain	?	Isobe and Okuyama (1975); Isobe, Nakajima and Okuyama (1977)
TCBP-10	1	10 000	4.5	cilium	cilia movement	Ohnishi and Watanabe (1983)
CBP-18	1	18 000	—	brain	?	Manalan and Klee (1984)
Caligulin	1	24 000	acidic	brain	?	Waisman, Muranyi and Ahmed (1983)
Gelsolin	2	91 000	6.1	ubiquitous	cytoskeleton rearrangement	Yin and Stossel (1980)
Villin	3	95 000	—	—	cytoskeleton rearrangement	Hesterberg and Weber (1983)

Table 10.2. Homology among troponin C superfamily.

```
                                Helix         Ca2+-binding loop      Helix
                                               X Y Z -Y -X            -Z

Calmodulin       CH3CO-A D Q L T E E Q I A E F K E A F S L F D K D G D G T I T T K E L G T V M R S L
Troponin C               E M I A E F A A F D M F D A D G G G D I S V K E L G T V M R M L
Calcineurin B          * S H F D A D E I K R L G K R F K K L D L D N S G S L S V E E F M S L P - E L

Calmodulin               G Q N P T E A E L Q D M I N E V D A D G N G T I D F P E F L T M M A R K
Troponin C               G Q T ? T K E E L D A I I E E V D E D G S E T I D F E E F L V M M V R Q
Calcineurin B            Q Q N P - - - L V Q R V I D I F D T D G N G E V D F K E F I E G V S Q F

Calmodulin               M K D T D S E E E I R E A F R V F D K D G N G Y I S A A E L R H V M T - - - - N L
Troponin C       M K E D A K G K S E E E L A E C F R I F D R N A D G Y I D A E E L A E I F R A S
Calcineurin B    S V K G D K E Q K L R F A F R I Y D M D K D G Y I S N G E L F Q V L K M M V G N N L

Calmodulin               G E K L T D E E V D E M I R E A N I D G D G Q V N Y E E F V Q M M T A K
Troponin C               G E H V T D E E I E S L M K D G D K N N D G R I D F D E F L K M M E G V Q
Calcineurin B            K D T Q L Q Q I V D K T I I N A D K D G D G R I S F E E F S A V V G G L D I H K K M V V D

                                                X   Y  Z -Y -X         -Z
                                                  Ca2+-binding loop
```

* = $CH_3(CH_2)_{12}CO$–G N E A S Y P L E M M

Calmodulin sequence is for the δ-subunit of phosphorylase kinase.
Calmodulin and calcineurin B are from Aitken et al. (1983), troponin C from Ebashi (1974).

specialized function.

10.3 Mechanism of calmodulin activation

The activation of calmodulin was originally viewed as a two-step process (Wang et al., 1975). In the first step calmodulin binds Ca^{2+} causing a conformational change to the activated form of calmodulin. This activated complex then binds to the target enzyme causing a conformational change, leading to activation of the enzyme. When multiple Ca^{2+}-binding and the energy coupling between Ca^{2+}-binding and Ca^{2+}-induced calmodulin–enzyme interactions are considered, a more global scheme for the mechanism of enzyme activation may be proposed (Chau et al., 1982):

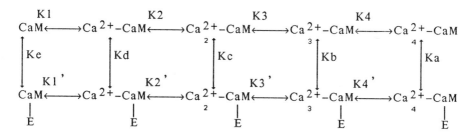

Scheme 1

where CaM and E stand for calmodulin and the enzyme respectively and the Ks represent dissociation constants for the respective reactions. Several investigators (Blumenthal and Stull, 1980; Burger, Stein and Cox, 1983; Cox, Malnoe and Stein, 1981; Huang et al., 1981) have carried out studies exploring the nature of multiple Ca^{2+} binding and energy coupling. Steady state kinetic analysis of the activation of phosphodiesterase by calcium and calmodulin has led to the suggestion that the enzyme activation requires the binding of at least three and most likely all four calcium ions per calmodulin (Chau et al., 1982). Using a specific Ca^{2+} indicator, 1,2-bis-(O-aminophenoxy)ethane-N,N,N',N'-tetraacetic acid (Tsien, 1980), the rate of Ca^{2+} dissociation from the complex E.CaM.Ca^{2+} has been determined and for the most rapidly dissociating Ca^{2+}, it is found to be about 3.5 to 4.6 sec^{-1}. Using a rapid chemical quenching method, the rate of inactivation of the fully activated phosphodiesterase by EGTA has been shown to be 4.5 sec^{-1}. The result supports the hypothesis that four Ca^{2+} per calmodulin are required to maintain the enzyme in its active state. Steady state kinetic studies of the activation of skeletal myosin light chain kinase (Blumenthal and Stull, 1980) have shown that four Ca^{2+} per calmodulin are also required for enzyme activation. On the other hand, Cox and coworkers (Cox, Malnoe and Stein, 1981; Burger, Stein and Cox, 1983) have studied the kinetics of activation of phosphodiesterase, phosphorylase kinase and Ca^{2+}/Mg^{2+} ATPase by Ca^{2+} and calmodulin, and concluded from their results that activation depends on the binding of three Ca^{2+} per calmodulin. Since kinetic studies involve certain assumptions which are difficult to test experimentally, the discrepancies in the different studies are not totally unexpected.

As has been thoroughly discussed by Huang et al. (1981), there are several advantages associated with the multiple Ca^{2+} binding. Analysis of the energy coupling between Ca^{2+}-binding and Ca^{2+}-induced protein–protein interaction revealed that multiple Ca^{2+} binding is essential for the reversible protein association system. The dissociation constants for calmodulin and most of the target enzymes in the absence and presence of saturating levels of Ca^{2+} are $>10^{-5}$ and $10^{-9}-10^{-10}$ respectively. Thus, the binding of Ca^{2+} to calmodulin results in an increase of over 100 000-fold in affinity towards its target enzymes. Since in most cells, Ca^{2+} concentrations fluctuate between 10^{-7} and 10^{-5} M, multiple Ca^{2+} binding to calmodulin is essential for protein–protein interactions. As is shown in Scheme 1, with the binding of four Ca^{2+}, each Ca^{2+} having an affinity toward the protein complex about one to two orders of magnitude higher than that for calmodulin alone, the affinity of calmodulin toward the enzyme can be increased more than 100 000-fold by Ca^{2+}.

In addition to being essential for the regulation via reversible protein–protein interaction, multiple Ca^{2+} binding may have other regulatory advantages. For example, it provides a highly cooperative activation of the enzymes by Ca^{2+} and it may also give more flexibility to the regulatory system. Thus different enzymes may be regulated by different form of the calcium–calmodulin complexes.

Although most of the target proteins are regulated by calmodulin according to the mechanism of Scheme 1, some of the Ca^{2+}-regulated enzymes may contain calmodulin as a subunit as phosphorylase kinase does. Calmodulin is the δ-subunit of phosphorylase kinase and remains bound in the presence or absence of calcium. Unlike the other calmodulin binding proteins, this calmodulin remains bound to the enzyme in the presence of EGTA (Cohen et al., 1978). This calmodulin appears to be bound to the γ-subunit. The activity of striated muscle phosphorylase kinase is dependent on calcium, however further stimulation can occur when exogenous calmodulin is added. This binding site has been termed the δ' site (Cohen, 1980; Picton, Klee and Cohen, 1980; Shenolikar et al., 1979). Cross-linking experiments indicate that this exogenous calmodulin binds to both the α and the β-subunits. There are isozymes of phosphorylase kinase in the red and white skeletal muscle and heart muscle. The red skeletal muscle and the heart muscle phosphorylase kinase are not stimulated by exogenous calmodulin (Sharma et al., 1980; Yoshikawa et al., 1983). The only difference in the subunits of these enzymes is that the fast (white) muscle has an α-subunit and the others α' which has a slighly different molecular weight on SDS gel electrophoresis (Burchell, Cohen and Cohen, 1976; Jennissen and Heilmeyer, 1974; McCullough and Walsh, 1979; Tam, Sharma and Wang, 1982). Chan and Graves (1982) found that the α-subunit was essential for calmodulin activation at the δ' site. This may explain in part why glycogen is retained in heart muscle and depleted in skeletal muscle by fasting, exercise or epinephrine administration (Evans, 1934; Evans and Bowie, 1936; Fisher and Lackey, 1925). Calmodulin also binds to a bovine brain protein (Andreasen et al., 1983) and others (Hooper and Kelly, 1984) in the presence of EGTA. This bovine brain protein is especially novel since it does not bind to calmodulin in the presence of calcium.

Levine and Dalgarno (1983) suggested that the Ca^{2+}-binding sites are near swivel points in the molecule. When calcium is bound it causes large movements in the helices. The helices contain hydrophobic residues which are oriented on one side and the binding of calcium causes these hydrophobic regions to be exposed creating a hydrophobic binding domain. The Ca^{2+}-induced formation of a hydrophobic region was confirmed by fluorescent probe analysis (Laporte, Wierman and Storm, 1980; Tanaka and Hidaka, 1980). Antipsychotic phenothiazine drugs bind to this hydrophobic domain preventing activated calmodulin from interacting with the target enzymes (Prozialeck, 1983). In the presence of Ca^{2+}, calmodulin, troponin C and S100b bind to phenothiazine-Sepharose, whereas, parvalbumin and the chicken intestinal vitamin D-dependent calcium-binding protein do not, suggesting that there may be a similar functional domain present in calmodulin, troponin C and S100b but absent in the other two (Marshak, Watterson and Van Eldik, 1981). Vogel, Thulin and Lindahl (1983) used trypsin to cleave calmodulin between domains II and III. They found that both the peptides formed showed hydrophobic bonding to phenyl-Sepharose in the presence of Ca^{2+}. When a similar experiment was performed using troponin C only the fragment with binding domains I and II showed Ca^{2+}-dependent hydrophobic binding. The other fragment with the Ca^{2+}/Mg^{2+} sites did not show this. Newton et al. (1984) used high performance liquid chromatography to purify five fragments of calmodulin after proteolysis. These fragments also bound to a phenothiazine affinity column in a Ca^{2+}-dependent manner. None of these fragments could stimulate calcineurin or cyclic nucleotide phosphodiesterase, however one fragment, containing residues 78—148 (binding domains III and IV) could stimulate phosphorylase kinase and prevent the stimulation of phosphodiesterase by calmodulin.

Along with proteins, several peptides have been found to bind to calmodulin in a calcium-dependent manner. These peptides may also be useful in determining the structure of the calmodulin-binding site. Vasoactive intestinal peptide (VIP), gastric inhibitory peptide, secretin (Malencik and Anderson, 1983), β-endorphin (Giedroc et al., 1983) and melittin (Maulet and Cox, 1983) have all been shown to bind to calmodulin. Malencik et al. (1982) recognised that these peptides contain sequences which are similar to the sequence recognised by the cyclic AMP-dependent protein kinase. They found that two peptides that could serve as substrates for cyclic AMP-dependent protein kinase also bound to calmodulin. They suggest the binding site may contain a strongly basic tripeptide sequence three positions away from a pair of bulky hydrophobic residues. While these findings are interesting, their resemblance to the calmodulin-binding site of known calmodulin-binding proteins is not known.

The competition between different calmodulin-binding proteins for calmodulin suggests the target proteins may contain a similar calmodulin-binding domain. From the subunit composition of the several purified calmodulin-binding proteins, it is concluded that there is no subunit common to them all that could be the basis for calmodulin binding. It is possible, however, that there may be similar domains on these proteins. From the fragmentation studies above, calcineurin and phosphodiesterase have stricter structural requirements for activation than phosphorylase kinase indicating some differences in the calmodulin-binding site. Some of the monoclonal antibodies developed in our laboratory have shown a cross-reactivity to calmodulin dependent proteins. suggesting there is some similarity in the calmodulin-binding site (Mutus et al., 1984; Wang et al., 1983a and b). One of these antibodies, produced by CR-B1 clonal line, has been analysed in some detail. This antibody reacts with purified phosphodiesterase, calcineurin and rabbit muscle phosphorylase kinase on ELISA assay. This antibody could immunoprecipitate phosphodiesterase and phosphorylase kinase and the immunoprecipitation of phosphodiesterase could be inhibited by calcineurin or phosphorylase kinase. The binding of this antibody was not inhibited by calmodulin nor does the antibody inhibit the activation of phosphodiesterase by calmodulin, suggesting that the antibody recognizes an extension beyond the calmodulin-binding domain. Two other cross-reactive antibodies are currently being studied in our laboratory.

Some of the target enzymes are themselves calcium-binding proteins. Calcineurin, phosphorylase kinase and ($Ca^{2+}+Mg^{2+}$)-ATPase may show two levels of calcium effects since they contain an integral calcium binding site on the enzyme and a site susceptible to regulation by Ca^{2+}-calmodulin.

10.4. Calmodulin-binding proteins

While calmodulin is found in most cells the calmodulin-binding proteins vary and it is these calmodulin-binding proteins which will determine the physiological effects of calmodulin in a tissue. These calmodulin-binding proteins can be divided into several categories according to their physiological function in the cell (Table 10.3). Calmodulin-binding proteins have been shown to be involved in the cytoskeletal and contractile elements of the cell, intermediate metabolism, regulation of calcium levels and regulation of the cyclic AMP second messenger system.

One function of particular interest in the interrelationship of the two second messenger systems, the Ca^{2+}-CaM system and the cAMP system (Rasmussen, 1981; Rasmussen and Kojima, 1983). Calmodulin was first discovered due to its stimulation of cycle nucleotide phosphodiesterase (Cheung, 1970; Kakiuchi and Yamazaki, 1970). Calmodulin does affect the cAMP system at two main levels: (1) it can affect cyclic AMP metabolism by regulation of adenylate cyclase and phosphodiesterase; (2) calmodulin can interact with the target enzymes of the cyclic AMP second messenger system by: (a) binding to substrates of cAMP-dependent protein kinase; (b) phosphorylation of the same substrates; (c) stimulation of a phosphatase that dephosphorylates many of the substrates of cAMP-dependent protein kinase. Thus calmodulin and cyclic AMP often modulate common biological processes.

10.4.1. *Regulation of cyclic AMP levels*

Calmodulin affects the levels of cyclic AMP in the cell by its effects on adenylate cyclase and cyclic nucleotide phosphodiesterase. Adenylate cyclase is a complex of proteins present in the plasma membrane of cells that synthesizes cyclic AMP under the control of hormonal regulation. In some cells and tissues calcium and calmodulin can stimulate adenylate cyclase activity. This has been shown in brain (Wolff and Brostrom, 1979), C6 astrocytoma (Brostrom et al., 1976), neonatal glial cells (Ebersolt, Perez and Bockaert, 1981), adrenal medulla (LeDonne and Coffee, 1979), pancreatic islet cells (Valverde et al., 1979), smooth muscle (Piascik, Babich and Rush, 1983) and intestinal epithelium (Amiranoff et al., 1983). The mechanism of this activation has not yet been clearly established. Some studies have found a relationship with GTP levels suggesting the involvement of the nucleotide binding subunit in calmodulin action (Brostrom, Brostrom and Wolff, 1978). Other studies have suggested that calmodulin binds to the catalytic subunit (Heideman, Weirman and Storm, 1982; Salter et al., 1981; Seamon and Daly, 1982). Most of the tissues that do show calmodulin stimulation also have a calmodulin-independent adenylate cyclase component that can be resolved from the dependent component by chromatography on calmodulin-Sepharose (Brostrom et al., 1975).

Like adenylate cyclase, cyclic nucleotide phosphodiesterase has both calmodulin-dependent and independent forms. The calmodulin-dependent enzyme has been purified from brain (Sharma et al., 1980) and heart (Laporte, Toscano and Storm, 1979) and shown to be dimeric. The heart enzyme has a

Table 10.3. Calmodulin-binding proteins.

(a) Cytoskeletal or contractile elements
  myosin light chain kinase — Adelstein (1982)
  dynein-microtubule ATPase — Blum et al. (1980)
  tau factor — Sobue et al. (1981a)
  microtubule associated factor 2 — Sobue et al. (1983)
  caldesmon — Sobue et al. (1981b)
  spectrin/fodrin — Sobue et al. (1980); Davies and Klee (1981)

  fodrin kinase — Sobue, Kanda and Kakiuchi (1982)
  tubulin/MAP 2 kinase — Goldenring et al. (1983)
  clathrin-coated vesicle proteins Mr 130 000, 93 000, 52 000 — Linden (1982)

(b) Intermediate metabolism
  phosphorylase kinase — Cohen et al. (1978)
  glycogen synthase kinase — Payne and Soderling (1980)
  phosphofructokinase — Mayr and Heilmeyer (1983)

(c) Cyclic AMP metabolism
  adenylate cyclase — Brostrom et al. (1975)
  phosphodiesterase — Cheung (1970); Kakiuchi and Yamazaki (1970)

(d) Calcium levels
  (Ca^{2+} + Mg^{2+}) ATPase — Sobue et al. (1979); LePeuch, Haiech and Demaille (1979);
  phospholamban — Louis and Jarvis (1982)

slightly smaller subunit (Hansen and Beavo, 1982) and shows less stimulation by calmodulin. The purified brain enzyme shows two bands of Mr 63 000 and 60 000 on Laemmli (1970) SDS polyacrylamide electrophoresis. Using monoclonal antibody-affinity chromatography partial separation has been achieved suggesting there are three forms of the enzyme, two homodimers and a heterodimer. It may appear futile that calmodulin would stimulate both the enzyme responsible for the synthesis of cyclic AMP and the enzyme responsible for its hydrolysis, but in the cell these effects may be sequential since calmodulin activates membrane-bound adenylate cyclase at a lower concentration of calcium than is required for activation of soluble phosphodiesterase.

10.4.2. Calmodulin binding to cyclic AMP-dependent protein kinase substrates
Several of the protein substrates for cyclic AMP-dependent protein kinase are known to be calmodulin-binding proteins, for example myosin light chain kinase (Adelstein, 1982) and phosphorylase kinase (Picton, Klee and Cohen, 1980). Contraction in smooth muscle and non-muscle cells requires the phosphorylation of the 20 000 dalton light chain of myosin to produce actin-activated ATPase activity. Calcium regulates this activity through the calmodulin-dependent activation of myosin light chain kinase (Adelstein, 1982). Cyclic AMP-dependent protein kinase can phosphorylate myosin light chain kinase at two sites (Conti and Adelstein, 1981). When calmodulin is present phosphorylation occurs at only one site and this has no effect on calmodulin binding. Phosphorylation in the absence of calmodulin occurs at two sites and causes the affinity for calmodulin to decrease 15 to 20-fold. Dephosphorylation restores calmodulin binding. This phosphorylation is a possible mechanism explaining the decrease in contractile activity in smooth muscle and non-muscle cells caused by β-adrenergic hormones.

Since calcium and calmodulin increase contractile activity requiring increased utilisation of ATP, it is logical that calmodulin would also be involved in coordinating intermediate metabolism to supply ATP for contraction. Phosphorylase kinase, a key regulatory enzyme in glycogenolysis, has long been known to be regulated by cyclic AMP-dependent protein kinase and by calcium. In 1978, Cohen et al (1978) showed that calmodulin was an integral subunit of phosphorylase kinase.

Other calmodulin-binding proteins have been shown to be substrates for cyclic AMP-dependent protein kinase including phospholamban (LePeuch, Hairch and Demaille, 1979; Louis and Jarvis, 1982), fodrin (Davies and Klee, 1981), MAP 2 (Sobue et al., 1983) and tau factor (Sobue et al., 1981).

10.4.3. Phosphorylation of common target proteins
As mentioned above, phosphorylase kinase is phosphorylated by cyclic AMP-dependent protein kinase. Calmodulin can also stimulate the autophosphorylation of this enzyme. Glycogen synthase is phosphorylated at several sites by several protein kinases including cyclic AMP-dependent protein kinase, phosphorylase kinase and calmodulin-dependent glycogen synthase kinase (Camici, 1984). Other proteins such as synapsin I (Kennedy and Greengard, 1981) and phospholamban (LePeuch, Hairch and Demaille, 1979) have been shown to be phosphorylated by both cyclic AMP-dependent and calmodulin-dependent protein kinases.

10.4.4. Calmodulin activation of a phosphatase
Calcineurin (Klee, Crouch and Krinks, 1979), a calmodulin binding protein, was initially discovered as an inhibitor of calmodulin stimulation of

phosphodiesterase (Wang and Desai, 1976). It consists of two subunits (Klee and Krinks, 1978; Klee, Crouch and Krinks, 1979), the α-subunit (Mr = 60 000), which binds calmodulin, and the β-subunit (Mr = 16 000), which binds Ca^{2+} and is a member of the troponin C superfamily, forming a structure similar to phosphorylase kinase. Recently (Stewart et al., 1982) this protein was shown to be a calmodulin stimulated protein phosphatase (protein phosphatase 2B). Calcineurin dephosphorylates inhibitor-1 of protein phosphatase a, phosphorylase kinase (α-subunit), myosin light chains, the type II regulatory subunit of cyclic AMP-dependent protein kinase, phosphohistone and phosphocasein which are phosphorylated by cyclic AMP-dependent protein kinase.

Recently it has been shown that calcineurin dephosphorylates the chromogenic substrate p-nitrophenyl phosphate (Pallen and Wang, 1983). When various non-protein physiological compounds were tested phosphoenolpyruvate and phosphotyrosine were the best substrates (Pallen and Wang, 1983) although as indicated above phosphothreonyl-proteins and phosphoseryl-proteins are good substrates for calcineurin. This led to the investigation to see if calcineurin had phosphotyrosyl protein phosphatase activity. It has been shown (Pallen et al., 1984) that calcineurin can dephosphorylate the EGF receptor from human placental membrane. Chernoff, Sells and Li (1984) have shown that phosphotyrosyl casein, histone and the EGF receptor from A431 membrane are substrates for calcineurin. The transforming proteins of several tumour viruses and the receptors for growth factors such as insulin, EGF and PDGF have been shown to possess tyrosine kinase activity. It is possible that calmodulin may have a function in coordinating the regulatory system involving phosphotyrosyl proteins with the other second messenger systems.

References

Adelstein, R.S. (1982). Calmodulin and the regulation of the actin-myosin interaction in smooth muscle and nonmuscle cells. *Cell* 30, 349—50.

Aitken, A., Klee, C.B., Stewart, A.A., Tonks, N.K. and Cohen, P. (1983). The structure of calcineurin B. In *Calcium-binding Proteins 1983*, pp.113—9. B. de Bernard, G.L. Sottocasa, G. Sandri, E. Carafoli, A.N. Taylor, T.C. Vanaman and R.J.P. Williams (eds.). Elsevier Science Publishers.

Alexis, M.N. and Gratzer, W.B. (1978). Interaction of skeletal myosin light chains with calcium ions. *Biochemistry* 17, 2319—25.

Amiranoff, B.M., Laburthe, M.C., Rouyer-Fessard, C.M., Demaille, J.G. and Rosselin, G.E. (1983). Calmodulin stimulation of adenylate cyclase of intestinal epithelium. *Eur. J. Biochem.* 130, 33—7.

Amphlett, G.W., Vanaman, T.C. and Perry, S.V. (1976). Effect of the troponin C-like protein from bovine brain (brain modulator protein) on the Mg^{2+}-stimulated ATPase of skeletal muscle actomyosin. *FEBS Lett.* 72, 163—8.

Andreasen, T.J., Luetje, C.W., Heideman, W. and Storm, D.R. (1983). Purification of a novel calmodulin binding protein from bovine cerebral cortex membranes. *Biochemistry* 22, 4615—8.

Becker, J.W., Recke, G.N., Wang, J.L., Cunningham, B.A. and Edelman, G.M. (1975). The covalent and three-dimensional structure of concanavalin A. *J. Biol. Chem.* 250, 1513—24.

Becker, J.W., Recke, G.N., Cunningham, B.A. and Edelman, G.M. (1976). New evidence on the location of the saccharide-binding site of concanavalin A.

Nature (Lond.) **259**, 406–9.
Blum, J.J., Hayes, A., Jamieson, G.A.Jr. and Vanaman, T.C. (1980). Calmodulin confers calcium sensitivity on ciliary dynein ATPase. *J. Cell. Biol.* **87**, 386–97.
Blumenthal, D.J. and Stull, J.T. (1980). Activation of skeletal muscle myosin light chain kinase by calcium (2+) and calmodulin. *Biochemistry* **19**, 5608–15.
Brostrom, M.A., Brostrom, C.O. and Wolff, D.J. (1978). Calcium-dependent adenylate cyclase from rat cerebral cortex: activation by guanine nucleotides. *Arch. Biochem. Biophys.* **191**, 341–50.
Brostrom, C.O., Huang, Y.C., Breckenridge, B. and Wolff, D.J. (1975). Identification of a calcium-binding protein as a calcium-dependent regulator of brain adenylate cyclase. *Proc. Natl. Acad. Sci. USA* **72**, 64–8.
Brostrom, M.A., Brostrom, C.O., Breckenridge, B.M. and Wolff, D.J. (1976). Regulation of adenylate cyclase from glial tumor cells by calcium and a calcium-binding protein. *J. Biol. Chem.* **251**, 4744–50.
Burchell, A., Cohen, P.T.W. and Cohen, P. (1976). Distribution of isozymes of the glycogenolytic cascade in different types of muscle fibre. *FEBS Lett.* **67**, 17–22.
Burger, D., Stein, E.A. and Cox, J.A. (1983). Free energy coupling in the interaction between Ca^{2+}, calmodulin, and phosphorylase kinase. *J. Biol. Chem.* **258**, 14733–9.
Camici, M., Ahmad, Z., DePaoli-Roach, A.A. and Roach, P.J. (1984). Phosphorylation of rabbit liver glycogen synthase by multiple protein kinases. *J. Biol. Chem.* **259**, 2466–73.
Chan, K.-F.J. and Graves, D.J. (1982). Rabbit skeletal muscle phosphorylase kinase: Interactions between subunits and influence of calmodulin on different complexes. *J. Biol. Chem.* **257**, 5956–61.
Chau, V., Huang, C.Y., Chock, P.B., Wang, J.H. and Sharma, R.K. (1982). Kinetic studies of the activation of cyclic nucleotide phosphodiesterase by Ca^{++} and calmodulin. In *Calmodulin and Intracellular Ca^{2+} Receptors*. pp.199–217. S. Kakiuchi, H. Hidaka and A.R. Means (eds.). Plenum Press, New York.
Chernoff, J., Sells, M.A. and Li, H.C. (1984). Characterization of phosphotyrosyl-protein phosphatase activity associated with calcineurin. *Biochem. Biophys. Res. Comm.* **121**, 141–8.
Cheung, W.Y. (1970). Cyclic 3',5'-nucleotide phosphodiesterase: Demonstration of an activator. *Biochem. Biophys. Res. Comm.* **38**, 533–8.
Cheung, W.Y. (1980a). Calmodulin plays a pivotal role in cellular regulation. *Science* **207**, 19–27.
Cheung, W.Y. (ed.) (1980b). *Calcium and Cell Regulation, Vol. 1, Calmodulin*, Academic Press, New York.
Cohen, P. (1980). The role of calcium ions, calmodulin and troponin in the regulation of phosphorylase kinase from rabbit skeletal muscle. *Eur. J. Biochem.* **111**, 563–74.
Cohen, P., Burchell, A., Foulkes, J.G., Cohen, P.T.W., Vanaman, T.C. and Nairn, A.C. (1978). Identification of the Ca^{2+}-dependent modulator protein as the fourth subunit of rabbit skeletal muscle phosphorylase kinase. *FEBS Letts.* **92**, 287–93.
Collins, J.H. (1976). Homology of myosin DTNB light chain with alkali light chains, troponin C and parvalbumin. *Nature (Lond.)* **259**, 699–700.
Conti, M.A. and Adelstein, R.S. (1981). The relationship between calmodulin binding and phosphorylation of smooth muscle myosin kinase by the catalytic subunit of 3':5' cAMP-dependent protein kinase. *J. Biol. Chem.* **256**, 3178–81.

Cox, J.A., Malnoe, A. and Stein, E.A. (1981). Regulation of brain cyclic nucleotide phosphodiesterase by calmodulin. *J. Biol. Chem.* 256, 3218–22.

Davies, P.J.A. and Klee, C.B. (1981). Calmodulin-binding proteins - A high molecular-weight calmodulin-binding protein from bovine brain. *Biochem. Int.* 3, 203–12.

Dedman, J.R. and Kaetzel, M.A. (1983). Calmodulin purification and fluorescent labeling. *Meth. Enzymol.* 102, 1–8.

Ebashi, S. (1974). Regulatory mechanism of muscle contraction with special reference to the Ca-troponin-tropomyosin system. *Essays Biochem.* 10, 1–36.

Ebersolt, C., Perez, M. and Bockaert, J. (1981). Neuronal, glial and meningeal localizations of neurotransmitter-senstive adenylate cyclase in cerebral cortex of mice. *Brain Res.* 213, 139–50.

Evans, G. (1934). The glycogen content of the rat heart. *J. Physiol. (Lond.)* 82, 468–80.

Evans, G. and Bowie, M.A. (1936). Cardiac glycogen in diabetic animals. *Proc. Soc. Exp. Biol. Med.* 35, 68–71.

Fisher, N.F. and Lackey, R.W. (1925). The glycogen content of the heart, liver and muscles of normal and diabetic dogs. *Am. J. Physiol.* 72, 43–9.

Giedroc, D.P., Puett, D., Ling, N. and Staros, J.V. (1983). Demonstration by covalent cross-linking of a specific interaction between β-endorphin and calmodulin. *J. Biol. Chem.* 258, 16–9.

Goldenring, J.R., Gonzalez, B., McGuire, J.S.Jr. and DeLorenzo, R.J. (1983). Purification and characterization of a calmodulin-dependent kinase from rat brain cytosol able to phosphorylate tubulin and microtubule-associated proteins. *J. Biol. Chem.* 258, 12632–40.

Goodman, M., Pechere, J.-F., Haiech, J. and Demaille, J.G. (1979). Evolutionary diversification of structure and function in the family of intracellular calcium-binding proteins. *J. Mol. Evol.* 13, 331–52.

Grand, R.J.A., Shenolikar, S. and Cohen, P. (1981). The amino acid sequence of the δ subunit (calmodulin) of rabbit skeletal muscle phosphorylase kinase. *Eur. J. Biochem.* 113, 359–67.

Haiech, J., Derancourt, J., Pechere, J.F. and Demaille, J.G. (1979). Magnesium and calcium binding to parvalbumins: Evidence for differences between parvalbumins and an explanation of their relaxing function. *Biochemistry* 18, 2752–8.

Han, M.H. and Benson, E.S. (1970). Fluorescence studies of tropocalcin, a troponin fraction undergoing Ca^{++}-induced conformational changes, *Biophys. J.* 10, 245a.

Hansen, R.S. and Beavo, J.A. (1982). Purification of two calcium/calmodulin-dependent forms of cyclic nucleotide phosphodiesterase by using conformation-specific monoclonal antibody chromatography. *Proc. Natl. Acad. Sci. USA* 79, 2788–92.

Hartshorne, D.J. and Dreizen, P. (1972). Studies on the subunit composition of troponin. *Cold Spring Harbor Symp. Quant. Biol.* 37, 225–34.

Heideman, W., Weirman, B.M. and Storm, D.R. (1982). GTP is not required for calmodulin stimulation of bovine brain adenylate cyclase. *Proc. Natl. Acad. Sci. USA* 79, 1462–5.

Hesterberg, L.K. and Weber, K. (1983). Demonstration of three distinct calcium-binding sites in villin, a modulator of actin assembly. *J. Biol. Chem.* 258, 365–9.

Hooper, J.E. and Kelly, R.B. (1984). Calmodulin is tightly associated with synaptic vesicles independent of calcium. *J. Biol. Chem.* 259, 148–53.

Huang, C.Y., Chau, V., Chock, P.B., Wang, J.H. and Sharma, R.K. (1981). Mechanism of activation of cyclic nucleotide phosphodiesterase:

requirements of the binding of four Ca^{2+} to calmodulin for activation. *Proc. Natl. Acad. Sci. USA* **78**, 871–5.
Isobe, T. and Okuyama, T. (1978). The amino-acid sequence of S100 protein (PAP I-b protein) and its relation to the calcium-binding protein. *Eur. J. Biochem.* **89**, 379–88.
Isobe, T. and Okuyama, T. (1981). The amino-acid sequence of the α subunit in bovine brain S-100a protein. *Eur. J. Biochem.* **116**, 79–86.
Isobe, T., Nakajima, T. and Okuyama, T. (1977). Reinvestigation of extremely acidic proteins in bovine brain. *Biochim. Biophys. Acta* **494**, 222–32.
Jennissen, H.P. and Heilmeyer, L.M.G. (1974). Multiple forms of phosphorylase kinase in red and white skeletal muscle. *FEBS Lett.* **42**, 77–80.
Kaibuchi, K., Takai, Y. and Nishizuka, Y. (1981). Cooperative roles of various membrane phospholipids in the activation of calcium-activated, phospholipid-dependent protein kinase. *J. Biol. Chem.* **256**, 7146–9.
Kakiuchi, S. and Yamazaki, R. (1970). Calcium dependent phosphodiesterase activity and its activating factor (PAF) from brain. *Biochem. Biophys. Res. Comm.* **41**, 1104–10.
Kennedy, M.B. and Greengard, P. (1981). Two calcium/calmodulin-dependent protein kinases, which are highly concentrated in brain, phosphorylate protein I at distinct sites, *Proc. Natl. Acad. Sci. USA* **78**, 1293–7.
Kishimoto, A., Takai, Y., Mori, T., Kikkawa, U. and Nishizuka, Y. (1980). Activation of calcium and phospholipid-dependent protein kinase by diacylglycerol, its possible relation to phosphatidyl inositol turnover *J. Biol. Chem.* **255**, 2273–6.
Klee, C.B. and Krinks, M.H. (1978). Purification of cyclic 3',5'-nucleotide phosphodiesterase inhibitory protein by affinity chromatography on activator protein coupled to Sepharose. *Biochemistry* **17**, 120–6.
Klee, C.B. and Vanaman, T.C., (1982). Calmodulin. *Adv. Prot. Chem.* **35**, 213–321.
Klee, C.B., Crouch, T.H. and Krinks, M.H. (1979). Calcineurin: a calcium- and calmodulin-binding protein of the nervous system. *Proc. Natl. Acad. Sci. USA* **76**, 6270–3.
Klee, C.B., Crouch, T.H. and Richman, P.G. (1980). Calmodulin, *Ann. Rev. Biochem.* **49**, 489–515.
Kretsinger, R.H. (1976). Calcium-binding proteins. *Ann. Rev. Biochem.* **45**, 239–66.
Kretsinger, R.H. (1980). Structure and evolution of calcium-modulated proteins. *CRC Crit. Rev. Biochem.* **8**, 119–74.
Laemmli, U.K. (1970). Cleavage of structural proteins during the assembly of the head of bacteriophage T4. *Nature (Lond.)* **227**, 680–5.
Laporte, D.C., Toscano, W.A. and Storm, D.R. (1979). Cross-linking of iodine-125-labeled, calcium-dependent regulatory protein to the Ca^{2+}-senstive phosphodiesterase purified from bovine heart. *Biochemistry* **18**, 2820–5.
Laporte, D.C., Wierman, B.M. and Storm, D.R. (1980). Calcium-induced exposure of a hydrophobic surface on calmodulin. *Biochemistry* **19**, 3814–9.
LeDonne, N.C.Jr. and Coffee, C.J. (1979). Inability of parvalbumin to function as a calcium-dependent activator of cyclic nucleotide phosphodiesterase activity. *J. Biol. Chem.* **254**, 4317–20.
LePeuch, C.J., Hairch, J. and Demaille, J.G. (1979). Concerted regulation of cardiac sarcoplasmic reticulum calcium transport by cyclic adenosine monophosphate dependent and calcium-calmodulin-dependent phosphorylations. *Biochemistry* **18**, 5150–7.
Levine, B.A. and Dalgarno, D.C. (1983). The dynamics and function of

calcium-binding proteins. *Biochim. Biophys. Acta* **726**, 187–204.
Linden, C.D. (1982). Identification of the coated vesicle proteins that bind calmodulin. *Biochem. Biophys. Res. Comm.* **109**, 186–93.
Louis, C.F. and Jarvis, B. (1982). Affinity labeling of calmodulin-binding components in canine cardiac sarcoplasmic reticulum. *J. Biol. Chem.* **257**, 15187–91.
MacLennan, D.H. and Holland, P.C. (1975). Calcium transport in sarcoplasmic reticulum. *Ann. Rev. Biophys. Bioeng.* **4**, 377–404.
MacManus, J.P. (1980). The purification of a unique calcium-binding protein from Morris hepatoma 5123 tc. *Biochim. Biophys. Acta* **621**, 296–304.
MacManus, J.P., Watson, D.C. and Yaguchi, M. (1983). The complete amino acid sequence of oncomodulin - a parvalbumin-like calcium-binding protein from Morris hepatoma 5123 tc. *Eur. J. Biochem.* **136**, 9–17.
Malencik, D.A. and Anderson, S.R. (1983). Binding of hormones and neuropeptides by calmodulin. *Biochemistry* **22**, 1995–2001.
Malencik, D.A., Huang, T.-S. and Anderson, S.R. (1982). Binding of protein kinase substrates by fluorescently labeled calmodulin. *Biochem. Biophys. Res. Comm.* **108**, 266–72.
Manalan, A.S. and Klee, C.B. (1984). Purification and characterization of a novel Ca^{2+}-binding protein (CBP-18) from bovine brain. *J. Biol. Chem.* **259**, 2047–50.
Marshak, D.R., Watterson, D.M. and Van Eldik, L.J. (1981). Calcium-dependent interaction of S100b, troponin C, and calmodulin with immobilized phenothiazine. *Proc. Natl. Acad. Sci. USA* **78**, 6793–7.
Maulet, Y. and Cox, J.A. (1983). Structural changes in melittin and calmodulin upon complex formation and their modulation by calcium. *Biochemistry* **22**, 5680–6.
Mayr, G.W. and Heilmeyer, L.M.G.Jr. (1983). Phosphofructokinase is a calmodulin binding protein. *FEBS Lett.* **159**, 51–7.
McCullough, T.E. and Walsh, D.A. (1979). Phosphorylation and dephosphorylation of phosphorylase kinase in the perfused rat heart. *J. Biol. Chem.* **254**, 7345–52.
Means, A.R. and Dedman, J.R. (1980). Calmodulin - an intracellular calcium receptor. *Nature (Lond.)* **285**, 73–7.
Moews, P.C. and Kretsinger, R.H. (1975). Refinement of the structure of carp muscle calcium-binding parvalbumin by model building and difference Fourier analysis. *J. Mol. Biol.* **91**, 201–28.
Mutus, B., Wang, K.C., Sharma, R.K., Lam, H.-Y.P. and Wang, J.H. (1984). A fluorescent probe and monoclonal antibody for the studies of the interation between calmodulin and cyclic nucleotide phosphodiesterase. *Adv. Cyc. Nucleotide Protein Phos. Res.* **16**, 65–76.
Nelsestuen, G.L., Broderius, M., Zytkovicz, T.H. and Howard, J.B. (1975). On the role of γ-carboxyglutamic acid in calcium and phospholipid binding. *Biochem. Biophys. Res. Comm.* **65**, 233–40.
Newton, D.L., Oldewurtel, M.D., Krinks, M.H., Shiloach, J. and Klee, C.B. (1984). Agonist and antagonist properties of calmodulin fragments, *J. Biol. Chem.* **259**, 4419–26.
Ohnishi, K. and Watanabe, Y. (1983). Purification and some properties of a new Ca^{2+}-binding protein (TCBP-10) present in *Tetrahymena* cilium. *J. Biol. Chem.* **258**, 13978–85.
Ostwald, T.J. and MacLennan, D.H. (1974). Effects of cation binding on the conformation of calsequestrin and the high affinity calcium-binding protein of sarcoplasmic reticulum. *J. Biol. Chem.* **249**, 5867–71.
Pallen, C.J. and Wang, J.H. (1983). Calmodulin-stimulated dephosphorylation

of p-nitrophenyl phosphate and free phosphotyrosine by calcineurin. *J. Biol. Chem.* **258**, 8550–3.

Pallen, C.J., Valentine, K., Wang, J.H. and Hollenberg, M. (1984). Calcineurin can dephosphorylate the epidermal growth factor receptor of human placental membrane. *Fed. Proc.* **43**, 1896.

Payne, M.E. and Soderling, T.R. (1980). Calmodulin-dependent glycogen synthase kinase. *J. Biol. Chem.* **255**, 8054–6.

Piascik, M.T., Babich, M. and Rush, M.E. (1983). Calmodulin stimulation and calcium regulation of smooth muscle adenylate cyclase activity. *J. Biol. Chem.* **258**, 10913–8.

Picton, C., Klee, C.B. and Cohen, P. (1980). Phosphorylase kinase from rabbit skeletal muscle: identification of the calmodulin-binding subunits, *Eur. J. Biochem.* **111**, 553–61.

Prozialeck, W.C. (1983). Structure-activity relationships of calmodulin antagonists. *Ann. Rep. Med. Chem.* **18**, 203–12.

Rasmussen, H. (1970). Cell communication, calcium ion, and cyclic adenosine monophosphate. *Science* **170**, 404–12.

Rasmussen, H. (1981). *Calcium and cAMP as Synarchic Messengers*, John Wiley & Sons, New York.

Rasmussen, H. and Kojima, I. (1983). Ca^{2+} and cAMP in the regulation of cell function. In *Calcium-Binding Proteins 1983*, pp. 417–31. B. de Bernard, G.L. Sottocasa, G. Sandri, E. Carafoli, A.N. Taylor, T.C. Vanaman and R.J.P. Williams (eds.) Elsevier Science Publishers.

Salter, R.S., Krinks, M.H., Klee, C.B. and Neer, E.J. (1981). Calmodulin activates the isolated catalytic unit of brain adenylate cyclase, *J. Biol. Chem.* **256**, 9830–3.

Seamon. K.B. and Daly, J.W. (1982). Calmodulin stimulation of adenylate cyclase in rat brain membranes does not require GTP. *Life Sci.* **30**, 1457–64.

Sharma, R.K., Tam, S.W., Waisman, D.M. and Wang, J.H. (1980a). Differential interaction of rabbit skeletal muscle phosphorylase kinase isozymes with calmodulin. *J. Biol. Chem.* **255**, 11102–5.

Sharma, R.K., Wang, T.H., Wirch, E. and Wang, J.H. (1980b). Purification and properties of bovine brain calmodulin-dependent cyclic nucleotide phosphodiesterase. *J. Biol. Chem.* **255**, 5916–23.

Shenolikar, S., Cohen, P.T.W., Cohen, P., Nairn, A.C. and Perry, S.V. (1979). The role of calmodulin in the structure and regulation of phosphorylase kinase from rabbit skeletal muscle. *Eur. J. Biochem.* **100**, 329–37.

Sobue, K., Kanda, K. and Kakiuchi, S. (1982). Solubilization and partial purification of protein kinase systems from brain membranes that phosphorylate calspectin. *FEBS Lett.* **150**, 185–90.

Sobue, K., Fujita, M., Muramoto, Y. and Kakiuchi, S. (1980). Spectrin as a major modulator binding protein of erythrocyte cytoskeleton. *Biochem. Int.* **1**, 561–6.

Sobue, K., Ichida, S., Yoshida, S., Yamazaki, R. and Kakiuchi, S. (1979). Occurrence of a Ca^{2+}- and modulator protein-activatable ATPase in the synaptic plasma membrane of brain. *FEBS Lett.* **99**, 199–202.

Sobue, K., Fujita, M., Muramoto, Y. and Kakiuchi, S. (1981a). The calmodulin-binding protein in microtubules is tau factor, *FEBS Lett.* **132**, 137–40.

Sobue, K., Muramoto, Y., Fujita, M. and Kakiuchi, S. (1981b). Purification of a calmodulin-binding protein from chicken gizzard that interacts with F-actin. *Proc. Natl. Acad. Sci. USA* **78**, 5652–5.

Sobue, K., Kanda, K., Adachi, J. and Kakiuchi, S. (1983). Calmodulin-binding proteins that interact with actin filaments in a Ca^{2+}-dependent flip-flop manner: survey in brain and secretory tissues. *Proc. Natl. Acad. Sci. USA* **80**, 6868–71.

Stewart, A.A., Ingebritsen, T.S., Manalan, A., Klee, C.B. and Cohen, P. (1982). Discovery of a Ca^{2+}- and calmodulin-dependent protein phosphatase: probable identity with calcineurin (CaM-BP$_{80}$). *FEBS Lett.* **137**, 80–4.

Szebenji, D.M.E., Obendorf, S.K. and Moffat, K. (1981). Structure of vitamin D-dependent calcium-binding protein from bovine intestine. *Nature (Lond.)* **294**, 327–32.

Takai, Y., Kishimoto, A., Iwasa, Y., Kawahara, Y., Mori, T. and Nishizuka, Y. (1979). Calcium-dependent activation of a multifunctional protein kinase by membrane phospholipid. *J. Biol. Chem.* **254**, 3692–5.

Tam, S.W., Sharma, R.K. and Wang, J.H. (1982). Phosphorylation and activation of red skeletal muscle phosphorylase kinase isozyme. *J. Biol. Chem.* **257**, 14907–13.

Tanaka, T. and Hidaka, H. (1980). Hydrophobic regions function in calmodulin-enzyme(s) interactions. *J. Biol. Chem.* **255**, 11078–80.

Tsien, R.Y. (1980). New calcium indicators and buffers with high selectivity against magnesium and protons: Design, synthesis, and properties of prototype structures. *Biochemistry* **19**, 2396–404.

Valverde, I., Vandermeers, A., Anjaneyulu, R. and Malaisse, W.J. (1979). Calmodulin activation of adenylate cyclase in pancreatic islets. *Science* **206**, 225–7.

Van den Bosch, H. (1980). Intracellular phospholipases A. *Biochim. Biophys. Acta* **604**, 191–246.

Van Eldik, L.J., Zendegui, J.G., Marshak, D.R. and Watterson, D.M. (1982). Calcium-binding proteins and the molecular basis of calcium action. *Int. Rev. Cytol.* **77**, 1–61.

Vogel, H.J., Thulin, E. and Lindahl, L. (1983). Purification of calcium binding proteins and peptides using phenyl-Sepharose. In *Calcium-Binding Proteins 1983*. B. de Bernard, G.L. Sottocasa, G. Sandri, E. Carafoli, A.N. Taylor, T.C. Vanaman and R.J.P. Williams (eds.) pp. 75–6. Elsevier Science Publishers.

Waisman, D., Stevens, F.C. and Wang, J.H. (1975). The distribution of the Ca^{++}-dependent protein activator of cyclic nucleotide phosphodiesterase in invertebrates. *Biochem. Biophys. Res. Comm.* **65**, 975–82.

Waisman, D.M., Muranyi, J. and Ahmed, M. (1983). Identification of a novel calcium binding protein from bovine brain. *FEBS Lett.* **164**, 80–4.

Wang, J.H. and Desai, R. (1976). A brain protein and its effect on the Ca^{2+}- and protein modulator-activated cyclic nucleotide phosphodiesterase. *Biochem. Biophys. Res. Comm.* **72**, 926–32.

Wang, J.H. and Waisman, D.M. (1979). Calmodulin and its role in the second messenger system. *Curr. Top. Cell. Regul.* **15**, 47–107.

Wang, J.H., Teo, T.S., Ho, H.C. and Stevens, F.C. (1975). Bovine heart protein activator of cyclic nucleotide phosphodiesterase. *Adv. Cyc. Nucleotide Res.* **5**, 179–94.

Wang, K.C., Mutus, B., Sharma, R.K., Lam, H.-Y.P. and Wang, J.H. (1983a). On the mechanism of interaction between calmodulin and calmodulin-dependent proteins. *Can. J. Biochem.* **61**, 911–20.

Wang, K.C., Wong, H.Y., Wang, J.H. and Lam, H.-Y.P. (1983b). A monoclonal antibody showing cross-reactivity toward three calmodulin-dependent enzymes. *J. Biol. Chem.* **258**, 12110–3.

Weeds, A.G. and McLachlan, A.D. (1974). Structural homology of myosin alkali light chains, troponin C and carp calcium binding protein. *Nature (Lond.)* **252**, 646–9.

Wolff, D.J. and Brostrom, C.O. (1979). Properties and functions of the calcium-dependent regulator protein. *Adv. Cyc. Nucleotide Res.* **11**, 27–88.

Yin, H.L. and Stossel, T.P. (1980). Purification and structural properties of Gelsolin, a Ca^{2+}-activated regulatory protein of macrophages. *J. Biol. Chem.* **255**, 9490–3.

Yoshikawa, K., Usui, H., Imazu, M., Takeda, M. and Ebashi, S. (1983). The effect of heart and skeletal muscle troponin complexes and calmodulin on the Ca^{2+}-dependent reactions of phosphorylase kinase isoenzymes. *Eur. J. Biochem.* **136**, 413–9.

THE CALCIUM MESSENGER SYSTEM

Howard Rasmussen, Itaru Kojima, Kumiko Kojima, Walter Zawalich, John Forder, Paula Barrett, David Kreutter, William Apfeldorf, Suzanna Park, Ann Caldwell, Dominique Delbeke and Priscilla Dannies

Department of Internal Medicine, Yale University, School of Medicine, New Haven/Connecticut 06520-8056 (USA)

Abstract

It has become increasingly evident that Ca^{2+} plays a nearly universal messenger role in stimulus-response coupling both in neural and endocrine-mediated responses. A crucial support for this conclusion was the discovery of calmodulin and its nearly universal distribution. However in an increasing number of systems in which the particular cell displays a sustained response to the sustained presence of a hormone, it has been found that the rise in intracellular free Ca^{2+} is transient, in spite of this sustained response. Investigating the possibility that calmodulin-dependent processes are responsible for only the initial phase of cell response, and that C-kinase-dependent processes are responsible for the sustained phase of response, we have examined the effects of A23187, an activator of calmodulin-regulated processes, and of phorbol ester (TPA), an activator of C-kinase, alone or in combination on the processes of vascular smooth muscle contraction, insulin secretion from islet cells, prolactin secretion from cultured pituitary cells, and aldosterone secretion from adrenal glomerulosa cell. In each system, A23187 produces a rapid but brief response, TPA a very gradually developing but submaximal response, and combined A23187-TPA a rapid and sustained response comparable to that produced by a natural agonist. These results support the hypthesis that the C-kinase pathway provides a type of gain control in the calcium messenger system.

11.1. Introduction

If one considers the biochemical make up of mammalian cells, one must conclude that mammalian cell types are more alike than unalike. Different cell types have each inherited a majority of the same ion channels, enzymes, and control systems. They have a number of universal attributes. However, each is also unique. This particularly is often achieved by adapting the components of a specific metabolic control system to a particular functional need (Rasmussen, 1981).

As evidence for the reality of this universal versus particular duality, one need only to cite the Krebs cycle enzymes and a group of ancillary proteins whose functions are coupled to the flux of metabolites through this enzyme sequence. It is clear that in one tissue, insect flight muscle, the Kreb cycle enzymes function almost exactly the way their function is depicted in a biochemistry textbook. On the other hand, in the case of mammalian liver mitochondria. the cycle rarely cycles as normally depicted. Rather, fluxes through parts of it are critical to processes such as lipogenesis, fatty acid oxidation, and gluconeogenesis. These enzymes, nearly identical to those in flight muscle, nevertheless express their function in quite different ways.

We believe that all nucleated mammalian cells posses, as one of their fundamental control systems, the calcium messenger system, that this system is as universal as the Krebs cycle (Campbell, 1983; Rasmussen, 1981; Rasmussen and Barrett, 1984), and that this universal system displays a number of particular variations as it serves to couple stimulus to response in cells with totally different response characteristics.

The vantage point from which we have approached our recent studies of the calcium messenger system is the problem of how it couples stimulus to *sustained* (hours) cellular responses (Rasmussen and Barrett, 1984). From this inquiry, we have developed a general model of how the calcium messenger system functions.

11.2. Recent experimental and conceptual progress

In the past 5 years there has been an explosive increase in interest in the calcium messenger system (Borle, 1981; Campbell, 1983; Cheung, 1980; Fleckenstein, 1983; Putney and Biancri, 1974; Rasmussen, 1981; Rasmussen and Barrett, 1984; Rubin, 1982; Wolheim and Sharp, 1981). This interest has been fuelled by the development of methods for the measurement of intracellular free calcium (Blinks *et al.*, 1982; Borle and Snowdowne, 1982; Feinstein *et al.*, 1983; Gershengorn and Thaw, 1983; Morgan and Morgan, 1983; Pozzan *et al.*, 1982; Scarpa *et al.*, 1978; Tsien, 1981; Tsien, Pozzan and Rink, 1981, 1982), refinements of methods for the analysis of phosphatidylinositol turnover (Berridge, 1983; Berridge *et al.*, 1983; Downes and Michell, 1982; Fain and Berridge, 1979; Irvine, Dawson and Freinkel, 1982), the discovery of a new type of protein kinase, the calcium-activated, phospholipid-dependent kinase, or C-kinase (Minakuchi *et al.*, 1981; Nishizuka, 1983; Takai *et al.*, 1981), and the development of the concepts of amplitude and sensitivity modulation (Rasmussen and Waisman, 1982). In addition, there has been an increased knowledge of all aspects of cellular calcium metabolism (Akerman and Nicholls, 1982; Baker, 1976; Borle, 1981; Drust and Martin, 1982; Foster and Rasmussen, 1983; Hansford and Castro, 1982; Hedeskov, 1980; Nicholls, 1978; Studer and Borle, 1983), new techniques for the study of calcium ion channels and calcium fluxes, and a detailed kinetic analysis of calcium action at a molecular level in the activation of specific proteins (Cox *et al.*, 1985; Wang *et al.*, 1980) including the calcium pump of Ca^{2+}/H^+ ATPase of the plasma membrane (Niggli, Sigel and Carafoli, 1982; Smallwood *et al.*, 1983).

The new insights gained have led to the recognition and general acceptance of the hypothesis that the calcium messenger system is a nearly universal means by which a whole class of extracellular stimuli are coupled to particular cellular responses (Campbell, 1983; Rasmussen, 1981; Rasmussen and Barrett, 1984), that calcium ion and cAMP often serve together as *synarchic* messengers in the regulation of cell function (Rasmussen, 1981), and that very small and transient changes in intracellular free Ca^{2+} are apparently sufficient to bring about sustained cellular responses (Rasmussen and Barrett, 1984).

11.3. The cellular milieu

The first question one might ask is why do cells use only a transient calcium message. Surrounded by an infinite sea rich in Ca^{2+}, the extracellular fluids containing 1000 μM free Ca, there is an overabundance of the Ca^{2+} necessary to increase the intracellular free $[Ca^{2+}]$ from 0.1 to 1.0 μM (Rasmussen and Barrett, 1984). There are even known pathways of Ca^{2+} influx across the plasma membrane capable of mediating this change (Campbell, 1983; Rasmussen and Barrett, 1984). Paradoxically, in spite of the fact that in cells displaying a sustained response to the sustained presence of extracellular messenger, there is a sustained increase in Ca^{2+} ion flux into the cell, the cytosolic $[Ca^{2+}]$, $[Ca^{2+}]_c$ is only 1.5–2-fold greater than that seen in the non-activated cell. The resolution of this apparent paradox is provided by the fact that during the sustained phase of cell activation, plasma membrane calcium efflux also remains high (Rasmussen and Barrett, 1984). Hence, during the sustained phase of cellular response, there is an increased rate of cycling of Ca^{2+} across the plasma membrane with efflux nearly balancing influx so that net Ca^{2+} accumulation by the cell is quite small.

11.4. Calcium as minatory messenger

The key to understanding the functioning of the calcium messenger system is to be found in an understanding of the relationship of cells to extracellular Ca^{2+}. Extracellular Ca^{2+} is a potential cellular poison. Any significant, sustained and unopposed influx of Ca^{2+} will lead to cell death. In order to minimize this possibility the cell has developed an elaborate and high effective system for maintaining calcium homeostasis. The various components of this system have been discussed in detail in other recent reviews (Borle, 1981; Campbell, 1983; Cheung, 1980; Fleckenstein, 1983; Rasmussen, 1981; Rasmussen and Barrett, 1984) and will not be considered in detail. The one component, which in many ways is the most important, is the calcium pump in the plasma membrane (Niggli, Sigel and Carafoli, 1982; Smallwood et al., 1983; Vincenzi, Hinds and Raess, 1980). This pump is activated (an increase in V_{max} and decrease in K_m for Ca^{2+}) by Ca^{2+}-calmodulin. Hence, any rise in cytosolic-free Ca^{2+} leads immediately to the association of Ca^{2+} with calmodulin and hence to an activation of this pump which acts as an autoregulatory device to prevent a rise in $[Ca^{2+}]_o$. As long as this sytem operates effectively, it is nearly impossible to induce a sustained rise in $[Ca^{2+}]_c$ from 0.1 to 1.0 μM under physiological conditions. This is the cellular milieu in which the calcium messenger system operates. Calcium is a minatory messenger. Hence, if this messenger system is to be employed to couple stimulus to sustained response, the cell must protect itself.

It is possible to consider at least three ways in which gain control might be achieved (Rasmussen and Barrett, 1984): (1) hysteresis in calmodulin-regulated enzymes; (2) sensitivity modulation of calmodulin-regulated enzymes by cAMP-dependent phosphorylation; or (3) a two-component calcium messenger system in which the different components possess different sensitivities to activation by Ca^{2+} and modulate different temporal phases of the response.

11.5. Gain control: hysteresis and/or cAMP-dependent sensitivity modulation

Analysis of the kinetics of association of Ca^{2+} and a response element (protein) such as phosphodiesterase shows that these association reactions are ordered and cooperative (Rasmussen and Waisman, 1982; Wang et al., 1980). In the usual scheme of the activation of a calmodulin-sensitive enzyme by Ca^{2+}, there is a direct correlation between the amplitude of the calcium signal, and the degree of activation of the enzyme, that is, the enzyme undergoes amplitude modulation. However, because of the cooperative nature of the binding of Ca^{2+} and CaM to the enzyme, the enzyme once activated will stay in the activated state at a lower $[Ca^{2+}]$ than that needed to place it in its active conformation - in other words, the system displays hysteresis. The possibility that this type of gain control operates within cells has not been widely explored, but may contribute in a small way to such a process.

In addition to amplitude modulation, a Ca^{2+}, CaM, response element system can display sensitivity modulation (Rasmussen and Waisman, 1982), a situation in which modification of the response element structure by, for example, cAMP-dependent phosphorylation leads to a change in its activity at a fixed $[Ca^{2+}]$, that is, the sensitivity of its activation by Ca^{2+} has been changed. Since in many cell types addition of an appropriate agonist leads to an increase in both Ca^{2+} and cAMP within the cell either in a cooperative, hierarchical, or sequential fashion (Rasmussen, 1981), it is possible to envision that initial cell activation occurs by the amplitude modulation of CaM-regulated enzymes, and that activation is sustained as a consequence of the positive sensitivity modulation, one or more of these response elements by cAMP-dependent phosphorylation (Rasmussen and Barrett, 1984). Under this circumstance, the enzyme will remain active even though the $[Ca^{2+}]_c$ falls back close to its original value.

11.6. Gain control: a two-component calcium messenger system

The concept that gain control is achived by the integrated operation of two branches of the calcium messenger system is supported by experimental evidence derived from several different cellular systems (Rasmussen and Barrett, 1984). In order to understand this evidence and this model, it is necessary to know of two recent developments. The first is the discovery of a calcium-regulated protein kinase (C-kinase) which is phospholipid-dependent and does not involve calmodulin as the calcium receptor protein, and the second the recent evidence as to the early transducing events in the calcium messenger system following hormone-receptor interaction (Minakuchi et al., 1981; Nishizuka, 1983; Rasmussen and Barrett, 1984; Takai et al., 1981).

Work by Nishizuka and co-workers led to the discovery of a new species of protein kinase, the C-kinase, which in its isolated form is a calcium-activated protein kinase which has a low V_{max} and a high K_{Ca} for activation. However, association of enzyme with diacylglycerol and a mixture of phospholipids causes the enzyme to undergo a conformational change to a state in which it has a high V_{max} and is highly sensitive to Ca^{2+} ($K_{Ca} \cong 0.1$ μM). In some cells at least, the non-sensitive form of the enzyme is found largely in the cytosol when the cell is not activated, and in its sensitive plasma membrane-bound form when the cell is activated. The enzyme can be activated either by synethetic diacylglycerol or phorbol esters, such as 12-0-tetradecanoyl-phorbol-13-acetate (TPA) (Castagna et al., 1982).

Older evidence had suggested that the turnover of phosphatidylinositides was related to the Ca^{2+} messenger system (Hokin and Hokin, 1955). Recent work suggests that receptor coupled to the calcium messenger system activate a specific phospholipase C when these receptors interact with their appropriate agonists (Berridge et al., 1983; Downes and Michell, 1982; Takai et al., 1981). This enzyme catalyses the hydrolysis of phosphatidylinositol 4,5-bisphosphate with the production of inositol triphosphate ($InsP_3$) and diacylglycerol (DG) rich in arachidonic acid. Each of these products is thought to have a messenger function: $InsP_3$ as a mobilizer of an intracellular Ca^{2+} pool (Streb et al., 1983); and DG as an activator of C-kinase. It is not yet clear how the third consequence of hormone-receptor interaction, the increase in plasma membrane Ca^{2+} influx is linked to hormone-receptor interaction.

The first data implying that the calcium messenger system might operate via two branches was provided by work done in platelets (Kaibuchi et al., 1982, 1983). When these cells are activated by thrombin there is a rise in the $[Ca^{2+}]_c$, the calmodulin-dependent phosphorylation of a protein, myosin light chain (MLC), the phosphorylation of a 47K protein which is a substrate for C-kinase, and a maximal serotonin secretory response. Addition of the calcium ionophore, A23187, to platelets leads to as great or even greater rise in $[Ca^{2+}]_c$ in MLC·P, but no phosphorylation of the 47K protein, and only a small secretory response induced by thrombin. Addition of TPA induces phosphorylation of the 47K protein but not of MLC, no rise in $[Ca^{2+}]_c$, and a submaximal secretory response. Combined A23187 and TPA induce a pattern of responses similar to those seen after thrombin addition.

Work in several other systems has extended these types of observations and led to the proposal that the flows of information through the two branches have unique temporal roles in mediating the sustained cellular responses see n in these systems (Delbeke, 1985; Kojima et al., 1983, 1985; Rasmussen et al., 1985; Zawalich, Brown and Rasmussen, 1983).

The system most extensively studied has been the action of angiotensin II in the control of aldosterone secretion from isolated, perifused adrenal glomerulosa cells (Foster and Rasmussen, 1983; Kojima et al., 1983, 1985). When angiotensin II interacts with its receptors on these cells, there is an immediate (10 s) hydrolysis of phosphatidylinositol 4,5-*bis*-phosphate (and phosphatidylinositol-4-phosphate) with the production of diacylglycerol (DG), rich in arachidonic acid, and inositol triphosphate ($InsP_3$), and inositol bisphosphate. The increase in DG peaks in 20 s but after a small fall remains elevated for at least 10 min. Likewise, the $[InsP_3]$ increases and remains elevated. Within 1–4 min after angiotensin II addition, there is a marked increase in Ca efflux from the cell which lasts for 5–8 min. This efflux is inhibited by prior addition of dantrolene to the cells. Saponin-permeabilized cells take up ^{45}Ca by an ATP-dependent process, into a non-mitochondrial pool. Calcium release from the pools is increased by $InsP_3$, and this $InsP_3$-induced release is blocked by dantrolene. Thus, the rise in $[InsP_3]$ causes the observed angiotensin II-induced release of Ca^{2+}. Following Ca^{2+} release (8 min), the aldosterone production rate increases monotonically to a sustained plateau which is maintained for 60 or more minutes. During this period the response is dependent on extracellular calcium, presumably because throughout this period there is a high rate of Ca^{2+} cycling across the plasma membrane.

Perfusion of cells with the calcium ionophore, A23187, leads to an increase in Ca entry and a release of intracellular Ca^{2+} without any hydrolysis of polphosphoinositides or production of DG and $InsP_3$. Shortly after ionophore addition (8 min), aldosterone secretion rate increases and rises monotonically to a value 80% of that seen after antiotensin II addition. However, this response is not sustained, but it falls gradually to a rate only 25–30% of the maximal rate. In contrast, perifusion with TPA, or a synthetic diacylglycerol, 1-oleoyl-2-acetyl-glycerol (OAG), leads to no immediate change in aldosterone secretory rate, but after 30 min the rate increases slowly to a value only 10–15% of that induced by angiotensin II. The perifusion of adrenal cells with the combination of A23187 and TPA leads to an aldosterone secretory response which is qualitatively and quantitatively similar to that induced by angiotensin II.

These data have been interpreted in terms of a model in which interaction of angiotensin II with its receptor leads to the generation of two messengers,

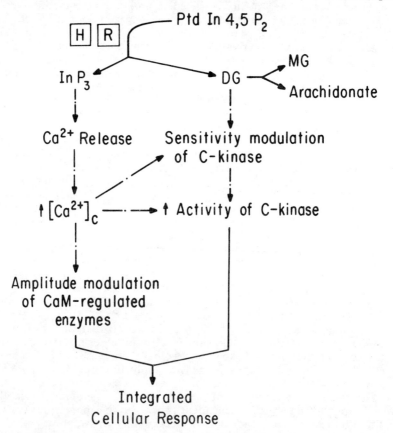

Fig.11.1. Operation of the calcium messenger system. Interaction of hormone H with receptor R leads to the generation of two messengers, inositol trisphosphate, $InsP_3$, and diacylglycerol, DG. The increase in cytosolic $InsP_3$ causes a release of Ca^{2+} from the endoplasmic reticulum. This released Ca^{2+} plus an increase in Ca^{2+} influx across the plasma membrane (mechanism unknown) leads to a transient rise in the free Ca^{2+} concentration in the cell cytosol, $[Ca^{2+}]_c$. The rise in $[Ca^{2+}]_c$ brings about the amplitude modulation of a group of calmodulin (CaM)-regulated enzymes (calmodulin branch) which are responsible for the initial phase of cellular response. Activation of one of these enzymes, the plasma membrane Ca^{2+} pump, leads to the subsequent fall in $[Ca^{2+}]_c$ and hence a decline in the activity of other CaM-regulated processes. The increase in the DG content of the plasma membrane, along with the rise in $[Ca^{2+}]_c$, leads to the sensitivity modulation of C-kinase. Once activated, this enzyme remains activated at a lower $[Ca^{2+}]_c$. Activation of this C-kinase branch is sustained and is largely responsible for the sustained phase of cellular response. The flow of information through the two branches of the system leads to the integrated cellular response.

InsP$_3$ and DG (Fig.11.1).

The increase in the [InsP$_3$] leads to the release of a trigger pool of Ca^{2+} from the endoplasmic reticulum. This Ca^{2+} along with that entering the cell via a receptor-operated channel, leads to a transient rise in [Ca^{2+}]$_c$ and thereby to the activation of calmodulin-regulated enzymes. One of these is the plasma membrane Ca^{2+} pump which is activated and plays a major role in bringing about the subsequent fall in the [Ca^{2+}]$_c$. Hence, the activation of this calmodulin branch of the calcium messenger system is transient and is largely responsible for only the initial phase of cellular response.

The increase in the DG content of the plasma membrane along with the transient rise in the [Ca^{2+}]$_c$ lead to the association of the C-kinase with the plasma membrane at which site it undergoes positive sensitivity modulation to its highly Ca^{2+}-sensitive form. In addition, it is exposed to a specific cellular domain of Ca^{2+}, the submembrane domain, the concentration of Ca^{2+} in which is maintained by the high rate of cycling across the plasma membrane. Activation of this C-kinase branch of the calcium messenger system is largely responsible for the sustained phase of cellular response. The temporal integration of events in the two branches leads to the observed cellular response.

One can bypass the receptor-mediated events and activate the two branches separately by the use of A23187 and TPA, respectively. The simultaneous use of these two agents produces an integrated response which closely mimics the response induced by the natural agonist.

Work in several other systems (Delbeke et al., 1985; Rasmussen et al., 1985; Zawalich, Brown and Rasmussen, 1983) provides support for this model. Both exocrine secretion from pancreatic acini (dePont and Fleuren-Jakobs, 1984) and prolactin secretion from pituitary cells (Delbeke et al., 1985) are induced by combined ionophore and TPA. Likewise, vascular smooth muscle contraction can be induced by TPA in a Ca^{2+}-dependent and reversible manner (Rasmussen et al., 1985). The secretion of insulin from rat pancreatic islets is an additional process in which the two branches appear to operate (Zawalich, Brown and Rasmussen, 1983). These studies in islets show a particular striking temporal separation of the roles of the two branches. Perifusion of islets with 7–10 mM glucose induces a biphasic insulin secretory response: an immediate sharp rise, an equally rapid fall of 50% or more, and then a slowly developing and sustained rate of insulin secretion. Perifusion with either A23187 or tolbutamide (in the absence of extracellular glucose) leads to only a first phase response equal or greater in mangnitude to that seen after glucose, but no second phase. Perifusion with TPA or OAG leads to no first phase, but a slowly developing and submaximal second phase. Perifusion with combined A23187 (or tolbutamide) and TPA leads to a biphasic secretory response qualitatively similar to that seen after glucose treatment.

It is known that agents which induce a rise in the cAMP content of islets enhance the effectiveness of a given increase in glucose concentration, and also that glucose in high concentration (8–10 mM or greater) induces cAMP formation (Hedeskov, 1980). By combining forskolin, a known activator of adenylate cyclase, with A23187 and TPA, it is possible to induce biphasic insulin secretory response qualitatively and quantitatively similar to that evoked by an increase in glucose concentration from 2.8 to 10 mM (a maximal physiological concentration). Thus, in this tissue, both cAMP-mediated process and the C-kinase branch of the calcium messenger system provide a means of achieving gain control.

11.7. Perspective

In discussing the applicability of our current model to a wide variety of cellular systems, it is important to point out that clear organizational differences exist in the way the calcium messenger system is organized in different cells. These differences imply a functional and temporal difference in the link between the calmodulin and C-kinase branches of the system.

A comparison of the properties of platelet activation by thrombin (Kaibuchi et al., 1983), and adrenal glomerulosa cells by angiotensin (Kojima et al., 1985) illustrate this type of difference. From the biological point of view, the platelet release reaction must occur rapidly (s or ms) and involves a large number of platelets, but must also be very tightly controlled by feedback loops both negative and positive. Activation of aldosterone secretion need not occur rapidly. It takes place rather slowly (5–15 min) but it must be capable of sustained response over a period of several hours or even under exceptional circumstances days. In the latter circumstance, a clear temporal role of the two branches of the calcium messenger system is apparent. It seems likely that in these cells, the C-kinase in the non-activated cell is located in the cytosol, and that its activation involves an association with the plasma membrane. In contrast, the C-kinase in platelets appears to be bound to the plasma membrane already. Hence, its activation by, for example, TPA, is nearly immediate even when the $[Ca^{2+}]_c$ does not change. In the platelets there is little temporal dissociation between events in the two branches; rather they provide supplemental information which is summated in the form of a synergistic response.

A similar arrangement appears to operate in the case of the neutrophil (Korchak et al., 1984; Pozzan et al., 1983). In this case, the two branches of the system appear to mediate separate phases of the neutrophil response to a single agonist. At low concentrations of the chemotactic peptide, FMLP, a rise in intracellular $[Ca^{2+}]$ occurs which mediates the movement of the cell, but no activation of C-kinase takes place, and hence no activation of superoxide production is seen. At higher concentrations, FMLP induces an activation of the C-kinase pathway and superoxide production. The latter is immediate and appears to be accounted for by the fact that the C-kinase and the system involved in superoxide production are already associated in the plasma membrane and the membranes of the phagocytic vesicles or granules.

Thus, it is already apparent that there are several ways in which the two branches of the calcium messenger system can be organized in mediating either rapid but brief or slower and more sustained cellular responses. Considerably more work is necessary to explore the particular variations of this universal two-branch theme. Nonetheless, one can predict that in different cell types one may find several distinct modes by which the C-kinase branch of this system operates.

Acknowledgement

This chapter was supported by grants from the National Institutes of Health (AM 19813 and HD 11487).

References

Akerman, K.E.O. and Nicholls, D.G. (1982). Physiological and bioenergetic aspects of mitochondrial calcium transport. *Rev. Physiol. Biochem. Pharmacol.* **95**, 149–210.

Baker, P.F. (1976). The regulation of intracellular calcium. *SEB Symposium* **30**, 67–8.

Berridge, M.J. (1983). Rapid accumulation of inositol triphosphate reveals that agonists hydrolyze polyphosinositides instead of phosphatidylinositol. *Biochem. J.* **212**, 849–58.

Berridge, M.J., Dawson, R.M.C., Downes, P., Heslop, J.P. and Irvine, R.F. (1983). Changes in the levels of inositol phosphates after agonist-dependent hydrolysis of membrane phosphoinostides. *Biochem. J.* **212**, 473–83

Blinks, J.R., Wier, W.G., Hess, P. and Prendergast, F.G. (1982). Measurement of Ca^{2+} concentrations in living cells. *Prog. Biophys. Molec. Biol.* **40**, 1–114.

Borle, A.B. (1981). Control, modulation and regulation of cell calcium. *Rev. Physiol. Biochem. Pharmacol.* **90**, 13–153.

Borle, A.B. and Snowdowne, K.W. (1982). Measurement of intracellular free calcium in monkey kidney cells with aequorin. *Science* **217**, 252–4.

Campbell, A.J. (1983). *Intracellular Calcium. Its Universal Role as Regulator.* John Wiley and Sons, Ltd., Chichester.

Castagna, M., Takai, Y., Kaibuchi, K., Sano, K., Kikkawa, U. and Nishizuka, Y. (1982). Direct activation of calcium-activated, phospholipid-dependent protein kinase by tumor promoting phorbol esters. *J. Biol. Chem.* **257**, 7847–51.

Cheung, W.Y. (1980). Calmodulin plays a pivotal role in cellular regulation. *Science* **207**, 19–27.

Cox, J.M., Comte, M., Malnoë, A., Burger, D. and Stein, E.A. (1985). Mode of action of the regulatory protein calmodulin.

Delbeke, D., Kojima, I., Dannies, P.S. and Rasmussen, H. (1985). Synergistic stimulation of prolactin release by phorbol ester, A23187, and forskolin. *Biochem. Biophys. Res. Comm.*

dePont, J.J.H.H.M. and Fleuren-Jakobs, A.M.M. (1984). Synergistic effect of A23187 and phorbol ester on amylase secretion from rabbit pancreatic acini. *FEBS Lett.* **170**, 64–8.

Downes, P. and Michell, R.H. (1982). Phosphatidylinositol 4-phosphate and phosphatidylinositol 4,5-bis-phosphate: lipids in search of a function. *Cell Calcium* **3**, 467–502.

Drust, D.S. and Martin, T.F.S. (1982). Thyrotropin-releasing hormone rapidly and transiently stimulate cytosolic calcium-dependent protein phosphorylation in GH_3 pituitary cells. *J. Biol. Chem.* **257**, 7566–73.

Fain, J.N. and Berridge, M.J. (1979). Relationship between hormonal activation of phosphatidylinositol hydrolysis, fluid secretion and calcium flux in the blowfly salivary gland. *Biochem. J.* **178**, 45–58.

Feinstein, M.B., Egan, J.J., Sha'afi, R.I. and White, J. (1983). The cytoplasmic concentration of free calcium in platelets is controlled by stimulators of cyclic AMP production (PGD_2, PGE, and Forskolin). *Biochem. Biophys. Res. Comm.* **113**, 598–604.

Fleckenstein, A. (1983). *Calcium Antagonism in Heart and Smooth Muscle*, New York, John Wiley and Sons.

Foster, R. and Rasmussen, H. (1983). Angiotensin-mediated calcium efflux from adrenal glomerulosa cells. *Am. J. Physiol.* **245** (*Endocrinol. Metab.* **8**), E281–8.

Gershengorn, M.C. and Thaw, C. (1983). Calcium influx is not required for TRH to elevate free cytoplasmic calcium in GH_3 cell. *Endocrinology* **113**, 1522–4.

Hansford, R.G. and Castro, F. (1982). Intramitochondrial and extramitochondrial free calcium ion concentrations of suspensions of heart mitochondria with very low, plausibly physiological, contents of total calcium. *J. Bioeng. Biomemb.* **14**, 361–76.

Hedeskov, C.K. (1980). Mechanism of glucose-induced insulin secretion. *Physiol. Rev.* **60**, 442–509.

Hokin, L.E. and Hokin, M.R. (1955). Effects of acetylcholine on the turnover of phosphoryl units in individual phospholipids of pancreas slices and brain cortex slices. *Biochim. Biophys. Acta* **18**, 102–10.

Irvine, R.F., Dawson, R.M.C. and Freinkel, N. (1982). Stimulated phosphatidylinositol turnover. A brief appraisal. In *Contemporary Metabolism, Vol. 2.* N.M. Freinkel (ed.), pp. 301–42. New York, Plenum Press.

Kaibuchi, K., Sano, K., Hoshijima, M., Takai, Y. and Nishizuka, Y. (1982). Phosphatidylinositol turnover in platelet activation; calcium mobilization and protein phosphorylation. *Cell Calcium* **3**, 323–35.

Kaibuchi, K., Takai, Y., Sawamura, M., Hoshijima, M., Fujikura, T. and Nishizuka, Y. (1983). Synergistic functions of protein phosphorylation and calcium mobilization in platelet activation. *J. Biol. Chem.* **258**, 6701–4.

Kojima, I., Lippes, H., Kojima, K. and Rasmussen, H. (1983). Aldosterone production: Effect of A23187 and TPA. *Biochem. Biophys. Res. Comm.* **116**, 555–62.

Kojima, I., Kojima, K., Kreutter, D. and Rasmussen, H. (1985). The temporal integration of the aldosterone secretory response to angiotensin occurs via two intracellular pathways. *J. Biol. Chem.*

Korchak, H.M., Vienne, K., Rutherford, L.E., Wildenfeld, C., Finkelstein, M.C. and Weissman, G. (1984). Stimulus response coupling in the human neutrophil. *J. Biol. Chem.* **259**, 4076–82.

Minakuchi, R., Takai, Y., Yu, B. and Nishizuka, Y. (1981). Widespread occurrence of calcium-activated phospholipid-dependent protein kinase in mammalian tissues. *J. Biochem.* **89**, 1651–4.

Morgan, J.P. and Morgan, K.G. (1983). Vascular smooth muscle: the first recorded Ca^{2+} transients. *Pflügers Arch.* **395**, 75–7.

Nicholls, D.G. (1978). The regulation of extramitochondrial free calcium ion concentration by rat liver mitochondria. *Biochem. J.* **176**, 463–74.

Niggli, V., Sigel, E. and Carafoli, E. (1982). The purified Ca^{2+} pump of human erythrocyte membranes catalyzes an electroneutral Ca^{2+}-H^+ exchange in reconstituted liposomal system. *J. Biol. Chem.* **257**, 2350–6.

Nishizuka, Y. (1983). A receptor-linked cascade of phospholipid turnover in hormone action. In *Endocrinology,* K. Shizume, H. Imura and N. Shimizu (eds.), pp. 15–24. Amsterdam, International Congress Series 598, Excerpta Medica.

Pozzan, T., Arslan, P., Tsien, R.Y. and Rink, T.J. (1982). Anti-immunoglobulin, cytoplasmic free calcium, and capping in lymphocytes-β. *J. Cell Biol.* **94**, 335–40.

Pozzan, T., Lew, D.P., Wollheim, C.B. and Tsien, R.Y. (1983). Is cytosolic ionized calcium regulating neutrophil activation? *Science* **221**, 1413–5.

Putney, J.W.Jr. and Biancri, C.P. (1974). Site of action of dantrolene in frog sartorius muscle. *J. Pharmacol. Exp. Ther.* **189**, 202–12.

Rasmussen, H. (1981). *Calcium and cAMP As Synarchic Messengers.* New York, John Wiley and Sons.

Rasmussen, H. and Barrett, P.Q. (1984). Calcium messenger system: An integrated view. *Physiol. Rev.* **64**, 938–84.

Rasmussen, H. and Waisman, D.M. (1982). Modulation of cell function in the calcium messenger system. *Rev. Physiol. Biochem. Pharmacol.* **95**, 111–48.

Rasmussen, H., Forder, J., Kojima, I. and Scriabine, A. (1985). TPA-induced contraction os isolated rabbit vascular smooth muscle. *Biochem. Biophys. Res. Comm.*

Rink, T.J., Smith, S.W. and Tsien, R.Y. (1982). Cytoplasmic free Ca^{2+} in human platelets: Ca^{2+} thresholds and Ca-independent activation for shape-change secretion. *FEBS Lett.* **148**, 21–6.

Rubin, R.P. (1982). *Calcium and Cellular Secretion.* New York, Plenum Press.

Scarpa, A., Brinley, F.J., Tiffert, T., and Dubyak, G.R. (1978). Metallochromic indicators of ionized calcium. *Ann. NY. Acad. Sci.* **307**, 86–112.

Smallwood, J., Waisman, D.M., Lafreniere, D. and Rasmussen, H. (1983). Evidence that the erythrocyte calcium pump catalyzes a $Ca^{2+}:nH^+$ exchange. *J. Biol. Chem.* **258**, 11092–7.

Streb, H., Irvine, R.F., Berridge, M.J. and Schulz, I. (1983). Release of Ca^{2+} from a non-mitochondrial store in pancreatic acinar cell by inositol-1,-4,5-triphosphate. *Nature (Lond.)* **306**, 67–9.

Studer, R.K. and Borle, A.B. (1983). Sex difference in cellular calcium metabolism of rat hepatocytes and in -adrenergic activation of glycogen phosphorylase. *Biochim. Biophys. Acta* **762**, 302–14.

Takai, Y., Kishimoto, A., Kawahara, Y., Minakuchi, R., Sano, K., Kikkawa, U., Mori, T., Yu, B., Kaibuchi, M.K. and Nishizuka, Y. (1981). Czlcium and phosphatidylinositol turnover as signally for transmembrane control of protein phosphorylation. *Adv. Cycl. Nucleotide Res.* **14**, 301–13.

Tsien, R.Y. (1981). A non-disruptive technique for loading calcium buffers and indicators into cells. *Nature (Lond.)* **290**, 527–8.

Tsien, R.Y., Pozzan, T. and Rink, T.J. (1981). Calcium homeostasis in intact lymphocytes: Cytoplasmic free calcium monitored with a new intracellular trapped fluorescent indicator. *J. Cell Biol.* **94**, 325–34.

Tsien, R.Y., Pozzan, T. and Rink, T.J. (1982). T-cell mitogens cause early changes in cytoplasmic free Ca^{2+} and membrane potential in lymphocytes. *Nature (Lond.)* **295**: 68–71.

Vincenzi, F.F., Hinds, T.R. and Raess, B.U. (1980). Calmodulin and the plasma membrane calcium pump. *Ann. NY Acad. Sci.* **356**, 232–44.

Wang, J.H., Sharma, R.K., Huang, C.Y., Chau, V. and Chock, P.B. (1980). On the mechanism of activation of cyclic nucleotide phosphodiesterase by calmodulin. *Ann. NY Acad. Sci.* **356**, 190–204.

Wolheim, C.B. and Sharp, G.W.G. (1981). Regulation of insulin release by calcium. *Physiol. Rev.* **61**, 914–73.

Zawalich, W., Brown, C. and Rasmussen, H. (1983). Insulin secretion: Combined effect of phorbol ester and A23187. *Biochem. Biophys. Res. Comm.* **117**, 448–55.

INTRACELLULAR MECHANISMS IN THE REGULATION OF CALCIUM METABOLISM BY SECRETAGOGUES IN EXOCRINE GLANDS

I. Schulz, H. Streb, K. Imamura, E. Bayerdörffer, T. Kimura

Max-Planck-Institut für Biophysik, 6000 Frankfurt (MAIN), West Germany

Abstract

Calcium is an important intracellular messenger for stimulation of enzyme secretion from the exocrine pancreas. Secretagogues induce Ca^{2+} release followed by Ca^{2+} influx and Mg-ATP-dependent Ca^{2+} uptake into intracellular stores. In isolated permeabilized rat pancreatic acinar cells, non-mitochondrial Ca^{2+} pools buffered the free $[Ca^{2+}]$ of the medium at a steady state of 4×10^{-7} mol/l. In isolated rough endoplasmic reticulum (RER) the Ca^{2+} uptake characteristics were the same as in 'leaky cells' suggesting that the RER plays an important role in the regulation of cytosolic free $[Ca^{2+}]$. Recently we have obtained evidence that the intracellular messenger for secretagogue-induced Ca^{2+} release is inositol-1,4,5-trisphosphate (IP_3) which is accumulated during receptor-activated hydrolysis of phosphatidylinositol-4,5 bisphosphate in a variety of tissues. In isolated subfractions of acinar cells IP_3 released Ca^{2+} only from the ER. Summarizing we propose the following model for stimulus–secretion coupling: secretagogue --→ phosphatidylinositol-4,5-bisphosphate↓ --→ inositol-1,4,5-trisphosphate↑ --→ Ca^{2+} release from ER --→ cytosolic $[Ca^{2+}]$↑ --→ enzyme secretion.

12.1. Introduction

Enzyme and fluid secretion from exocrine glands is stimulated by neurotransmitters and hormones via an increase in the cytosolic free calcium and/or cyclic adenosine 5'monophosphate (cAMP) concentrations (Gardner, 1979; Petersen, 1982; Schulz, 1980; Williams, 1980). In the exocrine pancreas the importance of cAMP for enzyme secretion varies in different animal species, but calcium is clearly the main intracellular trigger (Gardner, 1979; Petersen, 1982; Schulz, 1980). Calcium also mediates an isotonic fluid secretion from pancreatic acinar cells with plasma-like electrolyte composition (Petersen and Ueda, 1976), whereas an isotonic fluid secretion rich in $NaHCO_3$ is stimulated in pancreatic duct cells by mediation of cAMP (Case, Harper and Scratcherd, 1968). In salivary glands enzyme secretion is mediated by cAMP, whereas an intracellular rise of cytosolic free Ca^{2+} concentration mainly elicits electrolyte and fluid secretion and also, to a limited extent only, enzyme release (Butcher and Putney, 1980).
  Similarly cAMP can stimulate enzyme secretion in lacrimal glands, whereas

electrolyte and water secretion is mediated by calcium (Dartt et al., 1984). In sweat glands no difference in electrolyte composition of the secretory fluid between cholinergic and β-adrenergic stimulation has been found, indicating that with both agonists the same mechanism for electrolyte and fluid secretion is stimulated. Alternatively cAMP might mobilize intracellular Ca^{2+}, which is the final trigger of exocytosis (Sato, 1982). However, cyclic AMP-induced Ca^{2+} release has not yet been observed in any system.

In the pancreas both cAMP and Ca^{2+}-dependent pathways have been investigated in more detail than in any other gland. It is likely that the cellular mechanisms by which these intracellular messengers regulate secretion in other exocrine or in endocrine glands, are very similar. We therefore use the exocrine pancreas as a model system to describe regulatory mechanisms involved in stimulus–secretion coupling.

Secretagogue-induced increase in cytosolic free Ca^{2+} concentration of pancreatic acinar cells is due to both Ca^{2+} release from intracellular organelles as well as Ca^{2+} influx into the cells (Schulz, 1980).

In this report we describe the role of Ca^{2+} in enzyme secretion and cellular events which regulate the cytosolic free Ca^{2+} concentration in pancreatic acinar cells. This includes the characterization of the site and mechanism of hormone-induced Ca^{2+} release and Ca^{2+} uptake into intracellular structures. The data suggest that both Ca^{2+} uptake and secretagogue-induced Ca^{2+} release occur in the endoplasmic reticulum. The intracellular messenger for secretagogue action is most likely inositol 1,4,5-trisphosphate (IP_3) (Streb et al., 1983), which is accumulated in pancreatic acini during receptor-activated hydrolysis of phosphatidylinositol-4,5-bisphosphate (Rubin et al., 1984).

12.2. Materials and methods

12.2.1. *Isolation and permeabilization of acinar cells*
Pancreatic acinar cells were prepared by controlled digestion of pancreatic tissue from rats (Amsterdam and Jamieson, 1972; Kondo and Schulz, 1976) using collagenase (Worthington, Freehold, NJ, USA). Isolated cells were then permeabilized by treatment with saponin (Wakasugi et al., 1982) or by washing cells twice in a nominally 'Ca^{2+}-free solution' (Streb and Schulz, 1983). Both methods rendered cells leaky as judged by trypan blue uptake into 90% of cells and release of the cytosolic enzyme lactate dehydrogenase (LDH) by 80% of total LDH present in cells before treatment with saponin or washing in 'Ca^{2+}-free solution' (Streb and Schulz, 1983).

12.2.2. *Labelling of secretory proteins by ^3H-leucine*
Pancreatic tissue (about 4 g wet weight) from four Wistar rats which had been fasted overnight was cut into several pieces (~1 mm^3) and preincubated for 30 min in 15 ml minimal essential medium (MEM, Eagle, 1x) with Earlie's salt without both L-leucine and L-glutamine to which 5 μCi/ml L-^3H-leucine had been added. ^3H-Leucine that had not been incorporated into the tissue was removed by washing the tissue three times in Krebs–Ringer Henseleit Hepes buffer (KRH). Preparation of cells was then followed as described.

12.2.3. *Measurement of free Ca^{2+} concentration by a Ca^{2+} electrode*
The free Ca^{2+} concentration of the incubation medium was recorded continuously with a Ca^{2+}-specific electrode (Affolter and Sigel, 1979), using Ca^{2+}-selective membranes which contained the neutral carrier N,N'-di ((11-ethoxycarbonyl) undecyl)-N,N'-4,5-tetramethyl-3,6-dioxactane amide

(Glasblaeserei W. Moeller, Zurich, Switzerland). Electodes were calibrated at free Ca^{2+} concentrations between 6×10^{-8} and 10^{-2} mol/l using the chelators nitrilotriacetic acid (NTA) or ethylene-diamine-tetra-acetic acid (EDTA) as described previously (Streb and Schulz, 1983).

12.2.4. Measurement of $^{45}Ca^{2+}$ uptake

Isolated acinar cells or membrane vesicles were incubated in a 130 mmol/l KCl-Hepes buffer at different $^{45}Ca^{2+}$ concentrations. At given time points, triplicate samples were filtered through polycarbonate or cellulose nitrate filter, the filters were washed and the radioactivity was quantitated by liquid scintillation counting (Bayerdörffer et al., 1984).

12.2.5. Preparation of subcellular fractions

Isolated acinar cells were homogenized in an isotonic mannitol buffer, the homogenate was centrifuged at 1000 g and 11 000 g for 12 and 15 min, respectively, and the resulting supernatant for 15 min at 27 000 g. The pellet of the last centrifugation step was mixed with 11% Percoll in mannitol buffer, and a density gradient was formed by spinning the tubes at 41 000 g for 40 min as described previously (Bayerdörffer et al., 1984). Purified rough endoplasmic reticulum (rer) was localized in the most dense fractions with an average density of 1.055 g/cm^3.

Alternatively a fluffy layer on the 11 000 g pellet was removed and the remaining pellet was subfractionated at 7 000 g for 15 min. The supernatant was separated, the pellet was resuspended and both fractions were spun down at 18 000 g for 15 min.

For preparation of membranes enriched in the plasma membrane marker ($Na^+ + K^+$)-ATPase, the fluffy layer of the 11 000 g pellet (see above) was brought to 1.25 mol/l of sucrose, and the sample was layered over a 2 mol/l sucrose cushion and overlayered with 0.3 mol/l sucrose. The step gradient was centrifuged at 150 000 g for 90 min, and the material banding at the interfaces between 0.3 M and 1.25 M sucrose ('S 1' enriched in plasma membranes) and 1.25 M–2 M sucrose ('S 2' enriched in endoplasmic reticulum) were collected. Alternatively total homogenate was suspended in a 280 mmol/l mannitol–Hepes buffer and $MgCl_2$ (11 mmol/l) was added. It was then spun at 400 g for 10 min, and after resuspension the resulting supernatant and precipitate were spun down at 25 000 g for 15 min.

12.2.6. Preparation and labelling of inositol 1,4,5-trisphosphate

Inositol 1,4,5-trisphosphate (IP_3) was prepared by incubating human red blood cell ghosts with $CaCl_2$ followed by a Dowex-formate column separation, and desalted by elution from a Dowex-Cl column with 1 M LiCl, followed by removal of the LiCl with ethanol (Downes, Mussat and Michell, 1982). ^{32}P-labelled IP_3 was prepared from human erythrocytes by the method of Downes, Mussat and Michell from ^{32}P-prelabelled erythrocyte ghosts (Downes and Michell, 1981). Both IP_3 and $I^{32}P_3$ were obtained from Dr. R.F. Irvine, Cambridge, U.K.

12.2.7. Protein-phosphorylation and protein-separation by polyacrylamide gel elctrophoresis

The phosphorylation procedure was similar to that described by Amory, Foury and Goffeau (1980). The phosphorylation reaction was started by adding purified rough endoplasmic reticulum membranes to a Hepes–Tris buffer containing 10 μCi of [$\gamma^{32}P$] ATP (5×10^{-6} mol/l) and ions as indicated. Free Ca^{2+} and Mg^{2+} concentrations were adjusted with 3 mmol/l EGTA or EDTA as described previously (Streb and Schulz, 1983). At different incubation times the reaction was stopped by addition of 10% trichloracetic acid (TCA) containing 10 mmol/l KH_2PO_4 and 1 mmol/l ATP and washed subsequently with the same solution, followed by twice washing with a solution of 50 mmol/l sucrose with 50 mmol/l tetradecyltrimethylammonium bromide (TDAB) and submitted to polyacrylamide gel elctrophoresis at acidic pH as described previously (Amory, Foury and Goffeau, 1980). After electorphoresis, the gel was incubated in 1% (v/v) glycerol, dried and exposed to Kodak X-Omat AR 5 film for 15–48 h at -70°C. ^{32}P-labelled proteins were excised from gels and their radioactivity was counted in a Mark III liquid scintillation system, Model 6880.

12.2.8. Measurement of Ca^{2+}-(Mg^{2+})–ATPase activity

Adenosine triphosphatase activity was measured according to the method of Bais (1975). Briefly, after stopping the phosphorylation reaction by addition of 10% TCA, samples were centrifuged for 5 min at 3500 rpm at 4°C. Ten μl aliquots of the resulting supernatant were mixed with 500 μl of a charcoal suspension (125 mg of activated charcoal/ml 1N HCl) and centrifuged at 2500 g for 10 min at 4°C. Aliquots of the supernatant were counted for radioactivity in a liquid scintillation system (Mark III, Model 6880). Radioactivity obtained in the absence of membranes were subtracted from each sample. Ca^{2+}-dependent $^{32}P_i$ liberation was obtained after subtraction of the value obtained in the absence of Ca^{2+}. Liberated $^{32}P_i$ is expressed as pmol/mg membrane protein x time of incubation.

12.3. Results

12.3.1. Effects of free $[Ca^{2+}]$ and cAMP on ^3H protein release

As shown in Fig.12.1, ^{3}H-protein release from isolated permeabilized pancreatic acinar cells increases with increasing free Ca^{2+} concentration in the surrounding medium, suggesting a direct function of Ca^{2+} in exocytosis. In the presence of both Ca^{2+} and cAMP enzyme release increases to higher levels than compared to the sum of effects obtained by each stimulant alone. This suggests that both cAMP and Ca^{2+} act on the same target and do not stimulate independent secretory pathways. If cAMP would act by releasing Ca^{2+} from intracellular stores one would not expect additive effects of cAMP at maximally effective free Ca^{2+} concentrations.

12.3.2. Determination of the steady state free $[Ca^{2+}]$ in leaky acinar cells

Figure 12.2 shows that leaky acinar cells take up Ca^{2+} from a medium that contains ATP and a regenerating system until a steady state of ~4 x 10^{-7} mol/l free $[Ca^{2+}]$ is reached. In the presence of mitochondrial inhibitors the same steady state is reached as in the control. In the presence of vanadate which inhibits Ca^{2+}-(Mg^{2+})-ATPases but does not affect mitochondria (Streb and Schulz, 1983) the initial rate of Ca^{2+} uptake is the same as in the control but the steady state $[Ca^{2+}]$ is not reached. This suggests that a non-mitochondrial Ca^{2+} pool regulates the free $[Ca^{2+}]$ at 4 x 10^{-7} mol/l.

12.3.3. Into what non-mitochondrial Ca^{2+} pool is Ca^{2+} transported?

The site of Ca^{2+} uptake has been localized by electron microscopy, showing Ca^{2+} oxalate precipitates in the rough endoplasmic reticulum (Wakasugi et al., 1982). These precipitates were absent when leaky acinar cells had been

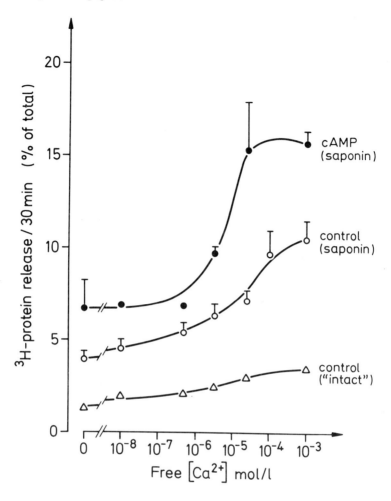

Fig.12.1. Effect of different free Ca^{2+} concentrations and cAMP (2×10^{-3} mol/l) on ^{3}H-protein release from isolated saponin-treated and saponin-untreated pancreatic acinar cells. Secretory enzymes were labelled with L-^{3}H-leucine as described. Cells were preincubated for 40 min in the presence of 45 μg/ml saponin (Sigma, St Louis, MO, USA) and indicated free Ca^{2+} concentrations. Incubation was for 30 min in the presence of saponin, cAMP and [Ca^{2+}] as indicated. Trypan blue uptake in saponin permeabilized cells was 70–80%, in untreated cells it was 3–5%. Points are mean values ± S.E. from at least three experiments and without S.E. from two experiments for saponin treated cells and from one experiment for 'intact' cells.

incubated without Mg-ATP or in the presence of the Ca^{2+} ionophore A23187.

This observation strongly suggests that the rough endoplasmic reticulum is the non-mitochondrial Ca^{2+} pool into which Ca^{2+} is taken up. Further support for this assumption came from studies in which the Ca^{2+} uptake properties into isolated leaky cells and into purified rough endoplasmic reticulum were compared (Bayerdörffer et al., 1984). All parameters tested such as [Ca^{2+}], [Mg^{2+}], cation and anion, ATP and pH dependencies of Ca^{2+} uptake were very similar for both preparations (Bayerdörffer et al., 1984).

12.3.4. Properties of the Ca^{2+} uptake mechanism

In the same preparation of rough endoplasmic reticulum which shows

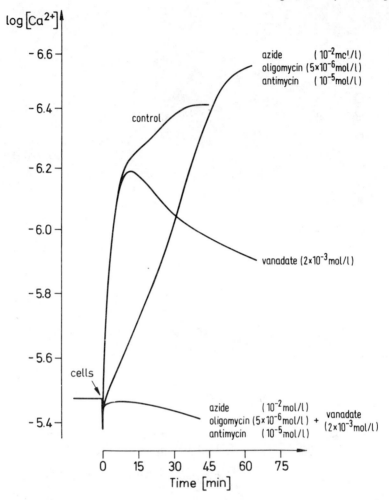

Fig.12.2. Effect of vanadate and mitochondrial inhibitors on Ca^{2+} uptake into leaky acinar cells. Cells were rendered leaky by washing them twice in a nominally 'Ca^{2+}-free solution' and were then incubated in a standard incubation buffer with 120 mmol/l KCl, 10 mmol/l Mg-ATP and an ATP regenerating system (Streb and Schulz, 1983) in the presence of indicated substances (from Streb and Schulz, 1983).

Mg-ATP-dependent Ca^{2+} uptake we could show the presence of a Ca^{2+}-(Mg^{2+})-ATPase and a phosphorylated intermediate of this enzyme, which were studied in more detail. Both the Ca^{2+}-(Mg^{2+})-ATPase and P_i incorporation were Ca^{2+} dependent in the same range of free $[Ca^{2+}]$ as for Ca^{2+} uptake into leaky cells and isolated rough endoplasmic reticulm with an optimum at 10^{-6} mol/l and decrease with higher free Ca^{2+} concentrations (Fig.12.3). The underlying mechanism for the observation that Ca^{2+} uptake into leaky acinar cells and isolated rough endoplasmic reticulum is cation dependent in the sequence $Rb^+ > K^+ > Na^+ > Li^+ >$ choline (Bayerdörffer et al., 1984), seems to be the cation dependency of the dephosphorylation step in the Ca^{2+}-(Mg^{2+})-ATPase intermediate. By use of different incubation temperatures at 4°C and at room temperature (that is: 22°C–25°C) it could be shown that ^{32}P-incorporation at 4°C was independent of different cations. At higher temperatures, however, at which both phosphorylation and dephosphorylation becomes more rapid, phosphorylation is kept at a higher level when K^+ in the incubation medium is replaced by choline (Fig.12.4). Similarly P_i liberation as an expression of the ATPase activity is small in the presence of choline as compared to K^+ (data not shown). Taken together these findings strongly indicate that the rough endoplasmic reticulum is the Ca^{2+} pool into which Ca^{2+} is transported by means of a Ca^{2+}-(Mg^{2+})-ATPase which regulates the free $[Ca^{2+}]$ to 4×10^{-7} mol/l.

Another question is: how do neurotransmitters and hormones release intracellular Ca^{2+} and from which structure is Ca^{2+} released. Recently we have obtained evidence (Streb et al., 1983) that the intracellular messenger for secretagogue-induced Ca^{2+} release is inositol-1,4,5-trisphosphate (IP_3) which is

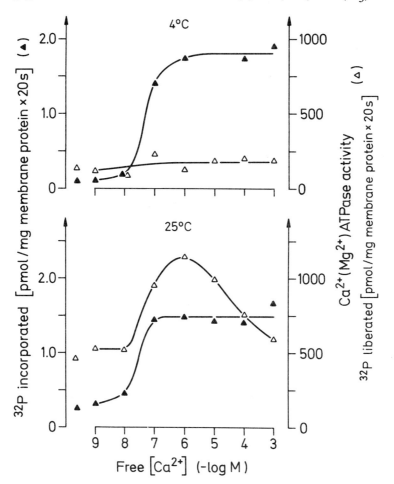

Fig.12.3. Dependence of ^{32}P-incorporation into the 130 kDa protein of rough endoplasmic reticulum (▲) and of Ca^{2+}-(Mg^{2+})-ATPase activity (Δ) on the free Ca^{2+} concentration. Incubation medium was Hepes–Tris buffer (pH 7.0) containing in mol/l: KCl 126×10^{-3}, free $[Mg^{2+}]$ 10^{-5}, $[\gamma^{32}P]$ATP 5×10^{-6}, ouabain 2×10^{-3}, NaN_3 5×10^{-3}, free $[Ca^{2+}]$ was adjusted with 3×10^{-3} mol/l EGTA (pH 7.0), 20 s incubation at 4°C and 25°C.

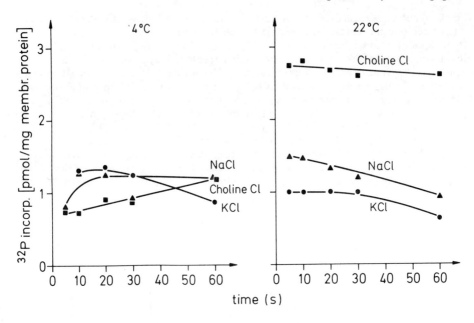

Fig.12.4. Dependence of ^{32}P-incorporation into the 130 kDa protein of rough endoplasmic reticulum on different cations at different temperatures. Free [Ca^{2+}] 10^{-6} mol/l, other conditions were as described in section 12.2 and in the legend to Fig.12.3.

accumulated during receptor-activated hydrolysis of phosphatidylinositol-4,5-bisphosphate in a variety of tissues (Abdel-Latif, Akhtar and Hawthorne, 1977; Kirk et al., 1981; Putney et al., 1983; Thomas et al., 1983).

12.3.5. *Ca^{2+} release from isolated leaky acinar cells by inositol 1,4,5-trisphosphate*
As shown in Fig.12.5 addition of inositol 1,4,5-trisphosphate (IP$_3$) to leaky acinar cells at steady state [Ca^{2+}] results in cellular Ca^{2+} release and transient increase in medium free Ca^{2+} concentrations followed by reuptake to the prestimulation value similar to the observation made with addition of secretagogues (Streb and Schulz, 1983). The dose-response relationship of IP$_3$-induced Ca^{2+} release showed maximal release at 5 x 10^{-6} mol/l IP$_3$ and an apparent K_m of 1.1 x 10^{-6} mol/l. The same result was obtained in the presence of mitochondrial inhibitors (Streb et al., 1983) suggesting that Ca^{2+} is released from a non-mitochondrial Ca^{2+} pool. If the acetylcholine analogue, carbachol, was added after IP$_3$, the carbachol-induced Ca^{2+} release was completely abolished. If, on the other hand, IP$_3$ was added after carbachol, the IP$_3$-induced release of calcium was decreased by about one-third, the sum of both effects, however, was constant (Streb et al., 1983). This suggests that both agents act on the same pool of releasable calcium.

12.3.6. *Inositol 1,4,5-trisphosphate production and breakdown in leaky acinar cells*
If IP$_3$ is a second messenger for the effects of secretagogues to release Ca^{2+} from an intracellular store, it should be accumulated during secretagogue stimulation in the same preparation of leaky cells from which addition of IP$_3$ causes Ca^{2+} release. As shown in Fig.12.6. stimulation of leaky acinar cells with carbachol results in increase in IP$_3$ as well as in the hydrolysis products

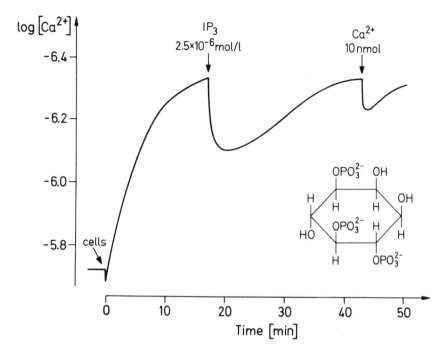

Fig.12.5. Effect of inositol 1,4,5-trisphosphate (IP$_3$, 2.5 μmol/l) on intracellular Ca^{2+} stores of leaky acinar cells. CaCl$_2$ (10 nmol) was added as calibration pulse where indicated (from Streb et al., 1983).

inositol 4,5-bisphosphate and inositol monophosphate. Figure 12.7 shows that the time-course of I^{32}P$_3$ breakdown which had been added to leaky cells coincides with the time-course of Ca^{2+} release. When I^{32}P$_3$ has reached its lowest value Ca^{2+} release stops and Ca^{2+} reuptake takes place.

If IP$_3$ is the second messenger for Ca^{2+} release, it should also be possible to release Ca^{2+} from isolated cell organelles by IP$_3$. This has indeed been demonstrated (Streb et al., 1984).

As shown in Fig.12.8 isolated endoplasmic reticulum responded to IP$_3$ (2.5 x 10^{-6} mol/l) with Ca^{2+} release in a way similar to isolated leaky cells (see Fig.12.5). There was either a negative correlation or none at all of IP$_3$-induced Ca^{2+} release and plasma membrane or mitochondrial marker enzymes (Streb et al., 1984).

12.4 Discussion

Our data show that the rough endoplasmic reticulum plays an important role in the regulation of cytosolic free Ca^{2+} concentration. The steady state value of 4 x 10^{-7} mol/l for [Ca^{2+}] as adjusted by the rough endoplasmic reticulum is close to the value for the cytosolic free Ca^{2+} concentration in intact cells at 1-2 x 10^{-7} mol/l as measured by the quin 2 method (Williams, 1984; and unpublished observations). It is therefore evident that the plasma membrane-bound extrusion mechanism regulates the final cytosolic free [Ca^{2+}] to the same or somewhat lower level than the steady state adjusted by the RER.

Ca^{2+} uptake into RER is promoted by a Ca^{2+}-(Mg^{2+})-ATPase with features

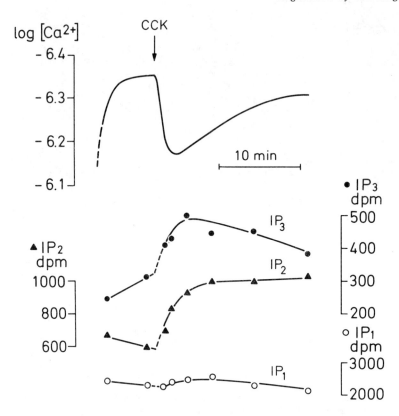

Fig.12.6. Release of Ca^{2+} and production of inositol trisphosphate (IP_3) inositol bisphosphate (IP_2) and inositol monophosphate (IP_1) in leaky acinar cells following stimulation with cholecystokinin (CCK, 3×10^{-6} mol/l), (from: Streb et al., 1985).

similar to that in the better investigated sarcoplasmic reticulum concerning the influence on monovalent cations and anions (Bayerdörffer et al., 1984; Duggan, 1977). In the presence of K^+, Ca^{2+} uptake into RER is highest and it is strongly inhibited when K^+ is replaced by choline (Bayerdörffer et al., 1984). Data on the phosphorylated intermediate of the Ca^{2+}-(Mg^{2+})-ATPase show that dephosphorylation is inhibited when K^+ is replaced by choline (see Fig.12.4). K^+ therefore seems to be rate limiting in the turnover of Ca^{2+}-(Mg^{2+})-ATPase in the Ca^{2+} transport cycle.

Our data provide strong evidence that inositol 1,4,5-P_3 functions as a second messenger for secretagogues to mobilize intracellular calcium from the endoplasmic reticulum, most likely from the same site into which Ca^{2+} is taken up. The events during stimulation could therefore include the following steps: secretagogues interact with their specific receptors on the plasma membrane. This induces hydrolysis of phosphatidylinositol-4,5-bisphosphate and an increase in the hydrolysis product inositol 1,4,5-trisphosphate. This increase causes release of Ca^{2+} from the endoplasmic reticulum. The increased cytosolic free Ca^{2+} concentration triggers enzyme secretion at a later step in stimulus–secretion coupling in which cAMP might also be involved. Detailed mechanisms for these events are not yet known.

It seems that during sustained stimulation of intact cells, after a transient

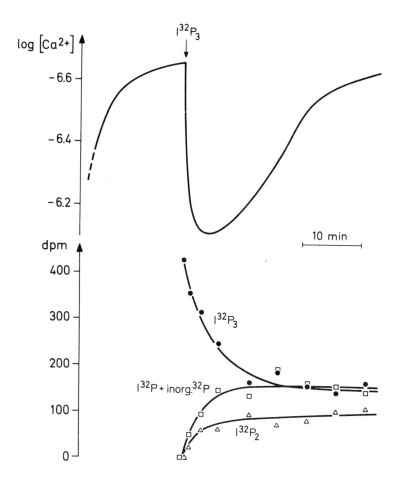

Fig.12.7. Time-course of ^{32}P-labelled inositol 1,4,5-trisphosphate (I^{32}P$_3$)-induced Ca^{2+} release and breakdown of added I^{32}P$_3$ in leaky acinar cells.

increase in the cytosolic free Ca^{2+} concentration, due to Ca^{2+} release from the endoplasmic reticulum, the steady state cytosolic [Ca^{2+}] is kept at a higher level than at rest due to increased Ca^{2+} influx into the cell (Kondo and Schulz, 1976), thus making sustained enzyme secretion possible (Petersen and Iwatsuke, 1978). Increased cytosolic [Ca^{2+}] also leads to increased Na$^+$ influx into the cell (Maruyama and Petersen, 1982a) which is likely to be the underlying mechanism for Ca^{2+}-regulated NaCl and fluid secretion (Petersen and Ueda, 1976). Increased Ca^{2+} influx, as seen in Ca^{2+}-tracer flux studies (Kondo and Schulz, 1976), does probably not involve IP$_3$ and its mechanism is not yet known. It is possibly, however, that Ca^{2+} influx into the cell occurs through the non-selective cation channel, which is opened up by increase in cytosolic Ca^{2+} in the plasma membrane of rat and mouse pancreatic acini (Maruyama and Petersen, 1982a and b). A leak of Ca^{2+} through this type of channel would be too small, however, to be noticeable even in high resolution current recording experiments (Petersen and Maruyama, 1983).

When stimulation ceases, Ca^{2+} influx decreases and the cytosolic [Ca^{2+}]

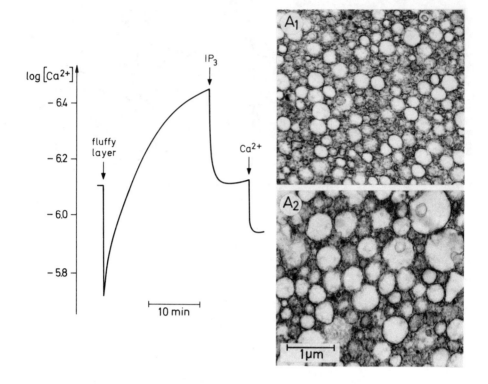

Fig.12.8. Inositol 1,4,5-trisphosphate (IP_3, 2.5×10^{-6} mol/l) induced Ca^{2+} release from isolated endoplasmic reticulum, obtained as 'fluffy layer' on an 11000 g pellet from centrifugation of total homogenate. For electronmicroscopy this 'fluffy layer' was resuspended and spun at 18000 g for 15 min and fixed as pellet with 5% (v/v) glutaraldehyde in 0.1 mol/l sodiumcacodylate buffer (pH 7.4) and postfixed with (1% w/v) osmic acid, dehydrated with alcohol, embedded in Spurr's resin, and stained with uranyl acetate and lead citrate. A_1 = upper layer, A_2 = basal layer of the 18000 g pellet. Where indicated $CaCl_2$ (10 nmol) was added (from Streb et al., 1984).

level is brought down to the prestimulation level by means of both active Ca^{2+} uptake into the rough endoplasmic reticulum and Ca^{2+} extrusion from the cell. The latter can be achieved by a Ca^{2+}-(Mg^{2+})-ATPase (Bayerdörffer et al., 1985; Kribben, Tyrakowski and Schulz, 1983) and a Ca^{2+}/Na^+ exchange mechanism located in the plasma membrane of pancreatic acinar cells (Bayerdörffer, in press).

References

Abdel-Latif, A.A., Akhtar, R.A. and Hawthorne, J.N. (1977). Acetylcholine increases the breakdown of triphosphoinositide of rabbit iris muscle prelabelled with (^{32}P) phosphate. Biochem. J. 162, 61–73.
Affolter, H. and Sigel, E. (1979). A simple system for the measurement of ion activities with solvent polymeric membrane electrodes. Anal. Biochem. 97, 315–9.
Amory, A., Foury, F. and Goffeau, A. (1980). The purified plasma membrane ATPase of the yeast schizosaccharomyces pombe forms a phosphorylated

intermediate. *J. Biol. Chem.* **255**, 9353–7.

Amsterdam. A. and Jamieson, J.D. (1972). Structural and functional characterization of isolated pancreatic exocrine cells. *Proc. Nat. Acad. Scio. USA* **69**, 3028–32.

Bais, R. (1975). A rapid and sensitive radiometric assay for adenosine triphosphatase activity using Cerenkov radiation. *Anal. Biochem.* **63**, 271–3.

Bayerdörffer, E., Eckhardt, L., Haase, W. and Schulz, I. (1985). Electrogenic calcium transport in plasma membrane of rat pancreatic acinar cells. *J. Membr. Biol.* **84**, 45–60.

Bayerdörffer, E., Streb, H., Eckhardt, L., Haase, W. and Schulz, I. (1984). Characterization of calcium uptake into rough endoplasmic reticulum of rat pancreas. *J. Membr. Biol.* **81**, 69–82.

Butcher, F.R. and Putney, J.W. (1980). Regulation of parotid gland function by cyclic nucleotides and calcium. *Adv. Cyclic Nucleotide Res.* **13**, 215–49.

Case, R.M., Harper, A.A. and Scratcherd, T. (1968). Water and electrolyte secretion by the perfused pancreas of the cat. *J. Physiol. (Lond.)* **196**, 133–49.

Dartt, D.A., Donowitz, M., Joshi, V.J., Mathieu, R.S. and Sharp, G.W.G. (1984). Cyclic nucleotide-dependent enzyme secretion in the rat lacrimal gland. *J. Physiol.* **352**, 375–84.

Downes, C.P. and Michell, R.H. (1981). The polyphosphoinositide phosphodiesterase of erythrocyte membranes. *Biochem. J.* **198**, 133–40.

Downes, C.P., Mussat, M.C. and Michell, R.H. (1982). The inositol trisphosphate phosphomonoesterase of the human erythrocyte membrane. *Biochem. J.* **203**, 169–77.

Duggan, P.F. (1977). Calcium uptake and associated adenosine triphosphatase activity in fragmented sarcoplasmic reticulum. *J. Biol. Chem.* **252**, 1620–7.

Gardner, J.D. (1979). Regulation of pancreatic exocrine function *in vitro*: initial steps in the action of secretagogues. *Ann. Rev. Physiol.* **41**, 55–66.

Kirk, C.J., Creba, J.A., Downes, P. and Michell, R.A. (1981). Hormone-stimulated metabolism of inositol lipids and its relationship to hepatic receptor function. *Biochem. Soc. Trans.* **9**, 377–9.

Kondo, S. and Schulz, I. (1976). Calcium ion uptake in isolated pancreas cells induced by secretagogues. *Biochim. Biophys. Acta* **419**, 76–92.

Kribben, A., Tyrakowski, T. and Schulz, I. (1983). Characterization of Mg-ATP-dependent Ca^{2+} transport in cat pancreatic microsomes. *Am. J. Physiol.* **244**, G480–90.

Maruyama, Y. and Petersen, O.H. (1982a). Single-channel currents in isolated patches of plasma membrane from basal surface of pancreatic acini. *Nature* **299**, 159–61.

Maruyama, Y. and Petersen, O.H. (1982b). Cholecystokinin activation of single-channel currents is mediated by internal messenger in pancreatic acinar cells. *Nature* **300**, 61–3.

Petersen, O.H. (1982). Stimulus–excitation coupling in plasma membranes of pancreatic acinar cells. *Biochim. Biophys. Acta* **694**, 163–84.

Petersen, O.H. and Iwatsuki, N. (1978). The role of calcium in pancreatic acinar cell stimulus–secretion coupling: an electrophysiological approach. *Ann. NY. Acad. Sci.* **307**, 599–617.

Petersen, O.H. and Maruyama, Y. (1983). What is the mechanism of the calcium influx to pancreatic acinar cells evoked by secretagogues? *Pflügers Arch.* **396**, 82–4.

Petersen, O.H. and Ueda, N. (1976). The importance of calcium for caerulein and secretin evoked fluid and enzyme secretion in the perfused rat pancreas. *J. Physiol. (Lond.)* **263**, 223P–4P.

Putney, J.W., Burgess, G.M., Halenda, S.P., McKinney, J.S. and Rubin, R.P.

(1983). Effect of secretagogues on (^{32}P) phosphatidylinositol 4,5-bisphosphate metabolism in the exocrine pancreas. *Biochem. J.* **212**, 483–8.

Rubin, R.P., Godfrey, P.P., Chapman, D.A. and Putney, J.W., Jr. (1984). Secretagogue-induced formation of inositol phosphates in rat exocrine pancreas. *Biochem. J.* **219**, 655–9.

Sato, K. (1982). Mechanism of eccrine sweat secretion. In *Fluid and Electrolyte Abnormalities in Exocrine Glands in Cystic Fibrosis*. P.M. Quinton, J.R. Martinez and U. Hopfer (eds.), pp. 35–52. San Francisco Press, San Francisco.

Schulz, I. (1980). Messenger role of calcium in function of pancreatic acinar cells. *Am. J. Physiol.* **239**, G335–47.

Streb, H., Bayerdörffer, E., Haase, W., Irvine, R.F. and Schulz, I. (1984). Effect of inositol-1,4,5-trisphosphate on isolated subcellular fractions of rat pancreas. *J. Membr. Biol.* **81**, 241–53.

Streb, H., Heslop, J.P., Irvine, R.F., Schulz, I. and Berridge, M.J. (1985). Relationship between secretagogue-induced Ca^{2+} release and inositol polyphosphate production in permeabilized pancreatic acinar cells. *J. Biol. Chem.* **260**, 7309–15.

Streb, H., Irvine, R.F., Berridge, M.J. and Schulz, I. (1983). Release of Ca^{2+} from a nonmitochondrial intracellular store in pancreatic acinar cells by inositol-1,4,5-trisphosphate. *Nature* **306**, 67–9.

Streb, H. and Schulz, I. (1983). Regulation of cytosolic free Ca^{2+} concentration in acinar cells of rat pancreas. *Am. J. Physiol.* **245**, G347–57.

Thomas, A.P., Marks, J.S., Coll, K.E. and Williamson, J.R. (1983). Quantitation and early kinetics of inositol lipid changes induced by vasopressin in isolated and cultured hepatocytes. *J. Biol. Chem.* **258**, 5716–25.

Wakasugi, H., Kimura, T., Haase, W., Kribben, A., Kaufmann, R. and Schulz, I. (1982). Calcium uptake into acini from rat pancreas: evidence for intracellular ATP-dependent calcium sequestration. *J. Membr. Biol.* **65**, 205–20.

Williams, J.A. (1980). Regulation of pancreatic acinar cell function by intracellular calcium. *Am. J. Physiol.* **238**, G269–79.

Williams, J.A. (1984). Regulatory mechanisms in pancreas and salivary acini. *Ann. Rev. Physiol.* **46**, 361–75.

MODULATION OF HUMAN PLATELET PHOSPHOLIPASE A_2 ACTIVITY

Leslie R. Ballou and Wai Yiu Cheung

Departments of Biochemistry, St. Jude Children's Research Hospital and University of Tennessee Center for the Health Sciences, Memphis, Tennessee, USA 38101

Abstract

Arachidonic acid serves as the sole substrate for prostaglandin biosynthesis. The intracellular level of free arachidonic acid is low; its release from phospholipids is a key control point in the biosynthesis of prostaglandins.

Phospholipase A_2 (PLA_2; phosphatide 2-acylhydrolase, EC 3.1.1.4) catalyses the hydrolysis of arachidonic acid from the 2 position of phosphatidylcholine, phosphatidylethanolamine, and to a lesser extent phosphatidylinositol and phosphatidylserine, producing free arachidonic acid and lysophospholipids. The enzyme requires Ca^{2+} for activity; half-maximum activation was obtained at approximately 1 μM Ca^{2+}.

The level of PLA_2 activity in platelet extracts is generally too low to account for the amount of arachidonic acid released in activated platelets. However, we have found that when a soluble or particulate fraction of human platelet homogenate was chromatographed on DEAE-cellulose and hydroxylapatite columns, the total PLA_2 activity increased sharply. PLA_2 activity detected in these partially purified preparations was 12 times greater than that associated with the original homogenate. Chromatography of the PLA_2 associated with the particulate fraction on a DEAE-cellulose column yielded one activity peak. Fractions eluted near the PLA_2 activity peak of the DEAE-cellulose column showed a trace of PLA_2 activity but inhibited PLA_2 activity stoichiometrically, suggesting the presence of an endogenous 'inhibitor' in the homogenate. The inhibitor appears to be associated mainly with a particulate fraction. Treatment of the inhibitor with trypsin, RNase, DNase or heat did not diminish its inhibitory activity, which was extractable by chloroform-methanol, indicating that the inhibitor is a lipid. Incubation of PLA_2 with phospholipids or various neutral lipids including saturated fatty acids had little or no effect on enzymic activity. In contrast, unsaturated fatty acids such as palmitoleic acid (16:1), oleic acid (18:1), linoleic acid (18:2), or arachidonic acid (20:4), all of which were detected in the particulate fraction, inhibited PLA_2 activity by 50% at approximately 5 x 10^{-7} M. The level of unsaturated fatty acids in the inhibitor fraction was equivalent to approximately 10^{-4} M, apparently sufficient to effectively inhibit PLA_2 activity. Inhibition by unsaturated fatty acids appears to be non-competitive with respect to phosphatidylcholine, the substrate of PLA_2; inhibition by unsaturated fatty acids was not reversed by excess Ca^{2+}.

The inhibition of PLA_2 by endogenous unsaturated fatty acids explains its apparent low level of activity in human platelet extract and its marked increase during the course of enzyme purification.

13.1. Introduction

In human platelets, as in other types of cells, arachidonic acid (20:4) serves as the sole substrate for the biosynthesis of prostaglandins and other prostanoid metabolites which possess a wide variety of biological activities. The intracellular level of free arachidonic acid is low, and its release from membrane phospholipids is a rate-determining step in the biosynthesis of prostanoids.

Several enzymatic pathways have been proposed to account for the rapid release of arachidonic acid following platelet activation. One involves a nonspecific phospholipase A_2 (PLA_2; phosphatide 2-acylhydrolase, EC 3.1.1.4) which catalyses the hydrolysis of arachidonic acid from the 2 position of phosphatidylcholine, phosphatidylethanolamine and phosphatidylinositol, producing free arachidonate and lysophospholipids. Another involves a phospholipase C specific for phosphatidylinositol, yielding a diglyceride which is then acted upon by a diglyceride lipase to produce glycerol and arachidonate (Bell et al., 1979). A third involves a specific phospholipase A_2 which catalyses the release of arachidonic acid from phosphatidic acid (Lapetina, Billah and Cuatrecasas, 1981). The relative contribution of each of these pathways to the release of arachidonate has yet to be determined but the dependence of the phospholipases for Ca^{2+} appears to be generally accepted (Feinstein and Sha'afi, 1983). It is believed that stimulation of the cell raises intracellular Ca^{2+}, which activates the phospholipases. However, other factors have been shown to affect phospholipase activity; they include various peptides (Argiolas and Pisano, 1983), and lipids (Ballou and Cheung, 1983, 1985). The phospholipases are presumably not activated in the unstimulated cell, and their activities would appear to be under stringent control mechanisms.

We have been interested in the role of Ca^{2+} in cell function and have extended our interest to a study on phospholipase A_2 of human platelets. Reviewed in this chapter are experiments showing that phospholipase A_2 activity is not fully expressed in human platelet extract and is inhibited by endogenous unsaturated fatty acids.

13.2. Materials and methods

13.2.1. Platelet preparation

Fresh human platelets were seperated from 20 units of platelet-rich plasma by centrifugation at 4850 g for 10 min at room temperature. The pelleted platelets, approximately 1.2×10^{12}, were resuspended in 200 ml of cold 50 mM Tris-HCl, pH 7.4, containing 1 mM ethylene glycol bis(β-aminoethyl ether)-N,N,N',N'-tetraacetic acid (EGTA) (buffer A). Subsequent operations were conducted at 0-4 °C. Then, 50 ml aliquots of the platelet suspension were sonicated for 1 min with a Branson sonifier (80 W) equipped with a microtip and then were centrifuged at 10 000 g for 10 min to remove cell debris (10 k pellet). The supernatant fraction was further centrifuged at 105 000 g for 1 h to give a high-speed supernatant fluid (HSS) and a 105 k high-speed pellet (HSP); these fractions were diluted in buffer A to a protein concentration of approximately 5 mg/ml.

13.2.2. Assay of PLA_2

PLA_2 activity was measured by the amount of [1-^{14}C] arachidonate released from L-α-1-palmitoyl 2-[1-^{14}C] arachidonyl phosphatidylcholine (54.5 mCi/mmol, New England Nuclear; 1 Ci = 3.7 x 10^{10} Bq). The substrate was dried under N_2 and dissolved in dimethyl sulfoxide (Me_2SO), and 10 µl (1.8 nmol) was added to a reaction mixture in a 12 x 75 mm polycarbonate tube (final volume of 100 µl) containing 50 mM Tris-HCl (pH 7.5 or 9.0 as indicated) and 100 µM $CaCl_2$. Incubations were at 30 °C for the times indicated. The reaction was terminated with 100 µl of 100% ethanol containing 2% acetic acid and 10 µg of arachidonic acid as carrier. After mixing, a 100 µl-fraction was applied to a silica gel G thin-layer chromatography (TLC) plate (5 x 20 cm, Analtech, Newark, DE) and developed in a solvent system of ethyl acetate and glacial acetic acid, 99:1 (v/v). Chromatography with authentic standards and autoradiography showed that free arachidonic acid migrated with the solvent front while the phospholipid remained at the origin; the only radioactive product of the reaction was arachidonic acid. After development of the TLC plate, the area corresponding to free arachidonate was scraped off and placed in a scintillation vial containing 500 µl of water. Then, 10 ml of an aqueous scintillant (ACS, Amersham, Bucks, UK) was added and the radioactivity was counted. One unit of activity is defined as 1 nmol of arachidonate released per min at 30°C.

13.3. Results

13.3.1. Inhibition of phospholipase A_2 activity in human platelet extract

Our initial study indicated a low level of PLA_2 activity in the extract of human platelets (Ballou and Cheung, 1983), (see Table 13.1). In separating the platelet extract into a high-speed supernatant (HSS) and a particulate fraction (HSP), we found that most of the activity was associated with the supernatant fraction. Chromatography on a DEAE-cellulose column and a hydroxylapatite column of the soluble and particulate fractions resulted in a 12-fold increase in total PLA_2 activity (see Table 13.1). More dramatically, the PLA_2 activity associated with the particulate fraction after column chromatography increased 140-fold. This indicated that a majority of the PLA_2 activity was actually associated with the particulate fraction. The marked increase of total PLA_2 activity after chromatography suggested the removal of an endogenous inhibitor during the course of enzyme purification. Mixing experiments indicated that the inhibitor was associated mainly with the particulate fraction. The presence of an inhibitor in human platelet extract suggests that the PLA_2 activity reported in the literature is probably underestimated.

Figure 13.1 shows the activity profile of PLA_2 of the particulate fraction on a DEAE-cellulose column. PLA_2 activity emerged from the column as a single peak. Fractions eluted from the column immediately preceeding the peak of PLA_2 activity (fractions 60–70, which contained low PLA_2 activity) possessed significant inhibitory activity. These fractions were pooled (designated as the inhibitor fraction) and analysed further.

Table 13.1. Demonstration of marked increase of human platelet PLA$_2$ activity during enzyme purification. Subcellular fractions from 6 x 10^{10} platelets were used in the purification; 23 mg of the 105 k HSP fraction, or 56 mg of the HSS fraction was applied to a DEAE-cellulose (Whatman DE-52) column (2 x 20 cm) equilibrated with 50 mM Tris-HCl, pH 8.0, containing 1 mM EGTA (buffer B). The column was washed with approximately 40 ml of buffer B and then eluted at a flow rate of 0.2 ml/min with a 160 ml linear NaCl gradient (0–0.5 M) in buffer B generated by an LKB 11300 Ultrograd gradient mixer; 2.1 ml fractions were collected and 20 μl of each fraction was assayed for PLA$_2$ activity. Active fractions were pooled and applied to a hydroxylapatite (Bio-Rad HT) column (3 x 8 cm) equilibrated with buffer B for further purification. This column was eluted with a 200 ml linear phosphate (K$_2$HPO$_4$) gradient (0-0.5 M) in buffer B at a flow rate of 0.2 ml/min; 1.6 ml fractions were collected; 20 μl of each fraction was assayed for PLA$_2$ activity at pH 9.0. Active fractions were pooled, and they could be stored at -20 °C in 10% (v/v) glycerol for several days without an appreciable loss of enzymatic activity. HSS, high-speed supernatant; HSP, high-speed pellet. (From Ballou and Cheung, 1983.)

Step	Protein mg	Total activity nmol/min	Yield %	Specific activity nmol/min per mg protein	Purification fold
Homogenate	99.0	3.7	100	0.04	1
10 000 g					
supernatant	80.5	2.8	75	0.03	1
pellet	15.1	0.3	8	0.02	—
105 000 g					
HSS	56.0	2.0	54	0.04	1
HSP	23.0	0.3	8	0.01	—
DEAE-cellulose					
HSS	14.2	8.4	227	0.6	15
HSP	8.0	8.0	216	1.0	25
Hydroxylapatite					
HSS	6.3	3.7	100	0.6	15
HSP	1.3	42.3	1143	32.5	812

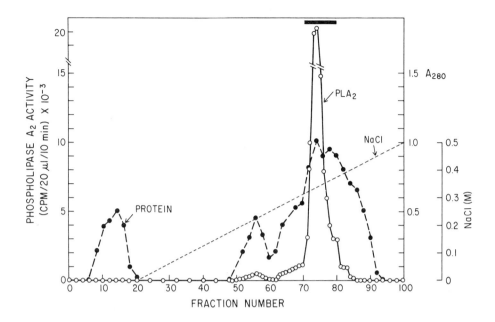

Fig.13.1. Elution profile of PLA$_2$ activity from a DEAE-cellulose column. The high-speed pellet from 6 x 10^{10} platelets was applied to the column which was eluted as described in the legend to Table 13.1. (From Ballou and Cheung, 1983.)

13.3.2. *Nature of the inhibitor*

To assess the nature of the inhibitor, we subjected aliquots of the inhibitor fraction to various treatments (Table 13.2). The inhibitory activity was insensitive to boiling, and to treatment with RNase, DNase, or trypsin, suggesting that the inhibitor was not associated with a protein or a nucleic acid-like substance (platelets are devoid of RNA or DNA). Extracting a sample of the inhibitor fraction with chloroform : methanol showed that all of the inhibitory activity was recovered in the organic phase, suggesting that the active component was associated with a lipid(s).

Analysis of the inhibitory fraction by thin-layer chromatography revealed the presence of phosphatidylcholine, phosphatidylethanolamine, phosphatidylinositol, phosphatidylserine and several neutral lipids, the majority being cholesterol. The phospholipids inhibited PLA$_2$ activity, but only at high concentrations; an aliquot of the inhibitor fraction that effectively suppressed PLA$_2$ activity did not contain a sufficient quantity of phospholipids to account for the observed inhibition, suggesting that the inhibitory component was not associated with the phospholipids.

13.3.3. *Effect of neutral lipids on phospholipase A$_2$ activity*

Since phospholipids present in the inhibitor fraction could not account for the inhibitory activity, we examined the effects of various neutral lipids on PLA$_2$ activity (Table 13.3). Diglyceride, cholesterol, cholesterol ester, phosphatidic acid, and alcohols of saturated fatty acids had little or no effect. Saturated fatty acids such as palmitic (16:0), stearic (18:0), or behenic (22:0) inhibited PLA$_2$ activity by 25%, while arachidic acid (20:0) inhibited enzymatic activity 40%. More effective, however, were the unsaturated fatty

Table 13.2. Nature of the inhibitor from human platelets. A standard reaction mixture (pH 7.5) contained 100 ng of RNase, DNase, trysin or 20 µl of inhibitor fraction. Tubes D and E were boiled for 30 minutes and cooled to 30 °C prior to addition of PLA_2; tubes F, H and J containing inhibitor fraction were treated as indicated for 30 min at 22 °C prior to the addition of PLA_2. Tube K containing trypsin received 10 µg of trypsin inhibitor prior to the addition of PLA_2. Tube L contained 5 µg of denatured protein obtained from the aqueous phase of a chloroform-methanol extract of the inhibitor fraction, and tube M contained 5 µg of lipid from the organic phase of the extract. In all cases, the PLA_2 assay was initiated by the addition of labelled phosphatidylcholine. The values repesent the means ± SEM of triplicate asssays from two separate experiments. (From Ballou and Cheung, 1985.)

Additions	PLA_2 activity (%)
A. PLA_2	100
B. Inhibitor	27 ± 4
C. Inhibitor + PLA_2	41 ± 2
D. Inhibitor (boiled)	0
E. Inhibitor (boiled) + PLA_2	20 ± 4
F. Inhibitor (RNase treated) + PLA_2	42 ± 3
G. RNase + PLA_2	90 ± 4
H. Inhibitor (DNase treated) + PLA_2	40 ± 3
I. DNase + PLA_2	90 ± 4
J. Inhibitor (trypsin treated) + PLA_2	15 ± 1
K. Trypsin + trypsin inhibitor + PLA_2	94 ± 3
L. Inhibitor 'protein fraction' + PLA_2	108 ± 5
M. Inhibitor 'lipid fraction' + PLA_2	20 ± 2

acids such as palmitoleic (16:1), oleic (18:1), linoleic (18:2), arachidonic (20:4), docosatetraenoic (22:4) and docosahexaenoic (22:6); they completely abolished the PLA_2 activity. These results show that of the neutral lipids tested, the unsaturated fatty acids were the most effective inhibitors, and suggested that they may be the active component in the inhibitor fraction.

13.3.4. *Fatty acid composition of the inhibitor fraction*
To determine whether unsaturated fatty acids were the endogenous inhibitors, we isolated them by thin-layer chromatography. The fatty acids were methylated and analysed by gas chromatography. Figure 13.2 shows a representative elution profile of the fatty acid derivatives obtained from the inhibitor fraction. Approximately 45% of the sample was comprised of unsaturated fatty acids: 30% oleic acid (18:1), 4% linoleic acid (18:2), 4% arachidonic acid (20:4), 3% of an unidentified long-chain fatty acid, 3% palmitoleic acid (16:1), and less than 1% tricosanoic acid (20:3) and linolenic acid (18:3). Saturated fatty acids made up approximately 55%: 29% palmitic acid (16:0), 23% stearic acid (18:0), and 3% myristic acid. These data demonstrated that the inhibitor fraction contained the unsaturated fatty acids that effectively suppressed PLA_2 activity.

To determine the amount of unsaturated fatty acids in the inhibitor fraction we applied a particulate fraction (HSP) prepared from 1×10^{11} platelets to a DEAE-cellulose column (see Table 13.1 and Fig.13.1). The

Table 13.3. Effect of neurtral lipids on human platelet PLA$_2$ activity. A standard reaction mixture contained 1 mM of each neutral lipid, added in 10 µl Me$_2$SO. The reaction was initiated by the addition of labelled substrate. The values represent the means ± SEM of triplicate assays from two separate experiments. The numbers in parenthesis refer to carbon number and number of double bounds in the fatty acid. (From Ballou and Cheung, 1985.)

Additions	PLA$_2$ activity (%)
1. None	100
2. Cholesteryl pentadecanoate	105 ± 0.1
3. Tricosanol	100 ± 10.7
4. Diglyceride	98 ± 3.1
5. Behenyl alcohol	93 ± 5.5
6. Phosphatidic acid	91 ± 2.2
7. Cholesterol	89 ± 0.4
8. Tricosanoic acid (23:0)	88 ± 4.3
9. Cholesteryl heptadecanoate	87 ± 1.2
10. Behenic acid (22:0)	80 ± 3.4
11. Stearic acid (18:0)	75 ± 4.4
12. Palmitic acid (16:0)	75 ± 4.9
13. Arachidic acid (20:0)	61 ± 3.2
14. Palmitoleic acid (16:1)	0
15. Oleic acid (18:1)	0
16. Linoleic acid (18:2)	0
17. Linolenic acid (18:3)	0
18. Arachidonic acid (20:4)	0
19. Docosatetraenoic acid (22:4)	0
20. Docosahexaenoic acid (22:6)	0

inhibitor fraction was pooled from the column (fractions 60–70), and the fatty acids were isolated by thin-layer chromatography. From this procedure, we obtained approximately 100 µg of fatty acids, 42 µg of which were unsaturated fatty acids. If we assume that 1 ml of packed platelets contained 1 x 10^{11} cells and that most of the platelet lipid was associated with the particulate fraction, the value of 42 µg unsaturated fatty acid/1 ml packed platelets would be equivalent to a concentration of approximately 1 x 10^{-4} M (the average molecular weight of fatty acid was taken as 300). This level of unsaturated fatty acid would appear to be more than sufficient to suppress most, if not all, PLA$_2$ activity in vitro, since 5 x 10^{-7} M unsaturated fatty acid inhibited PLA$_2$ activity by 50% (see Fig.13.3). We should point out that this estimation probably does not accurately reflect their cellular concentrations; some fatty acids may have been released from the particulate fraction during the course of PLA$_2$ purification, giving an apparently higher value. Nevertheless, these findings strongly support the notion that the active components of the inhibitor fraction consisted of known unsaturated fatty acids.

To explore further the active components of the inhibitor fraction, a lipid extract of the inhibitor fraction was methylated (Table 13.4). Methylation not only rendered the inhibitor inactive, but caused it to become slightly stimulatory. Saponification of the methylated sample to remove the methyl

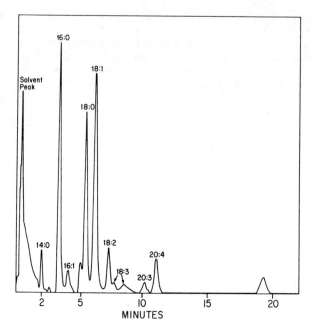

Fig.13.2. Fatty acid composition of the inhibitor fraction. The inhibitor fraction obtained from a DEAE-cellulose column (fractions 60–70, Fig.13.1) was analysed by gas chromatography. This profile is a representative of the fatty acid composition of the inhibitor fraction. The compound with a retention time of 19 min has not been identified. (From Ballou and Cheung, 1985.)

groups restored most of the inhibitory activity. In another experiment, we isolated the fatty acids from the inhibitor fraction by thin-layer chromatography. This sample, which contained a mixture of saturated and unsaturated fatty acids, inhibited PLA_2 activity significantly. When methylated, the inhibitory activity in the mixture was lost. When oleic acid (18:1) was methylated it not only lost its inhibitory activity, but became stimulatory. Stearic acid (18:0), which was only slightly inhibitory, became less inhibitory following methylation. These results show that the carboxyl group of the unsaturated fatty acids must remain free to be inhibitory.

13.3.5. *Effectiveness of unsaturated fatty acids as inhibitors of phospholipase A_2*
Unsaturated fatty acids present in the inhibitor fraction were examined individually for their ability to inhibit PLA_2 activity. Figure 13.3 shows that all unsaturated fatty acids tested inhibited PLA_2 activity, with 50% inhibition ranging from 2×10^{-6} M for oleic acid to 2×10^{-7} M for arachidonic acid. In increasing order of potency, their effectiveness was oleic acid (18:1) < docosatetraenoic acid (22:4) < linoleic acid (18:2) < docosahexaenoic acid (22:6) < linolenic acid (18:3) < palmitoleic acid (16:1) < arachidonic acid (20:4). These data clearly indicate that unsaturated fatty acids are effective inhibitors of PLA_2 activity and that the presence of these fatty acids could account for the low PLA_2 activity in the platelet homogenate.

Separate experiments showed that unsaturated fatty acids also inhibited pancreatic PLA_2 activity, suggesting that the inhibition may be a general phenomenon.

Human platelet phospholipase A_2 activity

Table 13.4. Loss of inhibitory activity by methylation. Samples of the inhibitor fraction (20 μl, tubes B, C and D), the fatty acids isolated from the inhibitor fraction (2 μM, tubes E and F), oleic acid (2 μM, tubes G and H), and stearic acid (2 μM, tubes I and J), were either left untreated or were methylated. All samples were evaporated to dryness under N_2, dissolved in 10 μl Me_2SO, and then assayed for their ability to inhibit PLA_2 activity. The fatty acids present in the inhibitor fraction were isolated by lipid extraction and thin-layer chromatography. Tube D contained a sample of inhibitor fraction which was methylated, evaporated to dryness, saponified in 33% KOH by reflux at 70 °C for 1 hour, and neutralized with 1 M HCl. All assays were initiated by the addition of PLA_2. Values represent the mean ± SEM of duplicate assays from two separate experiments. (From Ballou and Cheung, 1985.)

Additions	PLA_2 activity (%)
A. None	100
B. Inhibitor fraction	26 ± 1
C. Inhibitor fraction (methylated)	121 ± 8
D. Inhibitor fraction (methylated then saponified)	52 ± 3
E. Fatty acid fraction	38 ± 0
F. Fatty acid fraction (methylated)	94 ± 3
G. Oleic acid (18:1)	18 ± 2
H. Oleic acid (methylated)	156 ± 4
I. Stearic acid (18:0)	81 ± 1
J. Stearic acid (methylated)	92 ± 2

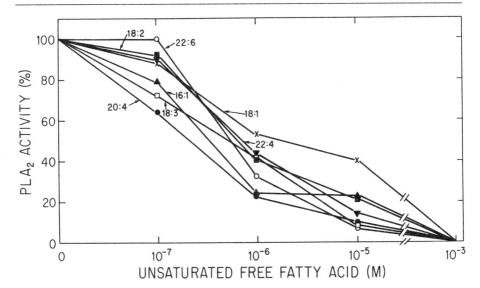

Fig.13.3. Effect of unsaturated fatty acids on PLA_2 activity. Each standard reaction mixture contained the indicated amounts of fatty acid. Oleic acid (18:1), x--x; palmitoleic acid (16:1), ▲--▲; linoleic acid (18:2), ■--■; linolenic acid (18:3), □--□, arachidonic acid (20:4), ●--●; docosahexaenoic acid (22:6), ○--○; and docosatetraenoic acid (22:4), ▼--▼. (From Ballou and Cheung, 1985.)

13.3.6. Mechanism of phospholipase A_2 inhibition by unsaturated fatty acids

PLA_2 activity was measured as a function of substrate concentration in the absence or presence of 10^{-6} M linolenic acid (18:3), a concentration that inhibited PLA_2 activity significantly (Fig.13.4). Without linolenic acid, PLA_2 activity increased rapidly as a function of substrate concentration, reaching 50% of maximal activity at 1 μM and 100% at 2 μM phosphatidylcholine. No further increase in PLA_2 activity was observed at higher substrate concentrations. In the presence of linolenic acid, PLA_2 activity was inhibited at all substrate concentrations tested; moreover, the enzymatic activity never exceeded 40% of maximum, which was reached at 5 μM phosphatidylcholine. The finding that inhibited of PLA_2 by linolenic acid was not diminished by increasing the substrate concentration suggests that the inhibition was noncompetitive, probably due to a direct interaction with the enzyme at an allosteric site(s).

13.3.7. 'Inhibitor-affinity' column chromatography of phospholipase A_2

Our finding that linolenic acid is a potent noncompetitive inhibitor of PLA_2 activity suggested that it interacts with the enzyme at a domain distinct from the substrate site, and that such an interaction could be exploited as a means of enzyme purification. Figure 13.5 shows a typical elution profile of PLA_2 activity from an inhibitor-affinity column, using ricinoleic acid as the ligand. A majority of the protein is excluded from the column, and further washing of the column with buffer A eluted the enzyme. The active fractions were pooled and applied to an Affi-Gel 501 column, which served to concentrate the enzyme. The PLA_2 thus obtained represented a 7000-fold purification, with a specific activity exceeding 200 mmol/min/mg protein, a significant improvement over our previous preparation (see Table 13.1). Another advantage of this procedure is that this enzyme preparation is rather stable,

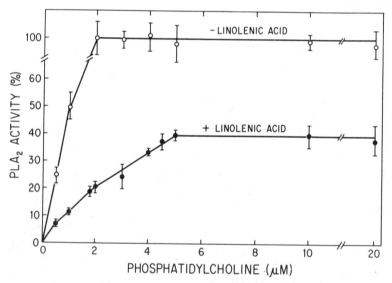

Fig.13.4. PLA_2 activity as a function of substrate concentration in the presence or absence of linolenic acid. Each reaction mixture contained the standard components and the indicated concentrations of labelled phosphatidylcholine. Linolenic acid (18:3) was used as an inhibitor at a concentration of 1 μM. (From Ballou and Cheung, 1985.)

but it does lose activity gradually over a period of several months in 50% glycerol (w/w) at -20°C.

13.3.8. Ca^{2+}-requirement of phospholipase A_2

Under our assay conditions, PLA_2 required approximately 1 μM Ca^{2+} for half-maximum activity (Fig.13.6), but enzyme activity could be detected at a concentration as low as 10^{-8} M. Full enzyme activity, however, required millimolar Ca^{2+}. Other investigators have found that optimal PLA_2 activity in membrane preparations requires millimolar Ca^{2+} (Bills, Smith and Silver, 1977; Derksen and Cohen, 1975; Rittenhouse-Simmons, Russell and Deykin, 1977). Detergents apparently enhanced Ca^{2+} sensitivity of the enzyme so that half-maximal activity could be attained at 60–100 μM Ca^{2+} (Billah, Lapetina and Cuatrecasas, 1980). Jesse and Franson (1979) noted that a partially purified PLA_2 from human platelets required 125 μM Ca^{2+} for 40% maximal activity.

Addition of calmodulin to the reaction mixture neither stimulated PLA_2 activity nor prevented its inhibition by unsaturated fatty acids. Inhibition of PLA_2 by unsaturated fatty acids was not overcome by excess Ca^{2+}.

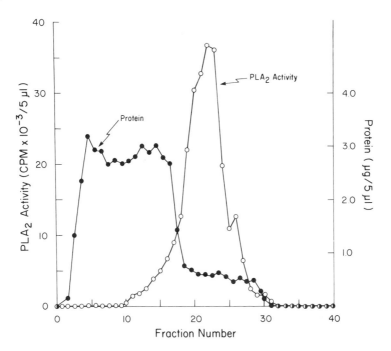

Fig.13.5. 'Inhibitor-affinity' column chromatography of PLA_2. Ricinoleic acid (18:1), an unsaturated fatty acid with a hydroxyl group at C-9, was coupled to sepharose 4-B via a 12-carbon spacer diglycidyl ether linkage through the hydroxyl group using a modification of the procedure of Sundberg and Porath (1974). A partially purified enzyme from the DEAE-cellulose column was loaded onto the 'inhibitor-affinity' column (1 x 3 cm), which was eluted with buffer A. Activity is expressed as CPM ^{14}C-arachidonate released/3 min at 30°C. The active fractions were pooled and applied to a small column (1 x 1 cm) of Affi-Gel 501, an organomercurial derivative of crosslinked Bio-Gel A-5m which readily forms covalent mercaptide bonds with free sulphydryl groups. The enzyme bound to the column, and was eluted with 5 ml of 10 mM dithiothreitol.

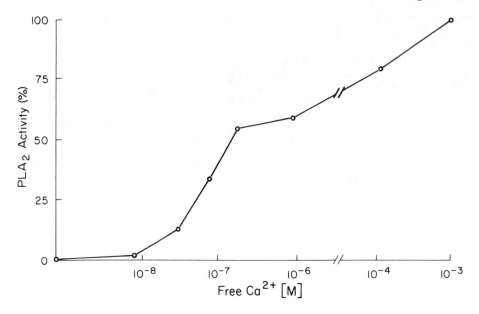

Fig.13.6. Stimulation of PLA$_2$ by Ca^{2+}. All PLA$_2$ assays were performed in a Ca^{2+}-EGTA, Tris-HCl buffer (pH 7.37) at 30°C, from which the concentration of free Ca^{2+} was calculated with the computer program of Perrin and Sayce (1967) using the following logarithmic association constants: H$^+$ to EGTA^{-4}, 9.32; H$^+$ to HEGTA^{-3}, 8.70; H$^+$ to H$_2$EGTA^{-2}, 2.66; H$^+$ to H$_3$EGTA^{-1}, 2.0; calcium to EGTA^{-4}, 10.76; and calcium to Ca-EGTA^{-2}, 3.85 (Piascik, Wisler and Johnson, 1980). The PLA$_2$ activity at 100% was 50.81 nmol/min/mg protein.

13.4. Discussion

The generally low level of PLA$_2$ activity reported in the literature for human platelet extract appears to be an underestimation. During the course of purification, the enzyme activity increased markedly due to the removal of an endogenous inhibitor from the enzyme. The inhibitor, associated mainly with a particulate fraction, has now been identified as a mixture of unsaturated fatty acids which inhibit PLA$_2$ noncompetitively. The inhibition by unsaturated fatty acids was not reversed by Ca^{2+}. A knowledge of how these unsaturated fatty acids modulate PLA$_2$ activity may shed light on the regulation of PLA$_2$ activity *in vivo*.

In the resting platelet, the level of free arachidonic acid is barely detectable (Bills, Smith and Silver, 1977; Marcus, Ullman and Safier, 1969). However, other unsaturated fatty acids are present in higher levels (Bills, Smith and Silver; Derksen and Cohen, 1975; Marcus, Ullman and Safier, 1969) and would appear sufficient to effectively inhibit PLA$_2$ activity *in vivo*, provided they are accessible to the enzyme in the milieu of the plasma membrane. Upon stimulation of the platelet, PLA$_2$ becomes activated, and arachidonic acid is cleaved from phospholipids to be converted into various prostanoids, potent modulators of a variety of physiological processes. In view of the important role of PLA$_2$ in regulating the intracellular level of free arachidonate, an understanding of the mechanism controlling PLA$_2$ activity would assume significance.

It has been proposed that stimulation of the platelet leads to the activation of a phospholipase C which catalyses the hydrolysis of phosphatidylinositol with the formation of diacylglycerol and inositol triphosphate (Berridge, 1984). Inositol triphosphate appears to trigger the release of Ca^{2+} from intracellular organelles, and diacylglycerol activates a protein kinase C (which requires Ca^{2+} and phosphatidylserine for activity). Protein kinase C catalyses the phosphorylation of specific proteins in the membrane, and may thus affect membrane permeability with respect ot Ca^{2+}. An increase of intracellular Ca^{2+} would then activate PLA_2. Various investigators (Bills, Smith and Silver, 1977; Derksen and Cohen, 1975; Jesse and Franson, 1979; Rittenhouse-Simmons, Russell and Deykin, 1977) have reported that PLA_2 requires millimolar Ca^{2+} for activity, a concentration far in excess of normal intracellular levels. It has been argued that the increase of intracellular Ca^{2+} associated with platelet activation may not be sufficient to activate PLA_2. However, we have found that human platelet PLA_2 is activated by submicromolar Ca^{2+} concentrations *in vitro*, well within the physiological range. This suggests that a slight increase in intracellular Ca^{2+} could serve as a potentially important signal in initiating PLA_2 activity *in vivo*.

Human platelet PLA_2 appears to be loosely associated with the membrane and is therefore present in the environment of membrane lipids, including its substrate and inhibitory unsaturated fatty acids. These unsaturated fatty acids appear to be present in sufficient quantities in platelets to effectively inhibit PLA_2 activity. How does the increase in the intracellular level of Ca^{2+} in activated platelets stimulate PLA_2 activity? And when the enzyme is activated, does the transient accumulation of arachidonate serve as a feedback inhibitor? Much more work is necessary before we can answer these questions.

References

Argiolas, A. and Pisano, J. (1983). Facilitation of phospholipase A_2 activity by mastoparans, a new class of mast cell degranulating peptides from wasp venom. *J. Biol. Chem.* **258**, 13697–702.

Ballou, L.R. and Cheung, W.Y. (1983). Marked increase of human platelet phospholipase A_2 activity *in vitro* and demonstration of an endogenous inhibitor. *Proc. Natl. Acad. Sci. USA* **80**, 5203–7.

Ballou, L.R. and Cheung, W.Y. (1985). Inhibition of human platelet phospholipase A_2 activity by unsaturated fatty acids. *Proc. Natl. Acad. Sci. USA* **82**, 371–5.

Bell, R.L., Kennerly, D.A., Stanford, N. and Majerus, P.W. (1979). Diglyceride lipase: A pathway for arachidonate release from human platelets. *Proc. Natl. Acad. Sci. USA* **76**, 3238–41.

Berridge, M.J. (1984). Inositol triphosphate and diacylglycerol as second messengers. *Biochem. J.* **220**, 345–60.

Billah, M.M., Lapetina, E.G. and Cuatrecasas, P. (1980). Phospholipase A_2 and phospholipase C activities of platelets. Differential substrate specificity, Ca^{2+} requirements, pH dependence and cellular localization. *J. Biol. Chem.* **255**, 10227–31.

Bills, T.K., Smith, J.B. and Silver, M.J. (1977). Selective release of arachidonic acid from the phospholipids of human platelets in response to thrombin. *J. Clin. Invest.* **60**, 1–6.

Derksen, A. and Cohen, P. (1975). Patterns of fatty acid release from endogenous substrates by human platelet homogenates and membranes. *J. Biol. Chem.* **250**, 9342–7.

Feinstein, M.B. and Sha'afi, R.I. (1983). Role of calcium in arachidonic acid metabolism and in the actions of arachidonic acid-derived metabolities. In *Calcium and Cell Functions*, Vol. IV. W.Y. Cheung (ed.), pp. 337—76. Academic Press, New York.

Jesse, R.L. and Franson, R.C. (1979). Modulation of purified phospholipase A_2 activity from human platelets by calcium and indomethacin. *Biochim. Biophys. Acta.* **575**, 467—70.

Lapetina, E.G., Billah, M.M. and Cuatrecasas, P. (1981). The phosphatidylinositol cycle and the regulation of arachidonic acid production. *Nature (Lond.)* **292**, 367—8.

Marcus, A.J., Ullman, H.L. and Safier, L.B. (1975). Lipid composition of subcellular particles of human blood platelets. *J. Lipid Res.* **10**, 108—14.

Michell, R.H. (1975). Inositol phospholipids and cell surface receptor function. *Biochim. Biophys. Acta* **415**, 81—147.

Perrin, D.D. and Sayce, J.G. (1967). Computer calculations of equilibrium concentrations in mixtures of metal ions and complexing species. *Talanta* **14**, 833—42

Piascik, M.T., Wisler, P.L., Johnson, C.L. and Potter, J.D. (1980). Ca^{2+}-dependent regulation of guinea pig brain adenylate cyclase. *J. Biol. Chem.* **255**, 4176—81.

Rittenhouse-Simmons, S., Russell, F.A. and Deykin, D. (1977). Mobilization of arachidonic acid in human platelets. Kinetics and Ca^{2+}-dependency. *Biochim. Biophys. Acta* **488**, 370—80.

Sundberg, L. and Porath, J. (1974). Preparation of adsorbents for biospecific affinity chromatography. I. Attachment of group-containing ligands to insoluble polymers by means of bifunctional oxiranes. *J. Chromatogr.* **90**, 87—98.

CONTROL OF THE INITIATION OF CHROMOSOME REPLICATION BY Ca^{2+}, Ca^{2+}·CALMODULIN AND cAMP: AN OVERVIEW

J.F. Whitfield, M. Sikorska, A.L. Boynton, J.P. Durkin, L. Kleine, R.H. Rixon and T. Youdale

Cell Physiology Group, Biological Sciences Division, National Research Council of Canada, Ottawa, Canada

and D. Bossi,

Insititute of General Pathology, School of Medicine, Catholic University, Rome, Italy

Abstract

The prokaryotes, which hang and replicate their naked chromosomes on cell membranes, invented cAMP to activate genes, but do not use Ca^{2+} to operate anything internally. During the 'Great Calcium Revolution', about 10^9 years ago, actomyosin filaments, microtubules, and the multipurpose Ca^{2+}-binding effector calmodulin were invented to operate a new set of functions which, in turn, forced the invention of the eukaryote nucleus and the placing of DNA replication and cell division under the dual control of Ca^{2+} and cAMP. Chromosome replication in cells starting from one of four possible proliferative states (newborn-cycling, quiescent-G_0, quiescent-dormant, quiescent-actively functioning) results from a gene activation programme which is triggered by a cAMP transient and burst of cAMP-dependent protein kinase activity and produces replication enzymes and initiators. Internal Ca^{2+}·calmodulin and enzyme(s) (such as, phospholipase A_2, protein kinase C) controlled by external Ca^{2+} might contribute to this prereplicative programme by causing the cAMP transient, stimulating mRNA translation, reducing protein degradation, stimulating entry of replication enzymes and initiators into the nucleus and finally by stimulating cyclic nucleotide phosphodiesterase and membrane Ca^{2+} pumps which prevent the cAMP and Ca^{2+}·calmodulin transients from lasting too long and disrupting subsequent S phase events.

14.1. Introduction

In the beginning was the accumulation of energy within emerging life forms in interfacial proton gradients, the prototypes of the transmembrane proton gradients which would power the growth, multiplication and movement of the sophisticated, DNA-coded prokaryotes (Wilson and Lin, 1980). These prokaryotes, which directly couple cell growth to chromosome replication by hanging and replicating their chromosomes on their membranes, did not, and still do not, use Ca^{2+} for anything important internally. However, they did invent a Ca^{2+} pump, driven by proton-motive force, to avoid loading themselves with Ca^{2+} from the outside which could activate dangerous degradative enzymes. Thus was created another transmembrane ion gradient which would someday become extremely important. While actively avoiding internal Ca^{2+} transients, they invented cAMP and the cAMP-binding/activated CAP protein (the ancestor of the eukaryotic cAMP·PK regulatory subunits) to stimulate certain genes (Weber et al., 1982).

Then came the 'Great Calcium Revolution' and from it the eukaryotes, which chose Ca^{2+} and used the old protective Ca^{2+} gradient to operate a set of new functions for feeding, proliferating, and reproducing. These new functions (such as, ciliary and flagellar beating, endocytosis, locomotion, phagocytosis, pinocytosis, cytoplasmic streaming) were based on a biotechnological breakthrough: arrays of membrane-anchored actomyosin and tubulin fibrils which could be shaped and moved by briefly opening potential-operated or receptor-operated membrane channels to admit triggering pulses of external Ca^{2+}.

Internal Ca^{2+} was still a very dangerous tool, which would kill cells by activating membrane-destroying phospholipases should its concentration in the cytosol exceed 10^{-5} mol/l for very long (Whitfield et al., 1982, 1985). So the first eukaryotes cleverly invented the ancestor of the small, Ca^{2+}-binding/activated multipurpose effector protein we now call *calmodulin* to enable the new processes to be triggered by harmlessly small internal Ca^{2+}

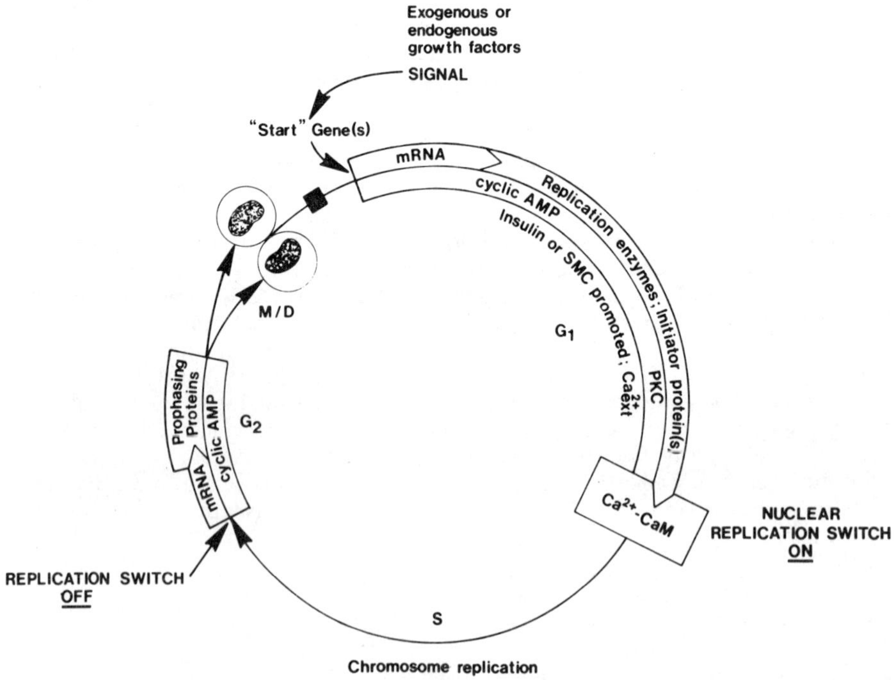

Fig.14.1. The type 1 prereplicative period of continuously cycling cells; this is the basic G_1 phase.

transients (Whitfield et al., 1982, 1985).

Mobilizing the cell membrane for endocytosis, phagocytosis and pinocytosis forced the transfer of the chromosome from its attachment to the cell membrane into a safe place, the nucleus, where covered with, and activated or silenced by, a host of new regulatory and structural proteins, it would not be endocytosed, dislodged by streaming cytoplasm, or battered by collisions with new organelles such as lysosomes and mitochondria. This, in turn, required new strategies to coordinate chromosome replication, cell growth and cell division which included a specific ordering of gene activations, moving gene transcripts from the nucleus into the cytoplasm for translation, delivering finished replication enzymes and initiator(s) to the nucleus at the right time, attaching multi-enzyme replication complexes to binding sites on the nuclear matrix, which are equivalent to the old prokaryotic cell membrane replication sites, and apportioning the replicated chromosomes between two new nuclei. cAMP, the powerful new cAMP·PKs (made by joining the old gene-activating cAMP-binding/activated CAP (Weber et al., 1982) to a protein kinase that could push aside the new access-limiting regulatory chromosomal proteins), external Ca^{2+} and Ca^{2+}·CaM were chosen to direct these new strategies.

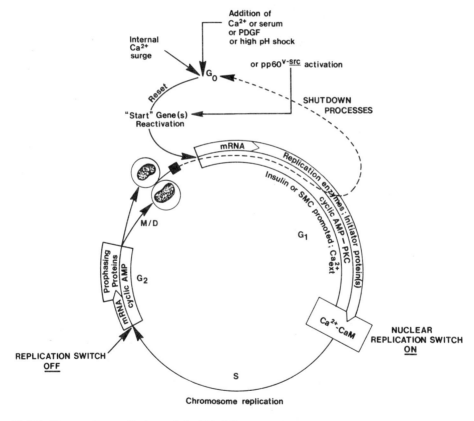

Fig.14.2. The type 2 prereplicative period which follows the stimulation of quiescent (G_0) cells in culture. This type consists of a competence-establishing pre-G_1 stage as well as the basic G_1 phase. All abbreviations are defined in the text.

Before discussing some of these strategies we must briefly describe the four kinds of prereplicative period (Boynton and Whitfield, 1983; Boynton et al., 1982).

The first is the basic G_1 phase which was originally defined by Howard and Pelc (1953). It starts almost immediately after the cell's birth and progresses uninterruptedly to the initiation of chromosome replication (Fig.14.1).

The second type (Fig.14.2) starts in quiescent cells, which had stopped (because of lack of external Ca^{2+} or serum) at or before a critical (or restriction) point in late G_1 phase, reverted to an early G_1 state by losing shortlived components and finally were obliged to activate a shutdown mechanism in order to enter an indefinitely maintainable quiescent, or G_0 state (Boynton and Whitfield, 1983). This type 2 prereplicative period is longer than the basic G_1 phase because there must be a pre-G_1 competence-re-establishing period during which prereplicative starter genes are reactivated, the translation machinery is replenished and switched on, and any late G_1 leftovers (such as excess calmodulin mRNA, excess cAMP, centriolar cilia) that might disrupt early G_1 events are erased.

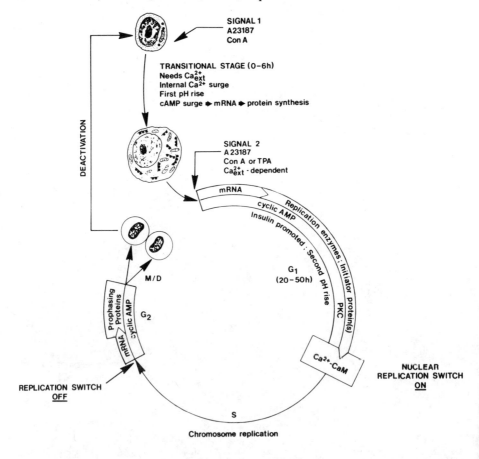

Fig.14.3. The type 3 prereplicative period which follows the activation of nearly dormant cells such as small lymphocytes. This type consists of a preparatory pre-G_1 stage as well as the basic G_1 phase, which are promoted by different external Ca^{2+}-dependent signals. All abbreviations are defined in the text.

The 50-hour type 3 period follows activation of nearly dormant cells such as small lymphocytes (Fig.14.3). These cells need a lot of time just to get ready for G_1 transit by decondensing chromatin, enlarging the nuclear matrix and associated transcript processing machinery, building a new protein-synthesizing apparatus and making enough mitochondria to fuel this new equipment.

The 18–20-hour type 4 period (Fig.14.4) occurs in already actively functioning cells such as hepatocytes, which, when stimulated by 70% (partial) hepatectomy (HPX), massively increase gene transcription, further increase their protein-synthesizing capability by lengthening ribosome half-life, and put more insulin receptors into their plasma membranes and increase the residence time of insulin-receptor complexes on the surface (to promote protein synthesis and reduce protein degradation), all in order to start growing (Moore, 1983; Walker and Whitfield, 1984; Whitfield et al., 1985; Whitfield, Rixon and Sikorska, 1984). Thus, the cells prime themselves for G_1 transit in a pre-G_1 transitional or growth stage (see Fig.14.4) while still servicing the rest of the body (Fausto, 1984; Whitfield, Rixon and Sikorska

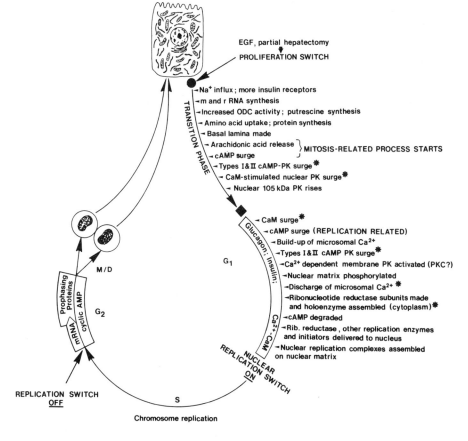

Fig.14.4. The events of the type 4 prereplicative period triggered by epidermal growth factor (EGF) or HPX in the actively functioning rat hepatocyte. (*)-labelled events fail to occur in the 72 h TPTX rat. All abbreviations are defined in the text

1984; Whitfield et al., 1985).

14.2. Terminology

'Prereplicative period' covers all of the periods extending from cell birth or stimulation to the onset of chromosome replication. Only the type 1 prereplicative period (see Fig.14.1) is equivalent to the original G_1 phase of Howard and Pelc (1953). The other three prereplicative periods (see Figs.14.2, 14.3 and 14.4) have pre-G_1 stages during which the cells ready themselves for transit through the basic G_1 phase. Therefore, we use the terms 'G_1' and 'G_1 phase' to refer either to whole type 1 prereplicative periods or to only the final common events in the other three prereplicative periods (see Figs.14.1, 14.2, 14.3 and 14.4).

14.3. Ca^{2+} and Ca^{2+}·CaM in G_1 phase

Most non-neoplastic parenchymal epithelial cells (such as hepatocytes) and mesenchymally derived cells (such as fibroblasts) need 1.0–1.8 mM external Ca^{2+} and Ca^{2+}·CaM activity (with or without an increase or redistribution of calmodulin) to transit G_1 (Armato, Andreis and Whitfield, 1983; Boynton and Whitfield, 1981, 1983, 1984; Boynton, Kleine and Whitfield, 1983, 1984, 1985; Durkin, Boynton and Whitfield, 1981; Durkin, MacManus and Whitfield, 1983; Swierenga et al., 1980; Whitfield, Rixon and Sikorska, 1984; Whitfield et al., 1976, 1979, 1980, 1982, 1985). Thus, dropping external Ca^{2+} to 0.01–0.02 mM, stripping calcium from the cell surface with EGTA, or exposure to Ca^{2+}·CaM blockers traps these cells for a while at a critical point in late G_1, but it does not generally affect transit through other stages of the cell cycle. For example, serum-stimulated confluent T51B rat liver cells need 1.25–1.8 mM external Ca^{2+} for the first 2 hours of their type 2 prereplicative period, do not need it for the next 6 hours, but need it again from shortly after 8 hours to the onset of DNA replication (Fig.14.5). If there is not enough external Ca^{2+} between 8 and 9 hours, the cells can go no further, because they cannot make and/or accumulate enough of some component(s) needed to trigger DNA replication. The arrested cells then rapidly ($t_{\frac{1}{2}}$= 1–2 h) revert to an early prereplicative, or even quiescent G_0, state. However, much of our knowledge of the possible Ca^{2+}-dependent G_1 control mechanisms has depended on the ability of these same T51B liver cells under different culture conditions to mark time for at least 48 hours in so advanced a G_1 state that they can start replicating DNA less than 1 h after external Ca^{2+} addition (Armato, Andreis and Whitfield, 1983; Boynton and Whitfield, 1981, 1983, 1984; Boynton, Kleine and Whitfield, 1983, 1984, 1985; Swierenga et al., 1980; Whitfield, Rixon and Sikorska, 1984; Whitfield et al., 1976, 1979, 1980, 1982, 1985).

A Ca^{2+}·CaM blocker, such as the naphthalene sulphonamide, W13, stops confluent serum-stimulated T51B liver cells (in medium containing 1.8 mM Ca^{2+}) much nearer the G_1/S transition than external Ca^{2+} deprivation (see Fig.14.5) (Boyton, Kleine and Whitfield, 1984, 1985; Whitfield et al., 1985). Moreover, the cells can then stay in this very late G_1 state for at least 10 hours (Boynton, Kleine and Whitfield, 1984, 1985; Whitfield et al., 1976), presumably by being able to retain enough of the external Ca^{2+}-controlled critical component(s) and maintain a cytosolic Ca^{2+} concentration high enough to generate the Ca^{2+}·CaM complexes needed to trigger DNA replication as soon as the blocker is removed. The possibility of there being a

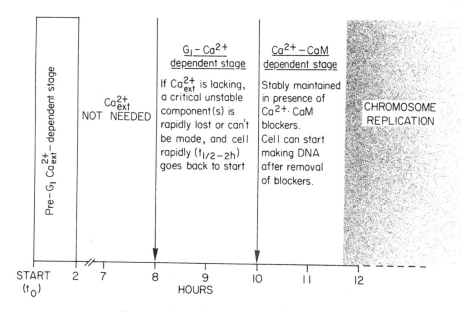

Fig.14.5. The external Ca^{2+}- and Ca^{2+}·CaM-controlled stages of the type 2 preplicative period triggered in quiescent confluent T51B liver cells by exposure to fresh medium (80% Eagle's basal medium + 20% bovine calf serum).

late G_1 internal Ca^{2+} transient to generate Ca^{2+}·CaM for the final G_1 event(s) is suggested by the facts that regenerating rat liver cells specifically load their endoplasmic reticulum (ER) with calcium at the beginning of G_1 and then release it just before the onset of DNA replication (Fig.14.6A), and that HPX-activated liver cells in thyropara-thyroidectomized (TPTX) rats still load their ER with calcium, but most of them cannot release it and cannot start replicating DNA (Fig.14.6B) (Whitfield, Rixon and Sikorska, 1984; Whitfield et al., 1980; Youdale et al., 1984).

Lining epithelial cells, such as epidermal cells, need 10 to 100 times less Ca^{2+}_{ext} than parenchymal epithelial cells, such as hepatocytes, to proliferate maximally and indefinitely (Boynton et al., 1982; Hennings, Holbrook and Yuspa, 1983; Whitfield et al., 1985). Indeed, the higher Ca^{2+}_{ext} concentrations needed by other cells cause the lining cells to stop proliferating and terminally differentiate (Hennings, Holbrook and Yuspa, 1983; Whitfield et al., 1985). These lining cells probably still need external Ca^{2+} to initiate DNA replication, but they make themselves hypersensitive to external Ca^{2+} at a certain point in their terminal differentiation programme by turning down their membrane Ca^{2+} pumps and/or putting special external Ca^{2+}-binding, desmosome-triggering, proliferation-inhibiting receptors onto their surfaces (Hennings, Holbrook and Yuspa, 1983; Whitfield et al., 1985).

Most, if not all, tumour cells also need about 100 times less Ca^{2+}_{ext} to transit G_1 and divide than the corresponding normal cells (Fig.14.7) (Durkin, Boynton and Whitfield, 1981; Swierenga et al., 1980; Whitfield et al., 1982). But they still need Ca^{2+}·CaM at the end of G_1. For example, NRK rat cells become able to proliferate normally in external Ca^{2+}-deficient medium when neoplastically transformed by the pp60$^{v\text{-}src}$ protein kinase of the avian sarcoma virus (Durkin, Boyton and Whitfield, 1981), but they are still stopped by Ca^{2+}·CaM blockers near the end of G_1 (Durkin, MacManus and Whitfield, 1983). Moreover, the ability of the activation of internal, thermolabile pp60$^{v\text{-}src}$ (by dropping the incubation temperature to a permissive level (36°C) to cause temperature-sensitive transformation-defective ASV (tsASV)-infected NRK cells, arrested in late G$_1$ by external Ca^{2+} deprivation at a restrictive incubation temperature (40°C), to initiate DNA replication *while still in the external Ca^{2+}-deficient medium* is blocked by any one of three putative Ca^{2+}·CaM blockers (R24571, trifluoperazine, W7) (unpublished observations). This, and the fact that external Ca^{2+} deprivation and Ca^{2+}·CaM blockers arrest T51B liver cells at different points in their G$_1$ transit, indicate two kinds of Ca^{2+}-dependent G$_1$ control mechanism, one operating at the cell surface under the control of external Ca^{2+} and the other operating inside the cell under the control of Ca^{2+} and Ca^{2+}·CaM transients.

The one controlled by external Ca^{2+} may be a surface protein kinase(s), because a Ca^{2+}-activatable cell membrane protein kinase is switched on when regenerating rat liver cells begin transiting G$_1$ (MacManus and Whitfield,

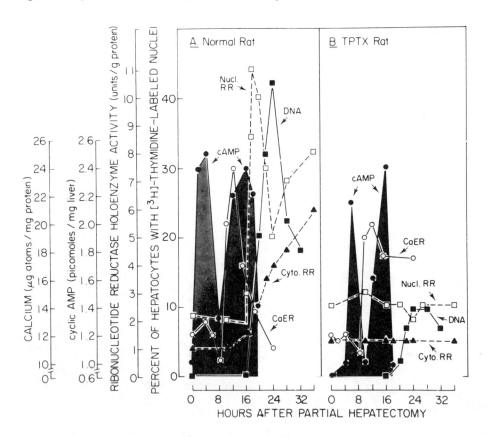

Fig.14.6. The temporal relations between the pre-G$_1$ and G$_1$ cAMP transients, the G$_1$ accumulation and release of Ca·ER, the slow rise of cytoplasmic ribonucleoside diphosphate reductase (RR) holoenzyme, the abrupt surge of RR into the nucleus, and the initiation of DNA replication following HPX in (A) normal and (B) 72 h TPTX rats. The sharp drop in nuclear RR activity (assayable by measuring the reduction of added cytidine 5'-diphosphate (CDP) to 2'-deoxycytidine 5'-diphosphate (dCDP) *before* the peak of DNA-synthetic activity is probably due to the locking of holoenzyme into substrate-inaccessible, matrix-bound multienzyme replication complexes (Youdale et al., 1984). All assay procedures are described in Rixon and Whitfield (1982), Whitfield et al. (1980), Youdale et al. (1984). All abbreviations are defined in the text.

1981). Moreover, while the phosphorylation of certain surface proteins and G_1 transit of normal T51B rat liver cells are both inhibited by lowering external Ca^{2+}, the same phosphorylations and G_1 transit of the neoplastic T51B-261B cells are not affected by lowering external Ca^{2+} (Boynton, Kleine and Whitfield, 1985; Kleine, Boynton and Whitfield, 1984).

We believe that the Ca^{2+}/diacylglycerol/phospholipd-dependent,

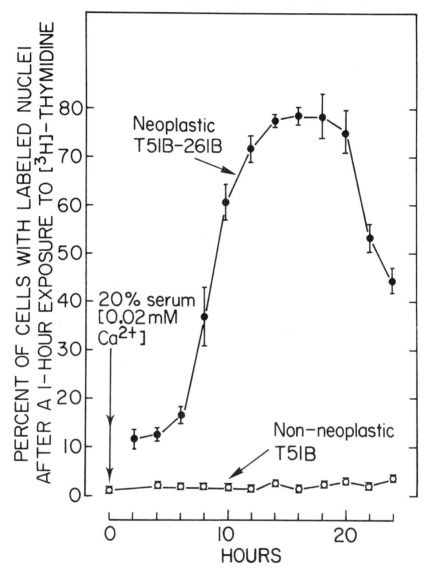

Fig.14.7. Quiescent T51B liver cells in confluent cultures cannot start replicating their DNA when stimulated with fresh medium (80% Eagle's basal medium + 20% mitogen-supplying bovine calf serum) containing only 0.02 mM external Ca^{2+}, but confluent neoplastic T51B-261B cells initiate DNA replication despite the external Ca^{2+} shortage. Cells were labelled by exposure to $[^3H]$-thymidine (2 μCi/medium) for 1 hour after which time they were prepared for autoradiography (Boynton, Kleine and Whitfield, 1983).

Ca^{2+}·CaM-independent protein kinase C is at least one of the external Ca^{2+}-sensitive G_1 controllers commonly altered by neoplastic transformation because:

(1) lowering external Ca^{2+} lowers protein kinase C activity and stops G_1 transit of normal T51B cells, but does not affect protein kinase C activity of G_1 transit of several preneoplastic and neoplastic derivatives of T51B liver cells (Boynton Kleine and Whitfield, 1983);
(2) protein kinase C is part of the receptor for TPA (Niedel, Kuhn and Vandenbank, 1983), a tumour promoter which makes the enzyme from T51B cells function without Ca^{2+} (unpublished observations) and makes normal T51B cells initiate DNA replication and proliferate like neoplastic cells in external Ca^{2+} deficient medium (Boynton, Kleine and Whitfield, 1984, 1985; Whitfield et al., 1982);
(3) external (that is, membrane receptor (protein kinase C)-binding) TPA, but not injected TPA, stimulates the initiation of replication of λ-phage DNA which has been injected into immature *Xenopus* frog eggs (Méchali, Méchali and Laskey, 1983);
(4) TPA stimulates G_1, but not pre-G_1, transit of small lymphocytes (Mastro and Smith, 1982);
(5) protein kinase C blockers (phenoxybenzamine, phentolamine) stop G_1 transit of regenerating rat liver cells (MacManus et al., 1973; Whitfield et al., 1985);
(6) protein kinase C activity rises during the external Ca^{2+}-dependent period just before the onset of DNA replication in serum-stimulated confluent T51B cells (Boynton, Kleine and Whitfield 1985).

No one knows how external Ca^{2+} and Ca^{2+}·CaM promote G_1 transit. But there are some clues. First are the Ca^{2+}- and Ca^{2+}·CaM-triggered events (Na^+/H^+ exchange and internal pH rise, cAMP transient, NAD kinase activation, release of sequestered mRNA, activation of the translation apparatus, mitochondrial activation) leading to the onset of DNA replication in fertilized eggs (Boynton and Whitfield, 1983; Boynton et al., 1982; Epel, 1982; Schatten, 1982). Then there are the rapid (within 1 h), external Ca^{2+}-triggered responses of many kinds of external Ca^{2+}-deprived cell, especially rat thymic lymphoblasts and T51B liver cells, analysis of which has uncovered a complicated web of interlocking Ca^{2+}- and cAMP-dependent processes which all funnel into mRNA production, syntheses of replication enzymes and initiator(s), and ultimately DNA replication (Fig.14.8) (Boynton and Whitfield, 1981, 1983; Boynton, Kleine and Whitfield, 1983, 1984, 1985; Boynton et al., 1982; Swierenga et al., 1980; Whitfield et al., 1976, 1979, 1980, 1982, 1985). Most importantly, stimulating any part of the Web (for example, with external Ca^{2+}, *external* Ca^{2+}·CaM, *external* cAMP, cAMP elevators, insulin, avian sarcoma virus $pp60^{v-src}$ protein kinase of the avian sarcoma virus) triggers the core events and starts DNA replication. That is why it is so hard to learn how external Ca^{2+} and Ca^{2+}·CaM control normal G_1 transit.

The most important clue to the web's core pathways to DNA replication is the observation of Swierenga et al. (1980) that external Ca^{2+}-stimulated T51B cells (which had been made to mark time for 2 days in late G_1 by external Ca^{2+} deprivation) responded promptly, but abortively, to external Ca^{2+} in the presence of actinomycin D and not at all in the presence of the translation inhibitor cycloheximide. This probably means that these 'old' late G_1 cells had retained just enough of certain mRNA transcripts to start, but not sustain, DNA replication and that external Ca^{2+} or any other web-triggerer ultimately stimulates the synthesis of *both* the mRNA transcripts and proteins needed for sustained DNA replication (see Fig.14.8).

While all of the possible external Ca^{2+}- and Ca^{2+}·CaM-dependent G_1-driving mechanisms are probably in the web of Fig.14.8, we still do not know which ones operate, and in what order they operate, during *normal* G_1 transit. An external Ca^{2+}-triggered explosion of the web events in cells which have been struggling for 2 days against shrinking pools of critical mRNA transcripts and proteins to stay in late G_1 is not the same as the stepwise unfolding of events as cells pass through G_1. Therefore, we cannot conclude from the external Ca^{2+}-deprived T51B model that external Ca^{2+} and/or Ca^{2+}·CaM directly or indirectly stimulate gene transcription and protein synthesis

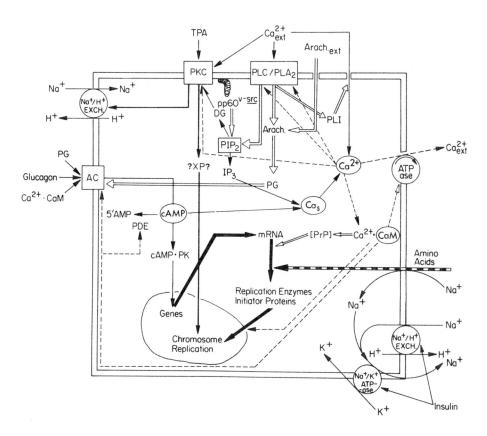

Fig.14.8. The web of interlocking Ca^{2+}-, Ca^{2+}·CaM- or cAMP-dependent processes which trigger chromosome replication in rat thymic lymphoblasts and T51B rat liver cells arrested in late G_1 by external Ca^{2+} deprivation. This complicated scheme can explain how such things as the tumour promoter TPA and the mitogenic/oncogenic membrane-bound $pp60^{v-src}$ protein kinase of avian sarcoma virus trigger chromosome replication when the cell is arrested in external Ca^{2+}-deficient medium. TPA triggers the web by causing its receptor, the normally Ca^{2+}-dependent PKC, to function without Ca^{2+}; $pp60^{v-src}$ converts membrane PIP to PIP_2, which is split into PI_3, which triggers the web by mobilizing *internal* Ca_S, and DG, which, along with the released Ca^{2+}, stimulates protein kinase C (Marx, 1984). Depending on the type and density of cell surface receptors, insulin or insulin-like factors (such as somatomedin C (SmC)) promote G_1 transit by: (a) maximizing system A amino acid transport by stimulating Na^+/K^+ ATPase; (b) stimulating Na^+/H^+ exchange which raises internal pH and thereby sensitizes Ca^{2+}-dependent enzymes and internal receptors of Ca^{2+} by reducing the number of H^+ competing with Ca^{2+} for binding sites; and (c) increasing the number of glucose transporters in the cell membrane (Moore, 1983; Whitfield *et al.*, 1985). All abbreviations are defined in the text.

during normal G_1 transit.

When serum-stimulated confluent T51B liver cells enter the final, external Ca^{2+}-dependent stage of their normal (that is, not 'explosive') G_1 transit (between 8 hours and the onset of DNA replication around 11 hours after serum addition) they rapidly become insensitive to actinomycin D, but important proteins must be made at this time because the cells stay sensitive

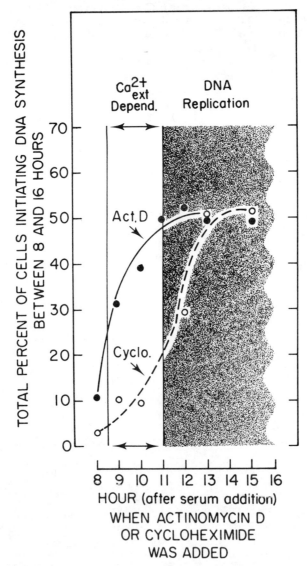

Fig.14.9. When serum-stimulated, confluent T51B rat liver cells enter the external Ca^{2+}-dependent G_1 period (compare Fig.14.5), they rapidly become insensitive to the gene transcription inhibitor actinomycin D (0.05 µg/ml), but they remain sensitive to the translation inhibitor cycloheximide (1 µg/ml) until the onset of DNA replication. Fresh medium (80% Eagle's basal medium + 20% bovine calf serum) was added at time '0', [^{3}H]-thymidine (2 µCi/ml) was added at 8 hours, the cells were prepared for autoradiography at 16 hours, and the inhibitors were added at various times.

to cycloheximide until the onset of DNA replication (Fig.14.9). Thus, external Ca^{2+} and/or $Ca^{2+}\cdot CaM$ might directly promote the accumulation of critical proteins. The $Ca^{2+}\cdot CaM$-dependent events in the last 2 hours of the G_1 phase of serum-starved tsLA-23 NRK rat cells (stimulated either by serum or activating the resident tsLA-23 ASV's thermolabile, mitogenic pp60$^{v\text{-}src}$ protein kinase) are also insensitive to actinomycin D, but, unlike those in T51B cells, they are relatively insensitive to cycloheximide (Durkin and Whitfield, 1984). Therefore, it seems that it is external Ca^{2+} which promotes (perhaps by supporting a diacyglycerol (DG)-induced burst of protein kinase C activity) the accumulation of essential proteins (by stimulating their synthesis and/or stabilizing them) while $Ca^{2+}\cdot CaM$ does something else.

Because of the lateness of its involvement in the G_1 transit of Chinese hamster CHO cells, NRK rat cells, and T51B rat liver cells (Boynton and Whitfield, 1981; Boynton, Kleine and Whitfield, 1985; Chafouleas *et al.*, 1984); Durkin, MacManus and Whitfield, 1983; Whitfield *et al.*, 1982), $Ca^{2+}\cdot CaM$ probably operates a 'non-synthetic' mechanism which activates DNA replication in the fully primed cell. This might be the delivery of accumulated replication enzymes and initiator(s) to the nucleus, as suggested by the abrupt entry of cytoplasmically accumulating ribonucleotide reductase holoenzyme into the regenerating liver cell's nucleus as Ca^{2+} is being discharged from the ER just before the onset of DNA replication (see Fig.14.6A) (Youdale *et al.*, 1984). By stimulating PDE and the cell membrane's Ca^{2+} pumps, $Ca^{2+}\cdot CaM$ may also prevent both itself and the G_1 cAMP surge from lasting too long and disrupting subsequent events (see Fig.14.8) (Boynton and Whitfield, 1983).

14.4. Ca^{2+} and $Ca^{2+}\cdot CaM$ in pre-G_1 stages

The importance of Ca^{2+} in type 2 pre-G_1 stages (see Fig.4.2) has been hard to assess, mainly because some quiescent cells (such as confluent T51B liver cells, serum-starved human WI-38 fibroblasts) need 1.0–1.8 mM external Ca^{2+} during their serum-triggered type 2 pre-G_1 stages (see Fig.14.2), but others (such as confluent or serum-starved BALB/c 3T3 and Swiss albino 3T3 mouse cells, *sparse* serum-starved T51B cells) can operate with 100 times less external Ca^{2+} (Boynton and Whitfield, 1984; Whitfield *et al.*, 1982, 1985). However, the fact that a large sharp rise in external Ca^{2+} arouses *all* kinds of G_0 cell (Whitfield *et al.*, 1982, 1985) suggests that an internal Ca^{2+} transient indeed may be a necessary part of the type 2 triggering mechanism. Moreover, we now know that cells such as BALB/c 3T3 and Swiss albino 3T3 mouse cells do not need external Ca^{2+} to trigger prereplicative development simply because when quiescent they build up an internal calcium reserve which is released into the cytosol upon stimulation by serum (Boynton and Whitfield, 1983; Boynton *et al.*, 1982; Engstrom, Zetterberg and Auer, 1982; Lopez-Rivas and Rozengurt, 1983). Serum also triggers large internal Ca^{2+} transients in serum-starved human cells (Cobbold and Goyns, 1983). An internal Ca^{2+} transient probably triggers PLA$_2$-PLC cascades, whose DG and prostaglandin (PG) products stimulate other enzymes such as protein kinase C (see Fig.14.8) (Whitfield *et al.*, 1985). Significantly, there are pre-G_1 bursts of protein kinase C activity in serum-stimulated confluent T51B cells (Boynton, Kleine and Whitfield, 1985) and the protein kinase C stimulator, TPA, replaces external Ca^{2+} and enables quiescent confluent T51B cells to respond to fresh external Ca^{2+}-deficient medium-serum (unpublished observation). These and/or the Ca^{2+} transient, which is caused by the internal Ca^{2+} transient, could be responsible for the pH rise in platelet-derived growth factor (PDGF)-stimulated or serum-stimulated Swiss albino 3T3 cells (Boynton et al., 1982; Cassel et al., 1983) which is known to be able to start the events leading to DNA replication all by itself without stimulating protein synthesis or growth (Engstrom, Zetterberg and Auer, 1982; Zetterberg and Engstrom, 1983).

An increase in internal Ca^{2+}, a Ca^{2+}-induced PLA_2-arachidonate cascade and a Ca^{2+}·CaM-induced pH rise appear also to function in the other two kinds (see Figs.14.3 and 14.4) of pre-G_1 stage (Mastro and Smith, 1982; Rixon, MacManus and Whitfield, 1979; Whitfield et al., 1982, 1985). While we do not know what they do in the pre-G_1, stage of the small lymphocyte, we now know what a transient PLA_2-arachidonate cascade does in the Pre-G_1, stage of the regenerating rat liver cell. Prostaglandins produced by this cascade are needed briefly between 2 and 3 hours after HPX to stimulate something which is concerned *not* with initiating DNA replication but with entry into mitosis *after* chromosome replication (Rixon, MacManus and Whitfield, 1979; Whitfield, Rixon and Sikorska, 1984).

It seems from all of this that some Ca^{2+}- or Ca^{2+}·CaM-triggered pre-G_1 events lead to DNA replication while others enable the cell to grow to a critical size needed for entry into mitosis.

14.5. cAMP and cAMP protein kinase (cAMP·PK) in G_1 phase

There is a transient cAMP surge in the G_1 phases of cells ranging from free-living unicellular diatoms, budding yeast and *Tetrahymena* to regenerating planarian cells, annelid blastema cells, rat aortic smooth muscle cells, rat hepatocytes and human lymphoma cells (Boynton and Whitfield, 1983; Franks, Plamondon and Hamet, 1984; Whitfield et al., 1985). This cAMP transient seems to result from a cycloheximide-insensitive burst of adenylate cyclase activity and a transient reduction of PDE activity (Boynton and Whitfield, 1983; Franks, Plamondon and Hamet, 1984). Its occurrence even in unicellular eukaryotes indicates that it is the work of an ancient mechanism which makes or activates an endogenous adenylate cyclase stimulator (such as a cholera toxin-like factor or a prostaglandin) and briefly activates a cyclic nucleotide phosphodiesterase (PDE) inhibitor or inactivates a PDE activator (such as Ca^{2+}·CaM). However, some cells in multicellular organisms have lost their independence. Thus, external glucagon seems to control the G_1 cAMP transient in regenerating rat liver cells, because these cells cannot transit G_1 without the normal HPX-induced rise in circulating glucagon, and cultured neonatal and adult rat hepatocytes need this hormone or help from external cAMP to transit G_1 (see Armato, Andreis, and Whitfield, 1983; Boynton and Whitfield, 1983; Whitfield et al., 1985). This transfer of control of the G_1 cAMP transient from an endogenous agent such as a prostaglandin to an external hormone explains the fact that blocking prostaglandin production with indomethacin stops the transient in mouse spleen lymphocytes *in vitro*, but does not stop it in regenerating rat liver cells (Whitfield, 1976, 1980).

Although we know very little about its cause, we do know that the G_1 cAMP transient and the accompanying burst of cAMP·PK activity are needed for G_1 transit, because:

(1) yeast (*Saccharomyces cerevisiae*) cells with defective adenylate cyclase cannot go beyond a G_1 start point without help from external cAMP;
(2) a yeast mutant making cAMP·PK with a reduced affinity for cAMP cannot transit G_1 without help from external cAMP;

(3) a mutation which permanently raises PK activity by preventing production of the cAMP·PK's regulatory (inhibitory) subunit enables yeast with defective adenylate cyclase to transit G_1 without help from external cAMP;
(4) yeast α-mating pheromone is a tridecapeptide adenylate cyclase inhibitor that stops cells of the opposite mating type at the G_1 start point;
(5) external cAMP shortens this α-pheromone-induced G_1 block;
(6) the prompt web (see Fig.14.8)-induced DNA-synthetic responses of external Ca^{2+}-deprived thymic lymphoblasts and T51B cells to external Ca^{2+} are preceded by cAMP transients;
(7) cAMP, cAMP elevators, type II cAMP·PK holoenzyme, or cAMP·PK catalytic subunits all get the same web (see Fig.14.8)-induced DNA-synthetic response from external Ca^{2+}-deprived rat thymic lymphoblasts and T51B cells as Ca^{2+}_{ext} addition;
(8) delaying or stopping the cAMP transient in these thymic lymphoblasts or T51B liver cells delays or prevents the onset of DNA replication;
(9) delaying or preventing the G_1 cAMP transient in mouse lymphocytes delays or prevents the onset of DNA replication;
(10) injecting propranolol in G_1 stops the G_1 cAMP transient and prevents the onset of DNA replication in rat parotid cells, but injecting cAMP along with the drug enables these cells to finish G_1 transit and initiate DNA replication;
(11) rat hepatocytes need external cAMP or a cAMP elevator (glucagon) to transit G_1;
(See Boynton and Whitfield (1983), Thorner (1982), Whitfield et al.(1982, 1985) for original references).

How do cAMP and cAMP·PK drive G_1 transit? This is difficult to answer because they affect so many things. Clues to this mystery are the facts that the prokaryotes invented cAMP and the cAMP-binding/activated CAP protein to stimulate genes and the cAMP/cAMP·PK combination of eukaryotes also stimulates genes such as the one coding for prolactin (Rosenfeld et al., 1983; Whitfield et al., 1985). Indeed, the cAMP/cAMP·PK couplet of Saccharomyces cerevisiae probably stimulates cdc 28, a master G_1 'start' gene, whose product is related both to cAMP·PK catalytic subunit and the putative protein kinase products of several viral oncogenes (Boynton and Whitfield, 1983; Edwards et al., 1985; Lorinez and Reed, 1984; Thorner, 1982; Whitfield et al., 1985). This cdc 28 product, in turn, triggers a cascade of other gene activations which lead to chromosome replication, budding and cell division (Fig.14.10) (Edwards et al., 1978; Thorner, 1982). The appearances of calmodulin and oncofetal c-ras gene transcripts and replication enzymes such as ribonucleotide reductase at the time of the G_1 cAMP transient in Chinese hamster CHO cells and regenerating rat liver cells is consistent with this (see Fig.14.6a) (Boynton and Whitfield, 1983; Chafouleas, 1984; Fausto, 1984; Whitfield et al., 1976, 1985; Youdale et al., 1984).

In mammalian cells, the type II cAMP·PK holoenzyme seems to be specially linked to the initiation of DNA replication, because: (a) it, but not the type I holoenzyme, can trigger DNA replication in external Ca^{2+}-deprived T51B liver cells; (b) the onset of DNA replication in calcitonin-stimulated T47D human mammary cells is preceded by activation of type II, but not type I, cAMP·PK; (c) Chinese hamster CHO cells, rat-1 fibroblasts and regenerating rat liver cells enrich their cAMP·PK pools with the type II isoenzyme in G_1 phase (see Boynton and Whitfield, 1983; Boynton, Kleine and Whitfield, 1985; Sikorska, Whitfield and Rixon, 1983; Whitfield et al., 1985). Thus, using the cAMP/cAMP·PK stimulation of a large block of genes in pituitary GH4 cells as a precedent (Rosenfeld et al., 1983), we suggest that the G_1 cAMP transient in mammalian cells stimulates the genes coding for replication enzymes and initiator protein(s) (see Fig.14.8) by activating a cAMP·PK (in this case the type II enzyme) which, in turn, phosphorylates a chromatin-associated regulator protein like the 23kDa basic protein that regulates the large block of genes in pituitary GH4 cells (Rosenfeld et al., 1983). The type II holoenzyme could be activated by cAMP in the nucleus or it could be transported into the nucleus during the G_1 cAMP transient as a

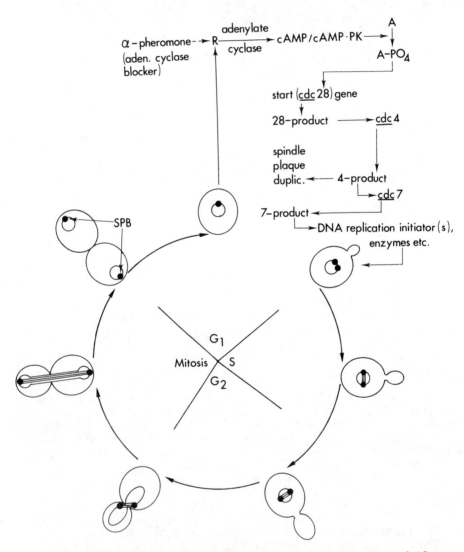

Fig.14.10. When a yeast (*Saccharomyces cerevisiae*) cell reaches the G_1 'start' point, a burst of AC activity and a cAMP transient turn on a master 'start' gene, *cdc* 28, whose product triggers a cascade of gene stimulations resulting in the production of replication enzymes and initiator(s) (Edwards et al., 1978; Thorner, 1982). All abbreviations are defined in the text.

cAMP·holoenzyme complex such as the one described by Cho-Chung (1980). An intriguing tale of relations between plasma Ca^{2+} (that is, external Ca^{2+}), $1,25(OH)_2D_3$, cAMP, cAMP·PK, replication enzymes and DNA replication in regenerating rat liver cells is beginning to unfold (Sikorska, Whitfield and Rixon, 1983 a and b; Whitfield, Rixon and Sikorska, 1984). In normal rat liver

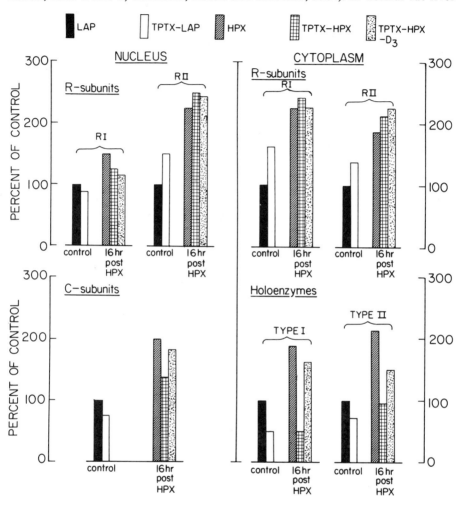

Fig.14.11. TPTX 72 h before HPX in the rat reduces or prevents the post-HPX G_1 cAMP·K holoenzyme transients in the cytoplasms and nuclei of the remaining liver cells by reducing or stopping the synthesis of the cAMP·PK catalytic (C) subunit without affecting the surges in the levels of the two regulatory (R_I and R_{II}) subunits. One intraperitoneal injection of $1,25(OH)_2D_3$ (600 ng/100 g of body weight) into the 72 h TPTX rat at the time of HPX restores, or nearly restores, the cytoplasmic and nuclear holoenzyme transients by restoring synthesis of their common C subunit. Tissue was homogenized and cytoplasmic and nuclear fractions were prepared according to Sikorska, Whitfield and Rixon (1983a). The assay of the R subunit contents was based on their abilities to bind avidly [^{3}H]-cAMP. Thus, the R_I and R_{II} subunits were resolved on a Mono Q column of the Pharmacia (Uppsala, Sweden) FPLC chromatography system with a linear NaCl gradient (from 0 to 0.6 mM in 20 mM TRIS/HCl buffer, pH 7.5) and eluted into a cAMP-containing assay mixture (5 μM [^{3}H]cAMP, 2 mM 3-isobutyl-1-methylxanthine, 10 mM $MgCl_2$, 20 mM TRIS/HCl (pH 7.5), and 5 mg of sample protein) and the cAMP binding activity assayed according to Gilman (see MacManus et al., 1983). The fractions' holoenzyme contents were measured according to Sikorska, Whitfield and Rixon (1983a).

cells type I cAMP·PK is mainly cytoplasmic while the type II enzyme is mainly nuclear. HPX triggers pre-G_1 and G_1 cytoplasmic type I and nuclear type II holoenzyme transients as well as a large G_1 type II holoenzyme transient in the cytoplasm, all of which coincide with the pre-G_1 and G_1 cAMP transients. Thyroparathyroidectomy (TPTX), which lowers the plasma Ca^{2+} concentration and causes a severe $1,25(OH)_2D_3$ shortage somehow inhibits synthesis of the common catalytic subunit of the two cAMP·PKs *without* reducing synthesis of the two regulatory subunits (Fig.14.11). Thus, in the $1,25(OH)_2D_3$-deficient 72 h TPTX rat, HPX still triggers pre-G_1 and G_1 transients of the two regulatory subunits, but not the holoenzyme transients because the cells cannot make more catalytic subunits. Under these restrictive conditions HPX still causes the pre-G_1 and G_1 cAMP transients, but now there are periodic excesses of cAMP-binding regulatory subunits and a shortage of holoenzymes for cAMP to activate which together render these transients impotent. This may be one of the reasons why HPX-stimulated liver cells cannot make ribonucleotide reductase subunits or initiate DNA replication in the TPTX rat (see Fig.14.12 and 14.6b) (Youdale et al., 1984).

Raising the plasma Ca^{2+} concentration, by injecting intraperitoneally a low dose (85 mg/100 g of body weight) of $1,25(OH)_2D_3$ and long-acting parathyroid extract (50 USP U/100 g of body weight) into the 72 h TPTX rat at the time of HPX, does not enable the remaining liver cells to initiate DNA replication. This cocktail still raises plasma Ca^{2+} but now enables the cells to start replicating their DNA and enter mitosis at the normal times when supplemented with calcitonin (100 mU/100 g of body weight) (Rixon, MacManus and Whitfield, 1979). Parathyroid extract is not needed when the

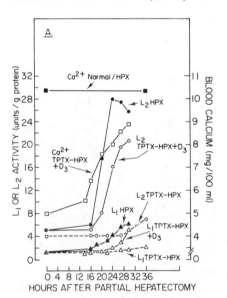
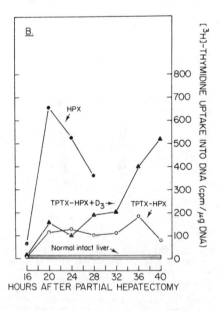

Fig.14.12. Most HPX-stimulated liver cells in the 72 h TPTX rat (A) do not accumulate the specific substrate (CDP)/effector-binding L_1 and the non-haem-iron-containing 'catalytic' L_2 subunits of ribonucleotide reductase (Youdale *et al.*, 1984) and (B) do not initiate DNA replication (see also Fig.14.6). However, just one intraperitoneal injection of $1,25(OH)_2D_3$ (400 ng/100 g of body weight) at the time of HPX (A) causes the TPTX rat's depressed plasma Ca^{2+} concentration to rise and enables the cells to make cytoplasmic L_1 and L_2 subunits and (B) initiate DNA replication. L_1 and L_1 subunits and holoenzyme activities were assessed from the rate of reduction of CDP to dCDP in an assay system described by Youdale *et al.* (1984).

TPTX animal is injected with calcitonin and a higher dose (170 ng/ 100 g of body weight) of $1,25(OH)_2D_3$ which cooperate to raise plasma Ca^{2+} and enable the cells to multiply (Rixon, MacManus and Whitfield). However, calcitonin serves simply to increase the effectiveness of $1,25(OH)_2D_3$, higher doses (400–600 ng/100 g of body weight) of which by themselves stimulate the synthesis of cAMP·PK catalytic subunits (see Fig.14.11), restore or at least partly restore the pre-G_1 and G_1 cAMP·PK holoenzyme transients in the cytoplasm and nucleus (see Fig.14.11; refs. Sikorska, Whitfield and Rixon, 1983a and b), and enable a delayed accumulation of ribonucleotide reductase subunits (see Fig.14.12a) and a delayed initiation of DNA synthesis (see Fig.14.12b). In other words, just one injection of a high dose of $1,25(OH)_2D_3$ at the time of HPX restores (albeit slowly) *all* of the essential prereplicative events which are missing in the TPTX rat.

We must note that actinomycin D stops $1,25(OH)_2D_3$ from stimulating synthesis of cAMP·PKs in the TPTX rat, which suggests that the steroid acts by stimulating the appropriate gene despite the fact that rat liver cells do not contain the specific cytosolic receptor with which the steroid combines to form gene-activating complexes in a more conventional target tissue such as intestine (Whitfield, Rixon and Sikorska, 1984). Thus, either the steroid acts through a different receptor, or the stimulation of the cAMP·PK catalytic subunit gene might be a byproduct of one of more conventional actions of the steroid such as the raising of plasma Ca^{2+}. However, the latter is unlikely, because the steroid restores cAMP·PK levels and the cAMP·PK responsiveness to HPX in less than 4 hours, which is long before plasma Ca^{2+} starts rising (see Fig.14.12a; Sikorska, Whitfield and Rixon, 1983a and b; Whitfield, Rixon and Sikorska, 1984).

14.6. cAMP and cAMP·PK in Pre-G_1 stages

While the cAMP/cAMP·PK couplet is a prime mover of G_1 transit, it functions differently in the three kinds of pre-G_1 stage.

There is no pre-G_1 cAMP transient in most type 2 periods. In fact, serum deprivation usually stops cells in the middle of the G_1 transient, and the excess cAMP is one of the things erased during the pre-G_1 stage (Boynton and Whitfield, 1983). However, external cAMP and cAMP elevators can arouse quiescent, serum-deprived Swiss albino 3T3 mouse cells (Boynton and Whitfield, 1983), possibly by mobilizing internal calcium.

A pre-G_1 cAMP transient and specific activation of type 1 cAMP·PK occur in lectin-activated lymphocytes, but we do not know their purpose if any (Boynton and Whitfield). However, the pre-G_1 cAMP surge, and presumably the large cytoplasmic pre-G_1 type I cAMP·PK holoenzyme transient (Sikorska, Whitfield and Rixon, 1983a) in regenerating liver cells probably cooperate with prostaglandins produced by the pre-G_1 Ca^{2+}-dependent PLA_2-arachidonate cascade (Rixon and Whitfield, 1982) to stimulate something (cell growth?) needed *not* for the initiation of DNA replication but for entry into mitosis. Thus, reducing the pre-G_1 cAMP transient by injecting propranolol intraperitoneally during the first hour after HPX does not affect the onset of DNA replication, but it prevents the stimulated cells from entering mitosis, an effect which can be remedied completely by injecting cAMP (but not 2',3'-cAMP) no earlier than 1.5 hours and no later than 2 hours after HPX (MacManus et al., 1973; and unpublished observations). Significantly, in view of the likelihood of the main role of the G_1 cAMP transient being stimulation of gene transcription, the pre-G_1 need for cAMP coincides exactly with the triggering of a 4-hour burst of mRNA production (Walker and Whitfield, 1984). Finally, it appears that the prostaglandins and cAMP/cAMP·PK function independently, because the PG-inhibiting indomethacin does not affect the pre-G_1 cAMP transient (Whitfield et al., 1976), the PG-dependent step occurs about 1 hour after the cAMP-dependent step, and injecting cAMP does not lift the blockage of the mitosis-determining process by hydrocortisone (which blocks the release of arachidonate from membrane phospholipids by stimulating the synthesis of the PLA_2 inhibitor, lipomodulin) (Rixon and Whitfield, 1982 and unpublished observations).

14.7. Conclusions

Although most of the pieces are still missing from this awesome puzzle, a picture of how Ca^{2+} and cAMP might coordinate and drive prereplicative development is starting to form. It seems that while cAMP and type I cAMP·PK are needed to start cell growth, it is cAMP and type II cAMP·PK which either directly stimulate the set of genes for replication enzymes or stimulate an oncogene-like master 'starter' gene(s) whose product, in turn, stimulates the genes for replication enzymes. External Ca^{2+}, internal Ca^{2+} and Ca^{2+}·CaM have a bewildering array of functions, the ultimate purposes of which may be to promote translation of the cAMP/cAMP·PK-evoked gene transcripts and accumulation of the finished products in the cytoplasm, to promote delivery of the finished replication enzymes and initiator(s) to the nucleus, and finally to switch off the G_1-specific cAMP and Ca^{2+}·CaM signals to avoid disrupting subsequent events.

Acknowledgements

One of us, D. Bossi, was supported by a NATO senior fellowship. We thank D. Gillan for preparing the illustrations. NRCC No.23447

Abbreviations

AC, adenylate cyclase; *ASV*, avian sarcoma virus; *Ca·ER*, calcium stored in endoplasmic reticulum (microsomes); Ca^{2+}_{ext}, external Ca^{2+}; Ca_S, internally sequestered and/or bound calcium; Ca^{2+}·*CaM*, calcium·calmodulin complex; *cAMP*, adenosine 3',5'-monophosphate; *2',3'-cAMP*, adenosine 2',3'-monophosphate, an inactive cAMP analogue; *cAMP·PK*, cAMP-dependent protein kinase; *CDP*, cytidine 5'-diphosphate; *dCDP*, 2'-deoxycytidine 5'-diphosphate; *DG*, diacylglycerol; *EGF*, epidermal growth factor; *EGTA*, a specific Ca^{2+} chelator; *ER*, endoplasmic reticulum (microsomes); *IIPX*, 70% (partial) hepatectomy; *IP₃*, inositol 1,4,5-*tris*phosphate; *IPR*, isoproterenol (a synthetic β-adrenergic agonist); *PDE*, cyclic nucleotide phosphodiesterase; *PDGF*, platelet-derived growth factor; *PG*, prostaglandin; *PIP*, phosphatidyl inositol 4-phosphate; PIP_2, phosphatidylinositol 4,5-*bis*phosphate; *PKC*, protein kinase C; PLA_2, the Ca^{2+}-activated phospholipase A_2; *PLC*, the Ca^{2+}-activated phospholipase C; *PLI*, phospholipid Ca^{2+}-ionophore (such as phosphatidic acid) produced by PLC activity; *PrP*, protein phosphatases; *RR*, ribonucleoside diphosphate reductase; *SmC*, somatomedin C; *SPB*, nuclear spindle plaque body of budding yeast; *TPA*, 12-O-tetradecanoyl phorbol-13-acetate, a PKC stimulator and tumor promoter; *TPTX*, thyroparathyroidectomy; *72 h TPTX* rat, a rat which

had been thyroparathyroidectomized 72 hours previously; *ts*ASV, a
temperature-sensitive, transformation-defective ASV; *1,25(OH)$_2$D$_3$* or *D$_3$*,
1,25-dihydroxyvitamin D$_3$; *XP*, unknown phosphorylated mediator(s).

References

Armato, U., Andreis, P. and Whitfield, J. (1983). The calcium-dependence of the stimulation of neonatal hepatocyte DNA synthesis and division by epidermal growth factor, glucagon and insulin. *Chem.-Biol. Int.* 45, 203–22.

Boynton, A. and Whitfield, J. (1981). Calmodulin and cyclic AMP-dependent protein kinase mediate calcium-induced stimulation of DNA synthesis by rat liver cells. *Adv. Cyclic Nucleotide Res.* 14, 411–9.

Boynton, A. and Whitfield, J. (1983). The role of cyclic AMP in cell proliferation: a critical assessment of the evidence. *Adv. Cyclic Nucleotide Res.* 15, 193–294.

Boynton, A. and Whitfield, J. (1984). The effect of serum on the calcium content of BALB/c 3T3 mouse cells. *Exp. Cell Res.* 152, 266–9.

Boynton A., Kleine, L. and Whitfield, J. (1983). Ca/phospholipid-dependent protein kinase activity correlates to the ability of transformed liver cells to proliferate in Ca^{++}-deficient medium. *Biochem. Biophys. Res. Commun.* 115, 383–90.

Boynton, A., Kleine, L. and Whitfield, J. (1985). The role of calcium in prereplicative development of non-neoplastic and neoplastic rat liver cells. In *Ions. Membranes and the Electrochemical Control of Cell Function*, A. Pilla and A. Boyton (eds.). Verlag-Chemie, New York.

Boynton, A., Kleine, L. and Whitfield, J. (1985). Cyclic AMP elevators stimulate the initiation of DNA synthesis by calcium-deprived rat liver cells. Boynton and Leffert, *Control of Animal Cell Proliferation, Vol. 1.* A. Boynton and H. Leffert (eds.). Academic Press, New York.

Boynton, A., McKeehan, W., Whitfield, J., et al. (1982). *Ions, Cell Proliferation and Cancer.* Academic Press, New York.

Cassel, O., Rothenburg, P. Zhuang, Y., Deuel, J. and Glaser, L. (1983). Platelet-derived growth factor stimulates Na$^+$/H$^+$ exchange and induces cytoplasmic alkalinization in NR6 cells. *Proc. Natl. Acad. Sci. USA* 80, 6224–8.

Chafouleas, J., Lagace, L., Bolton, W., Boyd, A. and Means, A. (1984). Changes in calmodulin and its mRNA accompany reentry of quiescent (GO) cells into the cell cycle. *Cell* 36, 73–81.

Cho-Chung, Y. (1980). Cyclic AMP and its receptor protein in tumor growth regulation *in vivo*. *J. Cyclic Nucleotide Res.* 6, 163–77.

Cobbold, P,. and Goyns, M. (1983). Measurements of the free calcium concentration of single quiescent human fibroblasts before and after serum addition. *Biosci. Rep.* 3, 79–86.

Durkin, J. and Whitfield, J. (1984). Partial characterization of the mitogenic action of pp60^{v-src}, the oncogenic protein product of the *src* gene of avian sarcoma virus. *J. Cell. Physiol.*

Durkin, J., Boynton, A. and Whitfield, J. (1981). The *src* gene product (pp60^{src}) of avian sarcoma virus rapidly induces DNA synthesis and proliferation of calcium-deprived rat cells. *Biochem. Biophys. Res. Commun.* 103, 233–9.

Durkin, J., MacManus, J. and Whitfield, J. (1983). The role of calmodulin in the proliferation of transformed and phenotypically normal *ts*ASV-infected rat cells. *J. Cell. Physiol.* 115, 313–9.

Edwards, D., Taylor, J., Wakeling, W., Watts, F. and Johnston, I. (1978). Studies on the prereplicative phase of the cell cycle in *Saccharomyces cerevisiae*. *Cold Spring Harbor Symp. Quant. Biol.* **43**, 577–86.

Engstrom. W., Zetterberg, A. and Auer, G. (1982). Calcium, phosphate and cell proliferation. *Ions, Cell Proliferation, and Cancer*, A. Boynton, McKeehan and J. Whitfield (eds.) pp. 341–57. Academic Press, New York.

Epel, D. (1982). The cascade of events initiated by rises in cytosolic Ca^{2+} and pH following fertilization in sea urchin eggs. In *Ions, Cell Proliferation, and Cancer*. A. Boynton, W. McKeehan and J. Whitfield (eds.) pp. 327–39. Academic Press, New York.

Fausto, N. (1984). Messenger RNA in regenerating liver: implications for the understanding of regulated growth. *Mol. Cell. Biochem.* **59**, 131–47.

Franks, D., Plamondon, J. and Hamet, P. (1984). An increase in adenylate cyclase activity precedes DNA synthesis in cultured vascular smooth muscle cells. *J. Cell. Physiol.* **119**, 41–5.

Hennings, H., Holbrook, K. and Yuspa, S. (1983). Factors influencing calcium-induced terminal differentiation in cultured mouse epidermal cells. *J. Cell. Physiol.* **116**, 265–81.

Howard, A. and Pelc, S. (1953). Synthesis of deoxyribonucleic acid in normal and irradiated cells and its relation to chromosome breakage. *Heredity, Supple.* **6**, 261–73.

Ingebritsen, T. and Cohen, P. (1983). Protein phosphatases: properties and role in cellular regulation. *Science* **221**, 331–8.

Kleine, L., Boynton, A. and Whitfield, J. (1985). A role for protein kinases in the initiation of DNA synthesis by rat liver cells. In *Ions, Membranes and the Electrochemical Control of Cell Function.* A. Pilla and A. Boynton (eds.) Verlag-Chemie, New York.

Lopez-Rivas, A. and Rozengurt, E. (1983). Serum rapidly mobilizes calcium from an intracellular pool in quiescent fibroblastic cells. *Biochem. Biophys. Res. Commun.* **114**, 240–7.

Lorinez, A. and Reed, S. (1984). Primary structure homology between the product of yeast cell division control gene CDC 28 and vertebrate oncogenes. *Nature (Lond.)* **307**, 183–5.

MacManus, J., Braceland, B., Youdale, T. and Whitfield, J. (1973). Adrenergic antagonists and a possible link between the increase in cyclic adenosine 3',5'-monophosphate and DNA synthesis during liver regeneration. *J. Cell. Physiol.* **82**, 157–64.

MacManus, J. and Whitfield, J. (1981). Stimulation of autophosphorylation of liver cell membrane proteins by calcium and partial hepatectomy. *J. Cell. Physiol.* **106**, 33–40.

Marx, J. (1984). A new view of receptor action. *Science* **224**, 271–4.

Mastro, A. and Smith, M. (1982). Two signal, calcium-dependent lymphocyte activation by an ionophore, A23187, and a comitogenic phorbol ester. *J. Cell Biol.* **95**, 21a.

Méchali, M., Méchali, F. and Laskey, R. (1983). Tumor promoter TPA increases initiation of replication on DNA injected into Xenopus eggs. *Cell* **35**, 63–9.

Moore, R. (1983). Effects of insulin upon ion transport. *Biochim. Biophys. Acta* **737**, 1–49.

Niedel, J., Kuhn, L. and Vandenbank, G. (1983). Phorbol diester receptor copurifies with protein kinase C. *Proc. Natl. Acad. Sci. USA* **80**, 36–40.

Rixon, R., MacManus, J. and Whitfield, J. (1979). The control of liver regeneration by calcitonin, parathyroid hormone and 1α, 25-dihydroxycholecalciferol. *Mol. Cell. Endocrinol.* **15**, 79–89.

Rixon, R. and Whitfield, J. (1982). An early mitosis-determining event in regenerating rat liver and its possible mediation by prostaglandins or thromboxane. *J. Cell. Physiol.* **113**, 281–8.

Rosenfeld, M., Amara, S., Birnberg, N., Mermond, J., Murdoch, G. and Evans, R. (1983). Calcitonin, prolactin, and growth hormone gene expression as model systems for the characterization of neuroendocrine regulation. *Rec. Prog. Horm. Res.* **39**, 305–51.

Schatten, G. (1982). Motility during fertilization. *Int. Rev. Cytol.* **79**, 59–163.

Sikorska, M., MacManus, J., Walker, P., Whitfield, J. (1980). The protein kinases of rat liver nuclei. *Biochem. Biophys. Res. Commun.* **93**, 1196–203.

Sikorska, M., Whitfield, J. and Rixon, R. (1983a). The effects of thyroparathyroidectomy and 1,25-dihydroxyvitamin D_3 on changes in the activities of some cytoplasmic and nuclear protein kinases during liver regeneration. *J. Cell. Physiol.* **115**, 297–304.

Sikorska, M., Whitfield, J. and Rixon, R. (1983b). 1,25-dihydroxyvitamin D_3 controls the synthesis of type I and II cyclic AMP-dependent protein kinase in normal and regenerating rat liver. *J. Cell Biol.* **97**, 147a.

Swierenga, S., Whitfield, J., Boynton, A., MacManus, J., Rixon, R., Sikorska, M., Tsang, B. and Walker, P. (1980). Regulation of proliferation of normal and neoplastic rat liver cells by calcium and cyclic AMP. *Ann. NY Acad. Sci.* **349**, 294–311.

Thorner, J. (1982). An essential role for cyclic AMP in growth control: the case for yeast. *Cell* **30**, 5–6.

Walker, P. and Whitfield, J. (1984). Colchicine prevents the translation of mRNA molecules transcribed immediately after proliferative activation of hepatocytes in regenerating rat liver. *J. Cell. Physiol.* **118**, 179–85.

Weber, I., Takio, K., Titani, K. and Steitz, T. (1982). The cAMP binding domains of the regulatory subunits of cAMP-dependent protein kinase and the catabolite gene activator protein are homologous. *Proc. Natl. Acad. Sci. USA* **79**, 7679–83.

Whitfield, J., Rixon, R. and Sikorska, M. (1985). Calcium, cyclic AMP and protein kinases combine to trigger chromosome replication in regenerating liver. In *Ions, Membranes and the Electrochemical Control of Cell Function*. A. Pilla and A. Boynton (eds.) Verlag-Chemie, New York.

Whitfield, J., Boynton, A., MacManus, J., Rixon, R., Sikorska, M., Tsang, B., Walker, P. and Swierenga, S. (1980). The roles of calcium and cyclic AMP in cell proliferation. *Ann. NY Acad. Sci.* **339**, 216–40.

Whitfield, J., Boynton, A., MacManus, J., Rixon, R., Walker, P. and Armato, U. (1976). The positive regulation of cell proliferation by a calcium-cyclic AMP control couplet. In *Cyclic Nucleotides and the Regulation of Cell Growth*. Abou-Sabe (ed.), pp. 97–130, Dowden, Hutchinson and Ross, Stroudsburg, PA.

Whitfield, J., Boynton, A., MacManus, J., Sikorska, M. and Tsang, B. (1979). The regulation of cell proliferation by calcium and cyclic AMP. *Mol. Cell. Biochem.* **27**, 155–79.

Whitfield, J., Boynton, A., Rixon, R. and Youdale, T. (1985). The control of cell proliferation by calcium, Ca^{2+}-calmodulin and cyclic AMP. In *Control of Animal Cell Proliferation, Vol.1*, A. Boynton and Leffert (eds.) Academic Press, New York.

Whitfield, J., MacManus, J., Boynton, A., Durkin, J. and Jones, A. (1982). Futures of calcium, calcium-binding proteins, cyclic AMP and protein kinases in the quest for an understanding of cell proliferation and cancer. In *Functional Regulation at the Cellular and Molecular Levels*, R. Corradino (ed.) pp. 61–87, Elsevier/North-Holland, New York.

Wilson, T. and Lin, E. (1980). Evolution of membrane bioenergetics. *J.*

Supramol. Struct. **13**, 421—46.

Youdale, T., Frappier, L., Whitfield, J. and Rixon, R. (1984). Changes in the cytoplasmic and nuclear activities of the ribonucleotide reductase holoenzyme and its subunits in regenerating liver cells in normal and thyroparathyroidectomized rats. *Can. J. Biochem. Cell Biol.*

Zetterberg, A. and Engstrom, W. (1983). Induction of DNA synthesis and mitosis in the absence of cellular enlargement. *Exp. Cell Res.* **144**, 199—207

PHOSPHOINOSITIDES AND CALCIUM MOBILIZING RECEPTORS

James W. Putney, Jr.

Medical College of Virginia, Division of Cellular Pharmacology, Richmond, Virginia 23298-0001, USA

Abstract

In a wide variety of tissues, Ca-mobilizing agonists stimulate the turnover of inositol lipids. Evidence now suggests that the initial reaction in this pathway is the phosphodiesteratic hydrolysis of the polyphosphoinositides which yields diacylglycerol and inositol phosphates. Each of these products may have a messenger role in cell activation. Diacylglycerol can activate a specific protein kinase (C-kinase) which phosphorylates specific intracellular substrates, leading to expression or modification of the appropriate response. The investigation of potential roles of the C-kinase in cell activation is facilitated by the use of the phorbol diester drugs which specifically activate the enzyme. Inositol 1,4,5-trisphosphate, the specific breakdown product of phosphatidylinhositol 4,5-bisphosphate, signals the release of Ca from internal pools, probably elements of the endoplasmic reticulum. In some cases, the elevation in cytosolic [Ca^{2+}] brought about by inositol 1,4,5-trisphosphate acts synergistically with the diacylglycerol and C-kinase pathway in bringing about the expression of an appropriate cellular response.

15.1. Introduction

In 1953, Hokin and Hokin (1953) reported that when pancreas slices were incubated in medium containing $^{32}PO_4$, muscarinic-cholinergic stimuli increased the incorporation of radioactivity into tissue phospholipids. Subsequently, it was found (Hokin and Hokin, 1954) that the lipids primarily involved were phosphatidylinositol (PI) and phosphatidic acid (PA). In the 30 years following the discovery of this phenomenon, research on this receptor-regulated phosphoinositide effect has centred on two basic questions: (1) What are the biochemical pathways involved? (2) What is the significance of these reactions to cell function? This chapter attempts to summarize current thinking and hypotheses in this area with emphasis on recent experimental findings.

15.2. Pathways of the phosphoinositide effect

Based on studies of the avian salt gland, Hokin and Hokin (1964) suggested that the PI and PA labelling response was actually a secondary consequence of receptor-activated breakdown of PI by a phosphodiesterase or phospholipase C. In his 1975 review, Michell (1975) proposed that this was the general case for activated PI turnover, which by this time had been shown to be a response characteristic of a wide variety of membrane receptors (Table 15.1). The diacylglycerol (DG) which would be formed following phospholipase C breakdown of PI would be rapidly phosphorylated by the actions of the plasmalemmal DG kinase, resulting in the synthesis of PA. This PA is then presumably transported to the endoplasmic reticulum (ER) where the enzymes necessary for resynthesis of PI are located. The newly synthesized PI is then transported, by a specific phospholipid transfer protein, to the plasma membrane thus completing the cycle. In labelling experiments, ^{32}P is introduced into the cycle from the γ-phosphate of ATP when PA is synthesized, and the radioactive phosphate is retained in the recycled PI. In support of this proposal, many investigators have found that associated with the increased labelling of PI and PA, there is a net decrease in PI, a net synthesis of PA, and a transient accumulation of DG (Laychock and Putney, 1982; Martin, 1983; Michell, 1975; Siess, Siegel and Lapetina, 1983).

Most cell types thus far examined contain specific kinases capable of phosphorylating the inositol ring on PI sequentially at the 4 and 5 positions. Specific phosphomonoesterases are also present which can sequentially remove the 5 and 4 position phosphates. These activities are quite rapid such that the resulting polyphosphoinositides, specifically PI-4-phosphate (PIP), and PI-4,5-bisphosphate (PIP_2) are turned over rapidly. Generally, they represent only about 10% or less of the total inositol lipids in the cell.

Abdel-Latif, Akhtar and Hawthorne (1977) first demonstrated that activation of muscarinic receptors in iris smooth muscle led to a decreased tissue content of PIP_2. Subsequently, Kirk et al. (1981) demonstrated breakdown of PIP and PIP_2 in hepatocytes in response to the same stimuli which they had previously shown to cause PI breakdown, and under similar experimental conditions. These findings raised the question as to which of the inositol lipids might be directly hydrolysed in response to receptor activation, and which of them might be depleted due to the rapid equilibration of the monoester phosphates, that is by a mass-action effect. Since the fractional rate of decline of PIP_2 was greater than that of PI, Kirk et al. (1981) suggested that a polyphosphoinositide might be the initial substrate for the receptor-activated phospholipase C (Kirk et al., 1981).

More definitive information on the initial reactions came with the development of a convenient radiochemical technique for studying the formation of the water soluble products of inositol lipid breakdown, the inositol phosphates (Berridge et al., 1983). Within the first 5 s following activation of 5-HT receptors in *Calliphora* salivary gland, inositol trisphosphate (IP_3) and inositol bisphosphate (IP_2) were substantially increased, but inositol phosphate (IP) did not increase until after a delay of 20—40 s (Berridge, 1983). Similar results have been obtained from the parotid gland (Aub and Putney, 1984; Downes and Wusteman, 1983) and liver (unpublished observation).

Once the inositol phosphates are formed, phosphatases can sequentially remove the monoester phosphates ultimately regenerating free inositol. Thus, a difficulty in interpreting cellular changes in IP_2, is that it could come from breakdown of either PIP or IP_3; similarly IP could come from PI or IP_2. By use of a computer-assisted kinetic modelling technique Aub and Putney (1984) estimated the steady-state rates of formation and breakdown of inositol phosphates in parotid acinar cells. The rate of breakdown of IP_3 was only about one-half of the estimated rate of formation of IP_2 implying that some IP_2 may be derived from direct hydrolysis of PIP. However, breakdown of IP_2 could more than adequately account for the rate of formation of IP. Thus it is likely that no appreciable phosphodiesteratic breakdown of PI occurs, at least under these experimental conditions.

The suggested pathways involved in the phosphoinositide cycle are depicted in Fig.15.1. The initial receptor regulated reaction is now believed to be a phospholipase C-mediated breakdown of polyphosphoinositides. The net decrease in PI characteristically seen in stimulated cells would thus be attributed to consumption by PI kinase in replenishing the polyphosphoinositides. The subsequent reactions may be as described above, with the DG backbone being carried through a somewhat more complex cycle than originally envisioned.

15.3. Significance of receptor-controlled phosphoinositide breakdown

Following the initial findings of Hokin and Hokin in the 1950s, a number of hypotheses were considered in an attempt to understand the role of phosphoinositide turnover in cell function. Most involved an involvement of the lipids in the expression of an appropriate end response; for example, enzyme secretion in exocrine pancreas or active Na^+ transport in epithelia. It soon became apparent, however (and inspection of Table 15.1 will support this idea) that there was no relationship at all between agonist-activated inositol lipid turnover and the end response of various tissues; but rather the common denominator was the messenger system involved. In each case, the receptors which caused phosphoinositide breakdown were ones believed to activate cellular responses through Ca mobilization. In his 1975 review, Michell (1975) noted this striking association of PI turnover and Ca mobilizing receptors. However, be also noted that enhanced PI metabolism could be experimentally

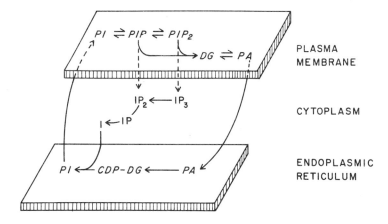

Fig.15.1. Proposed pathways of the phosphoinositide cycle in rat parotid gland. The initial reactions are believed to be the hydrolysis of PIP and PIP_2 to IP_2 and IP_3, respectively, and DG. Further details of the scheme are described in the text. (From Putney et al. (1985), with permission.)

dissociated from Ca mobilization. Depletion of cellular Ca with Ca deficient media inhibited the various cellular responses dependent on Ca mobilization with little effect on PI turnover; further, when cytosolic Ca was artificially increased by use of divalent cationophores, a variety of Ca-dependent cellular responses were activated, but PI turnover was increased very little (Michell, 1975). Thus, Michell (1975) argued that receptor-activated phosphoinositide breakdown did not appear to be a reaction mediated by Ca mobilization, but appeared more directly coupled to receptor activation. He further proposed that as such, this inositol lipid cycle might in some way serve as a biochemical coupling mechanism linking receptor activation to cellular Ca mobilization (Michell, 1975). Subsequent research has tended to support Michell's original idea of a direct link between receptor and inositide breakdown (Berridge, 1981; Putney, 1981), but exceptions have been noted (Cockcroft, 1981; Hawthorne, 1982).

Table 15.1. A partial listing of receptors that activate phosphoinositide cycling.

Receptor type	*Tissue*
Amine receptors	
Muscarinic-cholinergic	Exocrine glands, heart, smooth muscle, brain (synaptosomes)
α_1-Adrenergic	Exocrine glands, smooth muscle, liver
H_1-Histamine	Smooth muscle, brain
$5-HT_1$	Fly salivary gland, smooth muscle
Peptide receptors	
V_1-Vasopressin	Smooth muscle, liver, hippocampus
Thrombin	Platelets
Angiotensin II	Smooth muscle, liver
Substance P	Salivary gland
Cholecystokinin	Exocrine pancreas, cerebral cortex

15.4. Phosphoinositides and Ca mobilization

It now seems clearly to be the general case that receptor-mediated Ca mobilization occurs in two phases (Putney, 1978; Putney, Poggioli and Weiss, 1981): a release of Ca^{2+} to the cytosol from internal pools, and a sequential or concomitant entry of Ca^{2+} to the cytosol from the extracellular space. Evidence has been presented for a role of phosphoinositide breakdown in both phases of Ca mobilization.

Under appropriate experimental conditions, PA has been shown to transport Ca across artificial membranes. Also, in some, but not all, test systems exogenous PA can mimic the effects of receptor activation (Putney (1981). Since PA is rapidly formed following phosphoinositide breakdown, and this formation apparently occurs at the plasma membrane (Putney, 1981), it has been suggested that PA could mediate the enhanced Ca entry across the plasma membrane (Putney et al., 1980; Salmon and Honeyman, 1980). Recently, some of the experiments in artificial systems have been questioned (Holmes and Yoss, 1983), and it is not altogether clear the extent to which the effects of exogenous PA may be explained by oxidized PA or other contaminants in commercially available preparations. Putney et al. (1980) reported that pharmacological agents which blocked receptor-regulated Ca entry in parotid gland also blocked Ca binding to PA, and with similar potency. Thus, PA may mediate Ca entry, but as a co-factor or counter ion for a more complex mechanism which simple model transport systems cannot faithfully mimic. More detailed discussions of the relative merits of the PA hypothesis have been published previously (Laychock and Putney, 1982; Putney, 1981; Putney, Poggioli and Weiss, 1981).

As mentioned above, Ca mobilizing receptors also cause the release of Ca from internal pools; the available evidence suggests that the pool involved is most likely a component of the endoplasmic reticulum (Burgess et al., 1983;

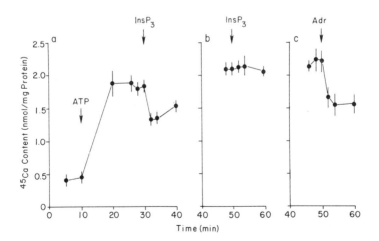

Fig.15.2. Release of sequestered Ca by IP_3 ($InsP_3$ in the figure) in guinea-pig hepatocytes.
(a) Hepatocytes made permeable with saponin sequester ^{45}Ca on addition of ATP into non-mitochondrial pools, presumably elements of the endoplasmic reticulum. The addition of 5 μM IP_3 at $t = 30$ min causes the rapid release of about 0.5 nmol/mg protein of the sequestered ^{45}Ca. (b) 5 μM IP_3 does not cause ^{45}Ca release from intact hepatocytes (not saponin-treated). (c) In intact hepatocytes, adrenaline (adr) releases a quantity of ^{45}Ca similar to that released by IP_3 from permeable cells. (For further details, see the text and Burgess et al. (1984a), from which this is taken, with permission.)

Fig.15.2). In order to couple receptor activation to internal Ca release, it has been suggested that a second messenger is formed at the plasma membrane which signals this release from the appropriate intracellular pool (Williamson, Cooper and Hoek, 1981). Berridge (1983) first suggested that the messenger might be a product of polyphosphoinositide breakdown, specifically, inositol 1,4,5-trisphosphate (IP$_3$), the water-soluble product of phosphodiesteratic breakdown of PIP$_2$. Recently, it has been shown for three different cell types, the rat pancreatic acinar cells (Rubin et al., 1984; Streb et al., 1983), the guinea pig hepatocyte (Burgess et al., 1984a), and the DMSO-differentiated HL-60 cell (a neutrophil-like tumour cell (Burgess et al., 1984b) that IP$_3$ is rapidly formed in response to receptor-activation, and is capable of releasing Ca from the endoplasmic reticulum.

Because IP$_3$ is a highly polar molecule, it must be applied to the intracellular environment either by direct injection or by use of experimental procedures that selectively permeabilize the plasma membrane. For the cell types mentioned above, saponin, a detergent with specificity for cholesterol enriched membranes such as the plasmalemma, was used to increase plasma membrane permeability. In the presence of added ATP, these permeable cells sequester Ca into internal organelles, primarily the mitochondria and endoplasmic reticulum. However, the general finding is that in the range of Ca^{2+} concentrations expected in the cytoplasm of cells (< 1 μM), Ca is sequestered almost exclusively by the endoplasmic reticulum. Under these conditions, micromolar concentrations of IP$_3$ induces a rapid release of sequestered Ca^{2+} (Burgess et al., 1984a,b; Streb et al., 1983). This response has been detected as a net release of Ca^{2+} to the medium with a Ca^{2+} electrode (Streb et al., 1983), or by loss of cell-associated ^{45}Ca in the presence of Ca-EGTA buffers (Burgess et al., 1984a,b; Fig. 15.2). Further, the concentrations IP$_3$ required to induce release (0.1–5 μM) are less than or equal to those estimated in stimulated cells (Burgess et al., 1984a; Rubin et al., 1984). More recently, it has been shown that the action of IP$_3$ has strict structural requirements (unpublished observation) which is consistent with the idea that IP$_3$ acts on a specific receptor (Burgess et al., 1984a) on a component of the endoplasmic reticulum to cause Ca release. The mechanism by which this IP$_3$ receptor functions is presently unknown.

15.5. Diacylglycerol and C-kinase

The other initial product of polyphosphoinositide breakdown, diacylglycerol (DG) is also believed to function as a cellular messenger. Nishizuka and his colleagues (Kawahara et al., 1980; Nishizuka, 1983) first described a protein kinase which was phospholipid dependent and Ca^{2+}-activated, and which they termed C-kinase. The sensitivity of the C-kinase to Ca^{2+} is greatly increased by DG which caused Nishizuka (Kawahara et al., 1980; Nishizuka, 1983) to suggest that DG derived from inositol lipid breakdown might be a physiological regulator of the enzyme. A number of experimental observations suggest that this view is correct, and that the DG and IP$_3$ (and perhaps PA), act in concert to produce a complete message necessary for the expression of an appropriate cellular response.

Two recent technical advances have provided the strongest support for this scheme. The first was the demonstration that the tumour-promoting drugs, the phorbol diesters, were potent and specific activators of the C-kinase (Castagna et al., 1982; Niedel, Kuhn and Vandenbark, 1983), apparently activating the enzyme in the same manner as DG. The second was the development of the intracellular Ca^{2+}-indicator, 'quin-2' (Tsien, Pozzan and Rink, 1982). The phorbol diesters provided a convenient means of demonstrating that activation of the C-kinase does in fact activate cellular responses in a variety of tissues (Gunther, 1981; Kojima et al., 1983; Putney et al., 1984; Rink, Sanchez and Hallam, 1983; Sha'afi et al., 1983). Studies with the quin-2 demonstrated that for Ca mobilizing receptors, the increase in cytosolic [Ca^{2+}] necessary to cause cellular activation is much less the increase necessary when Ca^{2+}-ionophores are used (Rink, Smith and Tsien, 1982). The interpretation of this latter finding is that receptor activation provides an additional signal which is missing when the ionophores are used; in other words, phosphoinositide breakdown and DG formation. Finally, these two lines of evidence can be brought together with the observation that when applied in combination, phorbol diesters and Ca ionophores are synergistic in activating many cellular responses (Kojima et al., 1983; Putney et al 1984). In other words, the combination of phorbol diester and ionophore serves to pharmacologically reconstitute the physiological signals which receptor-activation provides with DG and Ca^{2+} (mobilized by IP_3 and PA).

15.6. Conclusions

After some 30 years of investigation, we are beginning to understand the significant role of the 'phospholipid effect' in the regulation of cellular Ca^{2+}. The available evidence suggests that a wide variety of Ca^{2+}-mobilizing receptors initiates a cycling of inositol lipids which begins with the phosphodiesteratic breakdown of polyphosphoinositides liberating DG and soluble inositol phosphates. The DG activates an enzyme, the C-kinase, which phosphorylates and thereby regulates other enzymes and subcellular components. One of the inositol phosphates, IP_3, has been shown to release Ca^{2+} from the endoplasmic reticulum. PA generated from phosphorylation of DG may also contribute to cellular Ca^{2+}-mobilization by facilitating Ca^{2+} entry at the plasma membrane. The increase in cytosolic Ca^{2+} regulates intracellular enzymes and structures, probably through interaction with calmodulin or analogous proteins. This dual signalling system probably provides for flexibility and control as part of the complex process involved in the regulated and integral response of a cell to a signal from the external environment.

Acknowledgement

Work from the author's laboratory described in this review was supported by grants from the NIH (USA), #DE-05764 and AM-32823.

References

Abdel-Latif, A.A., Akhtar, R.A., Hawthorne, J.N. (1977). Acetylcholine increases the breakdown of triphosphoinositide of rabbit iris muscle prelabelled with [^{32}P] phosphate. *Biochem. J.* 162, 61–73.

Aub, D.L. and Putney, J.W., Jr. (1984). Metabolism of inositol phosphates in parotid cells: Implications for the pathways of the phosphoinositide effect and for the possible messenger role of inositol trisphosphate. *Life Sci.* 34,

1347–55.
Berridge, M.J. (1981). Phosphatidylinositol hydrolysis: a multifunctional transducing mechanism. *Mol. Cell. Endocrinol.* **24**, 115–40.
Berridge, M.J. (1983). Rapid accumulation of inositol trisphosphate reveals that agonists hydrolyse polyphosphoinositides instead of phosphatidylinositol. *Biochem. J.* **212**, 849–58.
Berridge, M.J., Dawson, R.M.C., Downes, C.P., Heslop, J.P. and Irvine, R.F. (1983). Changes in the levels of inositol phosphates following agonist-dependent hydrolysis of membrane phosphoinositides. *Biochem. J.* **212**, 473–82.
Burgess, G.M., McKinney, J.S., Fabiato, A., Leslie, B.A. and Putney, J.W., Jr. (1983). Calcium pools in saponin-permeabilized guinea-pig hepatocytes. *J. Biol. Chem.* **258**, 15336–45.
Burgess, G.M., Godfrey, P.P., McKinney, J.S., Berridge, M.J., Irvine, R.F. and Putney, J.W., Jr. (1984a). The second messenger linking receptor activation to internal Ca release in liver. *Nature (Lond.)* **309**, 63–6.
Burgess, G.M., McKinney, J.S., Irvine, R.F., Berridge, M.J., Hoyle, P.C. and Putney, J.W. Jr. (1984b). Inositol 1,4,5-trisphosphate may be a signal for intracellular Ca mobilisation in human leucocytes. *FEBS Lett.* **176**, 193–6.
Castagna, M., Takai, Y., Kaibuchi, K., Sano, K., Kikkawa, U. and Nishizuka, Y. (1982). Direct activation of calcium-activated, phospholipid-dependent protein kinase by tumor-promoting phorbol esters. *J. Biol. Chem.* **257**, 7847–51
Cockcroft, S. (1981). Does phosphatidylinositol breakdown control the Ca^{2+}-gating mechanism? *Trends Pharmacol. Sci.* **2**, 340–2.
Downes, C.P. and Wusteman, M.M. (1983). Breakdown of polyphosphoinositides and not phosphatidylinositol accounts for muscarinic agonist-stimulated inositol phospholipid metabolism in rat parotid glands. *Biochem. J.* **216**, 633–40.
Gunter, G.R. (1981). Effect of 12-0-tetradecanoly-phorbol-13-acetate on Ca^{2+} efflux and protein discharge in pancreatic acini. *J. Biol. Chem.* **256**, 12040–5.
Hawthorne, J.N. (1982). Is phosphatidylinositol now out of the calcium gate? *Nature (Lond.)* **295**, 281–2.
Hokin, M.R. and Hokin, L.E. (1953). Enzyme secretion and the incorporation of P^{32} into phospholipides of pancreas slices. *J. Biol. Chem.* **203**, 967–77.
Hokin, M.R. and Hokin, L.E. (1954). Effects of acetylcholine on phospholipides in the pancreas. *J. Biol. Chem.* **209**, 549–58.
Hokin, M.R. and Hokin, L.E. (1964). Interconversions of phosphatidylinositol and phosphatidic acid involved in the response to acetylcholine in the salt gland: In *Metabolism and Physiological Significance of Lipids*, Dawson and Rhodes (eds.), pp. 423–34. John Wiley and Sons, London.
Holmes, R.P. and Yoss, N.L. (1983). Failure of phosphatidic acid to translocate Ca^{2+} across phosphatidylcholine membranes. *Nature (Lond.)* **305**, 637–8.
Kawahara, Y., Takai, Y., Minakuchi, R., Sano, K. and Nishizuka, Y. (1980). Phospholipid turnover as a possible transmembrane signal for protein phosphorylation during human platelet activation by thrombin. *Biochem. Biophys. Res. Comm.* **97**, 309–17.
Kirk, C.J., Creba, J.A., Downes, C.P. and Michell, R.H. (1981). Hormone-stimulated metabolism of inositol lipids and its relationship to hepatic receptor function. *Biochem. Soc. Trans.* **9** 377–9.
Kojima, I., Lippes, H., Kojima, K. and Rasmussen, H. (1983). Aldosterone secretion: Effect of phorbol ester and A23187. *Biochem. Biophys. Res. Comm.* **116**, 555–62.
Laychock. S.G. and Putney, J.W., Jr. (1982). Roles of phospholipid metabolism

in secretory cells. In *Cellular Regulation of Secretion and Release,* Conn (ed.), pp. 53–105. Academic Press, New York.

Martin, T.F.J. (1983). Thyrotropin-releasing hormone rapidly activates the phosphodiester hydrolysis of polyphosphoinositides in GH_3 pituitary cells. Evidence for the role of a polyphosphoinositide-specific phospholipase C in hormone action. *J. Biol. Chem.* 258, 14816–22.

Michell, R.H. (1975). Inositol phospholipids and cell surface receptor function. *Biochim. Biophys. Acta* 415, 81–147

Niedel, J.E., Kuhn, L.J. and Vandenbark, G.R. (1983). Phorbol diester receptor copurifies with protein kinase C. *Proc. Natl. Acad. Sci. USA* 80, 36–40.

Nishizuka, Y. (1983). Calcium, phospholipid turnover and transmembrane signalling. *Phil. Trans. R. Soc. Lond. B* 302, 101–12.

Putney, J.W., Jr. (1978). Stimulus–permeability coupling: role of calcium in the receptor regulation of membrane permeability. *Pharmacol. Rev.* 30, 209–45.

Putney, J.W., Jr. (1981). Recent hypotheses regarding the phosphatidylinositol effect. *Life Sci.* 29, 1183–94.

Putney, J.W., Jr. Poggioli, J. and Weiss, S.J. (1981). Receptor regulation of calcium release and calcium permeability in parotid gland cells. *Phil. Tran. R. Soc. Lond. B* 296, 37–45.

Putney, J.W., Jr. Weiss, S.J., VanDeWalle, C.M. and Haddas, R.A. (1980). Is phosphatidic acid a calcium ionophore under neurohumoral control? *Nature (Lond.)* 284, 345–7.

Putney, J.W., Jr., McKinney, J.S., Aub, D.A. and Leslie, B.A. (1984). Phorbol ester-induced protein secretion in rat parotid gland. Relationship to the role of inositol lipid breakdown and protein kinase C activation in stimulus-secretion coupling. *Mol. Pharmacol.* 26, 261–6.

Putney, J.W., Jr., Burgess, G.M., Godfrey, P.P. and Aub, D.L. (1985). Messages of the phosphoinositide effect. In Inositol and Phosphoinosities, Bleasdale, Eichberg and Hauser (eds.), pp. 337–49. Humana Press, Clifton, NJ.

Rink, T.J., Smith, S.W. and Tsien, R.Y. (1982). Cytoplasmic free Ca^{2+} in human platelets: Ca^{2+} thresholds and Ca-independent activation for shape-change and secretion. *FEBS Lett.* 148, 21–6.

Rink, T.J., Sanchez, A. and Hallam, T.J. (1983). Diacylglycerol and phorbol ester stimulate secretion without raising cytoplasmic free calcium in human platelets. *Nature (Lond.)* 305, 317–9.

Rubin, R.P., Godfrey, P.P., Chapman, D.A. and Putney, J.W., Jr. (1984). Secretagogue induced formation of inositol phosphates in rat exocrine pancreas: Implications for a messenger role for inositol trisphosphate. *Biochem. J.* 219, 655–9.

Salmon, D.M. and Honeyman, T.W. (1980). Proposed mechanism of cholinergic action in smooth muscle. *Nature (Lond.)* 284, 344–5.

Sha'afi, R.I., White, J.R., Molski, T.F.P., Shefcyk, J., Volpi, M., Naccache, P.H. and Feinstein, M.B. (1983). Phorbol 12-myristate 13-acetate activates rabbit neutrophils without an apparent rise in the level of intracellular free calcium. *Biochem. Biophys. Res. Comm.* 114, 638–45.

Siess, W., Siegel, F.L. and Lapetina, E.G. (1983). Arachidonic acid stimulates the formation of 1,2-diacylglycerol and phosphatidic acid in human platelets: degree of phospholipase C activation correlates with protein phosphorylation, platelet shape change, serotinin release and aggregation. *J. Biol. Chem.* 258, 11236–42.

Streb, H., Irvine, R.F., Berridge, M.J. and Schulz, I. (1983). Release of Ca^{2+} from a nonmitochondrial intracellular store in pancreatic acinar cells by inositol-1,4,5-trisphosphate. *Nature (Lond.)* 306, 67–8.

Tsien, R.Y., Pozzan, T. and Rink, T.J. (1982). Calcium homeostasis in intact lymphocytes: cytoplasmic free calcium monitored with a new, intracellularly trapped fluorescent indicator. *J. Cell. Biol.* **94**, 325–34.

Williamson, J.R., Cooper, R.H. and Hoek, J.B. (1981). Role of calcium in the hormonal regulation of liver metabolism. *Biochim. Biophys. Acta* **639**, 243–95.

CALCIUM ACTIVATION IN SKELETAL MUSCLE

G. Cecchi, P.J. Griffiths, J.R. Lopez, S. Taylor and L.A. Wanek

Department of Pharmacology, Mayo Foundation, Rochester, Minnesota 55905, USA

Abstract

A rise in intracellular Ca^{2+} is followed by an increased number of attached cross-bridges, and the development of force becomes perceptible only after cross-bridge attachment. Contraction is then modulated by several factors. For example, cell length prior to stimulation influences the degree of mechanical activation induced by Ca^{2+}, and length changes during activation temporarily alter intracellular Ca^{2+} binding. Chemical reactions affected by Ca^{2+} not only change mechanical events, but mechanical events evidently change the driving chemical reactions. The temporal relations among these events is different during relaxation. Free Ca^{2+} falls first, then force, and finally the number of attached cross-bridges. The initiation, regulation, and termination of contractile activity in skeletal muscle appears to be governed by a combination of many factors.

16.1. Introduction

Intracellular Ca^{2+} regulation in skeletal muscle has been regarded as an important event in excitation–contraction coupling (ECC) for at least the third of a century since Sandow (1952) coined this phrase to describe the link between action potentials and contraction. At that time it seemed that a single mechanism, the release of ionized calcium throughout the cytoplasm, could account for the relation between membrane depolarization and mechanical activation as well as the effects of agents that potentiate mechanical activity. These ideas have recently been surveyed thoroughly, and the reader who requires information omitted from this necessarily brief and selective overview will be well-served by the *Handbook of Physiology* (Peachey, Adrian and Geiger, 1983).

Those already familiar with these publications will recognize the priority we have assigned to results obtained from amphibian skeletal muscle. In what follows we will frequently compare these results with those obtained from giant muscle cells of the barnacle *Balanus nubilus*, for several reasons. Single cells isolated from barnacle or frog muscle are particularly well suited for a number of critical experiments that cannot or have not yet been performed on mammalian muscle. In addition, both preparations are from cross-striated muscles with contractile filaments composed of actin and myosin. Most models of the physiological behaviour of these muscles are based upon the sliding filament model in which motion or tension development are produced by interactions among identical sites that are uniformly distributed in the zone of each half sarcomere where actin and myosin filaments overlap. In models where each site of interaction is associated with an independent force generator it is also supposed that cross-bridges form extensible links prior to force production (H.E. Huxley, 1979; A.F. Huxley, 1980). Hence one can measure the stiffness of a single muscle cell, for example, under conditions where a large part of the stiffness is presumably produced by the cross-bridges. A final reason is that 'No one muscle can possibly provide all the answers to the core question: What is truly general about muscles?' (Hoyle, 1983).

16.2. Methods

For more than a decade it has been technically possible to study many aspects of ECC by directly monitoring changes in intracellular Ca^{2+} associated with individual cycles of contraction and relaxation. One of the first and most productive methods was to introduce the calcium-sensitive luminescent protein aequorin into the myoplasm of giant barnacle fibres (Ridgway and Ashley, 1967; Ashley and Ridgway, 1970). While such measurements have yielded information about the role of Ca^{2+} in ECC, it has also become evident that control of an entire contraction—relaxation cycle in amphibian skeletal muscle is a complex system of related circumstances that is not always correlated with changes in myoplasmic Ca^{2+} (Blinks, Rudel and Taylor, 1978; Cecchi, Griffiths and Taylor, 1984a and b; Taylor et al., 1982). An overview of ECC in skeletal muscle must include the consideration not only of Ca^{2+} transients, but characteristics of several events. Many of these conditions and events can presently be measured during excitation or contraction in living muscle. However, only a few will be considered here, namely (1) cell length prior to activation, (2) those contractile events which themselves may change intracellular Ca^{2+} binding, and (3) the possible influence of multiple cross-bridge states.

The emphasis of this overview will be on studies carried out by the authors on single fibres that were isolated intact from leg or toe muscles of the frog *Rana temporaria*. The fibres were usually injected with the calcium-sensitive photoprotein aequorin, which enabled us to simultaneously monitor relative changes in intracellular free calcium in addition to contractile force. In some experiments small tabs of tendon at each end of a fibre were covered with aluminium clips to reduce extrinsic compliance. One clip was connected to a capacitance gauge transducer; the other clip was attached to a length-step generator that made it possible to impose rapid step-length changes (complete in less than 200 μs) or high frequency sinusoidal length changes (up to 9 kHz). Application of these quick length changes allowed us to estimate the kinetics of cross-bridge attachment and detachment from the calculated moment-to-moment stiffness. Additional details of these methods have already been described (Blinks, Rudel and Taylor, 1978; Cecchi, Griffiths and Taylor, 1984a and b).

16.3. Fibre length prior to stimulation influences calcium activation

Gordon, Huxley and Julian (1964, 1966) described with sharp definition the relation of the average length between the striations in a frog skeletal muscle fibre and the force a fibre can actively produce. They observed a linear fall in tension with extension beyond slack length and marked changes in slope of the length–tension relation at striation spacings near the peak and below slack length. The changes they observed corresponded remarkably well to the best estimates at the time for thick and thin filament lengths, and also agreed closely with Ramsey and Street's original data from isolated, uninjured muscle fibres (1940). These results could lead one to infer that most, if not all, of the dependence of active tension on muscle length could be accounted for by the following factors: (1) the degree of overlap between adjacent thick and thin filaments; (2) the collision of thin filaments at the middle of the sarcomere; (3) the beginning of overlap of each thin filament with bridges at the other half of the thick filament; and (4) the contact between thick filaments and the Z lines. The first factor pertains to tension development in stretched fibres. The last three were among several other possibilities suggested to account for the decline of tension in shortened fibres (A.F. Huxley, 1965; Gordon, Huxley and Julian, 1966). It still remains to be shown that these potential sources of resistance to shortening are small enough to be completely disregarded in intact muscle. Isolated bundles of myofibrils apparently have virtually no intrinsic resistance to shortening when activated maximally by Ca^{2+} (see, for example, Sugi, Ohta and Tameyasu, 1983). Several laboratories have tried to determine whether or not the normal degree of activation decreases as a living muscle shortens, following a possibility suggested by A.F. Huxley (1957).

Direct observation of isolated frog muscle fibres during contractions below slack length led to the suggestion that shortening somehow interrupted the spread of the action potential to all parts of a cell (Taylor and Rüdel, 1970; Rüdel and Taylor, 1971). If one merely monitors tension at different fibre lengths and compares the effects of different stimulus frequencies the results are also compatible with this suggestion. This might have been noticed many years ago except for the presumptive evidence that Ramsey and Street (1940, 1941) were primarily concerned with producing contractions that were propagated along the whole length of a fibre and had the longest possible period of rhythmic activity. They placed their stimulating electrodes at either end of a fibre and typically spaced the stimuli to be just under the optimal frequency in order to determine length–tension diagrams that covered a range from extreme lengthening to extreme shortening. On the other hand, Fig.16.1 shows part of one of our experiments in which the electrodes were transverse to the whole length of the fibre and shortening was limited to the initial two-thirds of the range shorter than slack length. The maximum force in a tetanic contraction was markedly enhanced by increasing the stimulus frequency at short fibre lengths, whereas the influence of stimulus frequency near the highest part of the length–tension relationship was much smaller. The frequency-dependent increase in tetanic force is not readily detected unless one works at moderately warm temperatures (about 10°C or higher for frogs indigenous to the temperate zones) and exceeds the lowest frequencies of stimulation that produce a fused plateau of force. Ramsey and Street performed their experiments at moderately warm temperatures wheras most of the subsequent determinations of the length–tension diagram were done in the cold. Similar results have previously been described briefly (Taylor et al., 1982).

Although experiments such as those shown in Fig.16.1 support the notion that a major factor limiting the ability of a fibre to develop force as it shortens below slack length is an impairment in the link between the

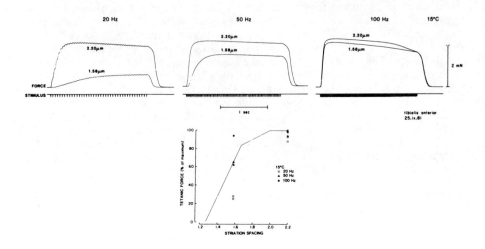

Fig.16.1. The influence of stimulus frequency on responses at short and optimum striation spacings. All records shown were made on the same fibre isolated from the tibialis anterior muscle of a frog. At least 5 min rest was allowed between contractions and the maximum force reached in each of the first eleven tetani is plotted. Up to this point, the striation spacing, measured by optical microscopy at three widely separated locations along the fibre, showed no detectable changes from the resting value. The average striation spacing measured prior to and after each tetanic contraction was 2.20 μM, which is indicated next to the top three records. The striation spacing during contraction at the short length — the three smaller contractions — was calculated by measuring the length of the cell from tendon to tendon, using the measured value of 2.20 μM for the average striation spacing at this cell length, and releasing the fibre to let it hang slack prior to stimulation. The solid line is redrawn from the work of Gordon, Huxley and Julian (1966), to facilitate comparison with our values measured at the peak of force production.

excitatory signal and calcium activation, this is not an effective way to examine the entire length–tension relation in the same fibre. Fibres stimulated repeatedly at high frequencies and at short lengths eventually begin to show irregularities in striation spacing, and spontaneous relaxation at the end of stimulation no longer remains uniform or complete (Ramsey and Street, 1940). Treatment with any one of a number of potentiating agents, substances known to increase the force of contraction in a twitch (Sandow, 1964), is a more enduring way to enhance the level of cytoplasmic Ca^{2+} in ECC. Hence we have used chemical potentiating agents to examine the possibility that decreased calcium release might be a factor in limiting force production below slack length. Figure 16.2 shows one such experiment in which it is evident that the enhanced release of Ca^{2+} in ECC does not produce the same effect on contraction at all lengths. For example, enhanced Ca^{2+} release has little or no effect on the descending limb of the length–tension relationship (that is, the four sets of records along the bottom of Fig.16.2). However, enhanced Ca^{2+} release has a marked effect on the ascending limb (records along the top row), which supports the idea that calcium activation is indeed less than maximal at short muscle lengths established soon after the development of even a single action potential (Taylor, 1974; Taylor et al., 1982).

This possibility is also supported by experiments in which the opposite approach was employed. Elevated extracellular calcium inhibits tetanus development, particularly at short lengths and at frequencies of stimulation below those necessary to produce complete fusion of a tetanus (Howell and Snowdowne, 1981). The author's interpretation of these results is that tension inhibition results from a failure of the action potential propagation within the transverse tubular system, and the inhibition is strongest at short muscle lengths. This factor evidently is not limited merely to skeletal muscle of so-called lower animals, as similar results have been reported for mammals including skeletal muscles in intact humans (Vandervoort, Quinlan and McComas, 1983).

There are several other factors associated with the influence of stimulus frequency and fibre length on tetanic force. They often are not fully described by those of us who experiment on isolated amphibian skeletal muscle fibres. Likewise, space restrictions prevent us from fully discussing every relevant detail of our experiments here. The muscles we use and the manner in which they are dissected eliminates fibre types 3,4, and 5 from our experiments (Lannergren, 1979; Smith and Ovalle, 1973). The types of fast fibres that remain may indeed be functionally independent (Smith and Ovalle, 1973), because we observe differences in force production that indicate more than one population of variables. Our frogs are usually cold-adapted and their fibres have twitch-to-tetanic-force ratios that range from about 0.1 to 0.5 at slack length and 15 °C (also see Lannergren, Lindblom and Johansson, 1982). Fibres with low, medium, and high twitch-to-tetanic-force ratios not only have unique length-twitch force relations, and potentiating agents have singular effects on their length—twitch force relations, but they have distinct length-*tetanic* force relations in the presence of potentiating agents or at different frequencies of stimulation (Taylor *et al.*, 1982).

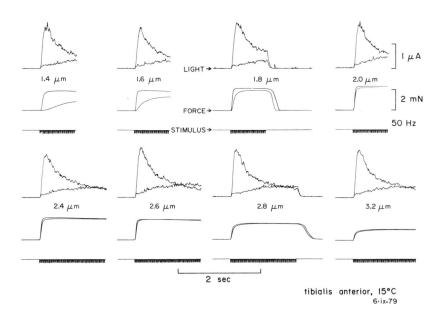

Fig.16.2. The effects of Zn^{2+} (50 μM) on fused tetanic responses at eight different striation spacings. The top trace is aequorin light emission, the middle trace is force development, and the bottom trace shows the applied stimuli. (This figure is reproduced from Taylor *et al.*, 1982 (Fig.4), with the publisher's permission.)

16.4. Length changes during activation influence calcium binding

An aequorin signal changes its overall size with changes in length prior to stimulation, roughly in the same manner that force measured simultaneously during stimulation changes its amplitude (Blinks, Rudel and Taylor, 1978). But quick length changes imposed during and immediately after stimulation can have different effects. The effects on aequorin signals from frog muscle fibres are influenced by several factors including the size and direction of the quick length change, and the length at which a contraction is begun. These results do not direct one towards a simple or unequivocal explanation (Cecchi, Griffiths and Taylor, 1984a). However, as our results with frog muscle fibres increase and are compared with those from barnacle muscle, the situation may become more open to understanding. Sudden shortening transiently decreases the apparent level of cytoplasmic free Ca^{2+} in frog muscle (Cecchi, Griffiths and Taylor, 1984a). On the other hand, in single muscle fibres from the barnacle, shortening during contraction trasiently increases aequorin light emission, whereas a quick stretch produces the opposite effect (Ridgway, Gordon and Martyn, 1983; Ridgway and Gordon, 1984). Why, despite different results, have the investigators reached similar conclusions from studies on either barnacle or frog?

Bremel and Weber (1972) and Weber and Murray (1973) found that rigour bonding between thick and thin filaments enhanced binding of Ca^{2+} to the low affinity sites on troponin, which is associated with the thin filaments, by almost 10-fold. Although there were already theoretical arguments that mechanical events might affect Ca binding to activating sites (see, for example, Hill, 1983 for references), Weber and co-workers provided the first biochemical evidence for this idea. They thus raised important questions for future workers to consider with intact muscle; for example, can other kinds of bonds such as those involved in force production also increase the Ca^{2+} affinity of troponin? Weber and Murray (1973) were also led by their evidence to speculate that a reduction in the number of force-generating bonds at non-saturating calcium concentrations might cause a fall in the binding constant of the regulatory proteins for calcium. If this occurs it might enhance removal of calcium from troponin after imposed shortening. In vertebrate skeletal muscle such as the frog, regulation of contraction by Ca^{2+} is uniquely restricted to the thin filament, whereas Ca^{2+} regulation of barnacle muscle involves Ca^{2+} binding to both thin and thick filaments (Lehman and Szent-Györgyi, 1975). This adds some uncertainty as to the source and the regulatory significance of the increased Ca^{2+} associated with length changes in barnacle fibres (Ridgway and Gordon, 1984).

There are other significant differences between barnacle and frog muscle cells, as well as significant differences in the way the individual experiments can be and were carried out. For example, the size of a length change in the barnacle experiments is limited by the presence of an electrode inside the fibre, which can damage the cell if the change is not limited. Furthermore, the effects of length changes cannot be examined in barnacle when the developed force is large because the connection between the fibre and the cannula tends to slip (Ridgway and Gordon, 1984). Recent results from frog skeletal muscle fibres that are related to these two differences will be outlined below. Other features, for example the relations among length changes during activation, the change in the calcium transient, and a possible change in the number of Ca^{2+} binding sites distributed uniformly along the myofilaments, are considered elsewhere (Cecchi, Griffiths and Taylor, 1984a).

Imposed shortening that is at least 4–5% of fibre length transiently

Calcium activation in skeletal muscle

decreases light emission during contraction of a frog muscle fibre and stretch has no detectable effect (Cecci, Griffiths and Taylor, 1984a). The effect of imposed shortening on a frog fibre is shown in Fig.16.3 where a constant velocity release caused a rapid drop in force and, after several ms, force began to redevelop as stimuli were continuously applied. Light emission did not begin to drop noticeably until after the release was complete if the release had a velocity of at least 3 fibre lengths/s. With continued application of stimuli, light emission ceased to decline and began to rise again at about the time that force had almost completely recovered to the prerelease level.

Paul Iaizzo of this laboratory has recently begun to further investigate the relation between the size of a length step and its effect on the aequorin signal. His results indicate that a reason we previously saw only a fall in the Ca^{2+} transient (Cecchi, Griffiths and Taylor, 1984a) might be because the length changes were limited to a range less than 10% of fibre length. Larger-size steps were not within the capability of the length-step generator at that time. The basic design of the length-step generator used by Iaizzo is the same as the one used by Cecchi, Griffiths and Taylor (1984a). But the ability to produce step changes of at least 15% of fibre length has allowed him to observe that the qualitative effects of a release become the opposite with a larger length change. Aequorin signals *rise* during an imposed release that is large, and the effects of a stretch are also detectable with large steps. A large stretch produces a fall in the aequorin signal whereas our earlier experiments on frog fibres indicated stretch had little or no effect (Cecchi, Griffiths and Taylor, 1984a). These new observations may result from the possibility that a very large stretch changes the ability of the sarcoplasmic reticulum (SR) to release Ca^{2+} or changes some other aspect of ECC. Our earlier experiments were explicitly designed to minimize the contribution of

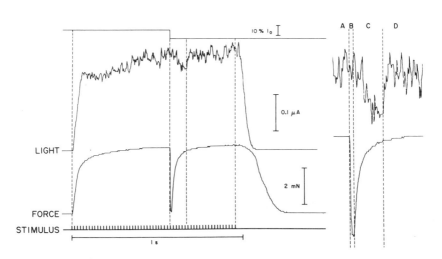

Fig.16.3. The relative change in intracellular Ca^{2+} and force production caused by a rapid stepwise decrease in length imposed during the plateau of a tetanic contraction. The complete records are shown on the left-hand side of the figure. An expanded view of the part of the records during the step decrease in length is shown on the right. The duration of period B is 6.41 ms (15°C). (This figure is reproduced from Cecchi, Griffiths and Taylor, 1984a (Fig.1) with the publisher's permission.)

such effects (Cecchi, Griffiths and Taylor, 1984a). As long as we continue to analyse our results on the basis of a sliding filament model with independent force generators uniformly distributed in the zone of each half sarcomere, frog fibres seem potentially more useful for further investigation of this question because the thick filament lengths and number of thin filaments surrounding each thick filament are essentially uniform in the resting cell compared to the variability intrinsic to most arthropod muscles (Franzini-Armstrong, 1970; Hoyle, McNeill and Selverston, 1973).

On the other hand, barnacle fibres can be voltage clamped by way of wires inserted along the length of the entire cell, and adequate voltage control of the membranes regulating ECC seems essential to study the effects of a length change. Ridgway, Gordon and Martyn (1983) found that a large, brief depolarizing pulse added to the early part of a contraction normally induced by a smaller, longer depolarizing pulse not only caused an enhanced initial release of Ca^{2+}, but maintained force at a higher level near the end of a stimulus pulse. Their calcium signal was about the same size at the end of the pulse in either case. They suggested that this hysteresis, the ability of Ca^{2+} to sustain more force during relaxation than during the rise of contraction, was due to a change in sensitivity of the regulatory calcium binding sites on the myofilaments. A given amount of Ca^{2+} produced a greater force in the 'quasi-steady state' if the mucle experienced a higher calcium level or higher force level in the initial stages of ECC (Ridgway, Gordon and Martyn, 1983). One of the possible explanations was that cross-bridge formation changed the calcium affinity of the calcium binding sites on the myofilaments. Their interpretation was based, in part, on the observed increase in light emission associated with shortening of barnacle fibres after a stimulus and at sub-tetanic force levels. Other explanations are not only possible, but our data on frog muscle fibres indicate that the hysteresis hypothesis may not account for some phenomena.

Figure 16.2 shows not only that potentiating agents enhance steady-state force most at short lengths, but enhance aequorin luminescence early in contraction. The aequorin signal declines with time despite continued stimulation. The control responses and the potentiated aequorin responses not only tend to approach one another but actually 'cross-over' if the contraction is sufficiently long. This is clear in the bottom four responses in Fig.16.2 here as well as in Fig.2 of Lopez, Wanek and Taylor (1981), and consistent with the idea of a hysteresis in the calcium—force relationship. However, when the time of the length change is varied through the contraction of a frog fibre, the calcium level in the initial stages of ECC is lowered and the force level in the steady state is increased. The opposite is true of barnacle fibres (Ridgway and Gordon, 1984, Figure 13).

Figure 16.4 shows part of such an experiment on a frog skeletal muscle fibre. We were also looking for an indication that the change in the aequorin transient during the tetanic plateau (see, for example 16.3) might be different at sub-tetanic force levels. A release imposed at the same time as the first stimulus of a tetanic train (Fig.16.4b) decreased the Ca^{2+} transient. Force had changed very little at that moment, but force response B (Fig.16.4b) on the plateau after the release was larger than A (Fig.16.4a) in the steady-state. Figure 16.4c shows a release imposed a short time later. Force was only about half its full tetanic value at the time of this release. This release also produced a transient depression of the aequorin signal, which can be detected in the slightly faster rate of decay of the first phase of the response. Hence we observed a transient drop in light whenever and wherever a release (less than 10% of fibre length) was made during tetanic contractions of frog fibres, and the size of the drop also appeared to be unrelated to the time of release.

All our experiments with quick length changes of frog muscle have thus far been performed on the descending limb of the length–tension relation, where the amount of Ca^{2+} seems to approach if not exceed the level required to saturate Ca^{2+} binding sites. Perhaps this might account for some of the different results between barnacles and frogs.

The time relations between our Ca^{2+} and force data (see, for examples, Fig.16.3) are supportive of the idea that the Ca^{2+} falls transiently only after appreciable redevelopment of tension, presumably owing to the reformation of force-generating actin–myosin complexes. This is consistent with the biochemical evidence that the formation of bonds between filaments enhances Ca^{2+} binding to troponin (Bremel and Weber, 1972). Our view is that the evidence for vertebrate skeletal muscle is largely explained by the idea tha Ca^{2+} binding to troponin not only instigates contraction, but there is feedback between mechanical events and Ca^{2+} binding to the regulatory sites, which in vertebrate muscle are associated only with the thin filaments.

'The possibility that there are combinations of force-generating and shortening mechanisms in the same cell and that there is a spectrum from non-myosin-operated systems to pure-myosin-operated ones, with various mixtures in between in different types, should not be shut off arbitrarily...' (Hoyle, 1983).

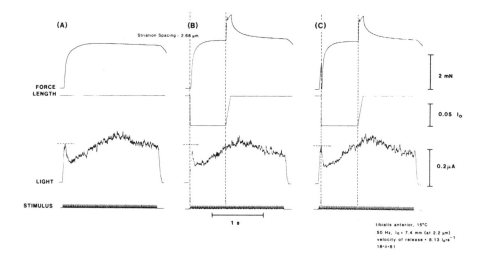

Fig.16.4. The influence of rapid, step-length changes on tetanic responses at different times during contraction. (A) Shows a contraction during which no change in length was imposed, that is, the control responses for comparison with (B) and (C). Light emission from the reaction of Ca^{2+} with injected aequorin did not follow a simple time-course as is often the case (Blinks, Rüdel and Taylor, 1978). Light initially rose rapidly than transiently decayed, then slowly rose and decayed again although the fibre was stimulated throughout the response and no extrinsic length change was imposed. In (B) a decrease in length was imposed at the same moment the first stimulus was delivered. The plateau of force at the end of the step was slightly higher in (B), which one would expect as the striation spacing was moved higher on the length–tension relationship. The relative change in intracellular Ca^{2+} was transiently decreased by the imposed shortening. A stretch to the original length during (B) had no detectable effect on aequorin light emission from the cell. However, force was transiently increased. In (C) a decrease in length was imposed at about the third stimulus in the tetanic train. Force dropped precipitously before redeveloping to about the same level as in (B). The transient depression of the aequorin signal was evident only as an increased rate of decay after the initial rise. These results were obtained from the same fibre shown in Figs. 2 and 3 of Cecchi, Griffiths and Taylor (1984a).

16.5. Calcium activation is influenced by multiple cross-bridge states

The idea that attached cross-bridges have more than one state was suggested in the model that A.F. Huxley and Simmons proposed as a basis for the redevelopment of tetanic tension after a release is imposed on the plateau, and also in the cross-bridge model based, in part, upon velocity transients that occur after a sudden change in load (Podolsky and Nolan, 1971). The idea of multiple cross-bridge states was introduced to accommodate the results of heat measurements (see, for example, A.F. Huxley, 1973; 1980). Various techniques for making fast time resolved measurements have revealed events on several time scales (see, for example, Kawai and Brandt, 1980; Kawai, Cox and Brandt, 1981).

To further explore the idea of different chemical states during Ca^{2+} activation and modulation of the cross-bridge cycle (A.F. Huxley, 1980), we measured the stiffness of intact fibres throughout entire tetanic contractions as a monitor of the proportion of attached cross-bridges. Some of our results are shown in Fig.16.5. This plot of stiffness and force as a function of time during tetani was made from data obtained on the same isolated fibre. Stiffness always increased more rapidly than force during the tetanus rise, and lagged well behind the fall of force at the end of stimulation. The

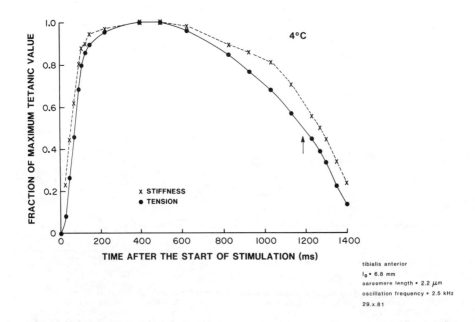

Fig.16.5. The relative stiffness (X) and relative tension (●) of a single fibre during the rise phase, the plateau, and relaxation of tetanic contraction. The arrow indicates the point at which segment length began to deviate from the length maintained during the tetanic plateau. The length of selected segments was monitored by a solid-state line scanner. The image of short strands of surgical silk attached to the fibre was focused onto the line scanner by way of a dissecting microscope. Segment movements less than 2 μM could be resolved and recorded in this manner. The conclusions drawn from these experiments have recently been strengthened by the finding that stiffness deviates from tension in the same manner shown here even when the central segment of a fibre is length-clamped by the output signal from a striation follower. This supports the idea (Cecchi, Griffiths and Taylor, 1981) that these differences are owing neither to tendon lengthening nor to sarcomere shortening (Cecchi et al., 1985). (This figure is reproduced from Cecchi, Griffiths and Taylor, 1984b (Fig.2), with the publisher's permission.)

difference was at least 10—15 ms, well within the resolution of our mechanical system which is capable of resolving differences on the order of microseconds (μs). The effects of transmission time (less than 15 μs) completely disappears at the end of a step and at the end of the first or second period of the imposed oscillations, and this is one of several reasons that makes it probable that the length changes were uniform in space and time along a fibre (Cecchi, Colomo and Lombardi, 1983). Conceivably compliance may add some unknown fraction of time to the delay between stiffness and force during the rise. However, there was no change in length for more than half a second, up to the arrow in Fig.16.5, that could account for the separation of events during the fall of tetanic tension. Recent results further support the validity of our suggestion that only a small part of the delay on the rise could be due to compliance extrinsic to a muscle fibre. When the length of a central segment of a fibre was under servocontrol, which eliminates tendon compliance as a potentially significant influence, stiffness still rose faster than tension (Cecchi et al., 1985). The conclusion we draw from these results is that cross-bridges may exist in a state generating relatively less force during the tetanus rise and during relaxation, compared to the prevailing state during the plateau of the tetanus (Cecchi, Griffiths and Taylor, 1981; 1984b). This essentially confirms a suggestion already made by H.E. Huxley (1979) from observing the intensity of the equatorial X-ray diffraction pattern change more rapidly than force. This pattern is presumed to reflect the movement of a substantial number of the cross-bridges to the vicinity of the thin filaments. But the X-ray pattern alone could not prove that the bridges were also attached. The mechanical evidence now supports the X-ray events. Similar conclusions have been drawn by a number of other investigators, each using different techniques. Hence the idea seems to be widely accepted at present (Mason and Hasan, 1980; Schoenberg and Wells, 1984; Tamamura et al., 1982)

The existence of multiple cross-bridge states might also explain why calcium transients fall long before relaxation begins (Blinks, Rudel and Taylor, 1978) and why stiffness lags even further behind. But this is only one of several possibilities. For example, we have not considered the likelihood that soluble Ca binding proteins can influence the relations among free and bound Ca and mechanical events during relaxation (see Ashley and Griffiths, 1983; Cannell and Allen, 1984; Gillis et al., 1982). Further evidence from coordinated measurements on the same preparations may help us to understand these things better. However, one thing seems clear. Changes in myoplasmic Ca^{2+} alone do not always correlate with contractile activation, and the modulation of an entire cycle of contraction and relaxation evidently involves several circumstances and events.

Acknowledgements

We are indebted to L.F. Wussow for preparing the figures and the manuscript, and to P. Iaizzo and J.G. Quinlan for criticizing the manuscript. Our work was supported in part by the USPHS (NS 14268), the Muscular Dystrophy Association, the Minnesota Heart Association, the Consejo Nactional de Investigaciones Cientificas y Tecnicas (CONICIT) of Venzuela, and the American Heart Association (77-983). One of us (SRT) also thanks the Dean and Staff of the University of Ulm, West Germany, for creating the opportunity for an extended visit, and for the hospitality and stimulation that aided the first draft of this manuscript

References

Ashley, C.C. and Griffiths, P.J. (1983). The effect of injection of parvalbumins into single muscle fibres from the barnacle *Balanus nubilus*. *Physiol. Soc.* 105P.

Ashley, C.C. and Ridgway, E.G. (1970). On the relationships between membrane potential, calcium transients and tension in single barnacle muscle fibres. *J. Physiol.* **209**, 105–30.

Blinks, J.R., Rüdel, R. and Taylor, S.R. (1978). Calcium transients in isolated amphibian skeletal muscle fibres. Detection with aequorin. *J. Physiol.* **277**, 291–323.

Bremel, R.D. and Weber, A. (1972). Cooperation within actin filament in vertebrate skeletal muscle. *Nature (Lond.)* **238**, 97–101.

Cannell, M.B. and Allen, D.G. (1984). Model of calcium movements during activation in the sarcomere of frog skeletal muscle. *Biophys. J.* **45**, 913–25.

Cecchi, G., Colomo, F. and Lombardi, V. (1983). Longitudinal distribution of step length changes in tetanized fibres from frog skeletal muscle. *J. Physiol.* **345**, 146P.

Cecchi, G., Colomo, F., Lombardi, V. and Piazzesi, G. (1985). Stiffness and isometric tension during tetanus rise in frog muscle fibre segments under length-clamp conditions. *J. Physiol.*

Cecchi, G., Griffiths, P.J. and Taylor, S.R. (1981). Decrease in aequorin response of skeletal muscle produced by quick release during tetanic contraction. *Proc. VII Internat, Biophys. Congr.* pp.177.

Cecchi, G., Griffiths, P.J. and Taylor, S.R. (1984a). Changes in intracellular Ca^{2+} induced by shortening imposed during tetanic contractions. In *Proceedings of the 2nd International Symposium on Cross-Bridge Mechanisms in Muscular Contraction, Seattle, 1982. Contractile Mechanisms in Muscle.* Pollack and H. Sugi (eds.), pp.455–72. Plenum Publishing Corp., New York.

Cecchi, G., Griffiths, P.J. and Taylor, S.R. (1984b). The kinetics of cross-bridge attachment and detachment studied by high frequency stiffness measurements. In *Proceedings of the 2nd International Symposium on Cross-Bridge Mechanisms in Muscular Contraction, Seattle, 1982. Contractile Mechanisms in Muscle.* Pollack and H. Sugi (eds.), pp. 641–56. Plenum Publishing Corp., New York.

Franzini-Armstrong, C. (1970). Natural variability in the length of thin and thick filaments in single fibres from a crab, *Portunus depurator*. *J. Cell Sci.* **6**, 559–92.

Gillis, J.M., Thomason, D., Lefevre, J. and Kretsinger, R.H. (1982). Parvalbumins and muscle relaxation: A computer simulation study. *J. Muscle Res. Cell Motil.* **3**, 377–98.

Gordon, A.M., Huxley, A.F. and Julian, F.J. (1964). The length–tension diagram of single vertebrate striated muscle fibres. *J. Physiol.* **171**, 28–30P.

Gordon, A.M., Huxley, A.F. and Julian, F.J. (1966). The variation in isometric tension with sarcomere length in vertebrate muscle fibres. *J. Physiol.* **184**, 170–92.

Hill, T.L. (1983). Two elementary models for the regulation of skeletal muscle contraction by calcium. *Biophys. J.* **44**, 383–96.

Howell, J.N. and Snowdowne, K.W. (1981). Inhibition of tetanus tension by elevated extracellular calcium contration. *Amer. Physiol. Soc.* C193–C200.

Hoyle, G. (1983). *Muscles and Their Neural Control.* John Wiley and Sons, New York.

Hoyle, G., McNeill, P.A. and Selverston, A.I. (1973). Ultrastructure of barnacle giant muscle fibers. *J. Cell Biol.* **56**, 74–91.

Huxley, A.F. (1957). Muscle structure and theories of contraction. *Prog. Biophys. Biophys. Chem.* **7**, 255–318.

Huxley, A.F. (1965). Muscle tension on the sliding-filament theory. *Proceedings of the XXIIIrd International Congress on Physiological Science, Tokyo, September 1965.* Excerpta Medica International Congress Series No. 87, pp. 383–7.

Huxley, A.F. (1973). A note suggesting that the cross-bridge attachment during muscle contraction may take place in two stages. *Proc. R. Soc. Lond. B.* **183**, 83–6.

Huxley, A.F. (1980). *Reflections on Muscle.* Princeton University Press, Princeton, NJ.

Huxley, H.E. (1979). Structure. In *The Molecular Basis of Force Development in Muscle.* Ingels (ed.), pp. 1–13. Palo Alto Medical Research Foundation, Palo Alto, California.

Kawai, M. and Brandt, P.W. (1980). Sinusoidal analysis: a high resolution method for correlating biochemical reactions with physiological processes in activated skeletal muscles of rabbit, frog, and crayfish. *J. Muscle Res. Cell. Motil.* **1**, 279–303.

Kawai, M., Cox, R.N. and Brandt, P.W. (1981). Effect of Ca ion concentration on cross-bridge kinetics in rabbit psoas fibers. Evidence for the presence of two Ca-activated states of thin filament. *Biophys. J.* **35**, 375–84.

Lannergren, J. (1979). An intermediate type of muscle fibre in *Xenopus laevis. Nature (Lond.)* **279**, 254–56.

Lannergren, J., Lindblom, P. and Johansson, B. (1982). Contractile properties of two varieties of twitch muscle fibres in *Xenopus laevis. Acta Physiol. Scand.* **114**, 523–35.

Lehman, W. and Szent-Györgi, A.G. (1975). Regulation of muscle contraction. Distribution of actin control and myosin control in the animal kingdom. *J. Gen. Physiol.* **66**, 1–30.

Lopez, J.R., Wanek, L.A. and Taylor, S.R. (1981). Skeletal muscle: length dependent effects of potentiating agents. *Science* **214**, 79–82.

Mason, P. and Hasan, H. (1980). Muscle crossbridge action in excitation and relaxation. *Experientia* **36**, 949–50.

Peachey, Adrian and Geiger, (1983). *Handbook of Physiology, Section 10: Skeletal Muscle,* **688**. Williams and Wilkins Co., Baltimore.

Podolsky, R.J. and Nolan, A.C. (1971). Cross-bridge properties derived from physiological studies of frog muscle fibers. In *Contractility of Muscle Cells.* R.J Podolsky (ed.). pp.247–60. Prentice-Hall, Englewood Cliffs, NJ.

Ramsey, R.W. and Street, S.F. (1940). The isometric length-tension diagram of isolated skeletal muscle fibers of the frog. *J. Cell. Comp. Physiol.* **15**, 11–34.

Ramsey, R.W. and Street, S.F. (1941). Muscle function as studied in single muscle fibers. *Biol. Symposia* **3**, 9–34.

Ridgway, E.B. and Ashley, C.C. (1967). Calcium transients in single muscle fibers. *Biochem. Biophys. Res. Comm.* **29**, 229–34.

Ridgway, E.B. and Gordon, A.M. (1984). Muscle calcium transients: effect of post-stimulus length changes in single fibers. *J.Gen. Physiol.* **83**, 75–103.

Ridgway, E.B., Gordon, A.M. and Martyn, D.A. (1983). Hysteresis in the force–calcium relation in muscle. *Science* **219**, 1075–7.

Rüdel, R. and Taylor, S.R. (1971). Striated muscle fibers: facilitation of contraction at short lengths by caffeine. *Science* **172**, 387–8.

Sandow, A. (1952). Excitation-contraction coupling in muscular response. *Yale J. Biol. Med.* **25**, 176–201.

Sandow, A. (1964). Potentiation of muscular contration. *Arch. Phys. Med. Rehabil.* **45**, 62–81.

Schoenberg, M. and Wells, J.B. (1984). Stiffness, force, and sarcomere shortening during a twitch in frog semitendinosus muscle bundles. *Biophys. J.* **45**, 389–97.

Smith, R.S. and Ovalle, W.K. Jr. (1973). Varieties of fast and slow extrafusal muscle fibres in amphibian hind limb muscles. *J. Anat.* **116**, 1–24.

Sugi, H., Ohta, T. and Tameyasu, T. (1983). Development of the maximum isometric force at short sarcomere lengths in calcium-activated muscle myofibrils. *Experientia* **39**, 147–8.

Tamamura, Y., Hatta, I., Matsuda, T., Sugi, H. and Tsuchiya, T. (1982). Changes in muscle stiffness during contraction recorded using ultrasonic waves. *Nature (Lond.)* **299**, 631–3.

Taylor, S.R. (1974). Decreased inactivation in skeletal muscle fibres at short lengths. In *Ciba Foundation Symposium 24: The Physiological Basis of Starling's Law of the Heart*, pp. 93–116. American Elsevier, New York.

Taylor, S.R. and Rüdel, R. (1970). Striated muscle fibers: inactivation of contraction induced by shortening. *Science* **167**, 882–4.

Taylor, S.R., Lopez, J.R., Griffiths, P.J., Trube, G. and Cecchi, G. (1982). Calcium in excitation-contraction coupling of frog skeletal muscle. *Can. J. Physiol. Pharmacol.* **60**, 489–502.

Vandervoort, A.A., Quinlan, J. and McComas, A.J. (1983). Twitch potentiation after voluntary contraction. *Exp. Neurol.* **81**, 141–52.

Weber, A. and Murray, J.M. (1973). Molecular control mechanisms in muscle contraction. *Physiol. Rev.* **53**, 612–73.

CALCIUM AND EXCITATION—CONTRACTION COUPLING IN SMOOTH MUSCLE

R. Casteels, G. Droogmans, F. Wuytack and L. Raeymaekers

Laboratorium voor Fysiologie, Campus Gasthuisberg K.U. Leuven, B-3000 Leuven, Belgium.

Abstract

Smooth muscle cells depend for a maintained force development on a continuous supply of calcium from the outside medium. Calcium is flowing into the cells through Ca channels which have been classified in both voltage-dependent and receptor-operated channels. A certain amount of calcium can also be released from an intracellular compartment, most likely the endoplasmic reticulum. It is proposed that for some vascular smooth muscle cells the receptor-operated channels and the endoplasmic reticulum form a functional unit. The action of an agonist would consist of a release of calcium from the store, and this depletion would then result in a refilling from the outside through the ER—plasma membrane connection.

The calcium extrusion system depends on a Ca-Mg-ATPase present in the plasma membrane. This enzyme can be distinguished from another Ca-Mg-ATPase which is present in the membrane of the endoplasmic reticulum.

17.1. Introduction

Smooth muscle cells are, as all other cells characterized by a large inwardly directed electrochemical gradient for Ca^{2+}. This gradient consists of a 10 000-fold concentration gradient and of a transmembrane potential difference of about 60 mV. Such a large electrochemical gradient of Ca^{2+} results even at a very low permeability of the plasma membrane in a continuous influx of this ion. The large changes of the cytoplasmic Ca^{2+} concentration, as required for the adaptation of the cell function, can be obtained by stimulus-dependent changes of the membrane permeability to Ca^{2+}, that is, by the formation of Ca-channels.

All these processes which increase the cytoplasmic concentration of Ca^{2+}, have to be neutralized by a reverse process, extruding Ca out of the cell. Such Ca extrusion is necessary to preserve the functional integrity of the cells. On the basis of these two properties of smooth muscle cells — Ca influx and Ca extrusion — one can consider smooth muscle cells as a one-way system for Ca movements. This picture is, however, further complicated by the existence in most smooth muscle cells of a cellular calcium compartment from which Ca can be released by an agonist and which under some conditions can also reaccumulate a limited amount of Ca.

We have to point out that the classification 'smooth muscle' embraces a large variety of contractile cells, which have different properties of both their plasma membrane and intracellular Ca stores. We will not describe those differences here, but rather review the mechanisms which are common to most types of smooth muscle. We will therefore discuss the agonist-sensitive Ca store, the transmembrane Ca influx and the transmembrane Ca extrusion.

17.2. The agonist-sensitive Ca store

The existence of an intracellular Ca store, which plays a role in excitation-contraction coupling in smooth muscle cells, has first been deduced from the observation that excitatory agonists can induce a transient contraction in tissues exposed to Ca-free solutions containing EGTA, or to solutions which had been supplemented with ions as La^{3+} or Mn^{2+}, which prevent the entry of Ca^{2+} across the cell membrane (Deth and van Breemen, 1974; Deth and Casteels, 1977; Deth and Lynch, 1981). The contractile response under these conditions is accompanied by a transient increase of the ^{45}Ca efflux, suggesting that the excitatory agonists induce a release of calcium from some intracellular store and thereby activate the contractile proteins. This Ca is then subsequently extruded out of the cells, causing the transient increased ^{45}Ca-efflux (Droogmans, Raeymaekers and Casteels, 1977; Deth and van Breemen, 1977). During a maintained exposure to Ca-free medium a supramaximal stimulus of the α-adrenoceptor will only elicit a phasic contraction and a ^{45}Ca efflux during its first application. Later identical stimuli in Ca-free medium only cause a small tonic force development, which is not accompanied by a ^{45}Ca release and which remains constant even after prolonged exposure to Ca-free medium. This tonic contraction is most likely independent of changes of $[Ca^{2+}]_{cyt}$, as has been described before (Casteels and Droogmans, 1981). We can therefore propose that under the present experimental conditions, the Ca released from the cellular store is extruded out of the cells, rather than being reaccumulated into the same structure.

An important aspect of the agonist-sensitive Ca store is that during exposure of the tissue to Ca-free solution the amount of calcium remaining in this store declines in time. This decline of calcium varies appreciably according to the smooth muscle tissue, being complete in less than 4 min in taenia coli (Casteels and Raeymaekers, 1979) and lasting over 50 min in the ear artery of the rabbit (Droogmans, Raeymaekers and Casteels, 1977). This rate of loss of Ca from the agonist-sensitive store also depends on temperature (Droogmans and Casteels, 1981). The maximal amount of Ca which can be released by an agonists amounts in the ear artery of the rabbit to 60 $\mu M/kg$ (Droogmans, Raeymaekers and Casteels, 1977), a quantity which is sufficient to fully activate the myofilaments. The observation that after stimulation with noradrenaline in Ca-free medium, the released Ca is not reaccumulated by this compartment does not alter the fact that under physiological conditions Ca is taken up by these organelles during relaxation. The Ca uptake activity of this store has been demonstrated by the finding that pretreatment with K-rich solutions or β-agonists in a Ca-containing solution augments the subsequent force development induced by noradrenaline in Ca-free medium. The most likely explanation is that under these conditions more calcium is taken up by the cellular store. This phenomenon has been demonstrated for the taenia coli of the guinea-pig (Casteels and Raeymaekers, 1979) and the ear artery of the rabbit (Van Eldere, Raeymaekers and Casteels, 1982). It is very likely that the agonist-sensitive Ca store corresponds to part of the endoplasmic reticulum (ER). It has been demonstrated that ER vesicles from smooth muscle can accumulate calcium in an ATP-dependent way after chemical skinning of the smooth muscle fibres (Raeymaekers, 1982). The release of Ca from the ER by caffeine has been extensively investigated in different smooth muscles before and after chemical skinning (Deth and Casteels, 1977; Itoh, Kuriyama and Suzuki, 1981). This effect of caffeine on an intracellular Ca-accumulating structure suggests a similarity between the endoplasmic reticulum of smooth muscle and that from skeletal muscle (Endo, 1977).

Two further aspects of this agonist-sensitive Ca store which is discussed below are the nature of the Ca-Mg-ATPase transporting Ca across its membrane, and the possible connections between this Ca store and the external medium.

17.3. Ca channels in the plasma membrane

In smooth muscle, as in many other tissues, Ca channels are opened during stimulation. As yet these presumed Ca channels have not been demonstrated in smooth muscle by the patch clamp technique, and only indirect evidence can be brought forward in favour of their existence and to describe their properties. It has been proposed that in smooth muscle cells two types of Ca channels occur; those which are activated by depolarization of the membrane are described as voltage-dependent channels, while those activated by the action of an agonist are called receptor-operated channels (Bolton, 1979). Up till now this distinction seems to be acceptable for some vascular smooth muscles, but the properties of Ca channels opened by K-depolarization or by agonist action in visceral smooth muscle look similar and it remains an acceptable hypothesis that in these tissues only one type of Ca channel would be present, which could be activated as well by membrane depolarization as by agonist action. The similar inhibitory activity of Ca antagonists on the force development induced by both stimuli in visceral smooth muscle supports this hypothesis (Casteels and Login, 1983). In contrast to these findings in visceral smooth muscle, Ca antagonists are not equally effective in inhibiting the contractions occurring in vascular smooth muscle. Ca antagonists were also in these smooth muscle tissues very efficient in inhibiting K-induced contractions, while their inhibitory action on responses elicited by agonists is less (Godfraind, 1976; Winquist and Baskin, 1983).

We have recently proposed that in some vascular smooth muscle cells the receptor-operated channels are an integral part of the agonist-sensitive Ca store and its connections with the plasma membrane (Casteels and Droogmans, 1981). An initial finding suggesting a direct relation between the external medium and the agonist sensitive Ca-store was that the rate of filling of this store after its depletion by stimulation with an agonist in Ca-free medium proceeds at a much faster rate than its depletion during exposure to Ca-free medium. In addition the total amount of Ca taken up by this store is a function of the external Ca concentration. Consistent with such a direct pathway from the extracellular medium is the observation that the loading of the store, as determined by the amplitude of contraction elicited by an agonist in Ca-free medium, depends much more on extracellular Ca concentration than on the cytoplasmic Ca concentration. This latter parameter which is increased during exposure to a K-rich solution is less efficient in increasing the Ca content of the ER than the extracellular Ca concentration. This direct pathway might be related to the close connection formed between the ER and the surface membrane as observed in electron microscopic pictures (Devine, Somlyo and Somlyo, 1971). At those specialized regions, the ER membrane is separated from the surface membrane by a 10–12 nm gap containing periodic electron opaque processes. Based on this presumed connection we have proposed that stimulation of these vascular smooth muscle cells with an agonist in a Ca-containing medium results in a release of Ca from the endoplasmic reticulum. This depletion of the store would facilitate the influx of Ca from the outside through the membrane–ER junction and thereby result in a continuous supply of Ca from the extracellular medium to the cytoplasm. This agonist-dependent release of calcium would by the continuous replenishment of the store be responsible for the tonic component of the contraction. The action of the agonist at its receptor would according to this hypothesis occur at only one site - the endoplasmic reticulum - while its seeming effect of opening receptor-operated channels in the plasma membrane could correspond to the continuous influx of Ca^{2+} across the membrane–ER junction and the endoplasmic reticulum as a consequence of the depletion of the agonist-sensitive store.

It was observed that K-rich media inhibit the filling of the agonist-sensitive store with Ca from the outside medium, and this experimental procedure has been used to obtain some evidence in favour of this hypothesis (Casteels et al., 1982a).

17.4. Ca transport mechanisms in smooth muscle

In order to maintain the force development of vascular smooth muscle during stimulation with an agonist there must be a continuous supply of Ca from the outside. This finding implies that in order to maintain the homeostasis of cellular calcium, Ca^{2+} must be extruded continuously against their electrochemical gradient. Two types of Ca-extrusion mechanism have been considered to exist in plasma membranes of smooth muscle - in other words, a transport depending on a Ca-Mg-ATPase and the Na–Ca exchange mechanism. In order to study the Na–Ca exchange the electrochemical gradient for Na^+ can be changed in smooth muscle cells by either exposure to Na-free solutions or by inhibiting the Na–K pump by ouabain or K-free solutions. Such prodecures can evoke a contractile response, whereas a restoration of the Na gradient induces relaxation (Ozaki and Urakawa, 1981). Although such observations are consistent with the Na–Ca exchange hypothesis there is as yet no demonstration of a direct causal relation between these two phenomena in smooth muscle. Moreover, contractions and relaxations can still be induced by various procedures in the complete absence of a Na gradient. Also after complete dissipation of the Na gradient by a prolonged exposure to K-free solutions, some smooth muscles remain relaxed (Casteels, Droogmans and Hendrickx, 1971). Another observation which argues against a functional importance of the Na–Ca exchange is that modifications of the Na gradient do not affect the rate of extrusion of ^{45}Ca from the rabbit ear artery during stimulation with noradrenaline (Droogmans and Casteels, 1981). For the guinea-pig taenia coli it was shown that the effect of Na-free solutions on the ^{45}Ca efflux could be largely a consequence of the increased diffusional pathway for Ca^{2+} caused by an increased Ca-binding capacity in the extracellular space (Raeymaekers, Wuytack and Casteels, 1975). An increased binding capacity of the extracellular matrix in Na-free solutions has also been proposed for vascular smooth muscle tissue (Villamil, Rettori and Yeyati, 1973). It is obvious that it is not possible to exercise sufficient control in the above-mentioned experimental conditions on the composition of the intracellular compartment of the smooth muscle cells. It was therefore necessary to investigate as well the Na-Ca exchange and the Ca-Mg-ATPase in simplified experimental systems such as microsomal preparations from smooth muscle. While the study of Na-Ca exchange in cardiac sarcolemmal vesicles has confirmed that such process could be quantitatively important for the Ca homeostasis in this tissue (Caroni and Carafoli, 1981), the data obtained for smooth muscle microsomes only indicate that the Na-Ca exchange is poorly developed in these tissues and could not play a significant role for the Ca extrusion under physiological conditions (Grover, Kwan and Daniel, 1981; Morel and Godfraind, 1982). The rate of Ca extrusion by Na-Ca exchange in plasmalemmal vesicles of smooth muscle is two orders of magnitude smaller than the value determined for cardiac muscle membranes.

Another argument against an important role of Na-Ca exchange in smooth muscle can be deduced from a comparison of the amount of the Na-K-ATPase and the Ca-Mg-ATPase activities in smooth muscle membranes (Casteels *et al.* 1984).

Because of the minimal requirement of 3 Na^+ for 1 Ca^{2+} in the Na-Ca exchange, and because of the 3 Na^+ for 1 ATP stoichiometry of the Na^+ pump, the Na-Ca exchange, which depends finally on the Na-K pump, could maximally extrude 1 Ca^{2+} for 1 ATP hydrolysed. Over the whole spectrum of isolated membrane fractions of smooth muscle separated on a density gradient, the activity of the Na-K-ATPase is always lower than that of the Ca-Mg-ATPase. In contrast it is observed that highly purified plasma membrane fractions from cardiac muscle contain much more Na-K-ATPase than the plasma membrane fractions from smooth muscle. In addition the Na-K-ATPase activity in cardiac plasmalemmal vesicles is higher than that of the Ca-Mg-ATPase. These differences in the activity of transport enzymes are certainly compatible with the important role of Na-Ca exchange in cardiac muscle and this hypothesis is further confirmed by the high value of V_{max} for Na-Ca exchange in plasmalemmal vesicles of cardiac muscle. Taking into account the activity in smooth muscle membranes of the Na-K-ATPase and of the Ca-Mg-ATPase, the activity of the pure enzymes and their molecular weight, we can calculate that the density of the Na-K-pump and Ca-pump sites in smooth muscle membranes would amount to 250–300 sites and to 300 sites per μm^2 respectively.

In recent years very important progress has been made in understanding how the ATP-dependent Ca transport works in smooth muscle membranes. It was quite easy to demonstrate that membrane vesicles prepared from different types of smooth muscle present a Ca^{2+} uptake with high affinity which depends on the presence of Mg-ATP (Carsten, 1969; Wuytack *et al.*, 1978). However, the demonstration of the presence of a corresponding (Ca^{2+} + Mg^{2+})-dependent ATPase activity was much more difficult (Wuytack and Casteels, 1980). We have in recent years succeeded in purifying these enzymes and we have obtained evidence that both types of the Ca-Mg-ATPases with transport functions are present in smooth muscle membranes. One of these enzymes is a calmodulin-stimulated Ca^{2+}-transport ATPase of 130 kD which has many characteristics in common with the well-known analogous enzyme from the plasma membrane of human erythrocytes (Wuytack, De Schutter and Casteels, 1981; Wuytack *et al.*, 1982). Also a second Ca transport ATPase of 100 kD has been isolated from smooth muscle microsomes. This transport protein has many properties in common with the Ca^{2+}-transport ATPase from the sarcoplasmic reticulum of skeletal muscle (Raeymaekers et al., 1983). The resemblances between the Ca-Mg-ATPases of smooth muscle and the enzymes from other tissues have recently been confirmed in terms of molecular weight, reaction to La^{3+} and in a similar proteolytic degradation pattern. The presence of the calmodulin-stimulated ATPase in mixed membranes from smooth muscle is confirmed by its binding of calmodulin and of antibodies against erythrocyte Ca^{2+}-transport ATPase, whereas such binding does not occur with proteins present in the presumed endoplasmic reticulum from smooth muscle (Wuytack et al. 1983, 1984).

Finally, we have succeeded in separating in a sucrose density gradient the plasma membranes and endoplasmic reticulum membranes by changing the density of the cholesterol-containing vesicles by adding digitonin. It was observed that most of the Ca-Mg-ATPase was shifted in parallel with the Na—K-ATPase. A limited fraction of the Ca-Mg-ATPase remained in the lower density fractions and most likely corresponds to the endoplasmic reticulum fraction. A further study of the transport enzymes present in the different fractions might allow us to show conclusive evidence on the subcellular localization of the different Ca transport enzymes.

References

Bolton, T.B. (1979). Mechanism of action of transmitters and other substances on smooth muscle. *Physiol. Rev.* **59**, 606—718.
Caroni, P. and Carafoli, E. (1981). The Ca^{2+}-pumping ATPase of heart sarcolemma. *J. Biol. Chem.* **256**, 3263—70.
Carsten, M.E. (1969). Role of calcium binding by sarcoplasmic reticulum in the contraction and relaxation of uterine smooth muscle. *J. Gen. Physiol.* **53**, 414—26.
Casteels, R. and Droogmans, G. (1981). Exchange characteristics of the noradrenaline sensitive calcium store in vascular smooth muscle cells of rabbit ear artery. *J. Physiol. Lond.* **317**, 263—79.
Casteels, R., Droogmans, G. and Hendrickx, H. (1971). Membrane potential of smooth muscle cells in K-free solution. *J. Physiol.* **217**, 281—95.
Casteels, R., Droogmans, G., Raeymaekers, L. and Wuytack, F. (1982a). Ca-exchange and transport in vascular smooth muscle cells. In *Advances in Pharmacology and Therapeutics II, Vol. 3. Cardio-Renal and Cell Pharmacology,* Yoshida, Hagihara and Ebashi (eds.), pp. 71—9. Pergamon Press, Oxford/New York.
Casteels, R., Raeymaekers, L., Suzuki, H. and Van Eldere, J. (1982b). Tension response and ^{45}Ca release in vascular smooth muscle incubated in Ca-free solution. *PflCgers Arch.* **395**, 81—3.
Casteels, R. and Login, I.S. (1983). Reserpine has a direct action as a calcium antagonist on mammalian smooth muscle cells. *J. Physiol.* **340**, 403—14.
Casteels, R. and Raeymaekers, L. (1979). The action of acetylcholine and catecholamines on an intracellular calcium store in the smooth muscle cells of the guinea-pig *taenia coli. J. Physiol.* **294**, 51—68.
Casteels, R., Raeymaekers, L., Droogmans, G. and Wuytack, F. (1984). Na^+-K^+ ATPase, Na-Ca exchange and excitation-contraction coupling in smooth muscle. A critical review. *J. Cardiovasc. Pharmacol.*
Deth, R. and Casteels, R. (1977). A study of releasable Ca^{2+} fractions in smooth muscle cells of the rabbit aorta. *J. Gen. Physiol.* **69**, 401—16.
Deth, R. and Lynch, C.J. (1981). Mobilization of a common source of smooth

muscle Ca^{2+} by norepinephrine and methylxanthines. *Amer. J. Physiol.* **240**, C239–47.

Deth, R. and Van Breemen, C. (1974). Relative contributions of Ca^{++} influx and cellular Ca^{++} release during drug-induced activation of the rabbit aorta. *Pflügers Arch.* **348**, 13–22.

Devine, C.E., Somlyo, A.V. and Somlyo, A.P. (1971). Sarcoplasmic reticulum and excitation-contraction coupling in mammalian smooth muscle. *J. Cell Biol.* **52**, 690–715.

Droogmans, G. and Casteels, R. (1979). Sodium and calcium interactions in vascular smooth muscle cells of the rabbit ear artery. *J. Gen. Physiol.* **74**, 57–70.

Droogmans, G. and Casteels, R. (1981). Temperature-dependence of ^{45}Ca fluxes and contraction in vascular smooth muscle cells of rabbit ear artery. *Pflügers Arch.* **391**, 183–9.

Droogmans, G., Raeymaekers, L. and Casteels, R. (1977). Electro- and pharmacomechanical coupling in the smooth muscle cells of the rabbit ear artery. *J. Gen. Physiol.* **70**, 129–48.

Endo, M. (1977). Calcium release from the sarcoplasmic reticulum. *Physiol. Rev.* **57**, 71–108

Godfraind, T. (1976). Calcium exchange in vascular smooth muscle, action of noradrenaline and lanthanum. *J. Physiol.* **260**, 21–35.

Grover, A.K., Kwan, C.Y. and Daniel, E.E. (1981). Na-Ca exchange in rat myometrium membrane vesicles highly enriched in plasma membranes. *Amer. J. Physiol.* **240**, C175–82.

Itoh, T., Kuriyama, H. and Suzuki, H. (1981). Excitation–contraction coupling in smooth muscle cells of the guinea-pig mesenteric artery. *J. Physiol.* **321**, 513–35.

Morel, N. and Godfraind, T. (1982). Na-Ca exchange in heart and smooth muscle microsomes, *Arch. Int. Pharmacodyn.* **258**, 319–321.

Ozaki, H. and Urakawa, N. (1981). Involvement of a Na–Ca exchange mechanism in contraction induced by low-Na solution in isolated guinea-pig aorta. *Pflügers Arch.* **390**, 107–12.

Raeymaekers, L. (1982). The sarcoplasmic reticulum of smooth muscle fibers. *Z. Naturforsch.* **37c**, 481–8.

Raeymaekers, L., Wuytack, F. and Casteels, R. (1975). Na-Ca exchange in taenia coli of the guinea-pig. *Pflügers Arch.* **347**, 329–40.

Raeymaekers, L., Wuytack, F., Eggermont, J., De Schutter, G. and Casteels, R. (1983). Isolation of a highly enriched plasma membrane faction from gastric smooth muscle. Comparison of the Ca-uptake to that in endoplasmic reticulum. *Biochem. J.* **210**, 315–22.

Van Eldere, J., Raeymaekers, L. and Casteels, R. (1982). Effect of isoprenaline on intracellular Ca uptake and on Ca influx in arterial smooth muscle. *Pflügers Arch.* **395**, 81–3.

Villamil, M.F., Rettori, V. and Yeyati, N. (1973). Calcium exchange and distribution in the arterial wall. *Amer. J. Physiol.* **224**, 1314–9.

Winquist, R.J. and Baskin, E.P. (1983). Calcium channels resistant to organic calcium entry blockers in a rabbit vein. *Amer. J. Physiol.* **245**, H1024–30.

Wuytack, F. and Casteels, R. (1980). Demonstration of a $(Ca^{2+} + Mg^{2+})$ ATPase activity probably related to Ca^{2+} transport in the microsomal fraction of porcine coronary artery smooth muscle. *Biochim. Biophys. Acta.* **595**, 257–63.

Wuytack, F., De Schutter, G. and Casteels, R. (1981). Purification of $(Ca^{2+} + Mg^{2+})$-ATPase from smooth muscle by calmodulin affinity chromatography. *FEBS Lett.* **129**, 297–300.

Wuytack, F., Raeymaekers, L., De Schutter, G. and Casteels, R. (1982). Demonstration of the phosphorylated intermediates of the Ca^{2+} transport ATPase in a microsomal fraction and in a (Ca^{2+} + Mg^{2+})-ATPase purified from smooth muscle by means of calmodulin affinity chromatography. *Biochim. Biophys. Acta* **693**, 45–52.

Wuytack, F., De Schutter, G., Verbist, J. and Casteels, R. (1983). Antibodies to the calmodulin-binding Ca^{2+}-transport ATPase from smooth muscle. *FEBS Lett.* **154**, 191–5.

Wuytack, F., Landon, E., Fleischer, R. and Hardman, J.G. (1978). The calcium accumulation in a microsomal fraction from porcine artery smooth muscle. A study of the heterogeneity of the fraction. *Biochim. Biophys. Acta.* **540**, 253–69.

Wuytack, F., Raeymaekers, L., Verbist, J., De Smedt, H. and Casteels, R. (1984). Evidence for the presence in smooth muscle of two types of Ca^{2+}-transport ATPase. *Biochem. J.* **224**, 445–51.

CONTRIBUTION OF CALCIUM CURRENT TO CONTRACTION IN HEART MUSCLE

Martin Morad and Lars Cleemann

Department of Physiology, University of Pennsylvania, Philadelphia, Pennsylvania, 19104, USA

Abstract

The mechanism of Ca^{2+} transport and its contribution to development of tension were examined by combining the voltage clamp method with rapid photo inactivation of Ca^{2+} antagonists (nifedipine or nisoldipine) and measurements of Ca^{2+} depletion using extracellular Ca^{2+} probes. In frog heart inactivation of Ca^{2+} antagonists with 100 μsec light pulse (430 nm) recovered the tension accompanied by a slow maintained component of I_{si}. Using antipyrylazo III as an extracellular Ca^{2+} probe we monitored a continuous Ca^{2+}-depletion signal in voltage-clamped frog ventricular strips. Ca^{2+}-depletion signal in muscle had the same wave length-dependence as the Ca-induced difference spectrum of the dye in the perfusate. Rate of Ca^{2+} depletion was maximal immediately after the upstroke of the action potential (long before the onset of tension) and reached a value of 10–200 μM at the end of the action potential. Ca^{2+} depletion was maintained for the duration of depolarization. Nifedipine blocked I_{si} and suppressed the signal and epinephrine enhanced I_{si} and increased Ca^{2+} depletion. But acetylstrophanthidin which potentiated the contraction failed to enhance Ca^{2+} depletion signal. Our studies show that in the frog heart Ca^{2+} transport across the membrane may be sufficient to activate contraction. The primary route for the transport of Ca^{2+} appears to be the Ca^{2+} channel.

18.1. Introduction

Since Ringer's original observation 100 years ago, it has been known that development of force in heart muscle depends critically on the extracellular Ca^{2+} concentration. In more recent years the role of various pools of Ca^{2+} in activation of tension in heart muscle has been more rigorously evaluated. It is now generally agreed that the vertebrate hearts may be divided into two major catagories when considering the sources and the pathways of Ca^{2+} transport system. One category, represented by the frog heart in Fig.18.1, includes amphibian, reptilian, some fish and elasmobranch hearts. The other category includes all mammalian hearts, terestrial and marine. Figure 18.1 compares the major steps of E-C coupling processes in frog and cat heart. Two major differences appears to exist between these two categories.

(1) In the frog heart the source of the activator Ca^{2+} appears to be primarily the extracellular space, while in the mammalian heart both extracellular and intracellular Ca^{2+} are used in activation of contraction.
(2) A significant part of Ca^{2+} sequestered in mammalian heart appears to recirculate to the Ca^{2+} release pools. Such a component of Ca^{2+} could not be identified in the frog heart (Klitzner and Morad, 1983a; Morad and Goldman, 1973).

In this chapter we focus on the frog heart in order to probe in greater detail the mechanism of Ca^{2+} transport and the role of different Ca^{2+} transport systems in activation of contraction.

18.2. Results

To examine the pathways of Ca^{2+} transport we combined voltage clamp technique with two novel optical procedures. One optical technique takes advantage of light sensitivity of dihydropyridine group of Ca^{2+} antagonists (nifedipine, nisoldipine) such that Ca^{2+} channel block may be removed with an intense and short flash of light (5–10 msec). The second technique uses a Ca^{2+}-sensitive dye (antipyrylazo III) in the extracellular space, and attempts to measure the time-course and the rate of Ca^{2+} depletion from the extracellular space.

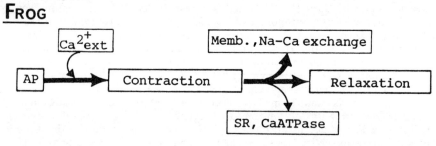

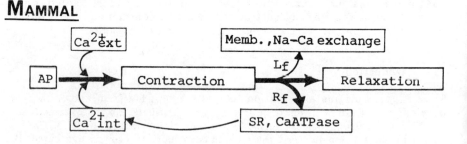

Fig.18.1. A schematic representation of the Ca^{2+}-cycle in frog and mammalian hearts. In frog heart Ca^{2+} for contraction comes from the extracellular pool. A Na^+-Ca^{2+} exchanger mediates in part the uptake of Ca^{2+} into the extracellular pool causing relaxation. Ca^{2+} sequestered by the sarcoplasmic reticulum (SR) does not reappear in the contractile cycle directly. In mammalian heart the primary sequestering system, the SR, helps to recycle Ca^{2+} back to internal pools of Ca^{2+}. This recirculating fraction (R_f) determines the amount of Ca^{2+} in the pool and the subsequent force of contraction. The lost fraction (L_f) of Ca^{2+} may be transported by the Na^+-Ca^{2+} exchanger and does not reappear directly in the releasable Ca^{2+} pools. (From Morad, 1982.)

18.2.1. Rapid photoinactivation of Ca^{2+} antagonists

Figure 18.2 shows the chemical reaction when nifedipine is exposed to light (at 350 nm). The pyridine ring is phenolized as a water molecule is released. This light exposed molecule shows a markedly different absorption spectrum than that of mother compound and fails to block Ca^{2+}-channel or suppress contraction. Figure 18.2 also shows the time-course of suppression of tension by nifedipine and its reversal by a single flash of light. Note that tension was potentiated to its original value within one beat after the light-induced inactivation of nifedipine. These results suggest that the nifedipine-suppressed tension may be reversed as soon as the nifedipine molecule is inactivated (the light-induced inactivation reaction in solution takes place in less than 1 μsec; Morad, Goldman and Trentham, 1983).

In order to examine whether the recovery of tension is accompanied by recovery of Ca^{2+}-current, frog ventricular strips were voltage clamped using a single-sucrose gap voltage clamp technique (Fig.18.3). In the presence of nisoldipine the developed tension and the inward Ca^{2+} current were suppressed (traces 2 and 3 in Fig.18.3). Tension and Ca^{2+} current recovered

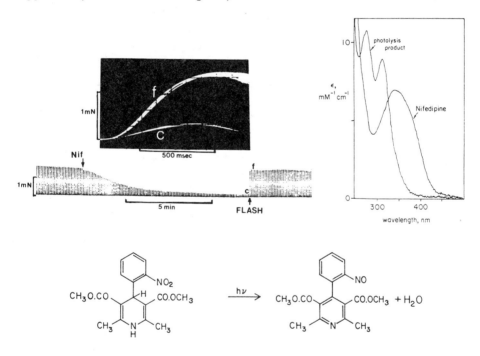

Fig.18.2. Photoconversion of nifedipine. Tension records show twitches of an electrically stimulated (12/min) frog ventricular strip bathed in Ringer solution containing (mM) Na^+, 115; Cl^-, 118; K^+, 3; HCO_3^-, 2; Ca^{2+}, 1; pH 7.1 at 20°C. Nifedipine (Nif) 1 μM, was added at the time indicated by the arrow. Twitch c (also shown in expanded time scale) is the suppressed twitch before exposure of the muscle to a 1 s pulse of 300–400 nm light. Twitch f is the first beat after the light pulse. In this experiment the bathing solution was not flowing. The photochemical reaction of nifedipine is shown together with the accompanying ultraviolet spectral change recorded in 100 mM TES buffer at pH 7.0. Photolysis of nisoldipine (which has an identical structure to nifedipine except that one of the methyl esters is replaced by an isobutyl ester) was also accompanied by a similar ultraviolet spectral change. The photoproduct of nisoldipine was crystallized and had an identical ultraviolet spectrum to the photoproduct in the mother liquor, indicating that the photolysis results in a single product plus water. The structure of the nisoldipine photoproduct was confirmed by proton NMR measured in deuterated chloroform. In particular, single protons 4.14 (NH) and 4.20 (both singlets) in the nisoldipine spectrum were absent in that of the photoproduct. (From Morad, Goldman and Trentham, 1983.)

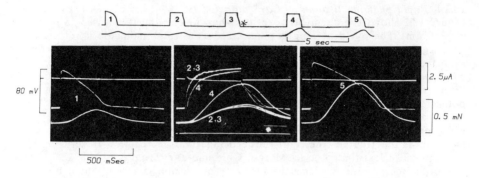

Fig.18.3. Recovery of inward current by photoinactivation of nisoldipine. The upper diagram shows the protocol of action potential and clamp pulses applied at 5 s intervals to a ventricular muscle equilibrated in 1 μM nisoldipine. The control (nisoldipine-suppressed) action potential and contraction are shown as traces 1 (left panel, original recordings). The sequence of stimulation is then interrrupted by a series of three clamp pulses before resumption of regular stimulation (upper diagram). After the second clamp pulse the muscle is exposed to a 200 ms light pulse (horizontal bar with *). Superimposed traces of membrane current, voltage and developed tension are shown in the middle panel and are labelled 2, 3 and 4 in correspondence with the diagram. Note that before the flash, tensions and membrane currents superimpose from clamp pulses 2 and 3 to 0 mV. However, after the flash, membrane current is more inward and developed tension is enhanced for a clamp pulse to the same potential (traces 4). The right panel shows the enhanced action potential and tension of the next normally stimulated beat (traces 5). (From Morad, Goldman and Trentham, 1983.)

(trace 4) when the preparation was exposed to an intense flash of light. The stimulated action potential was also markedly enhanced when the drug was inactivated (compare traces 1 and 5). The clamp protocol and the timing of the flash is indicated as an inset of Fig.18.3. Since recovery of tension and the inward current occur simultaneously it may be tentatively suggested that the recovery of Ca^{2+} current may be responsible for the immediate activation of tension.

It can be argued, however, that dihydropyridines may inhibit a number of processes which suppress contraction in addition to blocking Ca^{2+} current. In chemically skinned ventricular strips (detergent-treated or EDTA-treated; Endo and Iino, 1980; Winegrad, 1971), we found that Ca^{2+}-antagonists had no significant effect on the Ca^{2+}-sensitivity of the myofilaments (Morad, Goldman and Trentham, 1983). Figure 18.4 further confirms that the primary site of action of Ca^{2+}-antagonists is the Ca^{2+} channel. In this experiment the preparation was first exposed to nifedipine which suppressed tension within 10–15 min. Addition of Ni^{2+}, although it further suppressed the rate of development of tension and reduced the level of plateau potential, potentiated the final tension because of the marked Ni^{2+}-induced increase in the action potential duration (Klitzner and Morad, 1983b). Photoinactivation of nifedipine with an intense flash of light failed to enhance tension or the action potential. This observation is consistent with the idea that the suppressive effects of nifedipine on tension and action potential are mediated by blockage of the Ca^{2+} channel. Since in Ni^{2+}-treated preparations inactivation of nifedipine did not result in significant recovery of tension or action potential plateau, it was concluded that a common site, possibly the Ca^{2+}-channel, was the likely site of action of the drug.

The rapid photoinactivation of nifedipine, also results in recovery of maintained or slowly inactivating inward current, with kinetics quite similar to developed tension. The simultaneous recovery of such a slow and 'maintained' component of membrane current (i_{mi}), simultaneous with the recovery of maintained tension may suggest that a slowly inactivating or maintained Ca^{2+} channel is responsible for development of tension in the frog heart. If nifedipine binds primarily to the Ca^{2+} channel thereby suppressing tension, then the results of photoinactivation of nifedipine suggest that well over 80% of contractile Ca^{2+} in the frog heart is transported through a Ca^{2+} channel system in the surface membrane (Fig.18.5).

18.2.2. Direct measurement of Ca^{2+} influx

The above results imply that Ca^{2+} for activation of tension may enter the cell continuously across the cell membrane during the action potential. However, no direct evidence was provided as to the time-course and the magnitude of such a Ca^{2+} influx. Using antipyrylazo III (a Ca^{2+}-indicator metalochromic dye; Scarpa, Brinley and Dubyak, 1978) in the extracellular space we attempted to monitor the time-course and the magnitude of Ca^{2+} influx in frog ventricular strips.

Figure 18.6a shows a schematic of the apparatus used. A beam of light passing through the experimental chamber was used to measure simultaneously the absorption signal of antipyrylazo-equilibrated muscle at three different wavelengths. The concentration of antipyrylazo III and Ca^{2+} was adjusted such that the free Ca^{2+} activity ranged between 0.05 and 0.2 mM as measured with a Ca-sensitive electrode or calculated from published

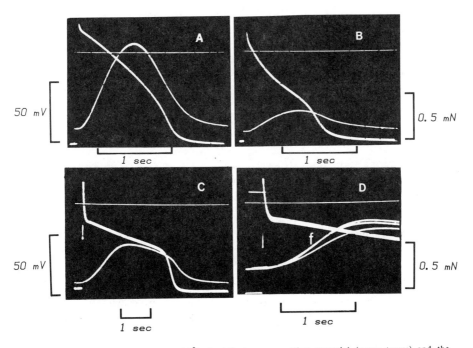

Fig.18.4. Interaction of nifedipine and Ni^{2+}. In (A), trans-gap action potential (upper traces) and the accompanying contraction (lower trace) are indicated. (B) Shows the steady-state suppression of 1 μM nifedipine on both action potential and contraction. In (C) the preparation has been exposed for 30 min to Ringer solution containing 1 μM nifedipine and 2 mM Ni^{2+}. Action potential is prolonged 3-fold to 4-fold, and although the initial rate of development of tension is suppressed (compare D, lower trace, and B), the peak tension is larger than in the presence of nifedipine (B). In (D) (expanded time base) three successive action potentials and contractions are superimposed in the presence of nifedipine and Ni^{2+}. The first, before the flash (control), was accompanied by the smallest twitch. The second and third, occurring after a 200 ms light pulse (horizontal bar), were accompanied by contractions with slightly higher initial rates of rise of twitch tension (f). (From Morad, Goldman and Trentham, 1983.)

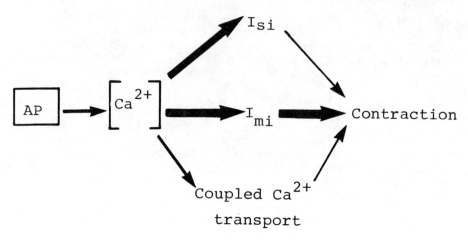

Fig.18.5. Possible sarcolemmal Ca^{2+} transport mechanisms in frog heart. The figure suggests three parallel transport systems for Ca^{2+} influx: the slow inward current (I_{si}), a maintained or slowly inactivated Ca^{2+} current (I_{mi}) and the Na^+-Ca^{2+} exchanger.

binding constants (Blinks et al., 1982). Motion-induced absorption changes were reduced by compressing the preparation between moveable transparent walls of the perfusion chamber (W_1 and W_2). The strip was illuminated from the left and the transmitted light was detected on the right. The transmitted light was measured at three different wavelengths (I_{788}, I_x, and I_{600}). Amplifiers with variable gains (G_1, G_2, G_3) were used to adjust the signals to a common reference level and the normalized signals ($\Delta I/I$) were summed with the predetermined weights (a_{788}, -1, and a_{600}) yielding,

$$\Sigma = a_{788}(\Delta I_{788}/I_{788}) - (\Delta I_x/I_x) + a_{600}(\Delta I_{600}/I_{600})$$

Only one of the light intensities, I_x, was used to detect Ca^{2+} depletion. This signal was assigned the weight of -1. The other two wavelengths were used to obtain information about accompanying motion-induced artifacts (antipyrylazo III is transparent at 788 nm and has an isosbestic point at 600 nm). The weights at 788 nm and 600 nm were calculated assuming that the motion-induced light scattering by the muscle or the dye at any given wavelength was directly proportional to the absorption at the wavelength. For instance, since the absorption of the dye containing perfusate at 600 nm was five times larger than at 718 nm the motion-induced light scattering should be eliminated if $a_{600} = 1/5 = 0.2$. The weight of the 788 signal was then chosen to compensate for the muscle motion seen at the two other wavelengths. The value found was $a_{788} = 0.85$, since the muscle absorption was found to be approximately proportional to the inverse square of the wavelength. $(0.2/(600)^2 - 1/(718)^2 + 0.85(788)^2 = 0)$. Figure 18.6b shows that this precedure, in the absence of any dye, produces a flat (Σ) trace. Similar procedures with Ca^{2+} insensitive dyes or with antipyrylazo III saturated with Ca^{2+} showed that this compensation procedure was effective in suppressing the motion-induced light scattering signal.

Figure 18.6c shows that in antipyrylazo-equilibrated muscles, large Ca^{2+} depletion signals (downward depletion) could be measured during the action potential. The wavelength dependence of Ca^{2+} signal was used routinely as a check on the accuracy of the signal and the adequacy of the compensation procedure. Figure 18.7 shows that the wavelength dependence of Ca^{2+} depletion signal in the extracellular space of the muscle was similar to that of the dye in cuvette. The shape of depletion signals was similar at different wavelengths, but was inverted at 540 nm (Fig.18.7b) and was larger at 660 nm (Fig.18.7c) than at 718 nm (Fig.18.7d). The peak depletion (during contraction) and the initial rate of depletion (occuring prior to the onset of tension) were in agreement with the spectrophotometric measurements when scaled by appropriate factors (see legends to Fig.18.7). These findings indicate that the absorption signal measured in antipyrylazo-equilibrated muscles represent in fact Ca^{2+}-depletion. The similarity of action spectrum during the peak of twitch confirms that the weighting procedures were effective in suppressing the motion-induced light scattering thus allowing measurement of Ca^{2+} depletion signal reliably during contraction. The finding that depletion signals had an action spectrum quite similar to that shown in Fig.18.7a in hypertonic solutions, where tension was suppressed, further confirmed the reliability of the substraction procedure.

Figure 18.8 shows that depletion of Ca^{2+} occurs prior to the onset of tension lasts for the duration of depolarization, and terminates upon premature repolarization of the action potential. Immediately after the upstroke the rate

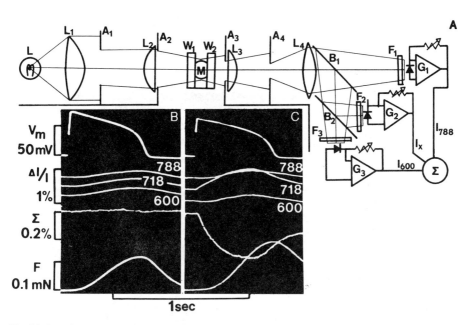

Fig.18.6. Experimental setup for antipyrylazo III measurement of extracellular Ca depletion. In (A) the strip (M) is compressed between the moveable transparent walls of the perfusion chamber (W_1 and W_2). The illuminator on the left consists of a lamp (L), two lenses (L_1 and L_2), two apertures (A_1 and A_2) and a heat-reflecting mirror (not shown); L_1 collimates the light, A_1 determines the illuminated areas since it is imaged by L_2 onto the strip, A_2 determines the numerical aperture of the illuminating beam. In the detector on the right the strip is first imaged by the lens (L_3, numerical aperture A_3) onto the plane of the aperture A_4; A_4 determines the visible, rectangular area within the central homogeneously compressed part of the strip and is imaged onto the light-sensitive areas of three photodiodes by means of the lens L_4 and the beam splitters B_1 and B_2. The wavelengths of the detected light are determined by interference filters. G_1, G_2, and G_3 are amplifiers with variable gain; Σ is the weighted average of the light intensities at three different wavelengths, I_{600}, I_x and I_{788}. The inset panels show sample records measured before (B) and after (C) admitting the dye-containing perfusate. The traces are the action potential measured through the sucrose gap (V_m), the light intensities at 788 nm, 718 nm and 600 nm ($\Delta I/I$), the weighted average (Σ), and the isometric force (F). (Modified from Pizarro, Cleemann and Morad, 1984.)

of Ca^{2+} depletion is typically 100–200 $\mu M/s$ corresponding to an inward Ca current of 0.5–1 $\mu A/cm^2$ (25% extracellular space, μm^{-1} surface-to-volume ratio; Page and Niedergerke, 1972). Thus it appears that depletion of Ca^{2+} from the extracellular space continues for the duration of depolarization, suggesting direct control of the process by the membrane potential. After repolarization the reaccumulation of Ca^{2+} often had a small rapid phase but the major part of Ca^{2+} repletion occurred long after relaxation was complete. These findings show that although Ca^{2+} leaves the extracellular space rapidly it returns to the extracellular space much slower, suggesting possible sequestration of Ca^{2+} by intracellular pools.

18.2.3. Possible mechanisms of Ca^{2+} transport

The voltage-dependence of Ca^{2+} signal was explored in order to examine the mechanism responsible for depletion of Ca^{2+}. The initial rapid rate of Ca^{2+}

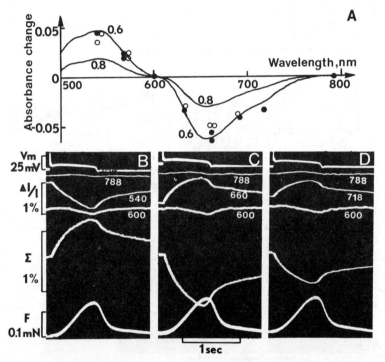

Fig.18.7. The action spectrum. Continuous curves in (A) were measured differentially with 0.15 mM cuvettes. The total Ca concentration was 0.6 and 0.8 mM compared with 1 mM in the reference solution. Points correspond to scaled values of peak depletion (o) and maximal rate of depletion (●). (B), (C), and (D) show sample records where the Ca depletion was measured at 540 nm, 660 nm, and 718 nm. From top to bottom each panel shows the action potential measured through the sucrose gap (V_m), the change in light intensity at 788 nm, at the wavelength in question, and at 600 nm, ($\Delta I/I$), the depletion signal (Σ) and the twitch force (F). Repeated measurements at 718 nm were used to bracket the measurement at other wavelengths. The measured Ca^{2+} depletion signals were calibrated by first determining the optical path through the extracellular space from the increase in absorption which resulted from admission of the dye-containing perfusate to the setup. The Ca^{2+}-induced difference spectrum of the perfusate was then normalized to represent a cuvette with the same optical path (0.015 cm). Finally the absorption changes measured during the beat were scaled to give the best fit with the curve (0.6) representing a 400 μM reduction in the total Ca^{2+} concentration. For instance, the values corresponding to peak depletion in normal Ringer solution (o) were multiplied by 15.5 to match this curve. This indicates a net Ca^{2+} depletion of 400 $\mu M/15.5$ = 26 μM or 32 μM when corrections are made for the non-linear Ca^{2+}-binding of antipyrylazo III. (Modified from Cleemann, Pizarro and Morad, 1984.)

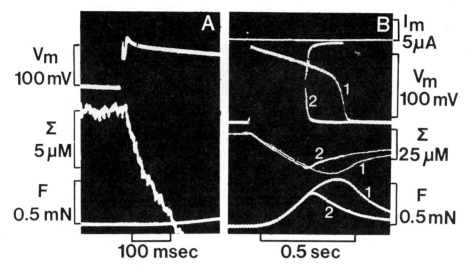

Fig.18.8. Continuous control of the extracellular Ca depletion by the membrane potential. (A) Shows that the rate of Ca depletion (Σ) is maximal within 10 ms of depolarization (V_m) and before any force development (F). In (B) V_m is the membrane potential measured with a microelectrode and I_m is the membrane current. (B) Shows that the depletion process is promptly terminated when the action potential (trace 1) is terminated by clamping the membrane potential back to the resting potential (trace 2). (Modified from Cleemann, Pizarro and Morad, 1984.)

depletion (Ca^{2+} influx) was measured at different potentials with clamps from rest and under conditions where Ca^{2+} current was inactivated by a conditioning pre-pulse to zero (Fig.18.9). The initial rate of Ca^{2+} depletion accompanying clamps from rest shows a bell-shaped voltage dependence similar to that of slow inward current (filled circles, Fig.18.9d). Addition of adrenaline enhanced the initial rate of Ca^{2+} depletion at all potentials, consistent with the enhancement of Ca^{2+}-current seen with this agent (triangles, Fig.18.9d). On the other hand the rate of Ca^{2+} depletion was markedly reduced when a 1.6 s depolarizing prepulse was used to inactivate the slow inward current (open circles, Fig.18.9d). These findings confirm that the major part of the Ca^{2+} influx occurs through the slow inward channel. The small fluxes observed after the inactivation of Ca^{2+} current, appears to have a voltage dependence similar to that predicted for Na^+-Ca^{2+} counter-transporter (Mullins, 1981). However, the magnitude of Ca^{2+} influx through such a system appears to be too small in the range of plateau potentials to contribute significantly to activation of tension. Consistent with this idea, strophanthidin (a rapidly acting cardiac glycoside) failed to enhance the rate of Ca^{2+} depletion accompanying the action potential as it potentiated tension (Fig.18.10b). Even though adrenaline and strophanthidin appear to be equally effective in enhancing the force of contraction (Fig.18.10), unlike strophanthidin, the positive inotropic effect of adrenaline is accompanied by a marked increase in Ca^{2+}-depletion signal, (compare Figs. 18.10a and 18.10b). The results of Figs. 18.9 and 18.10, therefore, strongly imply that the influx of Ca^{2+} through Na^+-Ca^{2+} counter-transport system may be neither large nor enhanced by suppression of Na^+ pump at the plateau potentials. These experiments, however, do not exclude the possibility that strophanthidin-induced potentiation of contraction is mediated by slower elevation of the resting intracellular Ca^{2+} concentration. Our results are

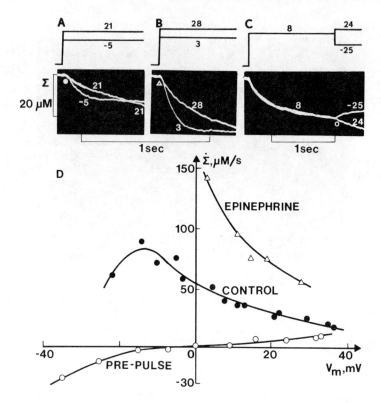

Fig.18.9. The voltage dependency of the Ca^{2+} transport. The rate of Ca^{2+} depletion was measured during voltage clamp pulses to different potentials under control conditions (A), after addition of 10^{-6} M epinephrine (B), and after a conditioning pre-pulse (C). Each of the panels (A), (B), and (C) show a schematic of the voltage-clamp procedure and records of the Ca^{2+} depletion signal (Σ). Traces are labelled with the membrane potential in mV. In (D) the rates of Ca^{2+} depletion (Σ) measured shortly after the clamp steps were plotted versus the membrane potential. The curves show that the rate of Ca^{2+} depletion has a voltage dependency similar to that of I_{si} (●), is enhanced by epinephrine (△) and suppressed by an inactivating pre-pulse (○). (From Pizarro, Cleemann and Morad, 1984.)

more consistent with the idea that Na^+-Ca^{2+} counter-transport system contributes significantly to the sequestration and efflux of Ca^{2+}. In support of such an idea the rate of Ca^{2+} repletion becomes significant at potentials negative to -30mV.

18.3. Summary and conclusions

Evidence from both rapid photoinactivation of Ca^{2+}-channel blockers or direct measurement of Ca^{2+} influx with Ca^{2+}-sensitive dyes suggest that Ca^{2+} for activation of tension enters the frog myocordial cells, primarily from the extracellular space, through a slowly inactivating Ca^{2+} channel. Contribution of other Ca^{2+}-transport systems, appear to be small during depolarization to plateau potentials, but may be significant at resting potential. This idea is consistent with the rate of Ca^{2+} influx and efflux measured at plateau and resting potentials and with the finding that cardiac glycosides do not enhance Ca^{2+} influx during the action potential. Thus it appears that while the I_{si} system regulates the beat-to-beat entry of Ca^{2+}, the Na^+-Ca^{2+}

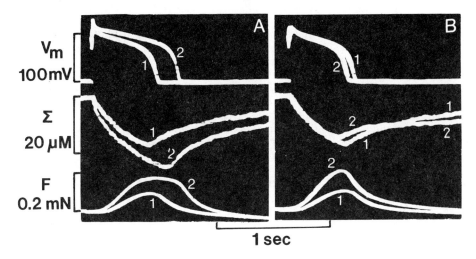

Fig.18.10. The effect of adrenaline (10^{-6}M, (A) and strophanthidin (10^{-6}M, (B)) on extracellular Ca^{2+} depletion. Each panel shows action potentials measured across the sucrose gap (V_m), the extracellular Ca^{2+} depletion measured at 718 nm (Σ) and the twitch force (F) recorded in the absence (1) and presence (2) of the drug. Note that although both drugs potentiated the twitch, only adrenaline enhanced Ca^{2+} influx. (Modified from Cleemann, Pizarro and Morad, 1984.)

counter-transporter serves not only to regulate the $(Ca)_i$, but also to help with the sequestration of Ca^{2+}.

Acknowledgements

This work was supported in part by W.W. Smith Foundation and N.H Grant no HL16152.

References

Blinks, J.R., Wier, W.G., Hess, P. and Prendegast, F.G. (1982). Measurement of Ca^{2+} concentrations in living cells. *Prog. Biophys. Molec. Biol.* **40**, 1–114.

Cleemann, L., Pizarro, G. and Morad, M., (1984). Optical measurements of extracellular calcium depletion during a single heart beat. *Science* **226**, 174–7.

Endo, M. and Iino, M. (1980). Specific perforation of cell membranes with preserved SR functions by saponin treatment. *J. Muscle Res. and Cell Motil.* **1**, 89–100.

Klitzner, T. and Morad, M. (1983a). Excitation–contraction coupling in frog ventricle: possible Ca^{2+} transport mechanism *Pflügers Arch.* **398**, 274–83.

Klitzner, T. and Morad, M. (1983b). The effect of Ni^{2+} on ionic current and tension generation in frog ventricular muscle. *Pflügers Arch.* **398**, 267–73.

Morad, M. (1982). Ionic measurements mediating the inotropic and relaxant effects of adrenaline on the hear muscle. In: *Catecholamines in the Non-ischaemic and Ischaemic Myocardium*. R.A. Riemersma and M.G. Oliver (eds.). Elsevier Biomedical, Amsterdam

Morad, M. and Goldman, Y.E. (1973). Excitation–contraction coupling in heart muscle: membrane control of development of tension. *Prog. Biophys. Mol. Biol.*

27, 257–313.

Morad, M., Goldman, Y.E. and Trentham, D.R. (1983). Rapid photochemical inactivation of Ca^{2+}-antagonists shows that Ca^{2+} entry directly activates contraction in frog heart. *Nature (Lond.)* **304**, 635–8.

Mullins, L.J. *Ion Transport in Heart*. Raven Press, New York.

Page, S.G. and Niedergerke, R. (1972). Structures of physiological interest in the frog heart ventricle. *J. Cell. Sci.* **11**, 179–203.

Pizarro, G., Cleemann, L. and Morad, M. (1985). Measurement of voltage-dependent trans-sacrcolemmal Ca-transport in *f*rog heart muscle with antipyrylazo III. *Proc. Natl. Acad. Sci. USA* **82**, 1864–88.

Scarpa, A., Brinley, F.J. Jr. and Dubyak, G. (1978). Antipyrylazo III, a 'middle range' Ca^{2+} metallochronic indicator. *Biochemistry* **17**, 1378–86.

Winegrad, S. (1971). Studies of cardiac muscle with a high permeability produced by treatment with ethylenediaminetetraacetic acid. *J. Gen. Physiol.* **58**, 295–303.

THE ROLE OF CALCIUM AND CALMODULIN IN HIGHER PLANTS

Dieter Marmé

Institute of Biology III, University of Freiburg, 7800 Freiburg, Germany; Department of Biophysics, Gödecke Research Institute, 7800 Freiburg, West Germany

Elisabeth Andrejauskas and Rainer Hertel

Institute of Biology III, University of Freiburg, 7800 Freiburg, West Germany

Abstract

Calmodulin which has been shown to be an ubiquitous protein in animals has been demonstrated in plants too. In plants so far the following enzymes are known to be regulated by Ca^{2+} and calmodulin: a membrane-bound ATPase, a soluble and two membrane-bound NAD kinases, a quinate: NAD^+ oxidoreductase and several protein kinases.

The activities of these enzymes can be regulated by changes in the free cytoplasmic Ca^{2+} concentration. In plant cells the low free, cytoplasmic Ca^{2+} concentration seems to be maintained mainly by the extrusion of Ca^{2+} out of the cell which is driven by a plasma membrane-bound, calmodulin-dependent Ca^{2+}-transport ATPase.

The activity of the plasma membrane Ca^{2+}-transport system is modulated by primary signals such as light and hormones. Specific binding sites for Ca^{2+} antagonists (such as verapamil) which are known to inhibit Ca^{2+} influx through the Ca^{2+} channel, have been shown to occur in plants too. We suggest that modulation of active and passive Ca^{2+} fluxes leads to an increase of the cytoplasmic Ca^{2+} concentration, to the formation of the active Ca^{2+}-calmodulin complex and thus to the activation of Ca^{2+}, calmodulin-dependent enzymes, such as NAD kinase. It could be demonstrated that the reaction product of the NAD kinase, NADP, increases significantly when intact plant seedlings are irradiated with light or when intact plant segments are exposed to calcium in the presence of the divalent cation ionophore A23187.

19.1. Introduction

Living organisms have established several mechanisms by which they transduce extracellular signals into an intracellular language that can be understood by the biochemical and biophysical machinery of the cell. One such mechanism has been discovered in animal cells almost 20 years ago by E.W. Sutherland and his co-workers and is called the second messenger principle, with cAMP as the intracellular second messenger. With the discovery of calmodulin it became evident that cytoplasmic Ca^{2+} ions play a similar role as second messengers in animal physiology.

Plant physiologists have tried to establish a similar role for cAMP in higher plants, expecting some insights into the molecular mechanisms of the transduction of extracellular signals, such as (non-photosynthetic) light, hormone-like substances, or gravity. There is certainly convincing evidence now for the existence of cAMP in higher plants (Brown and Newton, 1981), but there is no physiological evidence for a role as second messenger similar to that in animal cells. In particular, no cAMP-dependent protein kinase - the primary target molecule for cAMP in animal cells - could be demonstrated so far (Salimath and Marmé, 1983).

In plants there are many physiological processes that have been reported to be under the control of Ca^{2+} (table 19.1). However, in no case has the

Table 19.1. Ca^{2+}-dependent processes in plants

Mitosis

Plant growth hormone actions

Secretion of cell wall material

Protoplasmic streaming

Chloroplast movement (*Mougeotia*)

Phototaxis (*Chlamydomonas*)

Leaf movement (*Mimosa*)

Abscission

Senescence

Membrane damage and leakiness

Glutamate dehydrogenase

Pyruvate kinase

Alkaline lipase

Phospholipid acylhydrolase

NAD kinase

Auxin transport

Auxin-mediated H^+-secretion

Cation transport into plant cells

Ca^{2+} release from microsomes

Microsomal Ca^{2+} transport

Microsomal Ca^{2+}-ATPase

---

molecular mechanism of the Ca^{2+} regulation been convincingly elucidated. One of the reasons is the difficulty in measuring the free, cytoplasmic Ca^{2+} concentration (Williamson, 1981). Another important reason was the lack of evidence for the existence of calmodulin in plants. It was only in 1978 that Anderson and Cormier presented evidence for a heat-stable protein that was able to activate bovine brain cAMP phosphodiesterase and that was identified as calmodulin shortly thereafter (Anderson et al., 1980).

In this chapter we present further evidence for Ca^{2+} being an intracellular second messenger in plants.

19.2. Plant calmodulin

A comparison of the physical and biochemical properties of calmodulin from various sources - animals and plants - shows that this protein has been well conserved throughout eukaryotic species. This suggest a fundamental role of calmodulin for Ca^{2+}-dependent processes in the cell (Burgess et al., 1983).

Plant calmodulin has been found in many different species (Marmé and Dieter,1983). The complete amino acid sequence of spinach calmodulin has been published recently (Van Eldik and Watterson, 1984). Only 13 out of 148 amino acids are exchanged when compared with the amino acid sequence of bovine brain calmodulin. Apparently this does not cause a dramatic functional change. The dependence of cyclic AMP phosphodiesterase activation on the calmodulin and Ca^{2+} concentration is comparable for calmodulin from bovine brain and from zucchini (Marmé and Dieter, 1983).

19.3. Calmodulin-dependent plant enzymes

Since the discovery of the first calmodulin-dependent enzyme in plants, the NAD kinase from peas, by Anderson and Cormier (1978) a few additional enzymes have been reported to be activatable by the Ca^{2+}-dependent regulator protein. Table 19.2 summarizes these enzyme activities.

19.3.1. NAD kinase
The NAD kinase is of interest because it depends not only on Ca^{2+} and calmodulin but has also been shown to be under the control of light (Marmé and Dieter, 1982). As is outlined below, the light-dependence is due to a light-mediated change of the cytoplasmic Ca^{2+} concentration. All these

Table 19.2. Ca^{2+}, calmodulin-dependent enzymes in plants

NAD kinase (cytoplasm)

NAD kinase (outer mitochondrial membrane)

NAD kinase (chloroplast envelope)

$Ca^{2+} + Mg^{2+}$) ATPase

Protein kinases

Quinate: NAD^+ oxidoreductase (light-grown cells)

NAD-kinases which have been reported to be localized in different intracellular compartments (see table 19.2) expose their regulatory site towards the cytoplasm and therefore are able to sense changes of the cytoplasmic Ca^{2+} concentration.

19.3.2. Protein kinase and quinate: NAD^+ oxidoreductase

A Ca^{2+} and calmodulin-dependent protein kinase activity has been discovered in a membrane fraction obtained from zucchini (Salimath and Marmé, 1983). In carrot cell cultures the quinate: NAD^+ oxidoreductase has been shown to be under control of Ca^{2+}, calmodulin-dependent phosphorylation (Ranjeva et al., 1983). Therefore it can be anticipated that Ca^{2+}-dependent, calmodulin-mediated enzyme regulation in plants occurs directly, as in the case of the NAD kinases and indirectly by Ca^{2+}, calmodulin-dependent phosphorylation.

19.3.3. $(Ca^{2+} + Mg^{2+})$ ATPase

The calmodulin-dependent $(Ca^{2+} + Mg^{2+})$ ATPase has been shown to occur in a plasma membrane fraction from many plants and is most probably the ubiquitous mechanism to control the cytoplasmic Ca^{2+} concentration in the non-excited cell (Dieter and Marmé 1980, 1981a). Recently evidence has been provided for a ATP-dependent Ca^{2+} transport into the endoplasmic reticulum (Gross, 1982; Buckhout, 1983) and also for an active Ca^{2+} accumulation into plant vacuoles (Gross, 1982). All those mechanisms keep the cytoplasmic concentration at a low level.

19.4. Modulation of cytoplasmic Ca^{2+}

When the cell becomes stimulated (for example, by light or plant hormones) cytosolic Ca^{2+} can be increased by either opening Ca^{2+} channels presumably in the plasma membrane or by changing the transport properties of the Ca^{2+} transport mechanisms. So far no electrophysiological evidence is available for the existence of Ca^{2+} channels. Saunders and Hepler (1982) have reported that the Ca^{2+} antagonist D 600 (which inhibits the function of Ca^{2+} channels) is able to decrease cytokinin-induced cell division in *Funaria*.

Binding experiments using the specific Ca^{2+} antagonist verapamil show that binding sites exist for this drug in plants. Figure 19.1 shows a displacement curve for verapamil. Plant membranes were prepared from zucchini by differential centrifugation and incubated with ^3H-verapamil and various concentrations of the non-radioactive compound. Scatchard analysis yields a $K_D = 10^{-7}$ M and B_{max} = 60 pmol/mg protein. These results support the assumption that Ca^{2+} channels in plants may exist which could be responsible for short-term effects such as light-dependent chloroplast movement in algae (see table 19.1).

Long-term effects, such as growth and development of plants could be initiated and maintained by slow changes of the active Ca^{2+} transport mechanisms. Dieter and Marmé (1981b) have shown for corn coleoptiles that the calmodulin-dependent Ca^{2+}-transport into plasma membrane-derived vesicles loses its calmodulin dependency when the plants are transferred from darkness to infrared light. Kinetic studies revealed that the maximal velocity of the Ca^{2+} transport is decreased (Dieter and Marmé, 1983). For the intact cell this means that the efflux rate for Ca^{2+} is reduced and consequently the free cytoplasmic Ca^{2+} concentration increased. Infrared light is not the only extracellular factor which is capable of changing the Ca^{2+} transport activity.

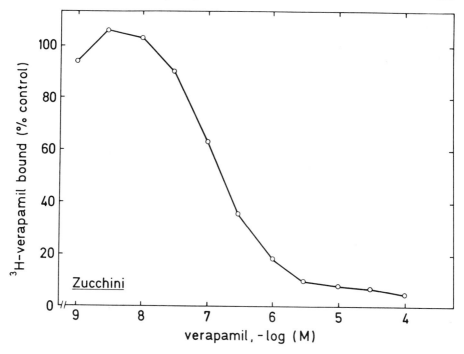

Fig.19.1. Concentration-dependent displacement of ^{3}H-verapamil by non-radioactive verapamil.

Blue and ultraviolet light, as well as the two plant hormones auxin and zeatin, when applied to the intact tissue are able to change the Ca^{2+} transport properties *in vitro* (Marmé, 1982).

19.5. Conclusion

Experimental evidence obtained so far suggests that plants use Ca^{2+} as an intracellular second messenger. In contrast to animals, when stimulated plants seem to be able to change cytoplasmic Ca^{2+} by long-term modulation of the active Ca^{2+} transport mechanisms. Recent evidence for Ca^{2+} antagonist binding sites gives the first indication for the existence of Ca^{2+} channels that may be involved in short-term regulation.

References

Anderson, J.M., Charbonneau, H., Jones, H.P., McCann, R.O. and Cormier, M.J. (1980). Characterization of the plant NAD kinase activator protein and its identification as calmodulin. *Biochemistry* **19**, 3113–20.

Anderson, J.M. and Cormier, M.J. (1978). Calcium-dependent regulator of NAD kinase. *Biochem. Biophys. Res. Commun.* **84**, 595–602.

Brown, E.G. and Newton, R.P. (1981). Cyclic AMP and higher plants. *Phytochemistry* **20**, 2453–63.

Buckhout, T.J. (1983). ATP-dependent calcium transport in endoplasmatic reticulum isolated from roots of *Lipidium satisum*. *Planta* **159**, 84–90.

Burgess, W.H., Schleicher, M., Van Eldik, L.J. and Watterson, D.M. (1983). Comparative studies of calmodulin. In *Calcium and Cell Function IV*. W.Y. Cheung (ed.), pp. 209–61. Academic Press, New York.

Dieter, P. and Marmé, D. (1980). Calmodulin activation of plant microsomal calcium uptake. *Proc. Natl. Acad. Sci. USA* **77**, 7311–14.

Dieter, P. and Marmé, D. (1981a). A calmodulin-dependent microsomal ATPase from corn (*Zea mays l*) *FEBS Lett.* **125**, 245–8.

Dieter, P. and Marmé, D. (1981b). Far red light irradation of intact corn seedlings affects mitochondrial and calmodulin-dependent microsomal calcium transport. *Biochem. Biophys. Res. Comm.* **101**, 749–55.

Dieter, P. and Marmé, D. (1983). The effect of calmodulin and infra-red light on the kinetic properties of the mitochondrial and microsomal calcium-ion. *Planta* **159**, 277–81.

Gross, J. (1982). Oxalate-enhanced active calcium uptake in membrane functions from zucchini squash. In *Plasmalemma and Tonoplast: Their Functions in the Plant Cell*. D. Marmé, E. Marré and R. Hertel (eds.), pp. 369–76. Elsevier/North-Holland Biomedical Press, Amsterdam, New York, Oxford.

Marmé, D. (1982). The role of calcium in signal transduction of higher plants. In *Plant Growth Substances 1982*. P.F. Wareing (ed.), pp. 419–26. Academic Press, London, New York.

Marmé, D. and Dieter, P. (1982). Calcium and calmodulin-dependent enzyme regulation in higher plants. In *Plasmalemma and Tonoplast: Their Functions in the Plant Cell*. D. Marmé, E. Marré and R. Hertel (eds.), pp. 111–18. Elsevier/North-Holland Biomedical Press, Amsterdam, New York, Oxford.

Marmé, D. and Dieter, P. (1983). Role of calcium and calmodulin in plants. In *Calcium and Cell Function IV*. W.Y. Cheung (ed.), pp. 264–311. Academic Press, New York.

Ranjeva, R., Refeno, G., Boudet, A.M. and Marmé, D. (1983). Activation of plant quinate: NAD^+ oxidoreductase by Ca^{2+} and calmodulin. *Proc. Natl. Acad. Sci. USA* **80**, 5222–4.

Salimath, B.P. and Marmé, D. (1983). Protein phosphorylation and its regulation by calcium and calmodulin in membrane fractions from zucchini hypocotyls. *The EMBO Journal* **2**, 535–42.

Saunders, M.J. and Hepler, P.K. (1982). Abstracts of the *Eleventh International Conference on Plant Growth Substances* **212**, 23.

Van Eldik, L.J. and Watterson, D.M. (1984). Calmodulin structure and function. In *Calcium and Cell Physiology*. D. Marmé (ed.), Springer Verlag, Berlin, Heidelberg, New York.

Williamson, R.E. (1981). Free calcium concentration in the cytoplasm: a regulator of plant cell function. *What's New Plant Physiol* **12**, 45–8.

CALCIUM ANTAGONISTS AND ATHEROSCLEROSIS. SUPPRESSION OF PLAQUE FORMATION AND REVERSAL OF ESTABLISHED LESIONS: AN OVERVIEW

Dieter M. Kramsch

Merck Sharp and Dohme Research Laboratories, P.O. Box 2000, Rahway, New Jersey, 07065-0914, USA.

and Anita J. Aspen

Cardiovascular Institute, Boston University, School of Medicine, Boston, Massachusetts 02118, USA.

Abstract

Recent mounting evidence in rabbits and monkeys indicates that Ca antagonists in the broad sense of the word suppress the formation of calcific fibrous atheromatous plaques by cellular and extracellular mechanisms. These Ca antagonists include the calcium-entry blockers lanthanum, nifedipine, verapamil and nicardipine; the diphosphonates ethanehydroxy diphosphonate (EHDP), aminohydroxypropane diphosphonate (APDP), azacycloheptane diphosphonate (AHDP), imminopyrolidone diphosphonate (IPDP); the cAMP phosphodiesterase inhibitor trimazosin; as well as the hormone thyrocalcitonin. Several of these Ca antagonists promote regression of established atherosclerotic lesions in rabbits. These *in vivo* studies by ourselves and others, as well as *in vitro* findings with cultured arterial smooth muscle cells exposed to anticalcium agents, support the hypothesis that focal increases in arterial calcium concentrations may facilitate: (a) excessive activation of cellular functions in response to atherogenic stimuli leading to lesion formation (cell migration, mitosis, secretion of cell products, endocytosis); (b) development of extracellular atherosclerotic abnormalities of connective tissue. Maintenance of Ca homeostasis may play a key role in prevention of atherosclerosis even in the presence of unmitigated atherogenic lipoprotein patterns. The mechanism for reversal of lesions with Ca antagonists are less clear. Possibly, the arrest of lesion progression allows catabolic processes to restitute diseased arterial tissue including removal of calcifications.

20.1. Introduction

The clinically important lesion of human atherosclerosis is the fibrous atheromatous plaque which often is calcified (National Heart, Lung and Blood Institute Task Force on Atherosclerosis, 1971). The connective tissue changes of fibrous lesions begin to develop in the second decade of life but it usually takes decades of slow growth before they produce clinical symptoms. However, it is these connective tissue changes which threaten health and life by seriously impairing arterial function through luminal stenoses by the raised fibrous intima, through loss of elasticity by elastica destruction and collagen accumulation, as well as through increased rigidity and brittleness by calcium mineralization. These impairments of arterial function in turn lead to insufficiency of blood flow, thrombosis, hypertension, aneurysms and vessel rupture.

In contrast to fibrous plaques, the often coexisting fatty streaks consist of lipid-filled intimal cells, largely without connective tissue changes. They do not raise the intima, do not impair arterial function, and appear to be reversible. There is evidence that some fatty streaks can progress into fibrous plaques although the majority presumably does not.

Fibrous lesions are present, with rare exceptions, in virtually all individuals over the age of 35 in western societies (National Heart, Lung and Blood Institute Task Force on Atherosclerosis, 1971). Moreover, there is no evidence that calcified collagenous plaques can be reversed once they are established. Quite to the contrary, recent studies at Bowman-Gray School of Medicine (Clarkson, personal communication, 1983) are showing that collagen accumulations of intimal plaques induced by an atherogenic diet in monkeys did not regress even after 4 years of the most drastic plasma cholesterol reduction, possible only in animals, that is, by total withdrawal of dietary cholesterol. However, any remaining abnormal collagen deposition in the intima raises the spectre of fatal thrombotic events by interaction of collagen with blood elements (Mustard et al., 1974). In view of the clinical importance of the fibrous plaque, the ultimate and critical questions appear to be: how does connective tissue accumulate and/or become altered in the intima of plaques, and what specific measures can be taken to prevent, arrest or reverse the formation of fibrous plaques?

Several recent studies (Committee of Principal Investigators, 1978; Lipid Research Clinics Coronary Primary Prevention Trial Results, 1984) have shown that lowering of markedly elevated serum cholesterol levels is capable of significantly reducing morbidity and mortality from coronary heart disease and, by implication, may beneficially influence atherosclerosis. However, as these studies were performed in patients with excessively high serum cholesterol concentrations, it still is not certain whether cholesterol lowering may have similar effects in subjects in whom the serum cholesterol concentrations are only mildly elevated or not at all, but who constitute the majority of coronary heart disease victims in western societies. Moreover, it is not even known whether the reductions in morbidity and mortality observed in those recent trials are indeed due to an amelioration of coronary atherosclerotic lesions by the lowered serum cholesterol or whether other mechanisms are responsible for these effects. Even if inhibition of lipid-filling of cells does occur (inhibition of fatty streak formation, essentially), it is not certain to what extent lipid-lowering treatment regimens may affect the formation of the clinically more important fibrous plaques or their removal.

It, therefore, appears to be also desirable to search for agents which are able to influence the cellular and connective tissue components of fibrous plaques directly without need to improve abnormal serum lipids or lipoproteins. Such agents could become very attractive for alternative or adjunctive treatment to lipid-lowering drugs. Of all agents known to affect the arterial wall components directly, the calcium antagonists in the broadest sense of the word are furthest advanced in research. These are agents which interfere with the deposition, flux and utilization of calcium ions in the vessel wall.

20.2. Atherogenesis and the role of calcium ions

According to the current concepts of atherogenesis depicted in Fig. 20.1, the main processes leading to plaque formation include:

(1) increased permeability of the endothelium to macromolecules such as low density lipoproteins (LDL), as well as to blood cells;
(2) migration of cells into the intima such as smooth muscle cells (SMC) from the media as well as monocyte-macrophages, other white blood cells and platelets from the blood stream;
(3) proliferation of these cells in the intima by mitosis;
(4) synthesis and secretion by the increased SMCs of excessive amounts of collagen, elastin and other connective tissue elements;

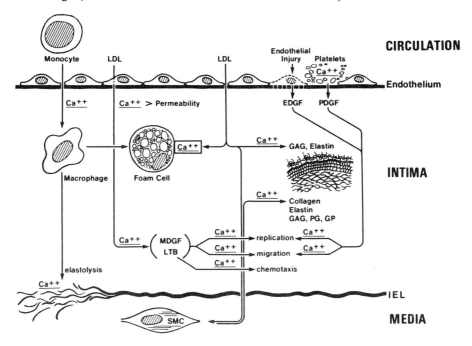

Fig.20.1. Schematic representation of the current concepts of atherogenesis and the role of Ca^{2+} in the main processes leading to plaque formation. Ca^{2+} = ionic calcium; EDGF = endothelium-derived growth factor; GAG = glycosaminoglycans; GP = glycoprotein; IEL = internal elastic lamina; LDL = low density lipoproteins; LTB = leucotriene B; MDGF = macrophage-derived growth factor; PDGF = platelet-derived growth factor; PG = proteoglycans; SMC = smooth muscle cell.

(5) increased internalization of lipids and lipoproteins by SMCs and macrophages, including by endocytosis;
(6) destruction of elastica by elastolysis; and
(7) gross calcium mineralization of connective tissue components in a late phase of plaque development.

As recently reviewed by Kramsch, Aspen and Apstein (1980) and Kramsch, Aspen and Rozler (1981), there has been mounting evidence that localized increases in arterial calcium content in ionic form may play a key role in early atherogenesis. Some of the basic cellular and extracellular processes of atherogenesis requiring readily-available ionic calcium (Ca^{2+}) are also depicted in Fig.20.1.

20.2.1. Cellular events
The cellular events requiring Ca^{2+} include:

(1) the contraction–relaxation cycle in contractile cells such as endothelium cells (contraction leads to increased endothelial permeability) as well as SMCs, macrophage-monocytes, other blood cells and platelets (leading to migration of SMCs and blood cells and the release reaction of platelets);
(2) directional movement of cells along a Ca^{2+} gradient (chemotaxis);
(3) the stimulus–division coupling (leading to cell replication by mitosis);
(4) the stimulus–secretion coupling (leading to excessive secretion by SMCs of collagen, elastin, glycosaminoglycans (GAG), proteoglycans (PG) and glycoproteins (GP), as well as by macrophages of leucotriene B (LTB) and elastase);
(5) platelet function (adhesion to subendothelial collagen, aggregation and release reaction releasing platelet-derived growth factor (PDGF) and proteolytic enzymes);
(6) endocytosis of low density lipoprotein (LDL), leading to foam cell formation; and
(7) energy production from Ca^{++}-dependent ATP for all of above cellular events.

20.2.2. Extracellular events
The extracellular events of atherogenesis requiring Ca^{2+} include:

(1) degradation of elastin (elastolysis) by macrophage elastase;
(2) configurational changes of elastin causing increased absorption of cholesterol by elastin;
(3) binding of extraneous proteins to elastin leading to increased binding of cholesterol esters by elastin–protein Ca^{2+} complexes;
(4) complexing of lipoproteins to GACs and PGs; and
(5) gross calcification (calcium apatite) of connective tissue, especially altered elastin.

20.2.3. Primary stimuli
The primary stimuli thought to induce these cellular and extracellular atherogenic responses are heterogeneous and include:

(1) hyperlipoproteinemia with increased total plasma cholesterol and LDL-cholesterol;
(2) inflammatory reactions of the arterial wall to certain viruses as well as to

immune complexes, including those resulting from autoimmune mechanisms;
(3) thrombotic events, such as macrothrombi and microthrombi leading to intimal deposition of platelets, monocytes and fibrin;
(4) mechanical injury to the arterial wall, such as shear stress, sustained hypertension, sudden bursts of blood pressure increases, tachycardia, catheterizations;
(5) endogenous chemical injury to the vessels, for example, by catecholamines, angiotensin, renin, testosterone, oestrogens, triglyceride-rich beta(β)-very low density lipoproteins (VLDL), certain apo-LDL (apo-B_1), which are thought to be 'toxic' even in the absence of plasma cholesterol elevations; and
(6) deranged carbohydrate metabolism leading to arterial accumulations of proteoglycans;

Most of these primary stimuli affect the arterial intima and elicit atherogenic responses independently of plasma lipid levels. One or more of the stimuli may be operative simultaneously. In the presence of increased plasma cholesterol concentrations, the atherogenic responses to most stimuli generally appear to be accelerated.

20.3. Prevention of fibrous-atheromatous plaque formation by anticalcium drugs

20.3.1. *Experimental*
We have been experimenting since 1973 with agents that have the capacity to prevent or even reverse calcium mineralization and fibrous atherosclerotic disease in rabbits and *Macaca* fascicular (cynomolgus) monkeys on fibrogenic atherogenic diets. These agents, which did not affect the elevated serum cholesterol levels, initially included ethane-1-hydroxy-1,1 diphosphonic acid (EHDP), an agent which is known to prevent the deposition of calcium into soft tissues, including into arteries (Fleisch *et al.*, 1970; Rosenblum, Flora and Eisenstein, 1975; Wagner and Clarkson, 1973). Other and newer diphosphonates of similar properties (Potocar and Schmidt-Dunker, 1978) were later employed and included the disodium salts of amino-1-hydroxy- propane-1,1 diphosphonic acid (APDP), azacycloheptane-2,2 diphosphonic acid (AHDP) and 2-immino-pyrrolidone-5,5 diphosphonic acid (IPDP). Successful treatment with these calcium-regulating agents suggested that atherosclerosis may be beneficially influenced by regulating the availability of ionic calcium. We, therefore, treated animals on the atherogenic diet with the direct calcium ion antagonist, lanthanum, at increasing doses. Lanthanum is a calcium entry blocker of the class that operates to inhibit calcium influx into cells at receptor-operated channels (Weiss, 1974; Weiss and Goodman, 1976). Most recently we also reported results with another anticalcium agent which inhibits the utilization of ionic tissue calcium - the cyclic AMP-phosphodiestrase inhibitor trimazosin (Kramsch, Aspen and Swindell, 1983). The fibrogenic atherogenic diet for rabbits contained 8% peanut oil and 2% cholesterol by weight (Kritchersky *et al.*, 1971); that for monkeys contained 10% butter and 0.1% cholesterol (Kramsch, Hollander and Renaud, 1973). The control diets were Purina Rabbit Chow and Purina Monkey Chow, respectively. Altogether, 12 groups of ten male New Zealand White rabbits each and six groups of generally eight adult male *Macaca* fascicular

(cynomolgus) monkeys were fed the fibrogenic atherogenic diets or the control diets: rabbits for 8 weeks, monkeys for 24 months. One group of rabbits was fed the control diet and one group the atherogenic diet alone, while ten groups were given the atherogenic diet with added EHDP, APDP, AHDP, IPDP, lanthanum (LaCl$_3$, at three dosage levels) and trimazosin (at three dosage levels), respectively. Likewise, one group of monkeys was fed the control diet (eight monkeys) and one group the atherogenic diet alone (eight monkeys), while four groups were given the atherogenic diet with added EHDP (eight monkeys), LaCl$_3$ (eight monkeys), APDP (four monkeys) and AHDP (four monkeys), respectively. The daily oral doses of drugs per kg bodyweight were: EHDP, APDP, AHDP and IPDP = 40 mg for rabbits; EHDP, APDP and AHDP for monkeys = 120 mg for the first 6 months followed by a maintenance dose

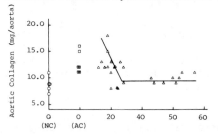

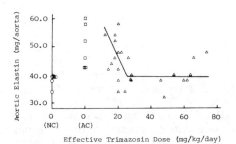

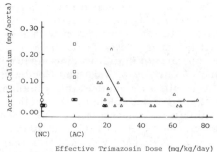

Fig.20.2. Content of collagen, elastin and calcium in intimamedia of whole aorta in rabbits fed the atherogenic diet without drugs (AC = atherogenic diet controls) or with trimazosin 20 mg, 30 mg and 80 mg/kg bodyweight/day. The cumulative effective dose was calculated from the amount of food, containing the trimazosin dose, that was actually consumed per day over the 8-week study period.

Table 20.1. Serum components of animals on control diets and on atherogenic diets with and without drug treatment.

Rabbits = 8 weeks; monkeys = 24 months; values in mg/100 ml (mean ± SD)

Experimental groups	Rabbits		Monkeys	
	cholesterol	calcium	cholesterol	calcium
Control diet	82 ± 23	13.2 ± 1.1	136 ± 42	11.0 ± 2.3
Atherogenic diet, untreated	2573 ± 568*	13.3 ± 1.2	452 ± 163*	10.8 ± 1.8
Atherogenic diet + EHDP	3535 ± 738*	13.4 ± 0.6	408 ± 147*	9.7 ± 1.6
Atherogenic diet + APDP	2488 ± 534*	14.5 ± 0.9	512 ± 189*	9.5 ± 1.4
Atherogenic diet + AHDP	2283 ± 664*	14.7 ± 0.7	468 ± 133*	9.2 ± 1.9
Atherogenic diet + IPDP	2056 ± 845*	14.2 ± 1.3	---	---
Atherogenic diet + LaCl3(20 mg)	2431 ± 489*	14.8 ± 1.0#	---	---
Atherogenic diet + LaCl3(30 mg)	2035 ± 921*	15.3 ± 1.3*	---	---
Atherogenic diet + LaCl3(40 mg)	2537 ± 621*	16.0 ± 1.2*	397 ± 126*	9.5 ± 2.0
Atherogenic diet + trimazosin(20 mg)	2905 ± 331*	---	---	---
Atherogenic diet + trimazosin(30 mg)	3077 ± 316*	---	---	---
Atherogenic diet + trimazosin(80 mg)	3524 ± 300*	13.6 ± 1.3	---	---

* Highly significant changes from controls ($p < 0.01$)
\# Significant changes from controls ($p < 0.05$)

Table 20.2. Components of intimamedia from rabbits on control diets and atherogenic diets with and without treatment. (absolute amounts in mg/whole aorta/kg body weight; mean ± SD)

Experimental groups	Calcium	Non-lipid phosphorus	Collagen	Elastin	Cholesterol
Control diet	0.03 ± 0.007	0.04 ± 0.02	2.8 ± 0.5	10.3 ± 2.8	0.3 ± 0.2
Atherogenic diet, untreated	0.08 ± 0.012*	0.15 ± 0.03*	6.5 ± 1.3*	27.1 ± 1.7*	3.5 ± 1.2*
Atherogenic diet + EHDP	0.06 ± 0.013*	0.11 ± 0.06#	4.3 ± 1.1#	21.5 ± 1.6*	1.3 ± 0.5*
Atherogenic diet + APDP	0.06 ± 0.015#	0.13 ± 0.08	3.7 ± 0.4#	15.3 ± 1.6*	0.7 ± 0.3#
Atherogenic diet + AHDP	0.04 ± 0.017	0.11 ± 0.05	2.9 ± 0.8	12.4 ± 1.4	0.5 ± 0.1#
Atherogenic diet + IPDP	0.02 ± 0.009	0.06 ± 0.05	2.6 ± 0.9	7.3 ± 2.3	0.3 ± 0.3
Atherogenic diet + LaCl$_3$ (20 mg)	0.04 ± 0.010#	0.06 ± 0.01#	4.2 ± 0.6*	17.3 ± 1.9*	1.9 ± 1.0*
Atherogenic diet + LaCl$_3$ (30 mg)	0.03 ± 0.010	0.04 ± 0.03	3.6 ± 0.6#	14.4 ± 0.7*	1.4 ± 0.7*
Atherogenic diet + LaCl$_3$ (40 mg)	0.03 ± 0.008	0.05 ± 0.02	2.9 ± 0.8	13.6 ± 4.7	0.6 ± 0.1#

* Highly significant changes from control values ($p < 0.01$)
\# Significant changes from control values ($p < 0.05$)

Table 20.3. Components of aortic intimamedia from monkeys on control diets and atherogenic diets with and without treatment.
(μm/cm length *in situ*/kg body weight; mean ± SD)

Aortic segment and treatment	Calcium	Non-lipid phosphorus	Collagen	Elastin	Cholesterol
Thoracic aorta					
Control diet	4 ± 2	3 ± 2	217 ± 31	544 ± 39	22 ± 8
Atherogenic diet, untreated	82 ± 29*	49 ± 14*	782 ± 183*	912 ± 165*	347 ± 127*
Atherogenic diet + EHDP	5 ± 2	4 ± 2	307 ± 97	560 ± 26	136 ± 47*
Atherogenic diet + APDP	6**	2**	392**	612**	152**
Atherogenic diet + AHDP	4**	3**	302**	581**	113**
Atherogenic diet + LaCl$_3$(40 mg)	6 ± 1#	5 ± 1#	388 ± 112*	671 ± 199*	178 ± 83*
Abdominal aorta					
Control diet	2 ± 1	1 ± 1	308 ± 46	312 ± 29	10 ± 3
Atherogenic diet, untreated	46 ± 18*	34 ± 12*	963 ± 105*	474 ± 102*	457 ± 198*
Atherogenic diet + EHDP	3 ± 1	3 ± 2	262 ± 88	267 ± 52	122 ± 62*
Atherogenic diet + APDP	2**	1**	286**	228**	83**
Atherogenic diet + AHDP	3**	2**	273**	233**	91**
Atherogenic diet + LaCl$_3$(40 mg)	2 ± 1	2 ± 1	225 ± 75	242 ± 80	143 ± 71*

* Highly significant changes from control values ($p < 0.02$)
\# Significant changes from controls ($p < 0.05$)
** Average values from four monkeys

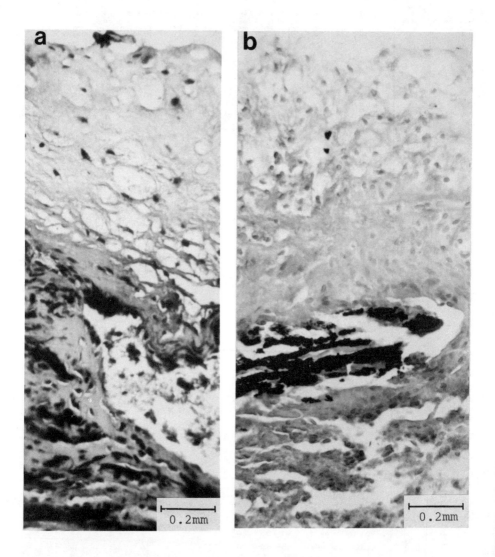

Fig.20.3. Sections through characteristic lesions of thoracic aorta of rabbits on the fibrogenic atherogenic diet for 8 weeks without drugs ((a) and (b) = sequential sections of the same lesion) or treated simultaneously with IPDP(c). (a) Note proliferation of cells and accumulation of collagen (grey) in intima and deranged elastica (black); Verhoeff's-Van Gieson, x85. (b) Note deposition of calcium (black) on deranged elastica; Yasue's-light green x85. (c) Note absence of collagen accumulation and calcium deposition, normal elastica, raising of intima by a thin layer of (lipid-rich) foam cells; Yasue's-light green, x70.

of 40 mg thereafter; LaCl$_3$ = 20, 30 and 40 mg for rabbits; and for monkeys = 120 mg followed by 40 mg as with diphosphonates; trimazosin = 20, 30 and 80 mg (rabbits only). The large initial doses in monkeys were given to saturate the tissues especially the large bone mass, in the case of the diphosphonates.

20.3.2. Biochemical findings

In all animals on the atherogenic diet, the serum cholesterol was markedly elevated to about the same level within each animal species, regardless whether or not any drugs were given (Table 20.1); the serum total calcium (and ionized calcium - not shown) remained at normal values for both species, regardless of drug treatment, with the exception of either dose of LaCl$_3$ in rabbits, where the total calcium was elevated.

As compared to animals on the control diets, rabbits (Table 20.2, Fig.20.2) and monkeys (Table 20.3) on the fibrogenic atherogenic diet revealed drastic changes in the components of aortic intima media: a marked increase in the arterial content of calcium was associated with drastic rises in the arterial content on non-lipid phosphorus, collagen, elastin and cholesterol. As noted previously (Kramsch, 1981), the arterial elastin fraction of untreated atherogenic-diet animals of both species also was markedly altered, revealing marked increases in calcium and phosphorus as well as of polar amino acids

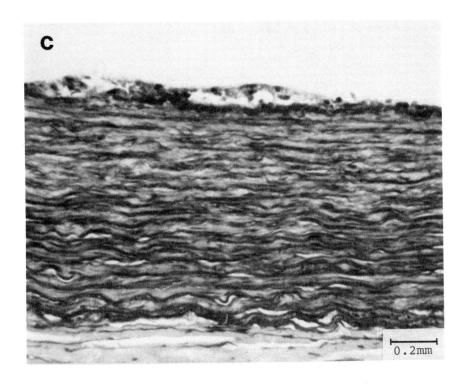

Table 20.4. Constituents of elastin from control rabbits and rabbits on the atherogenic diet with and without lanthanum (LaCl$_3$ doses in mg/kg body weight; cholesterol, calcium and phosphorus in mg/g elastin; amino acids in residues/100 residues; mean ± SD).

Experimental groups	Total cholesterol	Total calcium	Non-lipid phosphorus	Aspartic acid*	Glutamic acid*	Lysine*	Arginine*
Control diet	10.4 ± 2.3	0.8 ± 0.3	0.8 ± 0.07	2.1 ± 0.5	13.5 ± 1.2	4.0 ± 0.2	5.1 ± 0.3
Atherogenic diet without LaCl$_3$	37.4 ± 8.5*	2.1 ± 0.7#	1.9 ± 0.24#	13.1 ± 2.3#	31.1 ± 7.6#	9.3 ± 1.8#	8.9 ± 0.9#
Atherogenic diet + 20 mg LaCl$_3$/kg	24.6 ± 9.7#	1.5 ± 0.3**	0.9 ± 0.1**	4.1 ± 0.9#	17.0 ± 1.8#	3.8 ± 0.3	5.8 ± 0.4**
Atherogenic diet + 30 mg LaCl$_3$/kg	11.5 ± 1.6	1.2 ± 0.3	0.8 ± 0.13	2.2 ± 0.4	14.6 ± 0.8	3.6 ± 0.6	5.4 ± 0.2
Atherogenic diet + 40 mg LaCl$_3$/kg	8.9 ± 2.5	0.7 ± 0.2	0.6 ± 0.29	2.4 ± ##	12.6 ##	3.9 ##	5.3 ##
Control diet + 40 mg LaCl$_3$/kg	8.5 ± 3.7	0.9 ± 0.1	0.5 ± 0.30	2.4 ± 0.3	13.9 ± 0.7	4.2 ± 0.4	5.5 ± 0.3

* The remaining elastin amino acids were not significantly different in all groups.
\# $p < 0.01$ compared to untreated control
** $p < 0.05$ compared to untreated control
\## Average of three elastin samples

Table 20.5. Constituents of elastin from control monkeys and monkeys on the atherogenic diet with and without drugs (μg/g elastin; mean ± SD).

Arteries	Calcium				Non-lipid phosphorus			
	Control diet	Atherogenic diet	Atherogenic and La^{3+}	Atherogenic and EHDP	Control diet	Atherogenic diet	Atherogenic and La^{3+}	Atherogenic and EHDP
Thoracic aorta	3.6 ± 0.3	97.1 ± 21.3*	5.9 ± 0.4*	7.0 ± 0.5*	3.1 ± 1.5	50.7 ± 16.3*	3.5 ± 0.9	3.0 ± 1.1
Abdominal aorta	2.8 ± 1.1	77.4 ± 19.0*	3.2 ± 1.3	4.2 ± 1.6	3.7 ± 1.3	14.0 ± 3.8*	2.2 ± 1.5	3.2 ± 0.8
Subclavian	3.9 ± 0.8	22.9 ± 6.5*	2.0 ± 1.2	5.3 ± 0.6#	2.8 ± 0.7	8.3 ± 2.2*	3.5 ± 1.4	4.5 ± 1.1
Carotid	3.7 ± 1.2	22.5 ± 7.7*	3.8 ± 0.4	5.4 ± 1.2	1.2 ± 0.4	10.2 ± 4.1*	1.2 ± 0.2	1.4 ± 0.3
Iliac	4.3 ± 1.5	16.9 ± 4.9*	3.8 ± 0.5	7.7 ± 2.6	5.3 ± 1.6	11.9 ± 2.6*	3.2 ± 1.7	5.4 ± 1.4
Femoral	2.6 ± 0.7	6.4 ± 1.8*	3.1 ± 0.6	3.9 ± 1.3	---	---	---	---

* Highly significant changes from control values ($p < 0.01$)
\# Significant changes from control values ($p < 0.05$)

and cholesterol (Tables 20.4 and 20.5). These elastin changes occur early, even before lipid-filled intimal cells can be detected, and are another important marker of early atherogenesis, involving the connective tissue.

In rabbits (see Table 20.2, Fig.20.2) simultaneous treatment with any of the drugs employed resulted in marked inhibition of the accumulations of aortic calcium which was associated with marked reductions in the content of the other arterial components despite continued presence of the atherogenic stimulus of unmitigated hypercholesterolaemia. Best results were obtained by treatment with AHD, IPDP and $LaCl_3$ (40 mg) which completely suppressed the aortic accumulations of calcium, phosphorus, collagen and elastin; the aortic accumulations of cholesterol were markedly reduced, with the remaining lipids presumably being contained in the thin layer of intimal foam cells (Fig.20.3c) that was observed as the only atherogenic alteration in these animals. The calcium, phosphorus, amino acid and cholesterol composition of elastin was also normal in rabbits treated with these three drugs (for $LaCl_3$-treated rabbits see Table 20.4). In atherogenic-diet rabbits treated with the phosphodiesterase inhibitor trimazosin, the effective dose (calculated from the amount of food actually consumed per day) for maximal suppression (normalization) of the accumulations of aortic calcium, collagen and elastin was 30 mg/kg bodyweight/day (Fig.20.2). The aortic cholesterol accumulations, however, were not appreciably suppressed by trimazosin treatment at any dosage level given.

In monkeys (see Table 20.3), treatment with lanthanum and diphosphonates resulted in similar marked reductions in the content of aortic calcium, as well as the other arterial components with the abnormalities of calcium, phosphorus, collagen and elastin being completely suppressed in abdominal aorta (and other major sytemic arteries - not shown). The exception again was the arterial cholesterol content, which was greatly reduced but not to normal, with the residual lipids presumably being mainly contained in the intimal cells of the few small cellular lesions present (Fig.20.4c). Treatment with EHDP and $LaCl_3$ also suppressed the atherogenic calcium and phosphorus changes in elastin of aorta as well as in other major systemic arteries of monkeys (Table 20.5). Arterial elastin of APDP and AHDP-treated monkeys was not analysed but presumably would have shown similar results as with EHDP; on microscopic inspection the elastica was normal in these animals.

20.3.3. Morphological findings

The biochemical results were in agreement with the morphological findings. Typical lesions are shown in Fig.20.3a to c for rabbits and Fig.20.4a to c for monkeys. In untreated rabbits, the fibrogenic atherogenic diet elicited numerous characteristic fibrous-atheromatous plaques (see Fig.20.3a), consisting of a markedly raised intima which contained large accumulations of collagen, many proliferated cells (some of which were lipid-filled foam cells), fragmented and deranged intimomedial elastica as well as depositions of calcium minerals predominantly on the altered elastica (see Fig.20.3b). Although the lesions do not entirely resemble a human fibrous plaque, they did exhibit the same plaque ingredients. In contrast, rabbits treated simultaneously with $LaCl_3$ (40 mg), EHDP, APDP, AHDP or IPDP (see Fig.20.3c) revealed a complete suppression of the calcium deposition, as well as marked suppression of lesion formation, leaving a few lesions which consisted of a small aggregate of intimal foam cells on top of an otherwise normal arterial wall (especially effective: $LaCl_3$, AHDP and IPDP). Trimazosin-treated atherogenic-diet rabbits revealed similar suppressions of arterial depositions of calcium, collagen and elastica.

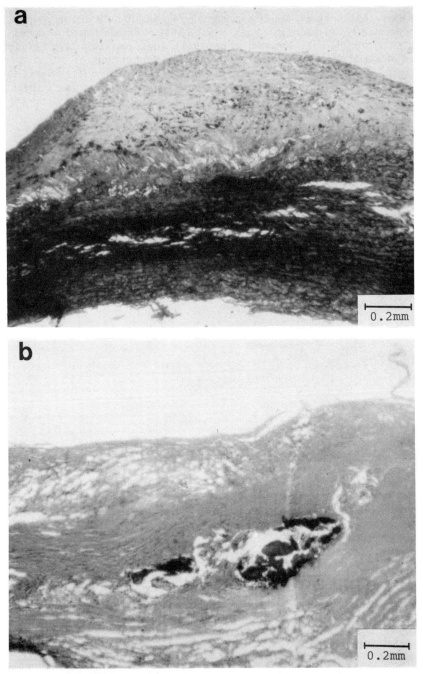

Fig.20.4. Sections through characteristic lesions of thoracic aorta of monkeys on the fibrogenic atherogenic diet for 24 months without drugs ((a),(b) = sequential sections through the same lesion) or treated simultaneously with AHDP(c). (a) Note collagenous plaque capsule (grey) surrounding a necrotic core with cholesterol clefts; the intimomedial elastica (black) is fragmented and deranged; Verhoeff's-Van Gieson, x70. (b) Note calcium deposition (black), including on intimomedial elastica; Yasue's -light green, x70. (c) The residual lesion consists of a few layers of (lipid-filled) foam cells in the intima. No connective tissue changes (or calcifications) were detectable in the lesion; Verhoeff's-Van Gieson, x70.

In untreated monkeys, the fibrogenic atherogenic diet elicited many of the same human-type fibrous plaques (see Fig.20.4a) as demonstrated in earlier studies (Kramsch, 1981; Kramsch et al., 1981). These markedly raised plaques (here of aorta) were characterized by massive accumulations of intimal collagen, especially a fibrous cap, necroses and cholesterol clefts deep in the raised intima, derangement of intimomedial elastica, but very few intimal foam cells; calcium mineralization was also present, including on deranged elastica (see Fig.20.4b). By contrast, no visible calcium deposits were detectable in aorta of monkeys on the same atherogenic diet for the same length of time but treated with either the calcium entry blocker lanthanum or any of the diphosphonates. In addition, the accumulations of collagen, the elastica alterations, necroses and cholesterol crystal depositions also were suppressed, the few small residual lesions consisted of a few layers of lipip-filled foam cells (see Fig.20.4c), giving the lesion the appearance of a mild human fatty streak, especially in monkeys treated with $LaCl_3$ and AHDP. Similar microscopic observations were made in other major systemic arteries, including coronary arteries of untreated monkeys and monkeys treated with anticalcium agents.

20.3.4. Studies with calcium antagonists by other authors
There are many reports from other authors on the effectiveness of calcium

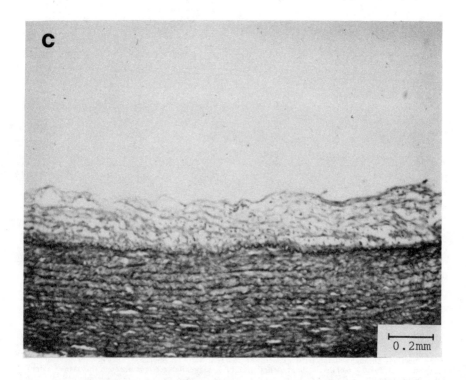

antagonists (in the broad sense of the word) as antiatherosclerotic agents. It is of particular interest that in many of these studies, as in our own experiments, the calcium antagonists displayed their most remarkable effects in the suppression of arterial calcium and connective tissue accumulations. Before our own first published studies in 1978 with EHDP in rabbits on an atherogenic diet (Kramsch and Chan, 1978), Wartman et al. (1967) already observed that the calcium chelator Mg-EDTA inhibited the rise in aortic calcium, collagen, and elastin in atherogenic-diet rabbits. Potocar and Schmidt-Dunker (1978) reported suppression of vitamin D-induced calcific-fibrous arterioatherosclerosis in rabbits using the newer diphosphonates APDP and AHDP.

It is of considerable interest in this context that calcium entry blockers, of the classes which have been shown to be effective for the teatment of other cardiovascular disorders in man, also have been shown to be antiatherosclerotic in various animal models. Fleckenstein and his coworkers (Fleckenstein, 1983, 1984; Fleckenstein, Frey and Leder, 1983; Fleckenstein, Frey and Witzleben, 1983; Fleckenstein-Grün, Frey and Fleckenstein, 1984; Frey, Keidel and Fleckenstein, 1980; Janke et al., 1972) have shown over the years since 1972 that verapamil, nifedipine and diltiazem suppress the calcium and connective tissue accumulations in rats with arterioatherosclerosis induced by nicotine, alloxan-diabetes, vitamin D excess or simply by advancing age. More recently marked antiatheroslerotic effects have been shown in rabbits on atherogenic diets for nifedipine, verapamil and diltiazem by Henry and Bentley (1981), and by Henry (1980); for verapamil by Rouleau et al. (1982, 1983), and for nicardipine by Nagel et al. (1984).

It should be emphasized that in several of the latter studies only lesion size was measured as outlined by the areas of yellowish discoloration caused by the presence of lipid-filled cells in the intima (or lipid staining thereof). However, as shown by many of the studies, calcium antagonists have their least marked effect in the prevention of lipid-loading of intimal cells. Since the arterial calcium and connective tissue was not measured, the true extent of the specific effect of the calcium antagonists on these important plaque components was missed. Frequently, in animals treated with calcium antagonists, lesion formation is inhibited to the extent that the lesions consist only of a few layers of lipid-filled intimal cells which, however, may be spread over a considerable area. This then may give rise to the erroneous impression that no lesion inhibition has been affected by the calcium antagonist. This presumably was the case in the study by Naito et al. (1984) in which only lesion size but no other plaque component was determined after treatment with nifedipine and diltiazem in atherogenic-diet rabbits. In another study reporting treatment failure, also with nifedipine (Stender et al., 1984) not even lesion formation was measured, but only influx of radioactive cholesterol into arteries. However, there never was an indication that calcium antagonists may affect this particular parameter. On the contrary, they have been shown to suppress the cellular responses in spite of the presence of the stimuli of increased arterial cholesterol influx and content.

For the specific effects of calcium antagonists it also appears to be particularly relevant that Robert et al. (1978) have earlier demonstrated that in rabbits with certain immunogenic arterial injury, the formation of atherosclerotic connective tissue lesions can be suppressed by daily injections of the hormone thyrocalcitonin which is known to play a pivotal role in calcium homeostasis. Similarly, previous studies by ourselves (Chan, Wells and

Kramsch, 1978) had shown a marked inhibition of fibrous rabbit atherosclerosis through administration of thiophenocarboxylic acid which is thought to exert a thyrocalcitonin-like effect on maintenance of calcium homeostasis (Lloyd et al., 1979).

20.3.5. *Mechanism of atherogenesis suppression by anticalcium agents*
Taken together, the evidence from the studies in rabbits and monkeys by ourselves and others led us to formulate the ionic calcium hypothesis which states that:

(1) Focal increases in arterial calcium ion concentrations may facilitate excessive activation of cellular functions in response to any atherogenic stimuli (Ca^{2+} as second messenger).
(2) A relative focal abundance of calcium ions in arteries also may facilitate the development of extracellular atherosclerotic abnormalities involving the connective tissue.
(3) The effects of a variety of calcium antagonists in the broad sense of the word support the concept that excessively available arterial calcium ions may play a key role in atherogenesis and that maintenance of calcium homestatis at the arterial tissue level may be pivotal in prevention of the disorder by reducing the abundance of ionic calcium
(4) Maintenance of calcium homeostasis by calcium antagonists (reduction of calcium overload) is especially effective in preventing the cell proliferation and connective tissue changes seen in fibrous plaques.

It is noteworthy that there is experimental evidence for the direct effect of calcium entry blockers on some of the basic processes of atherogenesis. Our laboratory has treated four normal rabbits on control diet with 40 mg lanthanum per kg bodyweight for 4 weeks at the end of which period several aortic explants were taken from each rabbit for cultures of arterial smooth muscle cells. These cultures then were compared to similarly obtained SMC cultures from four untreated control rabbits. Aortic explants from untreated normal rabbits after 7 days in culture revealed normal outgrowth of SMCs (Fig.20.5a). In contrast, explants from aortas of normal rabbits pretreated *in vivo* with lanthanum revealed a marked retardation of the cellular outgrowth at day 7 (Fig.20.5b), with the slower cell growth persisting even up to day 4 of the second passage. These results indicate that the calcium antagonists may affect *in vivo* arterial smooth muscle cell migration or mitosis or presumably both.

In another experiment (Table 20.6) we added lanthanum (La^{3+}) in increasing concentrations to confluent cultures of arterial SMCs from three untreated normal rabbits. The cultures than were pulse-labelled with ^{14}C-proline and the appearance of the radiolabel was measured in the hydroxyproline fraction as an indicator of collagen biosynthesis and secretion. As indicated by the increasing reduction of the hydroxyproline radioactivity in the incubation media, increasing doses of the calcium entry blocker caused increasing inhibition of collagen secretion from the cells into the media. On the other hand, as indicated by the unchanged normal hydroxyproline radioactivity at all dosage levels in the washed cell layers, collagen synthesis in the cells appeared to be unaffected by the La^{3+} exposure. It is of interest that studies by others (Minkin, Rabadjia and Goldhaber, 1974) have demonstrated that diphosphonates also inhibit secretion of collagen from collagen-producing cells *in vitro*. These *in vitro* studies suggest that suppression of excessive

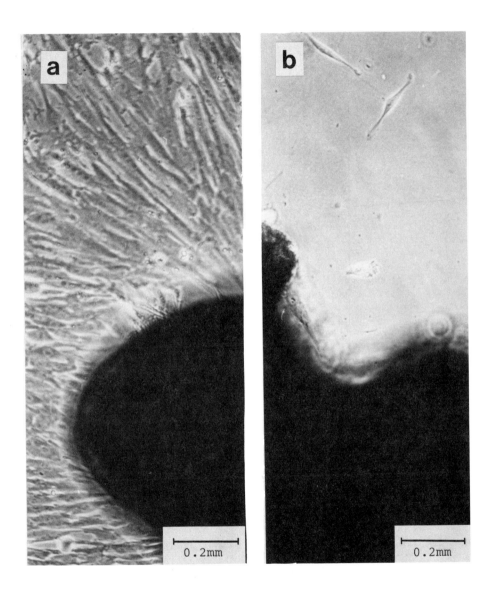

Fig.20.5. Phase contrast micrographs of smooth muscle cell cultures from aortic explants of normal rabbits fed a normal control diet without (a) and with lanthanum 40 mg/kg bodyweight (b) for 4 weeks; explants (large dark area in each micrograph) in culture for 7 days. (a) Note streaming out of smooth muscle-type cells from the aortic explant of an untreated rabbit to form a confluent cell layer, x100. (b) Note the paucity of cells grown out of the explants cultured for the same period of time from the aorta of rabbits treated *in vivo* with lanthanum, x100.

Table 20.6. Effect of lanthanum (La^{3+}) on collagen and total protein synthesis and secretion by arterial SMCs. (cpm x 10^4/mg N; averages of cultures from three rabbits)

La^{3+} Concentration in media (nMol)	A. Incubation media			Ratio Proline/OH-proline
	Proline	OH-Proline	Inhibition (OH-proline)	
None	14.4	7.4	0%	2.0
0.001	9.4	4.1	-45%	2.3
0.010	5.5	2.7	-64%	2.1
0.100	4.1	1.6	-78%	2.5
	B. Cell layer			
None	28.0	5.5	--	5.1
0.001	27.6	4.5	--	6.2
0.010	29.8	5.6	--	5.3
0.100	32.8	4.8	--	6.8

Calcium antagonists and atherosclerosis

secretion of connective tissue macromolecules by arterial SMCs may be one of the prominent beneficial effects of all calcium antagonists, supporting the *in vivo* treatment results with a variety of these anticalcium agents in animals with developing atherosclerosis.

20.4. Reversal of pre-established fibrous atheromatous plaques by anticalcium drugs

While a preventive effect of calcium antagonists on the development of atherosclerosis may not seem too surprising, the critical question remains as to whether these agents are capable of promoting reversal of the established disease. In some recent experiments our laboratory was able to shed some light on this question.

20.4.1. *Experimental*

Five groups of ten male New Zealand White rabbits each were studied for 16 weeks; one group on the control diet of Purina Rabbit Chow for the entire period and four groups on the fibrogenic atherogenic diet (described above for induction of atherosclerosis) for 8 weeks after which time all four groups were again placed on the control diet for another 8 weeks. After cessation of the atherogenic diet, one of these four groups received the control diet alone while the remaining three groups were fed the control diet with addition of (40 mg/kg bodyweight) the diphosphonates APDP and AHDP as well as the calcium entry blocker lanthanum. These five groups of rabbits studied for 16 weeks were compared to one additional group of ten rabbits which received the atherogenic diet without drug treatment for 8 weeks only (induction of atherosclerosis, atherogenic control group).

20.4.2. *Biochemical findings*

In all animals on the atherogenic diet the concentrations of total serum cholesterol rose to about the same levels as described above in section 20.3 (~2500–3000 mg/dl); and the serum cholesterol concentrations fell to normal control levels after cessation of the atherogenic diet.

As compared to animals receiving the control diet for the entire experimental period, rabbits on the atherogenic diet for 8 weeks and then sacrificed showed the usual marked increases in aortic collagen, elastin, cholesterol, calcium and phosphorus (Table 20.7; compare Table 20.2). After cessation of the atherogenic diet without drug treatment (see Table 20.7) there were further substantial increases in these aortic components which is the general experience in rabbits and other animals. Treatment with APDP, after the induction diet was stopped, resulted in the prevention of further increases of the atherosclerotic parameters measured, indicating that APDP had an arresting effect on further progression of atherosclerosis (see Table 20.7). In marked contrast, treatment with AHDP or lanthanum after cessation of atherosclerosis induction resulted in complete normalization of the aortic content in collagen, elastin and phosphorus while the content in cholesterol and especially calcium were greatly reduced from atherogenic levels, albeit not to normal (see Table 20.7). These findings suggest that both AHDP and lanthanum are capable of removing induced arterial calcification in addition to reversing fibrosis.

Changes in the composition of arterial elastin followed similar patterns as those for atherosclerosis involvement in the experimental groups (Table 20.8).

Table 20.7. Components of aortic intimamedia from control rabbits and rabbits on the atherogenic diet for 8 weeks on the control diet with and without drugs.
(absolute amounts in mg/kg body weight; mean ± SD)

Experimental drugs	Collagen	Elastin	Cholesterol	Calcium	Phosphorus
Control diet (16 weeks)	4.1 ± 0.8	15.6 ± 1.4	0.6 ± 0.2	0.01 ± 0.007	0.07 ± 0.03
Atherogenic diet (8 weeks)	6.2 ± 0.7*	21.4 ± 1.6*	4.6 ± 0.9*	0.06 ± 0.013*	0.14 ± 0.03*
Atherogenic diet followed by control diet without drugs	7.5 ± 1.0*	27.5 ± 3.0*	4.8 ± 1.8*	0.10 ± 0.018*	0.26 ± 0.06*
Atherogenic diet followed by control diet + APDP	5.6 ± 0.9#	19.4 ± 1.2#	5.3 ± 3.3*	0.07 ± 0.016*	0.11 ± 0.04#
Atherogenic diet followed by control diet + AHDP	3.6 ± 1.2	15.1 ± 1.8	2.5 ± 1.5*	0.02 ± 0.010#	0.06 ± 0.03
Atherogenic diet followed by control diet + LaCl3	4.2 ± 1.5	15.6 ± 2.4	1.5 ± 1.00#	0.03 ± 0.018#	0.07 ± 0.04

* Highly significant changes from control values ($p < 0.01$)
\# Significant changes from control values ($p < 0.05$)

Table 20.8. Constituents of elastin from aortic intimamedia of control rabbits and rabbits on the atherogenic diet for 8 weeks followed by 8 weeks on the control diet with and without drugs. (mg/g elastin; mean ± SD)

Experimental groups	Calcium	Non-lipid phosphorus	Total Cholesterol	Percentage Ester Cholesterol
Control diet (16 weeks)	0.6 ± 0.03	0.8 ± 0.07	10.4 ± 2.3	41 ± 9
Atherogenic diet (8 weeks)	2.1 ± 0.73*	1.9 ± 0.40*	34.6 ± 6.6*	81 ± 3*
Atherogenic diet followed by control diet without drugs	2.9 ± 0.50*	2.3 ± 0.44*	42.7 ± 8.2*	85 ± 7*
Atherogenic diet followed by control diet + APDP	1.7 ± 0.68*	1.1 ± 0.06*	36.7 ± 13.2*	78 ± 9#
Atherogenic diet followed by control diet & AHDP	0.8 ± 0.04#	0.7 ± 0.25	16.7 ± 6.1#	56 ± 8#
Atherogenic diet followed by control diet + LaCl$_3$	1.1 ± 0.05#	0.9 ± 0.32	14.2 ± 8.3	67 ± 9*

* Highly significant changes from control values ($p < 0.01$)
\# Significant changes from control values ($p < 0.05$)

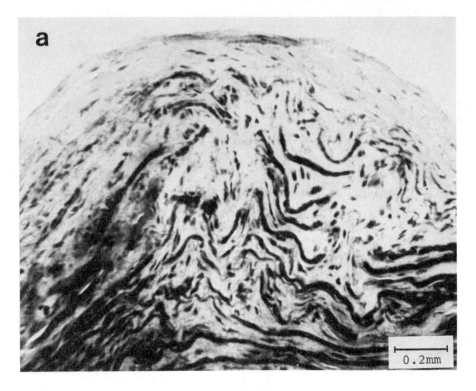

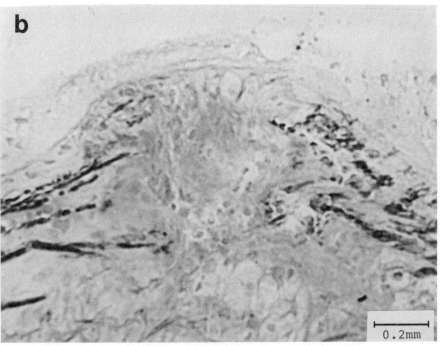

Calcium antagonists and atherosclerosis

As compared to normal controls, elastin of rabbits receiving the atherogenic diet alone for 8 weeks showed the usual increases in calcium, phosphorus and cholesterol, which still further increased after an 8-week cessation period of that diet with drug treatment. However, these additional increases were halted at atherosclerosis induction levels by treatment with APDP. In marked contrast, treatment with AHDP or lanthanum reduced the elastin content of phosphorus to normal and that of calcium to near-normal levels. Likewise, the elastin cholesterol content was greatly reduced in AHDP-treated animals and was normal in the lanthanum group. The results indicate that the elastin abnormalities can also be reversed by treatment with calcium antagonists.

20.4.3. Morphological findings

Again, the biochemical results were in agreement with the morphological findings. Characteristic lesions of thoracic aorta after cessation of the atherosclerosis induction diet with and without drug treatment are shown in Fig. 20.6 a to c. Typically, after cessation of the atherogenic diet without drug treatment, the pre-established lesions became even more fibrotic with dense collagenous tissue in the markedly raised intima which contained essentially no foam cells; the derangement and fragmentation of the intimomedial elastica became even more pronounced (Fig.20.6a; compare Fig.20.3a) and, in addition, even more extensive deposition of calcium minerals were seen,

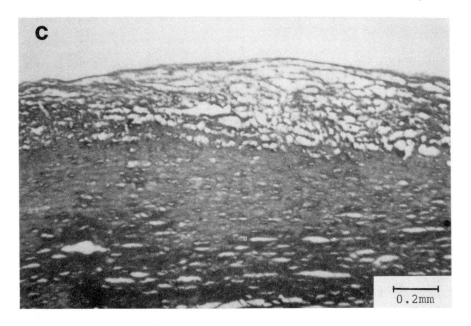

Fig.20.6. Sections through characteristic lesions of thoracic aorta of rabbits on the fibrogenic atherogenic diet for 8 weeks followed by 8 more weeks on the control diet without drugs; ((a),(b) = sequential sections of the same lesion) or treated simultaneously with lanthanum (c). (a) Note dense accumulations of collagen (grey) in the markedly raised intima, the extensively fragmented and deranged intimomedial elastica (black) as well as the absence of lipid-filled foam cells; this transformation into even more fibrous plaques is typical of atherosclerotic lesions of untreated rabbits after cessation of the atherogenic diet (compare Fig.20.3a), Verhoeff's-Van Gieson, x85.
(b) Note marked deposition of calcium (dark grey-black) on deranged elastica; Alizarin red-light green, x85. (c) Note that the lesion is composed of a few layers of lipid-rich foam cells in the intima overlying an essentially normal arterial wall with absence of calcium and collagen accumulations and essentially normal elastica; Yasue's-light green, x70.

especially on the severely deranged elastica (Fig.20.6b; compare Fig.20.3b). In marked contrast, when the animals were treated with either AHDP or lanthanum during the 'regression' period, only small lesions were seen characterized by absence of collagen accumulations, repair of the intimomedial elastica, and a moderate number of intimal foam cells remaining as the only abnormality together with some minor residual calcium mineral depositions (barely visible) on the largely intact elastica (Fig.20.6c).

20.4.4. *Mechanism of lesion reversal by anticalcium agents*

The precise mechanisms by which the diphosphonate AHDP and the calcium entry blocker lanthanum cause reversal of pre-existing calcific fibrous atherosclerosis is less clear than the mechanisms by which they may operate to cause its prevention. It is possible that once the further progression of atherosclerotic lesions is arrested, the regular catabolic processes of the organism are capable of slowly restituting the diseased arterial tissue, including reversing tissue calcification. It should be recalled in this context that lanthanum is known to actually displace tissue bound calcium (Weiss, 1974; Weiss and Goodman, 1976) as well as, like the diphosphonates (Fleisch *et al.*, 1970; Rosenblum, Flora and Eisenstein, 1975; Wagner and Clarkson, 1973), to prevent new calcium from being bound to receptor sites.

20.5. Conclusion

From the results of these studies by ourselves and other workers, it is concluded that calcium antagonists capable of regulating functional calcium levels in arteries may be suitable therapeutic agents for the prevention, arrest and even reversal of the most important and most life-threatening aspect of human atherosclerosis: the calcified fibrous plaque. These agents may be useful as adjunct treatment to the treatment of atheroslcerotic arterial disease with lipid-lowering drugs, or they may even be useful for such treatment by themselves as single entities. It should be emphasized again that these agents appear to exert their beneficial effects without altering unfavourable serum lipid and lipoprotein concentrations and patterns as are presumably present in most western populations.

Acknowledgements

The work in this chapter was supported by US Public Health Service Grants HL 15512 and HL 13262.

References

Chan, C.T., Wells, H. and Kramsch, D.M. (1978). Suppression of fibrous-fatty plaque formation in rabbits by agents not affecting elevated serum cholesterol levels. The effect of thiophene compounds. *Circ. Res.* **43**, 115–25.

Committee of Principal Investigators. (1978) Report: A cooperative trial in the primary prevention of ischemic heart disease using clofibrate. *Br. Heart J.* **10**, 1069–1118.

Fleckenstein, A. (1983). *Calcium Antagonism in Heart and Smooth Muscle. Experimental Facts and Prospects* (monograph). John Wiley and Sons, New York.

Fleckenstein, A. (1984). Introduction: drugs acting through calcium channels-calcium antagonists. *IUPHAR 9th International Congress of Pharmacology, London*, pp. S19—1.

Fleckenstein, A., Frey, M. and Leder, O. (1983). In *New Calcium Antagonists. Recent Developments and Prospects*. A. Fleckenstein, K. Hashimoto, M. Herrmann, A. Schwartz and J. Seipel (eds), pp. 15—31. Gustav Fischer Verlag, Stuttgart.

Fleckenstein, A., Frey, M. and V. Witzleben, H. (1983). Vascular calcium overload - a pathogenic factor in arteriosclerosis and its neutralization by calcium antagonists. In *Proceedings of the 5th Adalat Symposium*. M. Kaltenbach and H.N. Neufeld (eds), pp. 36—52. Exerpta Medica, Amsterdam.

Fleckenstein-Grün, G., Frey, M. and Fleckenstein, A. (1984). Calcium antagonists: mechanisms and therapeutic uses. *Trends Pharmacol. Sci.* 5, 283—6.

Fleisch, H., Russel, R.G.G., Bisaz, S., Muehlbauer, R.C. and Williams, D.A. (1970). The inhibitory effect of diphosphonates on the formation of calcium phosphate crystals in vitro and on aortic and kidney calcification in vivo. *Eur. J. Clin. Invest.* 1, 12—18.

Frey, M., Keidel, J. and Fleckenstein, A. (1980). Verhütung experimenteller Gefäss-Verkalkungen (Mönckeberg's Typ der Arteriosklerose) durch Calcium-Antagonisten. In *Calcium-Antagonismus*. A. Fleckenstein and H. Roskamm (eds), pp. 258—68. Springer-Verlag, Berlin.

Henry, P.D. (1980). Comparative pharmacology of calcium antagonists-nifedipine, verapamil and diltiazem. *Amer. J. Cardiol.* 46, 1047—58.

Henry, P.D. and Bentley, K.I. (1981). Suppression of atherogenesis in cholesterol-fed rabbit treated with nifedipine. *J. Clin. Invest.* 68, 1366—9.

Janke, J., Hein, B., Packinger, O., Leder, O. and Fleckenstein, A. (1972). Hemmung arteriosklerotischer Gefässprozesse durch prophylaktische Behandlung mit $MgCl_2$, KCl und organischen Ca^{++} Antagonisten (Quantitative Studien mit Ca^{45} bei Ratten). In *Vascular Smooth Muscle*. E. Betz (ed.), pp. 71—2. Springer-Verlag, Berlin.

Kramsch, D.M. (1981). Biochemical changes of the arterial wall in atherosclerosis with special reference to connective tissue: promising experimental avenues for their prevention. In *Connective Tissue in Arterial and Pulmonary Disease*. T.F. McDonald and A.B. Chandler (eds), pp. 95—151. Springer-Verlag, New York.

Kramsch, D.M. and Chan, C.T. (1978). The effect of agents interfering with soft tissue calcification and cell proliferation on calcific fibrous-fatty plaques in rabbits. *Circ. Res.* 42, 562—71.

Kramsch, D.M., Hollander, W. and Renaud, S. (1973). Induction of fibrous plaques versus foam cell lesions in Macaca fascicularis by varying the composition of dietary fats. *Circulation* 48 (**Suppl. IV**); 41.

Kramsch, D.M., Aspen, A.J. and Apstein, C.S. (1980). Suppression of experimental atherosclerosis by the Ca^{++}-antagonist lanthanum. Possible role of calcium in atherogenesis. *J. Clin. Invest.* 65, 967—81.

Kramsch, D.M., Aspen, A.J. and Rozler, L.J. (1981). Atherosclerosis: Prevention by agents not affecting abnormal levels of blood lipids. *Science* 213, 1511—2.

Kramsch, D.M., Aspen, A.J. and Swindell, A.C. (1983). Trimazosin suppreses fibrosis of atherosclerotic plaques. *Fed. Proc.* 42, 808.

Kramsch, D.M., Aspen, A.J., Abramowitz, B.M., Kreimendahl, T. and Hood, Jr., W.B. (1981). Reduction of coronary atherosclerosis by moderate conditioning exercise in monkeys on an atherogenic diet. *N. Engl. J. Med.* 305, 1483—9.

Kritchevsky, D., Tepper, S.A., Vesselinovitch, D. and Wissler, R.W. (1971). Cholesterol vehicle in experimental atherosclerosis. Part II (peanut oil). *Atherosclerosis* **14**, 53–64.

Lipid Research Clinics Coronary Primary Prevention Trial Results (1984). I. Reduction in incidence of coronary heart disease. II. The relationship of the reduction in incidence of coronary heart disease to cholesterol lowering. *J. Am. Med. Assoc.* **251**, 351–74.

Lloyd, W., Fang, W.S., Wells, H. and Tashijan, A.H. (1979). 2-Thiophene carboyxlic acid. A hypocalcemic, antilipolytic agent with hypocalcemic and hypophosphatemic effects in rats. *Endocrinology* **85**, 763–8.

Minkin, C., Rabadjia, L. and Goldhaber, P. (1974). Bone remodeling *in vitro*: the effects of diphosphoantes on osteoid synthesis. *Calcif. Tissue Res.* **14**, 161–8.

Mustard, J.F., Packham, M.A., Moore, S. and Kinlough-Rathbone, R.L. (1974). Thrombosis and Atherosclerosis. In *Atherosclerosis III*, G. Schettler and A. Weizel (eds). Springer-Verlag, Berlin.

Nagel, B., Mahmud, I., Churchill, V., Whyte, M., Smith, D.L. and Willis, A.L. (1984). Nicardipine and nifedipine inhibit atherosclerosis and aortic deposition of cholesterol and triglycerides. *Fed. Proc.* **43**, 843.

Naito, M., Kuzuwa, F., Asai, K., Shibata, K. and Yoshimine, N. (1984). Ineffectiveness of Ca^{2+}-antagonists nicardipine and diltiazem on experimental atherosclerosis in cholesterol-fed rabbits (ATH 03474). *Atherosclerosis* **51**, 343–4.

National Heart, Lung and Blood Institute Task Force on Arteriosclerosis (1971). *DHEW Publ. (NIH)* **1**, 72–137.

Potocar, M. and Schmidt-Dunker, M. (1978). The effect of new diphosphonic acids on aortic and kidney calcifications *in vivo*. *Atherosclerosis* **30**, 313–20.

Robert, A.M., Moczar, M., Brechemier, D., Godeau, G., Miskulin, M. and Robert, L. (1978). Biosynthesis and degradation of matrix molecules of the arterial wall. Regulation by drug action. In *International Symposium: State of Prevention and Therapy in Human Atherosclerosis and in Animal Models*. W.H. Hauss, R.W. Wissler and R. Lehmann (eds), pp. 301–12. Abh. Rhein.-Westf. Akad. Wiss. Vol. 3, Westdeutscher Verlag, Opladen.

Rosenblum, I.Y., Flora, L. and Eisenstein, R. (1975). The effect of sodium ethane-1-hydroxy-1,1 diphosphonate (EHDP) on a rabbit model of arterio-atherosclerosis. *Atherosclerosis* **22**, 411–21.

Rouleau, J-L., Parmley, W.W., Stevens, J., Wikman-Coffelt, J., Sievers, R., Mahley, R.W. and Havel, R.J. (1982). Verapamil suppresses atherosclerosis in cholesterol-fed rabbits. *Amer. J. Cardiol.* **49**, 889.

Rouleau, J-L., Parmley, W.W., Stevens, J., Wikman-Coffelt, J., Sievers, B.S., Mahley, R.W. and Havel, R.J. (1983). Verapamil suppresses atherosclerosis in cholesterol-fed rabbits. *J. Amer. Coll. Cardiol.* **1**, 1453–60.

Stender, S., Stender, I., Nordestgaard, B. and Kjeldsen, K. (1984). No effect of nifedipine on atherogenesis in cholesterol-fed rabbits. *Arteriosclerosis* **4**, 389–94.

Wagner, W.D. and Clarkson, T.B. (1973). Slowly miscible cholesterol pools in progressing and regressing atherosclerotic aortas. *Proc. Soc. Exp. Biol. Med.* **143**, 804–9.

Wartman, A., Lampe, A.L., McCann, D.S. and Boyle, A.J. (1967). Plaque reversal with MgEDTA in experimental atherosclerosis: elastin and collagen metabolism. *J. Atheroscler. Res.* **7**, 331–41.

Weiss, G.B. (1974). Cellular pharmacology of lanthanum. In *Annual Review of Pharmacology, Vol. 14*. H.W. Elliot, R. Okun and R. George (eds), pp. 343–54.

Annual Reviews, Palo Alto.
Weiss, G.B. and Goodman, F.R. (1976). Distribution of lanthanide [^{174}Pm] in vascular smooth muscle. *J. Pharmacol. Ther.* **198**, 366–74.

CALCIUM - ITS PRESUMPTIVE ROLE IN HYPERTENSION

J. Rosenthal

Centre of Internal Medicine, University of Ulm, West Germany

Abstract

Epidemiological, clinical and experimental data reveal that close relationships exist between primary hypertension and intracellular electrolytes. Recent observations have underlined the apparent roles of calcium and of sodium–calcium exchange mechanisms in the regulation of smooth muscle contractility. In this minireview current aspects are appraised and put into perspective with relevant clinical entities.

2.1. Introduction

High blood pressure is the consequence of a number of alterations and derangements which involve a large sequence of events including haemodynamic, endocrine and neural systems to mention only the apparently most important ones. In the following review, possible relations between ion transport across membranes and high blood pressure will be evaluated. Membrane abnormalities in essential hypertension in man and several forms of experimental hypertension in animals have been discussed for years (Table 21.1) (Blaustein, 1977). Various aspects of ouabain-insensitive cation transport in red cells have been investigated. Na-cotransport, K-cotransport and Na–Na exchange, as measured by Na–Li countertransport, have also received considerable attention. Other reports have described abnormal maximum rates of Na–Li countertransport in members of different generations of families. These data provide particularly important pictures of possible genetic transmission of this transport system (see Table 21.1) (Friedman *et al.*, 1977).

Other investigations explored ion movements in vascular smooth muscle in various models of experimental hypertension and in some cases positive correlations between ion movements and concentration and increased vascular smooth muscle tone are reported, but in some other cases no correlation between intracellular Na and vascular resistance was observed.

Additional investigations call particular attention to the possibility of a humoral factor (see Table 21.1) that inhibits Na–K pump being present in increased concentrations during volume expansion hypertension (Garay *et al.*, 1980). Important hypotheses are those that see connections between abnormal Na transport and high blood pressure in connection with intracellular Ca, where it is argued that any transport abnormality that leads to increased

Table 21.1 Transport across membranes and hypertension

---

(1) Membrane abnormalities

   Ouabain-insensitive cation transport in red cells
   Na—K cotransport
   Na—Na exchange (measured by Na—Li countertransport) - genetic transmission of this transport system?

(2) Ion movements in vascular smooth muscle

   positive correlations between ion movements and concentrations and increased vascular smooth muscle tone

(3) (Endogenous?) humoral factor(s) inhibit(s) Na—K pump
   increased concentration in volume-expansion hypertension

(4) Abnormal Na transport and hypertension via intracellular calcium
   increased intracellular Na pre-disposes to hypertension by leading to reduced extrusion of calcium via the Na—Ca exchange system

---

intracellular Na would predispose to high blood pressure by leading to reduced extrusion of Ca by the Na—Ca exchange system (see Table 21.1) thus increasing intracellular Ca concentration and causing increased shortening and tension in vascular smooth muscle fibres (Horackova and Vassort, 1979). Finally, there is growing evidence, as mentioned above, favouring the existence of an endogenous Na—K pump inhibitor. It is quite possible that such a hormone may play an important role in the regulation of blood pressure and other physiological processes.

It is obvious that Na cannot be dissociated from Ca when discussing blood pressure regulation and this point will be reiterated in this chapter (Potnov, Orlov and Poludin, 1979).

21.2. Factors predisposing to hypertension

There appear to be three key factors in the development of high blood pressure: a genetic factor, a humoral factor and an environmental factor.

The inherited factor may be a defect in the body's ability to excrete Na, but such a defect has not been identified in humans. However, in the Dahl Na-sensitive strain of rat undergoing chronic renal transplantation, a primary defect appears to reside in the kidneys: transplantation of kidneys from salt-sensitive to salt-resistant rats renders the latter salt-sensitive; conversely, replacement of kidneys in a salt-sensitive rat with those of a salt-resistant rat renders the former insensitive to salt. These findings suggest that even if a genetic defect - a transport abnormality - is present in tissues other than the kidney its presence must be inconsequential for the expression of high blood pressure (DeMendonea et al., 1980).

With an inherited renal defect in Na excretion, excessive Na ingestion tends to increase extracellular fluid volume. The normal homeostatic responses may then be reduced renin and aldosterone secretion, and increased secretion of a hormone that promotes Na excretion, the so-called natriuretic hormone. Clearly, with a restricted Na intake the stimulus for secretion of a natriuretic agent is removed. A large Na intake may be necessary to raise the level of natriuretic hormone to above normal in hypertension-prone individuals who would otherwise be unable to excrete this salt load because of an inherited renal defect.

Clinical studies, however, have demonstrated that hypertensive patients rarely exhibit overt fluid retention and expanded extracellular fluid volume. However, many hypertensives do have low renin levels and this may be further evidence that normal homeostatic mechanisms are effectively compensating to prevent net Na and fluid retention.

Under normal conditions Na retention is associated with development of high blood pressure, but there are cases where marked Na retention is not associated with high blood pressure, for instance in so-called 'essential hypernatraemia' in man which concurs with anteroventral hypothalamic lesions in the third ventricle of animals. It seems that some patients with essential hypernatraemia have intracranial lesions in the region of the hypothalamus. These observations appear to be supported by other evidence that natriuretic hormone is secreted in the brain and probably in the hypothalamus (Brody *et al.*, 1978). Natriuretic hormone inhibits Na transport in non-renal cells: the Na content of arterial smooth muscle cells of erythrocytes and leucocytes was found to be elevated in patients with essential hypertension, and also blood plasma from hypertensive patients contains a substance that inhibits Na transport in leucocytes from normal individuals (Table 21.2). All of these studies are consistent with the idea that natriuretic hormone levels may be elevated in plasma of hypertensive patients. This hormone may be the agent responsible for the high intracellular Na concentrations in various types of cells including vascular smooth muscle cells in these individuals (MacGregor *et al.*, 1981).

The natriuretic hormone and its relationship to increased peripheral vascular resistance requires further elaboration. If there is indeed a genetic renal defect as shown by excessive Na intake and elevation of natriuretic hormone in plasma of hypertensives a question remains: how is this translated into increased peripheral resistance? Two hypotheses have been put forward to explain this inter-relationship between elevated natriuretic hormone levels and increased peripheral vascular resistance (Table 21.3):

Table 21.2 Hypertension - Genetic (inherited) factor(s)

---

(1) Dahl sodium-sensitive rats - transport anomaly - natriuretic hormone may inhibit sodium reabsorption; with restricted sodium intake the stimulus for secretion is removed

(2) anteroventral hypothalamic lesions in 'essential hypernatraemia without hypertension'

Table 21.3 Hypertension - inhibition of sodium transport in non-renal cells by natriuretic hormone

(1) Sodium content of arterial smooth muscle, of erythrocytes and of leucocytes elevated in hypertension

(2) Blood plasma from hypertensives contains substance that inhibits sodium transport in leucocytes from normotensives

(3) Hence: Elevation of natriuretic hormone in plasma of hypertensives leading to high intracellular sodium concentrations in tissues

---

(1) The main manifestation of the circulating Na pump inhibitor may be a steady depolarization of the vascular smooth muscle cells. Vascular smooth muscle cells have voltage-regulated Ca-channels and depolarization should enhance their permeability to Ca. As a result, Ca influx would increase because a large electrochemical gradient for Ca across the plasma membrane favours Ca entry; this causes the cytoplasmic Ca concentration to rise, thereby promoting muscle contraction. One attributes the depolarization to inhibition of the electrogenic Na pumps in the smooth muscle cells.

(2) The alternative hypothesis is that the rise in cytoplasmic Na concentration *per se* contributes to the increased peripheral vascular resistance. Because of the operation of the Na–Ca exchange mechanism (Table 21.4), the Ca concentration gradient across plasma membrane will be tightly linked to the Na concentration gradient. Therefore a reduction in the Na gradient - for example as a result of rise in cell Na - will cause the cytoplasmic Ca concentration to rise as well. Then, because neural

Table 21.4 Hypertension - natriuretic hormone and increased peripheral vascular resistance - genetic (renal) defect leads to excessive sodium intake and increased peripheral resistance - mechanism(s)?

(1) Circulating sodium-pump inhibitor evokes steady depolarization of vascular smooth muscle cells which have voltage-regulated calcium channels, and depolarization may enhance their permeability to calcium

(2) Rise in cytoplasmic sodium concentration, *per se*, contributes to increased peripheral vascular resistance

(3) Natriuretic hormone also inhibits sodium pump in sympathetic neurones leading to reduced sodium gradient across plasma membrane of sympathetic nerve terminals

(4) Sodium and calcium (Na and Ca) transport are coupled by a counterflow transport system: sodium ions enter in exchange for exiting calcium

transmitter release is triggered by rise in the cytoplasmic Ca concentration one might expect both tonic, spontaneous as well as depolarization-evoked catecholamine release to be enhanced (Fig.21.1).

Figure 21.1 illustrates the parallel Na pump (in the lower part) and Na–Ca exchange (in the upper part), transport systems that function in the nerve terminals and in vascular smooth muscle. The Na pump utilizes energy from ATP hydrolysis to accumulate K and extrude Na, thereby maintaining a large electrochemical gradient for Na. The energy from the latter gradient can then be used to power Ca extrusion via the Na–Ca exchange mechanism. Because of this the limiting Ca electrochemical gradient will depend upon the Na

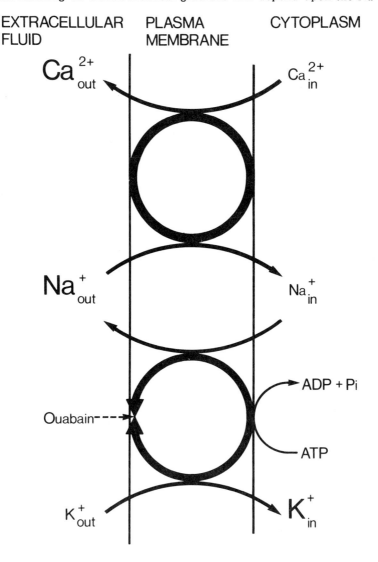

Fig.21.1. Sodium and calcium (Na and Ca) exchange mechanisms (for details see text).

gradient: the precise relationship will depend upon the Na—Ca stoichiometry (the number of Na ions exchanged for each Ca ion). In many types of cells the stoichiometry appears to be roughly 3 Na to 1 Ca, which would be sufficient to maintain the free intracellular Ca concentration in the physiological range. It is noteworthy that in these transport systems sardiotonic steroids such as ouabain selectively inhibit the Na pump and do not directly affect the Na—Ca exchange mechanism (Walter and Distler, 1980).

But all of this cannot be the only mechanism because there is considerable evidence that vascular smooth muscle reactivity, the contractile reponse to a given stimulus, is enhanced in patients with hypertension, and thus conceivably the smooth muscle cell itself may be altered.

Na—Ca exchange and vascular smooth muscle tone is a further subtopic: as has been demonstrated frequently vascular smooth muscle also possesses a Na—Ca exchange transport system (Table 21.4). Therefore in this tissue, too, a rise in the cytoplasmic steady state Na concentration will produce a concomitant rise in cytoplasmic Ca concentration. As a result there will also be increased storage of Ca in the sarcoplasmic reticulum which means both free and stored Ca levels will rise. A diagrammatic model of the cell illustrating the main pathways of Ca movement across the plasma membrane and limiting membrane of the sarcoplasmic reticulum, and the main intracellular Ca storage organelle, is shown here in Fig.21.2.

The route of Ca entry through the plasma membrane (shown as a gated pathway with rate coefficient) can be inhibited by the Ca-channel blocker verapamil.

EXTERNAL
MEDIUM

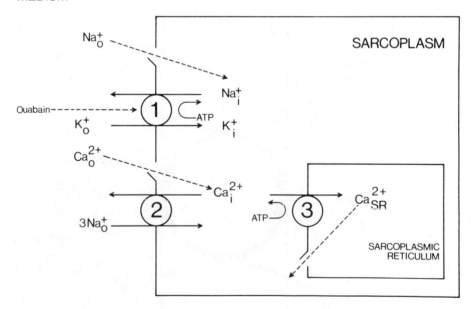

Fig.21.2. Model of vascular smooth muscle cell and mechanisms apparently involved in regulation of calcium (for details see text).

Some authors have adopted the view that Na—Ca exchange does not play an important role in cell Ca regulation in vascular smooth muscle, and favour the view that a Ca-dependent ATPase may be involved in the dominant Ca transport system. But they appear to ignore the fact that a portion of the Ca efflux is directly dependent on external Na; moreover, partial inhibition of the Na pump, with K reduction raising intracellular Na, sensitizes vascular smooth muscle to agonists such as norepinephrine (noradrenaline). These data support the view that Na—Ca exchange indeed plays an important role in the regulation of intracellular Ca in vascular smooth muscle.

In smooth muscle (Table 21.5), as in other types of muscle, the immediate trigger for contraction is appropriate elevation of the free ionized Ca concentration in the cytoplasm.

However, vascular smooth muscle is known to maintain constant tension or tone and two mechanisms can account for this:

(1) The steady state free Ca concentration may be maintained above contraction threshold so that the muscle fibres are always partially contracted.
(2) There may be spontaneous asynchronous activation of the smooth muscle fibres so that at any given moment some of the fibres are partially or completely contracted.

With either mechanism the increased cytoplasmic Ca that would result from the Na pump inhibition and the rise in cell Na would be expected to increase vascular smooth muscle tone and contractility. As mentioned, the increased tone is manifested as an increase in peripheral resistance and observed as an elevation of blood pressure. Vascular smooth muscle tension is a graded function of the Ca concentration. Therefore, if the Ca concentration is constantly maintained above the threshold any increase in the Ca level will be immediately translated into an increase in tension. If the second explanation for tone is correct then the Ca that enters the cytoplasm from the extracellular fluid and/or intracellular stores during depolarization would be superimposed on an elevated baseline. Moreover, if the stores are more saturated than normal more Ca will be released from the intracellular stores in response to a given stimulus. The net result will be a higher free Ca level and therefore greater tension than normal when the smooth muscle fibres are activated under these circumstances.

Agents such as verapamil that block Ca entry into vascular smooth muscle are effective in treating hypertension. This may be additional evidence that

Table 21.5 Hypertension-sodium—calcium (Na—Ca) exchange and vascular smooth muscle tone

---

Vascular smooth muscle possesses sodium—calcium (Na—Ca) exchange transport system

Therefore: Rise in cytoplasmic steady-state sodium concentration produces concomitant rise in cytoplasmic calcium concentration together with increased storage of calcium in sarcoplasmic reticulum

---

Ca plays a central role in the genesis of essential hypertension.

Some queries must be elucidated in the future. Work must still be done to: prove a specific inherited effect in renal tubular handling of Na; identify and isolate the natriuretic hormone (recent results have indicated that the hormone exists and is elevated in the plasma of patients with essential hypertension); prove that natriuretic hormone inhibits Na pumps including those in vascular smooth muscle cells; find more direct evidence that the Na—Ca exchange mechanism functions in vascular smooth muscle cells. Answers to all of these questions, however, would not explain why the normal circulatory reflex control mechanism no longer maintain blood pressure in normal range when the vascular smooth muscle cell tone is increased. Obviously the control mechanisms are reset to a higher dynamic range of blood pressure and this may occur when the primary increase in peripheral vascular tone due to a Ca gain by the cells prevents adequate reflex relaxation of the arterial smooth muscle by reduction of sympathetic discharge.

References

Blaustein, M.P. (1977). Sodium ions, blood pressure regulation and hypertension: a reassessment and a hypothesis. *Am. J. Physiol.* **232**, C165—73.

Brody, M.J., Fink, G.D., Buggy, J. et al. (1978). The role of the anteroventral third ventricle (AV3V) region in experimental hypertension. *Circ. Res.* **43** *(suppl. I)*, 12—3.

DeMendonca, M., Grichois, M.L., Garay, R.P. et al (1980). Abnormal net Na^+ and K^+ fluxes in erythrocytes of three varieties of genetically hypertensive rats. *Proc. Nat. Acad. Sci. USA* **77**, 4283—6.

Friedman, S.M., Nakashima, M., McIndoe, R.A. et al. (1976). Increased erythrocyte permeability to Li and Na in the spontaneously hypertensive rat. *Experientia* **32**, 476—8.

Garay, R.P., Dagher, G., Pernollet, M.G., Devynck, M.A. and Meyer, P. (1980). Inherited defect in a Na^+-K^+ co-transport system in erythrocytes from essential hypertensive patients. *Nature* **284**, 281—3.

Horackova, M. and Vassort, G. (1979). Na—Ca exchange in regulation of cardiac contractility. *J. Gen. Physiol.* **73**, 403—24.

MacGregor, G.A., Fenton, S., Alaghband-Zadeh, J. et al. (1981). Evidence for a raised concentration of a circulating sodium transport inhibitor in essential hypertension. *Br. Med. J.* **282**, 1267—9.

Potnov, Y.V., Orlov, S.N. and Poludin, N. (1979). Decrease of calcium binding in the red blood cell membrane in spontaneously hypertensive rats and in essential hypertension. *Pflügers Arch.* **379**, 181—95.

Walter, U. and Distler, A. (1980). Effects of ouabain and furosemide on ATPase activity and sodium transport in erythrocytes of normotensives and of patients with essential hypertension. In *Intracellular electrolytes and hypertension,* H. Zumkley and H. Losse (eds.), pp. 170—81. Thieme, Stuttgart.

INTRACELLULAR CALCIUM AND CARDIAC ARRHYTHMIAS

U. Borchard, D. Hafner, B. Henning and C. Hirth

Institute of Pharmacology, University of Düsseldorf, West Germany

Abstract

The various types of cardiac arrhythmias may be explained by the following pathophysiological mechanisms based on disturbances of either excitation and/or conduction:

(1) *Automaticity of ectopic pacemakers* requiring phase 4 depolarization of action potentials. Increase in $[Ca^{2+}]_i$ (for example, after intracellular injection of Ca^{2+}) enhances K^+ permeability, pacemaker current and slows phase 4 depolarization and beat frequency. Intracellular injection of Ca^{2+} chelating agents (EGTA) or application of slow channel inhibitors depress or abolish pacemaker current.
(2) *Depolarizing afterpotentials* leading to sustained rhythmic activity, for example triggered activity (brought about by hypoxia, elevated $[Ca^{2+}]_o$, removal of $[K^+]_o$, cardiac glycosides, catecholamines or by the toxins veratrine and aconitine), result from abnormal transient inward current, which appears to be initiated by a rise in $[Ca^{2+}]_i$, followed by an increase in electrogenic Na^+, Ca^{2+} exchange and/or activation of a non-specific cation conductance. This mechanism might be involved in fibrillation.
(3) *Decremental conduction* inducing circus movement of excitation (re-entry) or reflection. Slow responses, which predominantly occur in the ischaemic myocardium, favour decremental conduction, unidirectional block and re-entry. The antiarrhythmic action of slow channel inhibitors like diltiazem may be explained with respect to their rate-dependent and potential-dependent action on cardiac Ca^{2+} channels, as shown by experiments on slow responses.

22.1. Introduction

Pathophysiological conditions leading to cardiac arrhythmias in many cases involve an increase in intracellular calcium. It is of particular clinical significance that reduction of oxygen supply to the working myocardium, which occurs during ischaemic attacks or after myocardial infarction, effects a rapid decrease in energy-rich phosphates (Janse et al., 1979; Murphy et al., 1983). As a consequence, ATP-dependent processes carrying Ca^{2+} and/or Na^+ out of the cell are diminished: decreased activity of the Na^+, K^+ pump is followed by an increase in intracellular Na^+ and, via the electrogenic Na^+, Ca^{2+}-exchange mechanism, an elevation of $[Ca^{2+}]_i$ takes place (Barry and Smith, 1984; Mullins, 1979; Reuter and Seitz, 1968). In addition, occlusion of coronary arteries as well as development of pain and anxiety increase sympathetic activity (Heusch and Deussen, 1983; Meesman et al., 1979) so that a further enhancement in $[Ca^{2+}]_i$ is brought about by endogenous catecholamines, which stimulate cardiac β-adrenoceptors and enhance transmembrane Ca^{2+}-current through voltage sensitive Ca^{2+}-channels (Reuter, 1984).

22.2. Mechanisms leading to an increase in $[Ca^{2+}]_i$

Intracellular Ca^{2+} homeostasis is maintained by passive and active Ca^{2+} transport into or out of the cell and by sequestration within the cell, that is, uptake into intracellular stores as shown in Fig.22.1. In the past, many investigations have been carried out which demonstrate the arrhythmogenic action of those conditions which lead to an increase in $[Ca^{2+}]_i$. These are summarized in Table 22.1. Elevated $[Ca^{2+}]_i$ may be due to an increased influx through Ca^{2+}-channels of the sarcolemma, which is either induced or independent of a preceding rise in intracellular cAMP. Stimulation of different cardiac receptors and/or inhibition of phosphodiesterase, for example, by xanthine-derivatives enhance $[Ca^{2+}]_i$. The role of epinephrine (adrenaline) or other β-sympathomimetic agents in producing cardiac arrhythmias, especially in the presence of high $[Ca^{2+}]_o$, is well known (Szekeres and Papp, 1971).

It is now generally accepted that catecholamine-induced arrhythmias play an important role during physical and emotional stress as well as directly after coronary artery occlusion. With respect to the differentiation between

Table 22.1 Mechanisms increasing $[Ca^{2+}]_i$

---

(1) Ca^{2+}-influx ↑ via $[cAMP]_i$ ↑:
   (a) receptor stimulation: β_1, β_2, H_2
   (b) inhibition of phosphodiesterase: IBMX, amrinone

(2) Ca^{2+}-influx ↑, $[cAMP]_i$-independent:
   (a) receptor stimulation: α, H_1, 5-HT
   (b) $[Ca^{2+}]_o$ ↑, stimulation rate ↑, depolarizing current

(3) Na^+, Ca^{2+}-exchange: $[Na^+]_o$ ↓ or $[Na^+]_i$ ↑:
   (a) veratrine, aconitine
   (b) Na^+, K^+-pump ↓: cardiac glycosides, BIIA, hypoxia
       $[K^+]_o$ ↓, temperature ↓

(4) Ca^{2+}-pump ↓: cardiac glycosides

(5) Ca^{2+}-stores ↓: (a) SR: caffeine, $[Ca^{2+}]_i$
                        (b) mitochondrial: DNP, CN^-

---

myocardial β_1- and β_2- receptors (Waelbroeck et al., 1983) the question arises as to whether stimulation of β_2-receptors might be still more arrhythmogenic because of a greater elevation of [cAMP]$_i$ and hence [Ca^{2+}]$_i$. But recent investigations on the receptor-related acidosis during ischaemia are in favour of a predominant role of β_1-receptors, as only β_1- and not β_2-specific antagonists were able to prevent acidosis to a great extent (Sakai and Abiko, 1984). Other mechanisms, leading to an increase in cellular cAMP (as for example, stimulation of H$_2$-receptors in the sinoatrial node, the atrioventricular node, and ventricular myocardium by histamine or H$_2$-agonists (Borchard, Hafner and Hirth, 1984) may also induce arrhythmias, especially in the presence of phosphodiesterase inhibitors. The underlying mechanism of action is based on the phosphorylation of membrane proteins which leads to an increase in the probability of Ca^{2+} channels being in their open state (Reuter, 1984), thus effecting an elevation in [Ca^{2+}]$_i$. Furthermore, [Ca^{2+}]$_i$ may be enhanced by increasing Ca^{2+} influx on application of high external Ca^{2+} concentrations, high stimulation rate or strong depolarizing current which has been shown to be arrhythmogenic (Antoni, Töppler and Krause, 1970; Borchard, Bösken and Greeff, 1982; Imanishi and Surawicz, 1976).

On the other hand, inhibition of Ca^{2+} extrusion via the Na$^+$, Ca^{2+} exchange mechanism is able to raise [Ca^{2+}]$_i$ and precipitate arrhythmias (Clusin et al., 1982; Cranefield, 1977). This occurs by a decrease in [Na$^+$]$_o$ or an increase in [Na$^+$]$_i$ which may be induced by the toxins aconitine or veratrine. Elevated Na$^+$, Ca^{2+} exchange is also involved in enhancing [Ca^{2+}]$_i$ due to inhibition of

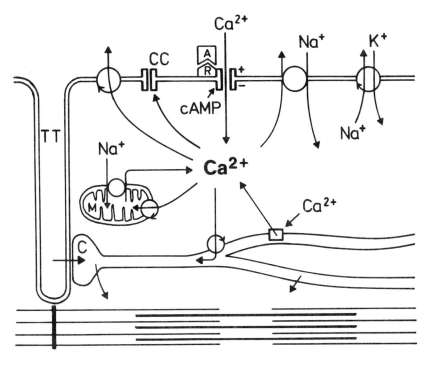

Fig. 22.1. Homeostasis of [Ca^{2+}]$_i$ in relation to transmembrane transport mechanisms and intracellular sequestration. C = terminal cisternae of the sarcoplasmic reticulum; M = mitochondrion; TT = transverse tubuli; CC = Ca^{2+}-dependent cation channel; A = agonist; R = receptor.

the Na^+, K^+-pump. Experimental conditions known to inhibit the Na^+, K^+-ATPase are a decrease in $[K^+]_o$, hypoxia or semi-competitive inhibition of the enzyme by cardiac glycosides with K^+ (Nayler, Poole Wilson and Williams, 1979; Orchard, Eisner and Allen, 1983; Vassalle and Mugelli, 1981).

Inhibition of this enzyme competitive with Na^+-binding has also been shown to be possible using the isoquinoline derivative BIIA (Borchard, Fox and Greeff, 1980). High concentrations of these inhibitors are able to provoke arrhythmias in all anatomical structures of the heart. A controversial discussion is currently taking place as to whether two binding sites exist for cardiac glycosides (Erdmann, Brown and Werdan, 1984), a higher affinity site which could be involved in the desired positive inotropic action and a low affinity site responsible for the arrhythmogenic side effects (Weingart, 1981), which are of clinical significance as they often limit the indication of these drugs. A further action of cardiac glycosides leading to an increase in $[Ca^{2+}]_i$ should be discussed. It concerns inhibition of a transsarcolemmal Ca^{2+} pump which has been derived from measurements on the activity of the Ca^{2+}, Mg^{2+}-ATPase of cardiac membrane preparations (Caroni and Carafoli, 1980; Will, Schirpke and Wollenberger, 1976). This enzyme is supposed to be stimulated by low concentrations but inhibited by high concentrations of cardiac glycosides (Lüllman, Peters and Preuner, 1982). Besides these mechanisms related to a transmembrane Ca^{2+} influx, enhancement of $[Ca^{2+}]_i$ may also be evoked by Ca^{2+} release from intracellular Ca^{2+} stores. The sarcoplasmic reticulum (SR) plays an important part in the regulation of $[Ca^{2+}]_i$ (see Fig. 22.1) because release of Ca^{2+} is increased if $[Ca^{2+}]_i$ is elevated (Fabiato, 1983). This phenomenon is of great importance with respect to cardiac arrhythmias as depolarizing afterpotentials and induction of rhythmic activity of cardiac cells in most pathophysiological conditions seem to be linked to the function of the SR, as will be discussed below. Still another intracellular store, the mitochondria, should also be taken into consideration. Metabolic poisons like DNP or cyanide effect a release of calcium from the mitochondria and are able to produce arrhythmias (Clusin, 1983). It should be mentioned that the role of increased $[Ca^{2+}]_i$ as an arrhythmogenic principle becomes even more pronounced if different pathophysiological mechanisms enhancing $[Ca^{2+}]_i$ occur simultaneously.

22.3. Electrophysiological actions of elevated $[Ca^{2+}]$

If $[Ca^{2+}]_i$ is enhanced due to the conditions described above (see Table 22.1) it may induce changes in transmembrane ionic currents as summarized in Table 22.2. Their significance with respect to cardiac arrhythmias will be discussed later. Furthermore, increase in $[Ca^{2+}]_i$ alters the passive electrical properties of the myocardium as shown by Ando et al. (1981) by the effects of ouabain on isolated rabbit atrial muscle. The authors observed a reduction in the space constant due to an increase in the axial resistance (electrical uncoupling) whereas the time constant remained unaffected. As a result a decrease in junctional conductance may be expected. Such an electrical uncoupling could inhibit local circuit current and interrupt impulse propagation. If $[Ca^{2+}]_i$ is raised to a great extent, a depletion of energy-rich phosphates is observed which, in turn, leads to an inhibition of the ion pumps. Intracellular enzyme release may destroy cellular membranes and induce irreversible damage (Janse et al., 1979; Murphy et al., 1983).

Intracellular calcium and cardiac arrhythmias

Table 22.2 Actions of elevated $[Ca^{2+}]_i$

---

(1) Ca^{2+}-release from SR: oscillations of $[Ca^{2+}]_i$
(2) Transient inward current: afterdepolarizations
(3) Electrogenic Na^+, Ca^{2+} exchange
(4) Opening of voltage-independent cation channels (patch clamp)
(5) Increase in outward current (Purkinje fibers): Effective refractory period ↓, inhomogeneity of repolarization (re-entry)
(6) Cycle length of automatic activity
(7) Increase in axial resistance: electric uncoupling
(8) ATP-depletion, enzyme release, cell injury

22.4. Arrhythmogenic actions of elevated $[Ca^{2+}]_i$

Three basic pathophysiological mechanisms in either excitation or conduction have turned out to be responsible for the induction of rhythm disorders: automaticity of ectopic pacemakers, depolarizing afterpotentials leading to sustained rhythmic activity (such as triggered activity) and decremental conduction facilitating circus movement of excitation (re-entry). These mechanism will be discussed with respect to the increase in $[Ca^{2+}]_i$.

22.4.1. Impulse generation

22.4.1.1. Automaticity of ectopic pacemakers. Partially depolarized cells of the working myocardium are able to develop spontaneous activity, which requires phase 4 depolarization of action potentials. As most of the conditions leading to automaticity of ectopic pacemakers may also induce delayed afterdepolarizations it is often difficult to attribute spontaneous activity to either gradual diastolic depolarization or afterdepolarization unless the latter phenomenon becomes visible. Figure 22.2. shows an example of both types of diastolic depolarization in human artrial fibers (Rosen and Hordof, 1980). The slope of phase 4 depolarization is enhanced by stimulation of β-adrenoceptors, elevation of $[Ca^{2+}]_o$ and depressed by acetylcholine, verapamil, lidocaine and TTX, but not altered by procaine amide (Rosen and Hordof, 1980), whereas delayed afterdepolarizations are enhanced by catecholamines (Mary-Rabine, Hordof, Danilo, Malm and Rosen, 1980) or toxic concentrations of digitalis (Wit, Rosen and Hoffman, 1974) and depressed by acetylcholine, verapamil (Wit, Rosen and Hoffman, 1974; Surawicz, 1980), manganese or high concentrations of caffeine (Allen, Eisner and Orchard, 1984; Karagueuzian and Katzung, 1982; Wier, Kort, Stern, Lakatta and Marban, 1983). This latter type of arrhythmia will be dealt with in section 22.4.1.2. One may conclude from the effects of the above mentioned drugs that the slope of diastolic depolarization in ectopic pacemakers at least partially depends on current through voltage sensitive Ca^{2+}-channels. Hyperpolarization of the membrane or decrease in slow inward current increase cycle length and conditions which lead to an increase in Ca^{2+}-influx decrease cycle length and enhance automaticity.

The role of elevated intracellular calcium with respect to the slope of diastolic depolarization in ectopic pacemakers has not yet been elucidated, as no experiments have so far been carried out in ectopic cells of the adult working myocardium, in which intracellular calcium has been increased prior to the change in transmembrane current, as e.g. by intracellular injection of Ca^{2+}. However, information on pacemaker activity may be obtained from the results of Trautwein and coworkers (Trautwein, Taniguchi and Noma, 1982; Fig.22.3) from clusters of AV nodal cells in which intracellular calcium was increased by injection through a micropipette or from experiments in which Ca^{2+} was applied to spontaneously beating cultured heart cells via Ca^{2+}-filled liposomes (Bkaily et al., 1983). In atrioventricular nodal cells there was no significant change in action potential duration, but a decrease in upstroke velocity, amplitude and slope of diastolic depolarization which might be explained by a decrease in driving force for the calcium inward current due to a reduction in the electrochemical gradient across the cell membrane. One may conclude from these results that cycle length in nomotopic or ectopic pacemakers is increased by $[Ca^{2+}]_i$ if the calcium inward current contributes to phase 4 depolarization. Similar alterations of the action potential and diastolic depolarizaton have been obtained by Bkaily et al. (1983) who raised $[Ca^{2+}]_i$ and $[Na^+]_i$ in cultured heart cells (reaggregates) prepared from

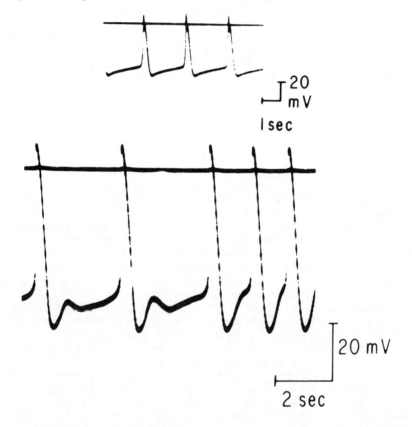

Fig.22.2. Action potential of human atrial fibres. Upper panel: gradual diastolic depolarization leads to impulse initiation. Lower panel: delayed afterdepolarizations are superimposed upon gradual diastolic depolarization. The third afterdepolarization reaches threshold and rate of spontaneous activity is increased. (From Rosen and Hordof, 1980.)

15-day-old embryonic chick ventricles.

As an increase in $[Na^+]_i$ also decreases the slope of phase 4 depolarization and firing frequency so diminution of the electrochemical gradient of both Ca^{2+} and Na^+ reduces inward background current during diastole. Enhanced depolarization upon elevation of $[Ca^{2+}]_i$ has been observed in atrioventricular nodal cells (Trautwein, Taniguchi and Noma, 1982) and in chick embryonic heart cells (Bkaily et al., 1983). It has been argued that this depolarization could be induced by the Ca^{2+}-activation of a non-specific Na^+, K^+-channel.

Isenberg (1977a and b) has injected Ca^{2+} into cardiac Purkinje fibers and observed a shift of the pacemaker potential in the hyperpolarizing direction and a shortening of the action potential. He attributed these effects to an increase in K^+ outward current. However, Trautwein, Taniguchi and Noma (1982) and Bkaily et al. (1983) did not observe a shift of the pacemaker potential after injection of Ca^{2+} into the atrioventricular nodal cells or cultured heart cells. Therefore, it seems to be unlikely that calcium controlled K^+ conductance plays an important role at least in these tissues.

Shortening of the action potential was also observed by Trautwein, Taniguchi and Noma (1982) after injection of Ca^{2+} into ventricular cells. The opposite effect occurred after intracellular application of EDTA. The

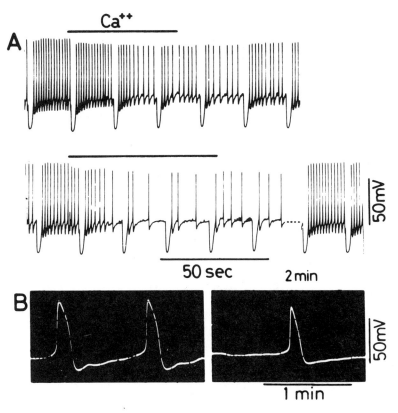

Fig.22.3. Pressure injection of Ca^{2+} into an atrioventricular-nodal cell belonging to a cluster isolated enzymatically from a rabbit atrioventricular node. Action potential registration at low speed in panel A and high speed in panel B. Elevation of $[Ca^{2+}]_i$ effects a decrease in maximal diastolic potential, amplitude of the action potential and a decrease in slope of diastolic depolarization and thereby an increase in cycle length. (From Trautwein, Taniguchi and Noma, 1982.)

significance of these observations will be discussed in section 22.4.2 in connection with the facilitation of re-entry mechanisms.

22.4.1.2. *Depolarizing afterpotentials.* Elevated intracellular calcium plays an important role in the initiation of depolarizing afterpotentials which are accompanied by aftercontractions and may lead to sustained rhythmic activity.

In 1962 Reiter described aftercontractions upon exposure of guinea-pig papillary muscles to toxic concentrations of cardiac glycosides. Lederer and Tsien (1976) carried out voltage clamp experiments on strophanthidin-treated cardiac Purkinje fibres in which they showed that afterdepolarizations are generated by the transient inward current. This is enhanced if $[Ca^{2+}]_o$ is increased, abolished if the bathing solution contains no calcium, and decreased on elevation of $[Mg^{2+}]_o$ (Fig.22.4).

Kass *et al.* (1978) measured the reversal potential of the transient inward current in an attempt to clarify the underlying ionic mechanism. It was found to be around -5 mV indicating the absence of selectivity for any of the common cations. The nature of the transient inward current in still obscure. At present, two mechanisms have to be discussed:

(1) A large increase in intracellular calcium is followed by an increase in the electrogenic Na^+, Ca^{2+}-exchange which transports 1 Ca^{2+} out of the cell and 3 Na^+ into the cell so that there is a net depolarizing inward current (Clusin, Fischmeister and DeHaan, 1983; Coraboeuf, Gautier and Guiraudou, 1981; Sheu and Fozzard, 1982). The abolition of the transient

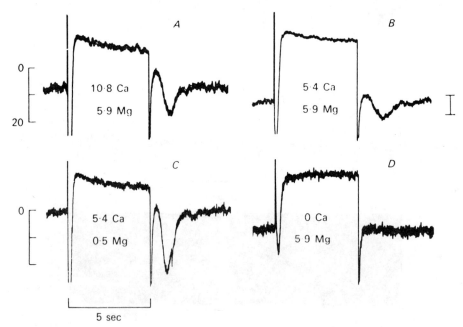

Fig.22.4. Decrease in transient inward current by reduction of $[Ca^{2+}]_o$ (A,B,D) or elevation of $[Mg^{2+}]_o$ (A,C) in a calf Purkinje fibre exposed to 1 μmol/l strophanthidin during the application of 5 s voltage-clamp pulses from -69 to -32 mV. (From Kass, Lederer, Tsien and Weingart, 1978).

inward current by replacing of Na^+ in the bathing solution by Tris or sucrose (Karagueuzian and Katzung, 1982) would be in agreement with an electrogenic Na^+, Ca^{2+}-exchange mechanism. However, no direct experimental evidence for this hypothesis has so far been presented.

(2) Increase in intracellular calcium opens Ca^{2+}-sensitive cation channels which are not appreciably affected by the membrane potential and non-selective for Na^+ or K^+. These channels have been detected the first time by Colquhoun and coworkers (1981) in patch clamp experiments on myocytes obtained from neonatal rat hearts. These currents have many features in common with the transient inward currents described by Kass et al. (1978) in strophanthidin-treated cardiac Purkinje fibres, with a reversal potential of -5 mV, also indicating poor selectivity for either Na^+ or K^+ ions. Replacement of extracellular sodium by choline or Tris or substitution of NaCl by sucrose induced a shift of the reversal potential by maximal -35 mV and there was no significant change in the reversal potential if extracellular calcium was varied from 2.7 to 16.2 mmol/l or extracellular potassium from 1 to 8 mmol/l. At present it is a matter of debate whether both mechanisms described above might be involved in the generation of the transient inward current.

The spontaneous rate of Purkinje fibres exposed to toxic concentrations of ouabain may be increased if the rate of a preceding overdrive suppression is enhanced (Rosen and Danilo, 1980).

Different mechanisms may be responsible for this effect: rapid stimulation increases intracellular calcium which has been shown to increase amplitude

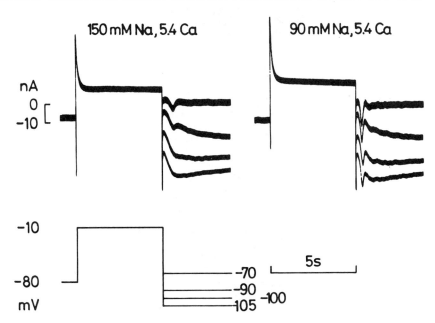

Fig.22.5. Voltage clamp steps from -80 to -10 mV and back to voltages ranging from -70 to -105 mV in canine Purkinje fibre. Upper panel, left: current traces show transient inward current which occurs during the clamp step back to -70 mV and disappears when returning to -100 mV, whereas pacemaker current is enhanced when repolarizing clamp steps are increased from -70 to -105 mV. Upper panel, right: decrease in $[Na^+]_o$ effects an increase in transient inward current, but pacemaker current is almost unaffected. (From Henning et al., unpublished observations.)

and frequency of afterdepolarizations so that threshold is reached more quickly. Increase in intracellular Ca^{2+} enhances the rate of intracellular Ca^{2+} oscillations, as shown by Allen, Eisner and Orchard (1984). Furthermore, rapid stimulation also leads to depolarization of the membrane which induces an increase in the amplitude (Ferrier, 1980; Henning and Wit, 1981) and rate (Kass and Tsien, 1982) of afterdepolarizations and brings membrane potential nearer to threshold. These phenomena are of clinical importance with respect to the spontaneous frequency of abnormal automaticity. The voltage dependency of the transient inward current, superimposed on the pacemaker current, is demonstrated by voltage clamp experiments in a canine Purkinje fibre which are shown in Fig.22.5. Depolarization or decrease in $[Na^+]_o$ enhance transient inward current, whereas pacemaker current is activated at more negative potentials and virtually unaffected by the decrease in $[Na^+]_o$.

Rate of triggered activity is linked to the Na^+, K^+ pump in a complex way (Kline et al., 1985). Bursts of triggered activity in atrial fibres of the canine coronary sinus exposed to catecholamines show an initial acceleration followed by slowing of the spontaneous rate and termination in parallel to

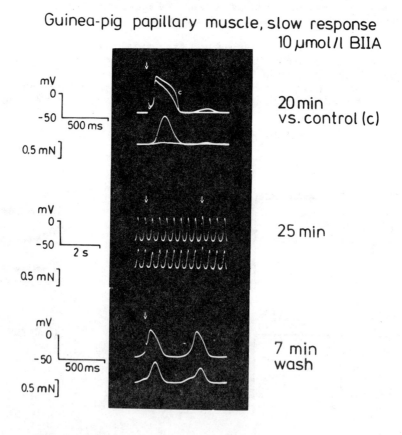

Fig.22.6. Upper trace: application of 10 μmol/l BIIA (isoquinoline derivative) leads to a decrease in amplitude and duration of the slow response (27 mmol/l $[K^+]_o$, 0.5 mmol/l $[Ba^{2+}]_o$) and large increase in force of isometric contraction (second trace). The afterdepolarization is paralleled by an aftercontraction. Middle trace: 25 min after application of BIIA the preparation beats spontaneously. Lower trace: 7 min after washout of BIIA cycle length of spontaneous activity increases slightly. (From Borchard et al., unpublished observations.)

hyperpolarization. Initial acceleration is accompanied by an increase in $[K^+]_o$ and depolarization, whereas subsequent decrease in spontaneous rate is brought about by a decline in $[K^+]_o$ as well as repolarization due to the enhanced Na^+, K^+ pump activation.

Various conditions leading to an increase in $[Ca^{2+}]_i$ may induce spontaneous activity in the working myocardium (see Table 22.1). Inhibition of the Na^+, K^+-ATPase does not only occur at the K^+ binding side by cardiac glycosides but also at the Na^+-binding side by the isoquinoline derivative BIIA. Therefore, it is not surprising that this substance is able to induce rhythmic activity in depolarized ventricular myocardium (Fig.22.6).

Another example of the induction of sustained rhythmic activity in depolarized ventricular myocardium is shown in Fig.22.7. Increase in slow inward current by 6 μmol/l isobutylmethylxanthine (IBMX) indicated by the increase in V_{max} of the slow response, raises contractile force and induces small subthreshold afterdepolarizations. Increase of IBMX to 10 μmol/l effects rhythmic depolarizations, which occur parallel to rhythmic oscillations of the contractile force.

It has been demonstrated in the last few years that afterdepolarizations, transient inward current and aftercontractions are due to oscillations of the intracellular Ca^{2+} concentration (Fig.22.8) which occur in close connection to the sarcoplasmic reticulum (Allen, Eisner and Orchard, 1984; Eisner, Orchard and Allen, 1984; Orchard, Eisner and Allen, 1983; Vassalle and Di Gennaro, 1983; Wier et al., 1983). Elevation of $[Ca^{2+}]_o$ increases amplitude and rate of the oscillations of $[Ca^{2+}]_i$. If Ca^{2+}-mediated Ca^{2+}-release from the sarcoplasmic reticulum is attenuated by caffeine or ryanodine, the oscillations are abolished (Allen, Eisner and Orchard, 1984; Hiraoka, Okamoto and Sano, 1979; Karagueuzian and Katzung, 1982; Kort and Lakatta, 1984; Wier et al., 1983).

In myocardial cell aggregates severely intoxicated with veratrine or strophanthin, spontaneous beating is replaced by a completely disorganized

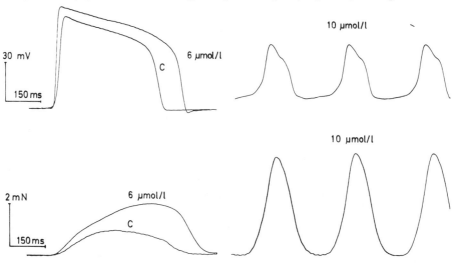

Fig.22.7. Upper trace: the phosphodiesterase inhibitor IBMX (isobutylmethylxanthine) increases amplitude, upstroke velocity and duration of the slow response (27 mmol/l $[K^+]_o$, 0.5 mmol/l $[Ba^{2+}]_o$) in a guinea-pig papillary muscle. At 10 μmol/l spontaneous activity is initiated. Lower trace: force of isometric contraction. (From Borchard et al., unpublished observations.)

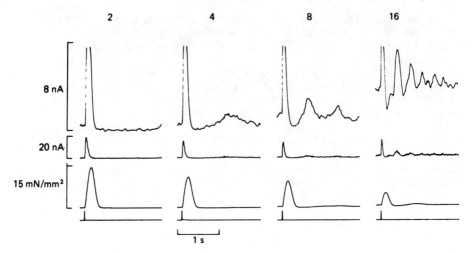

Fig.22.8. Dependency of stimulated $[Ca^{2+}]_i$ oscillations on $[Ca^{2+}]_o$ in ferret ventricular muscle exposed to 10 µmol/l strophanthidin and stimulated at 0.33 Hz. From top to bottom: numbers of $[Ca^{2+}]_o$ in mmol/l, aequorin light signal (high and low gain), tension, stimulus marker. (From Allen, Eisner and Orchard, 1984.)

movement which resembles that of the intact heart during experimentally induced fibrillation (Clusin, 1981). It cannot arise from re-entrant conductance as the aggregates are too small for significant conduction delays to occur as shown by DeHaan and Fozzard (1975). Fibrillatory activity arising from the transient inward current mechanism could be the result of spatial and temporal inhomogeneities in the Ca^{2+}-release process thus leading to asynchronous inward current in discrete regions throughout the tissue. This hypothesis is attractive in so far as experimental conditions that produce transient inward current also precipitate ventricular fibrillation in intact animals. Furthermore, Ca^{2+}-antagonists which protect the myocardium from Ca^{2+} overload are able to suppress oscillatory afterpotentials as well as ventricular fibrillation following acute coronary ligation (Clusin et al., 1984). It has been pointed out by Schechter, Freeman and Lazzara (1984) that afterdepolarizations are likely to occur also in patients, especially after infusion of epinephrine or during emotional stress (Fig.22.9). These afterpotentials are strikingly large in patients with the long QT-syndrome (Schechter, Freeman and Lazzara, 1984).

22.4.2. Impulse conduction

The third arrhythmogenic principle which has to be considered, with respect to increase in intracellular calcium, is circus movement of excitation.

The essential conditions which facilitate circus movement are unidirectional block resulting in part from inhomogeneity of repolarization, shortening of the refractory period and slow impulse propagation. As has been demonstrated by the experiments of Trautwein, Taniguchi and Noma (1982) and Isenberg (1977a), elevation of intracellular calcium concentration effects a shortening of the action potential and therefore a shortening of the refractory period in ventricular myocardium and Purkinje fibres. If this shortening occurs in separate areas of the heart, inhomogeneity of repolarization may be induced. If at the same time membrane potential is decreased, for example, by toxic concentrations of cardiac glycosides or ischaemia, ideal conditions for the initiation of circus movement are fulfilled. In addition, increase in intracellular Ca^{2+} may lead to an increase in axial resistance and electrical uncoupling, as shown in cardiac Purkinje fibres by De Mello (1975) after intracellular injection of Ca^{2+}, and by Weingart (1981) and Ando *et al.* (1981) after application of high concentrations of cardiac glycosides to Purkinje fibres or strips of right atrial muscle.

Decremental conduction in limited regions of atrial or ventricular myocardium may cause a circus movement of excitation involving regions of the heart with different conduction velocity. Although the investigation of such re-entry or reflection processes is hampered by their spatial and temporal complexity, it is generally agreed that calcium-dependent slow action potentials play an important role in these types of arrhythmias. Increasing degrees of hypoxia which are caused by ischaemia of the myocardium lead to an increasing inactivation of the fast inward current, while the slow inward current is not markedly affected within the potential range of -80 to -50 mV. Consequently, the transition between fast and slow action potentials is gradual, and under certain conditions the fast and the slow component can be dissociated during the upstroke of the action potential.

It might be argued that 'pure' slow responses initiated from a resting potential of -40 mV occur rather rarely since the depolarization during the majority of ischaemic episodes is less substantial. However, the contribution of the slow inward current to the upstroke of the action potential in

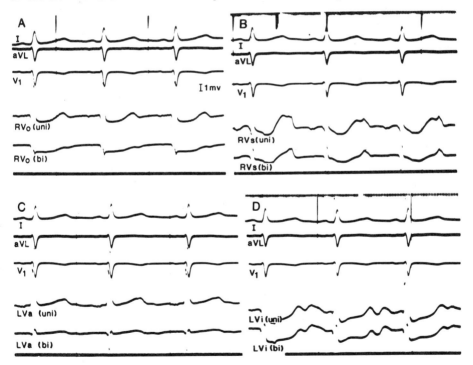

Fig.22.9. Surface electrocardiographic leads I, aVL and V_1 from a 23-year-old woman. Unipolar (uni) and bipolar (bi) endocardial recordings were taken from the right ventricular outflow tract (RV_o), right ventricular septum (RV_s), left ventricular apex (LV_a) and left ventricular inflow tract (LV_i). Diastolic afterdepolarizations occur in the right ventricular septum and left ventricular inflow tract. (From Schechter, Freeman and Lazzara, 1984.)

depolarized myocardium is enhanced by catecholamines: they shift the inactivation curve of the fast Na^+ system to more negative potentials and increase the slow inward current (Kiyosue and Arita, 1982; Windisch and Tritthart, 1982). This mechanism might be of importance in ischaemic attacks during which both depolarization and increased catecholamine-release may occur.

22.5. Calcium channel inhibitors

Inhibition of Ca^{2+}-entry through slow channels should be useful to prevent Ca^{2+} overload of the myocardial cells as well as re-entry arrhythmias by blocking the pathway of decremental conduction. Slow responses are very effectively diminished by slow channel blockers: 3×10^{-8} mol/l nifedipine or 10^{-6} mol/l diltiazem are sufficient to reduce V_{max} by 50% in human myocardium (Borchard et al., unpublished data) and to induce a substantial shortening of the slow response. This high sensitivity is most likely due to the facilitation of the drug-induced slow channel blockade by depolarization as was shown in voltage clamp experiments by McDonald, Pelzer and Trautwein (1984a). After block was established, I_{si} could be restored by hyperpolarizing prepulses, whereas depolarizing prepulses increased the block of I_{si}. Obviously, opening of the calcium channels is necessary for the blocking action of these drugs, and the dissociation of the drug from the channel is facilitated by repolarization. Thus, the blockade of slow calcium channels by calcium antagonists resembles the action of local anaesthetics on the Na^+ channel (Hondeghem and Katzung, 1984). Another important feature of the

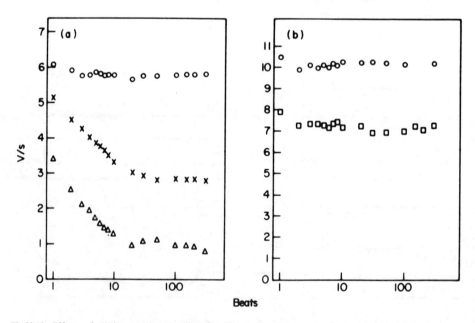

Fig.22.10. Effects of diltiazem (a) and nifedipine (b) on the time-course of the maximum upstroke velocity (V/s) of the slow response in guinea-pig papillary muscle after a stimulation pause of 30 min. Abscissa: number of stimuli (stimulation rate 0.17 Hz) after the end of the pause. ○, control, X, 1 μmol/l and Δ, 3 μmol/l diltiazem, □, 0.01 μmol/l nifedipine. The drugs were added at the beginning of the stimulation pause. (From Hirth, Borchard and Hafner, 1983.)

action of the verapamil-type calcium antagonists is their frequency dependence ('use-dependent' block) as shown in voltage clamp experiments with D600 (McDonald, Pelzer and Trautwein, 1984b). In contrast, nifedipine within the range of 0.34-0.017 Hz shows no use-dependent effect on V_{max} of slow response in human and guinea-pig myocardium, while the effects of diltiazem are greatly enhanced by increasing the stimulation frequency (Fig.22.10). The use-dependence is closely related to changes in the recovery kinetics of V_{max} of the slow response: 1 μmol/l diltiazem delays the recovery after stimulus-induced inactivation while comparable concentrations of nifedipine do not change the recovery process in guinea-pig papillary muscle

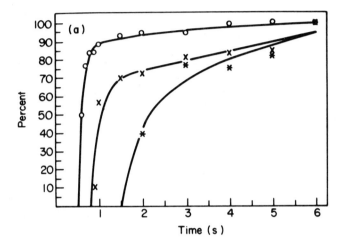

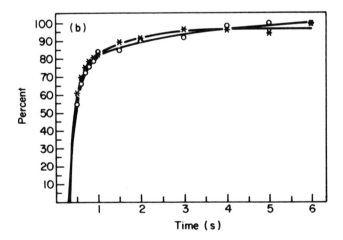

Fig.22.11. Effects of diltiazem (a) and nifedipine (b) on the recovery of $\dot{V}_{max}$ of the slow response from stimulus-induced inactivation in guinea-pig papillary muscle stimulated at a frequency of 0.17 Hz. Extra stimuli were applied at varying intervals from the last regular stimulus. Ordinate: $\dot{V}_{max}$ of the slow response elicited by the extra stimulus as percentage of $\dot{V}_{max}$ of the regular slow response. Abscissa: stimulus interval between regular stimulus and extra stimulus; curves were fitted to the experimental data using a double exponential equation. (From Hirth, Borchard and Hafner, 1983.)

(Fig.22.11). Patch clamp experiments have revealed that calcium antagonists decrease the probability of the Ca^{2+}-channels being open. Lifetime and frequency of channel openings are reduced, shut times between the shorter bursts of openings are prolonged and the number of traces without channel openings (blanks) is increased (Pelzer, Cavalié and Trautwein, 1984).

The beneficial role of calcium antagonists in preventing ventricular fibrillation has been demonstrated in coronary artery occlusion experiments in dogs (Clusin et al., 1984). Diltiazem reduced diastolic injury potential, which is supposed to be the trigger for automatic activity. Delay in ventricular fibrillation and suppression of early arrhythmias might be attributed to a diminution of ischaemia-induced conduction delay most probably due to increased Na^+-channel availability as a result of reduced depolarization and/or a relative increase in membrane resistance during ischaemia by preventing Ca^{2+} overload of the myocardial cells, thereby decreasing membrane resistance via the closure of Ca^{2+}-activated ion channels (Clusin et al., 1984). However, it is questionable if calcium antagonists show similar beneficial effects in patients with myocardial infarction as their ability to penetrate into the ischaemic zone seems to be limited, and their action on atrioventricular conduction may be disadvantageous.

22.6. Conclusion

Experimental results suggest that an increase in intracellular Ca^{2+} is involved in a slowing of beat frequency in nomotopic or ectopic pacemakers, in facilitating depolarizing afterpotentials which may lead to sustained rhythmic activity, and in favouring circus movement which results from inhomogeneity of repolarization or electrical uncoupling. Ca^{2+} overload of myocardial cells may be prevented by application of calcium channel blockers which effect a potential-dependent and/or use-dependent inhibition of Ca^{2+} influx by reducing the probability of calcium channels being in their open state.

References

Allen, D.G., Eisner, D.A. and Orchard, C.H. (1984). Characterization of oscillations of intracellular calcium concentration in ferret ventricular muscle. *J. Physiol.* 352, 113–28.
Ando, S., Kodama, J., Ikeda, N., Toyama, J. and Yamada, K. (1981). Effects of ouabain on electrical coupling of rabbit atrial muscle fibres. *Jap. Circ. J.* 45, 1050–5.
Antoni, H., Töppler, J. and Krause, H. (1970). Polarizing effects of sinusoidal 50-cycle alternating current on membrane potential of mammalian cardiac fibres. *Pflügers Arch.* 314, 274–91.
Barry, W.H. and Smith, T.W. (1984). Movement of Ca^{2+} across the sarcolemma: effects of abrupt exposure to zero external Na concentration. *J. Mol. Cell. Cardiol.* 16, 155–64.
Bkaily, G., Sperelakis, N., Elishalom, Y. and Barenholz, Y. (1983). Effect of Na^+- or Ca^{2+}-filled liposomes on electrical activity of cultured heart cells. *Am. J. Physiol.* 245, H756–61
Borchard, U., Fox, A.A.L. and Greeff, K. (1980). The positive inotropic, antiarrhythmic and Na^+, K^+-ATPase inhibitory effects of the isoquinoline derivative, BIIA. *Naunyn-Schmiedeberg's Arch. Pharmacol.* 312, 187–92.
Borchard, U., Bösken, R. and Greeff, K. (1982). Characterization of

antiarrhythmic drugs by alternating current induced arrhythmias in isolated heart tissues. *Arch. Int. Pharmacodyn. Ther.* **256**, 253–68.

Borchard, U., Hafner, D. and Hirth, C. (1984). Electrophysiological investigations of cardiac actions of H_1- and H_2-receptor agonists and antagonists. *Naunyn-Schmiedeberg's Arch. Pharmacol. Suppl.* **325**, 185.

Caroni, P. and Carafoli, E. (1980). An ATP-dependent Ca^{2+}-pumping system in dog heart sarcolemma. *Nature (Lond.)*, **283**, 765–7.

Clusin, W.T. (1981). Normal and abnormal automaticity in cultured embryonic cardiac cells. In *Cardiac Arrhythmias; A Decade of Progress*. Harrison (ed), pp. 69–85. G.K. Hall Medical Publishers, Boston.

Clusin, W.T., Bristow, M.R., Karagueuzian, H.S., Katzung, B.G. and Schroeder, J.S. (1982). Do calcium dependent ionic currents mediate ischemic ventricular fibrillation? *Am. J. Cardiol.* **49**, 606–12.

Clusin, W.T., Fischmeister, R. and DeHaan, R.L. (1983). Caffeine-induced current in embryonic heart cells: time course and voltage dependence. *Heart Circ. Physiol.* **14**, H528–32.

Clusin, W.T. (1983). Mechanism by which metabolic inhibitors depolarize cultured cardiac cells. *Proc. Natl. Acad. Sci. USA* **80**, 3865–9.

Clusin, W.T., Buchbinder, M., Ellis, A.K., Kernoff, R.S., Giacomini, J.C. and Harrison, D.C. (1984). Reduction of ischemic depolarization by the calcium channel blocker diltiazem. *Circ. Res.* **54**, 10–20.

Colquhoun, D., Neher, E., Reuter, H. and Stevens, C.G. (1981). Inward current channels activated by intracellular Ca in cultured cardiac cells. *Nature (Lond.)* **294**, 752–4.

Coraboeuf, E., Gautier, P. and Guiraudou, P. (1981). Potential and tension changes induced by sodium removal in dog Purkinje fibers: role of an electrogenic sodium–calcium exchange. *J. Physiol. (Lond.)* **311**, 605–22.

Cranefield, P.F. (1977). Action potentials, afterpotentials, and arrhythmias. *Circ. Res.* **41**, 415–23.

DeHaan, R.L. and Fozzard, H.A. (1975). Membrane responses to current pulses in spheroidal aggregates of embryonic heart. *J. Gen. Physiol.* **65**, 207–22.

DeMello, W.C. (1975). Effect of intracellular injection of calcium and strontium on cell communication in heart. *J. Physiol.* **250**, 231–45.

Eisner, D.A., Orchard, C.H. and Allen, D.G. (1984). Control of intracellular ionized calcium concentration by sarcolemmal and intracellular mechanisms. *J. Mol. Cell. Cardiol.* **16**, 137–46.

Erdmann, E., Brown, L. and Werdan, K. (1984). Two receptors for cardiac glycosides in the heart. In *Symposium München, 1983: Cardiac Glycoside Receptors and Positive Inotropy*. E. Erdmann (ed.), pp. 21–6. Steinkopff Verlag, Darmstadt.

Fabiato, A. (1983). Calcium-induced release of calcium from the cardiac sarcoplasmic reticulum. *Am. J. Physiol.* **245** (*Cell Physiol.* 14); C1–14.

Ferrier, G.R. (1980). Effects of transmembrane potential on oscillatory afterpotentials induced by acetylstrophanthin in canine ventricular tissues. *J. Pharmacol. Exp. Ther.* **215**, 332–41.

Henning, B. and Wit, A. (1981). Action potential characteristics control afterdepolarization amplitude and triggered activity in canine coronary sinus. *Circulation* **64**, IV–50.

Heusch, G. and Deussen, A. (1983). The effects of cardiac sympathetic nerve stimulation on perfusion of stenotic coronary arteries in the dog. *Circ. Res.* **53**, 8–15.

Hiraoka, M., Okamoto, Y. and Sano, T. (1979). Effects of Ca^+ and K^+ on oscillatory afterpotentials in dog ventricular muscle fibers. *J. Mol. Cell. Cardiol.* **11**, 999–1015.

Hirth, C., Borchard, U. and Hafner, D. (1983). Effects of the calcium antagonist diltiazem on action potentials, slow response and force of contraction in different cardiac tissues. *J. Mol. Cell. Cardiol.* **15**, 799–809.

Hondeghem, L.M. and Katzung, B.G. (1984). Antiarrhythmic agents: the modulated receptor mechanism of action of sodium and calcium channel-blocking drugs. *Ann. Rev. Pharmacol. Toxicol.* **24**, 387–423.

Imanishi, S. and Surawicz, B. (1976). Automatic activity in depolarized guinea-pig ventricular myocardium. *Circ. Res.* **39**, 751–9.

Isenberg, G. (1977a). Cardiac Purkinje fibres. Resting, action, and pacemaker potential under the influence of $[Ca^{2+}]_i$ as modified by intracellular injection techniques. *Pflügers. Arch.* **371**, 51–9.

Isenberg, G. (1977b). Cardiac Purkinje fibres. $[Ca^{2+}]_i$ controls steady state potassium conductance. *Pflügers Arch.* **371**, 71–6.

Janse, J.M., Cinca, J., Morena, H., Fiolet, J.W.T., Kleber, A.G., De Vries, P.G., Becker, A.E. and Durrer, D. (1979). The 'border zone' in myocardial ischemia. An electrophysiological, metabolic, and histochemical correlation in the pig heart. *Circ. Res.* **44**, 576–88.

Karagueuzian, H.S. and Katzung, B.G. (1982). Voltage-clamp studies of transient inward current and mechanical oscillations induced by ouabain in ferret papillary muscle. *J. Physiol.* **327**, 255–71.

Kass, R.S., Lederer, W.J., Tsien, R.W. and Weingart, R. (1978). Role of calcium ions in transient inward currents and aftercontractions induced by strophanthidin in cardiac Purkinje fibers. *J. Physiol.* **281**, 187–208.

Kass, R.S. and Tsien, R.W. (1982). Fluctuations in membrane current driven by intracellular calcium in cardiac Purkinje fibers. *Biophys. J.* **38**, 259–69.

Kiyosue, T. and Arita, M. (1982). Magnesium restores high K-induced inactivation of the fast Na channel in guinea-pig ventricular muscle. *Pflügers Arch.* **395**, 78–80.

Kline, R.P., Siegal, M.S., Henning, B. and Wit, A.L. (1985). Triggered activity in atrial fibers of the canine coronary sinus: the role of extracellular K^+ accumulation and depletion. *Circ. Res.*, in press.

Kort, A.A. and Lakatta, E.G. (1984). Calcium-dependent mechanical oscillations occur spontaneously in unstimulated mammalian cardiac tissues. *Circ. Res.* **54**, 396–404.

Lederer, W.J. and Tsien, R.W. (1976). Transient inward current underlying arrhythmogenic effects of cardiotonic steroids in Purkinje fibers. *J. Physiol. (Lond.)* **263**, 73–100.

Lüllmann, H., Peters, T. and Preuner, J. (1982). Mechanism of action of digitalis glycosides in the light of new experimental observations. *Eur. Heart J.* **3 (Suppl.)**, 45–51.

Mary-Rabine, L., Hordof, A.J., Danilo, P.Jr., Malm, J. and Rosen, M.R. (1980). Mechanisms for impulse initiation in isolated human atrial fibers. *Circ. Res.* **47**, 267–77.

McDonald, T.F., Pelzer, D. and Trautwein, W. (1984). Cat ventricular muscle treated with D600: effects on calcium and potassium currents. *J. Physiol.* **352**, 203–16.

McDonald, T.F., Pelzer, D. and Trautwein, W. (1984). Cat ventricular muscle treated with D600: characteristics of calcium channel block and unblock. *J. Physiol.* **353**, 217–41.

Meesmann, W., Gülker, H., Krämer, B., Stephan, K., Menken, U. and Wiegand, V. (1979). Ventrikuläre Arrhythmien und Vulnerabilität des Herzens in der Frühphase nach akutem experimentellem Koronarverschluss und nach Reperfusion - mögliche Mechanismen. In *3. Wiener Symposion 1978. Herzrhythmusstörungen.* H. Antoni *et al.* (eds.), pp. 27–43. F.K. Schattauer

Verlag, Stuttgart, New York.
Mullins, L.J. (1979). The generation of electric currents in cardiac fibers by Na/Ca exchange. *Am. J. Physiol.* **236**, C103–10.
Murphy, E., Aiton, J.F., Horres, R. and Lieberman, M. (1983). Calcium elevation in cultured heart cells: its role in cell injury. *Am. J. Physiol.* **245**, *(Cell Physiol.* **14)**, C316–21.
Nayler, W.G., Poole Wilson, P.A. and Williams, A. (1979). Hypoxia and calcium. *J. Mol. Cell. Cardiol.* **11**, 683–706.
Orchard, C.H., Eisner, D.A. and Allen, D.G. (1983). Oscillations of intracellular Ca^{2+} in mammalian cardiac muscle. *Nature (Lond.)* **304**, 735–8.
Pelzer, D., Cavalié, A. and Trautwein, W. (1984). Modulation of the gating properties of single calcium channels in cardiac cell membranes by D600. *INSERM Symposia.*
Reiter, M. (1962). Die Entstehung von 'Nachkontraktionen' im Herzmuskel unter Einwirkung von Calcium und von Digitalisglykosiden in Abhängigkeit von der Reizfrequenz. *Naunyn-Schmiedeberg's Arch. Exp. Pathol. Pharmakol.* **242**, 497–507.
Reuter, H. and Seitz, N. (1968). Dependence of calcium efflux from cardiac muscle on temperature and external ion composition. *J. Physiol.* **195**, 451–70.
Reuter, H. (1984). Ion channels in cardiac cell membranes. *Ann. Rev. Physiol.* **46**, 473–84.
Rosen, M.R. and Danilo, P. (1980). Effects of tetrodotoxin, lidocain, verapamil, and AHR-2666 on ouabain-induced delayed afterdepolarizations in canine Purkinje fibers. *Circ. Res.* **46**, 117–24.
Rosen, M.R. and Hordof, A.J. (1980). The slow response in human atrium. In *The Slow Inward Current and Cardiac Arrhythmias*. Zipes et al. (eds.), pp. 295–308. Martinus Nijhoff Publishers, The Hague, Boston, London.
Sakai, K. and Abiko, Y. (1984). Beta-1 but not beta-2 adrenoceptors contribute to myocardial acidosis after coronary occlusion in dogs. *Abstr. 9th Int. Congr. Pharmacol.*, 856 P.
Schechter, E., Freeman, C.C. and Lazzara, R. (1984). Afterdepolarizations as a mechanism for the long QT syndrome: electrophysiologic studies of a case. *JACC* **3 (No 6)**, 1556–61.
Sheu, S.-S. and Fozzard, H.A. (1982). Transmembrane Na^+ and Ca^{2+} electrochemical gradients in cardiac muscle and their relationship to force development. *J. Gen. Physiol.* **80**, 325–51.
Surawicz, B. (1980). Depolarization-induced automaticity in atrial and ventricular myocardial fibers. In *The Slow Inward Current and Cardiac Arrhythmias*. Zipes et al. (eds.), pp. 375–96. Martinus Nijhoff Publishers, The Hague, Boston, London.
Szekeres, L. and Papp, G.J. (1971). *Experimental Cardiac Arrhythmias and Antiarrhythmic Drugs.* Akadémiai Kiadó, Budapest.
Trautwein, W., Taniguchi, J. and Noma, A. (1982). The effect of intracellular cyclic nucleotides and calcium on the action potential and acetylcholine response of isolated cardiac cells. *Pflügers Arch.* **392**, 307–14.
Vassalle, M. and Mugelli, A. (1981). An oscillatory current in sheep cardiac Purkinje fibers. *Circ. Res.* **48**, 618–31.
Vassalle, M. and Di Gennaro, M. (1983). Caffeine eliminates the oscillatory current in cardiac Purkinje fibers. *Eur. J. Pharmacol.* **94**, 361–2.
Waelbroeck, M., Taton, G., Delhaye, M., Chatelain, I., Camus, J.C., Pochet, R., Leclerc, J.L., De Smet, J.M., Robberecht, P. and Christophe, J. (1983). The human heart beta-adrenergic receptors. *Mol. Pharmacol.* **24**, 174–82.
Weingart, R. (1981). Influence of cardiac glycosides on electrophysiologic processes. In *Cardiac Glycosides.* K. Greeff (ed.), pp. 221–54. Springer-Verlag,

Berlin, Heidelberg, New York.
Wier, W.G., Kort, A.A., Stern, M.D., Lakatta, E.D. and Marban, E. (1983). Cellular calcium fluctuations in mammalian heart: direct evidence fron noise analysis of aequorin signal in Purkinje fibers. *Proc. Natl. Acad. Sci. USA* **80**, 7271–367.
Will, H., Schirpke, B. and Wollenberger, A. (1976). Stimulation of Ca^{2+}-uptake by cAMP and protein kinase in sarcoplasmic reticulum-rich and sarcolemma-rich microsomal fractions from rabbit heart. *Acta Biol. Med. Ger.* **35**, 529–41.
Windisch, H. and Tritthart, H.A. (1982). Isoproterenol, norepinephrine and phosphodiesterase inhibitors are blockers of the depressed fast Na^+-system in ventricular muscle fibers. *J. Mol. Cell. Cardiol.* **14**, 431–4.
Wit, A.I., Rosen, M.R. and Hoffman, B.F. (1974). Relationship of normal and abnormal electrical activity of canine fibers to the genesis of arrhythmias. I. Automaticity. *Am. Heart J.* **88**, 515–24.

CA^{2+} REGULATION OF CELLS FROM BONE MARROW:

Overview Including Polycythaemia Vera and Other Blood Diseases

Ole Scharff and Birthe Foder

Department of Clincial Physiology, Finsen Institute, Rigshospitalet, Copenhagen, Denmark

Abstract

Blood diseases, such as polycythaemia vera and various leukaemias, reveal disturbances of the balance between proliferation and differentiation of bone marrow cells. This balance seems to be affected by specific regulatory factors, such as erythropoietin, and more unspecific signal substances, such as Ca^{2+} and 1α, 25-dihydroxyvitamin D$_3$. Several drugs, whose mode of action is better known, affect the differentiation of leukaemia cell lines in culture, contributing to the understanding of the signal-dependent cellular responses. The role of Ca^{2+} is reviewed, and a disorder of the Ca^{2+} homeostasis in polycythaemia vera cells is suggested and illustrated with an experimentally provoked Ca^{2+} transient in polycythaemia vera red cells.

23.1. Haematopoiesis

It is well established that the various blood cell types originate from a common pluripotent stem cell, mainly located in the bone marrow. The pluripotent stem cell can produce (1) identical copies of itself by cell proliferation (self-renewal), or (2) daughter cells which are committed to differentiation along a haemopoietic lineage. Loss of potential for self-renewal seems to occur concomitantly with the irreversible commitment to differentiation (Till, 1982).

Figure 23.1 shows a simplified model of the differentiation of myelopoietic stem cells. The differentiation leads to progenitor cells with the capacity to proliferate extensively: each progenitor cell can produce 5000-10 000 progeny cells (Metcalf, 1981). The progenitors are committed to one of three lineages: erythropoiesis, granulopoiesis and megakaryocytopoiesis, and the divisions produce daughters of increasing maturity, leading to the final cells that are functionally effective but incapable of division. B and T lymphocytes are also derived from pluripotent stem cells from bone marrow, and some stem cells may be common to both lymphopoiesis and myelopoiesis (McCulloch, 1983).

The pluripotent stem cells in rodents can be detected by their capacity to produce spleen colonies composed of erythroid, myeloid and megakaryocytic elements in irradiated, syngeneic recipients (Till and McCulloch, 1961). The committed progenitor cells are able to form colonies in culture under appropriate conditions. The cell giving rise to a particular type of colony is called a colony-forming unit (CFU), and a suffix designates the type of colony (S, spleen (stem); E, erythroid; GM, granulocyte/macrophage; M, megakaryocyte). Cells giving rise to very primitive erythroid colonies are called burst-forming units (BFU-E) (Spivak and Graber, 1980).

The maintenance of homeostasis requires that a strict balance is maintained between self-renewal, differentiation, and maturation, and sustained alterations in the ratios of the individual cell populations lead to hypoproliferation or hyperproliferation diseases, such as anaplastic anaemia, polycythaemia vera or leukaemia. Accordingly, great interest has been focused on trying to define what controls stem cell proliferation and differentiation.

The majority of pluripotent stem cells are in a quiescent state (G_0-cells). However, the proportion of pluripotent stem cells replicating DNA may be raised *in vitro* from 10% to about 30% by raising the extracellular calcium ion concentration from 0.65 to 1.26 mM, or by addition of the divalent cation ionophore A23187 at 0.65 mM Ca^{2+} (Gallien-Lartigue, 1976), similar to the responses of other cell types (Whitfield *et al.*, 1980).

Several factors controlling proliferation and differentiation have been described, for instance, erythropoietin, granulocyte/macrophage colony stimulating factor (GM-CSF), and a factor called BPA (burst-promoting activity) which is a lineage-indifferent regulator (Iscove *et al.*, 1982). The latter have suggested that the number of BPA receptors decreases during maturation down the different pathways while the number of receptors

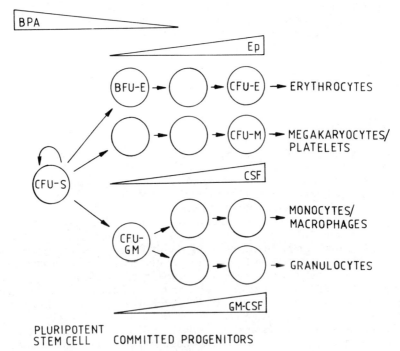

Fig.23.1. Simplified model of the differentiation of myelopoietic stem cells. The curved arrow symbolizes capacity of stem cell (CFU-S) for self-renewal. Abbreviations: CFU, colony-forming unit; BFU, burst-forming unit; E, erythrocyte; M, megakaryocyte; GM, granulocyte and macrophage; BPA, burst-promoting activity; Ep, erythropoietin; CSF, colony-stimulating factor.

recognizing lineage specific factors increases (see Fig.23.1).

Erythropoietin is acting exclusively on red cell precursors, the sensitivity to erythropoietin of primitive and mature BFU-E being approximately 15 and five times lower than the sensitivity of CFU-E (Eaves and Eaves, 1978). Erythropoietin binding to receptors on the progenitor cells generates a signal which initiates RNA synthesis, DNA synthesis, and synthesis of erythroid specific proteins, for instance, glycophorin, band-3 protein (the anion transporter), and haemoglobin (Goldwasser, 1981). GM-CSF stimulates GM progenitor cells to proliferate and to form colonies consisting preferentially of granulocytes at high GM-CSF concentrations and macrophages at low GM-CSF concentrations (Metcalf, 1981). Theories about generation of receptor-mediated signals are discussed below.

23.1.1. *Polycythaemia vera*

Polycythaemia vera is a relatively uncommon disease (annual incidence rate: 4 to 5 per million population) and characterized by panhyperplasia of the bone marrow with increased numbers of circulating erythrocytes together with some degree of excessive production of granulocytes and platelets. Later in the course of illness myelofibrosis with myeloid metaplasia and acute leukaemia may occur (Weinstein, 1973).

The involvement of three haemopoietic lines in polycythaemia vera suggests that this disease arises in a pluripotent stem cell (Adamson *et al.*, 1976). The characteristics of pluripotent stem cells from polycythaemia vera patients have been examined by use of their ability to produce mixed colonies containing granulocytic, erythroid, macrophage, and megakaryocytic elements in culture (CFU-GEMM). These studies show an increased number of detectable CFU-GEMM in blood and bone marrow samples from polycythaemia vera patients, and an increased megakaryocyte formation by the CFU-GEMM (Ash, Detrick and Zanjani, 1982). Furthermore, the CFU-GEMM from polycythaemia vera peripheral blood has proliferative activity, in contrast to the usual quiescent behaviour of CFU-GEMM from normal blood (Fauser and Messner, 1981). The erythroid progenitors (BFU-E and CFU-E) and CFU-GEMM obtained from bone marrow or peripheral blood from polycythaemia vera patients are able to form colonies without the addition of erythropoietin, in contrast to cells obtained from normal subjects or patients with secondary polycythaemia (Ash, Detrick and Zanjani, 1982). This property may be related to a tenfold increase in the sensitivity to erythropoietin of polycythaemia vera erythroid progenitors (Casadevall *et al.*, 1982).

23.1.2. *Other blood diseases*

Together with polycythaemia vera, such diseases as acute and chronic myeloblastic leukaemia are considered to originate in pluripotent stem cells (McCulloch, 1983). In leukaemia the normal relationship between proliferation and differentiation is uncoupled (Olsson, 1983). Since a new approach to the therapy of leukaemia may be based on the induction of normal differentiation in malignant cells (Olsson, 1983), the investigation of the mechanism of action of physiologically produced inducing factors and other inducing agents is of great importance.

23.2 Induction of cell differentiation

23.2.1. *Erythropoietin*

In a CFU-E assay using murine marrow cells, Misiti and Spivak (1979) have demonstrated the role of calcium. Erythropoietin-induced colony formation is increased by addition of the divalent cation ionophore A23187, and Ca^{2+} but not Mg^{2+}, Mn^{2+}, Zn^{2+}, or Fe^{2+} is the effective cation. One function of calcium seems to be a promotion of the binding of erythropoietin to its target cells.

In the colony assays the effects on differentiation are not separate from those on proliferation. However, another study shows that erythropoietin has a separate role in the differentiation of erythroid progenitors (Koury et al., 1982). Proliferating erythroblasts, obtained from mouse bone marrow cells infected *in vitro* with Friend virus, respond to erythropoietin by differentiating to erythrocytes, and the addition of erythropoietin increases the cellular uptake of $^{45}Ca^{2+}$ within 1 min, being one of the earliest effects of erythropoietin yet described (Sawyer and Krantz, 1984). Interestingly, A23187 and dimethylsulphoxid (DMSO; Fig.23.2) increase the uptake of $^{45}Ca^{2+}$ in the virus-infected erythroblasts and enhance the differentiation induced by suboptimal levels of erythropoietin (Sawyer and Krantz, 1984), indicating that Ca^{2+} increases the sensitivity to the natural regulatory factor.

These investigations clearly reveal a significant interaction between erythropoietin and calcium in the induction of differentiation of erythroid

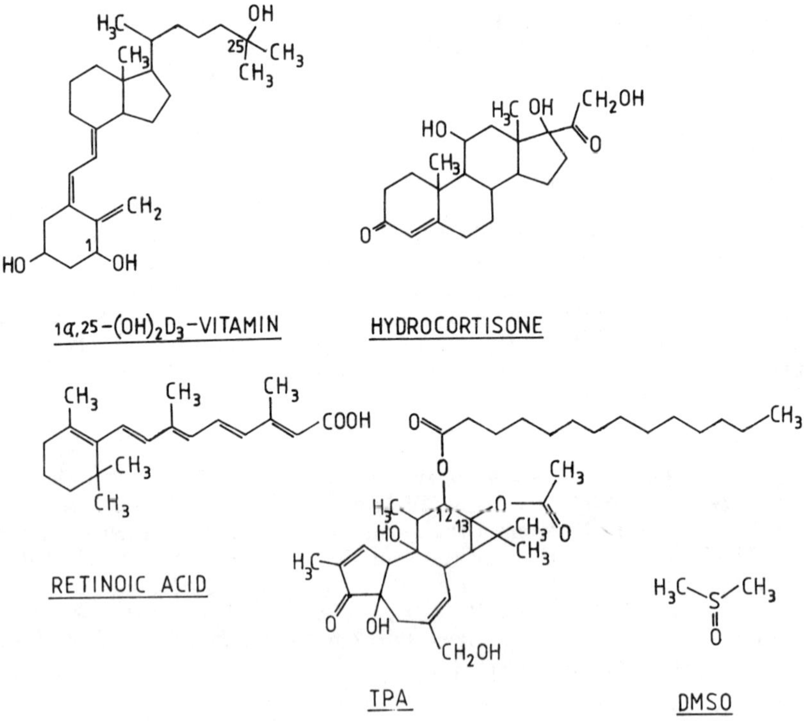

Fig.23.2. Chemical structure of various inducers and inhibitors of cell differentiation. Abbreviations: TPA, 12-0-tetradecanoyl phorbol-13-acetate; DMSO, dimethylsulphoxide.

progenitors.

23.2.2. Calcium, DMSO, and butyrate

More detailed information concerning the effect of calcium on cell differentiation has been obtained from studies on the continuously growing leukaemia cell lines (Table 23.1), especially the murine erythroleukaemia cells (MEL). Morphologically, the MEL cells are similar to proerythroblasts and undergo erythroid differentiation after exposure to various chemicals (Marks and Rifkind, 1978; Table 23.1) but do not respond to erythropoietin (Koury et al., 1982).

The differentiation of MEL cells induced by butyric acid or DMSO depends on the presence of Ca^{2+}. The onset of differentiation after DMSO addition is normally delayed 10–15 hours but in the presence of the Ca^{2+} ionophore A23187 DMSO induces differentiation with no lag (Bridges et al., 1981). Since A23187 alone does not initiate differentiation, another factor besides the A23187-induced Ca^{2+} influx into the cells must be required (Bridges et al., 1981).

23.2.2.1. Ca^{2+} channel and Na^+/Ca^{2+} antiport.
The lack of differentiation in the presence of the Ca^{2+} channel blocker D600 (Chapman, 1980) suggests the involvement of a plasma membrane Ca^{2+} channel (Fig.23.3) during the differentiation of the MEL cell.

On the other hand, the work of Cantley and co-workers (Smith et al., 1982) indicates that the increased Ca^{2+} influx may occur via a plasma membrane Na^+/Ca^{2+} antiport (Fig.23.3), since the differentiation is blocked by amiloride. The amiloride block of differentiation can be overcome either by the addition of A23187 or by addition of the Na^+ ionophore monensin and/or

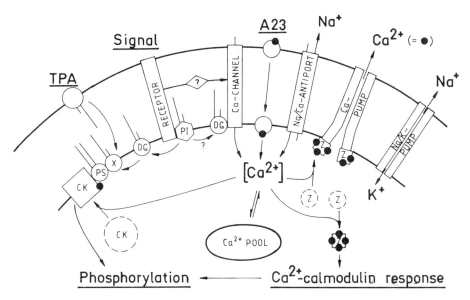

Fig.23.3. Cellular activation by extraneous signals of protein kinase C and the Ca^{2+}-calmodulin system mediated by a plasma membrane receptor (see text for further details). Abbreviations: PI, phosphatidylinositol; DG, diacylglycerol; PS, phosphatidylserine; CK, protein kinase C; TPA, tumour-promoting phorbol ester; X, DG or TPA; A23, Ca^{2+} ionophore A23187; Z, calmodulin with four Ca^{2+}-binding sites.

Table 23.1. Leukaemic cell lines grown in culture.

Cell line	Phenotype	Cell differentiation inducers	Cell differentiation inhibitors	Induced cell	References
HL-60 human	promyelocyte	DMSO		granulocyte	Collins et al. (1978)
		retinoic acid		granulocyte	Breitman, Selonide and Collins (1980)
		$1,25-(OH)_2D_3$		macrophage	Bar-Shavit et al. (1983); McCarthy et al. (1983)
		TPA		macrophage-like	Lotem and Sachs (1979); Murao et al. (1983); Rovera, O'Brien and Diamond (1979)
		(TPA)	3-deaza-adenosine		Feuerstein and Cooper (1984)
Erythroleukaemia murine					
Friend (MEL)	proerythroblast	DMSO		erythroid	Marks and Rifkind (1978)
		butyrate		erythroid	Marks and Rifkind (1978)
		(butyrate)	D600		Chapman (1980)
		(DMSO)	amiloride		Smith et al. (1980)
		(DMSO, butyrate)	$1,25-(OH)_2D_3$		Suda et al. (1984)
		(DMSO, butyrate)	hydrocortisone		Scher et al. (1978)
		(DMSO, butyrate)	TPA		Rovera, O'Brien and Diamond (1977); Yamasaki et al. (1977)
Rauscher	proerythroblast	erythropoietin		erythroid	Sytkowski et al. (1980)
		DMSO		erythroid	Sytkowski et al. (1980)
		TPA (high concentration)		erythroid	Miao, Fieldsteel and Fodge (1978)
human HEL	proerythroblast (similar to K-562; related to CFU-GEMM?)	spontaneous		erythroid	Rimmer and Horton (1984)
		haemin		erythroid	Rimmer and Horton (1984)
		TPA		macrophage-like	Papayannopoulou et al. (1983)
K-562	proerythroblast (also granulo-cyte/monocyte-specific markers)	butyrate		erythroid	Rimmer and Horton (1984)
		haemin		erythroid	Rimmer and Horton (1984)
		TPA		megakaryocyte	Tetteroo et al. (1984)

the sodium pump inhibitor ouabain.

23.2.2.2. *Sodium pump.* DMSO and other agents which induce differentiation in the MEL cells cause a 30–40% inhibition of the cation pumping activity of the Na^+, K^+-ATPase during a period of about 12 hours (Mager and Bernstein, 1978; Smith *et al.*, 1982). Cantley and co-workers suggest that this slow pump inhibition causes the required increase of Ca^{2+} influx via the Na^+/Ca^{2+} antiport during the lag period (10–15 hours) of the DMSO-induced differentiation (compare above).

23.2.3. *Vitamin D_3 and retinoic acid*

The effective concentrations of DMSO (about 150 mM) and sodium butyrate (0.5 mM) are much higher than those of two very potent agents, $1\alpha,25$-dihydroxyvitamin D_3 ($1,25$-$(OH)_2D_3$) and retinoic acid (see Fig.23.2), affecting differentiation at concentrations from 10^{-10} to 10^{-6} M (Bar-Shavit *et al.*, 1983; Breitman, Selonick and Collins, 1980); $1,25$-$(OH)_2D_3$ is the most active metabolite of vitamin D_3 with a major role in maintaining the calcium homeostasis in the organism (Wasserman and Fullmer, 1983).

The agent $1,25$-$(OH)_2D_3$ inhibits the DMSO-induced erythroid differentiation but not the growth of MEL cells (Suda *et al.*, 1984), and similar effects on MEL cells are exerted by hydrocortisone (see Fig.23.2), dexamethasone and other steroids (Scher *et al.*, 1978).

The inhibitory effect of $1,25$-$(OH)_2D_3$ on erythroid differentiation is remarkable because this drug, on the contrary, induces differentiation of various myeloid leukaemia cells, for example, HL-60 cells (Bar-Shavit *et al.*, 1983; McCarthy *et al.*, 1983; see Table 23.1), preferentially along the monocyte-macrophage pathway (Suda *et al.*, 1984). This suggests a role of $1,25$-$(OH)_2D_3$ in the differentiation of cells in the haematopoiesis: stimulation of granulopoiesis and suppression of erythropoiesis (see Fig.23.1).

Retinoic acid, a derivative of vitamin A, induces HL-60 cells to differentiation into granulocytes instead of macrophages (Breitman, Selonick and Collins, 1980; see Table 23.1).

23.2.4. *Tumour-promoting phorbol esters*

It is not clear how DMSO and the other agents mentioned exercise their specific effects on cell differentiation. More is known about the reaction mechanism of tumour-promoting phorbol diesters, especially 12-0-tetradecanoyl phorbol-13-acetate (TPA, see Fig.23.2), as described below. TPA is very potent, being effective in low concentrations, 10^{-9}–10^{-7} M (Kraft and Anderson, 1983; Nishizuka, 1983). TPA and other phorbol diesters have pleiotropic effects which seem to depend primarily on the target cells (compare Table 23.1).

For instance, Sieber, Stuart and Spivak (1981), using an *in vitro* colony assay of murine haemopoietic cells, found that TPA and other phorbol esters stimulate the formation of monocyte–macrophage colonies by CFU-GM but inhibit the formation of erythroid bursts by BFU-E, whereas CFU-E is not affected by the phorbol esters. Analogously, the phorbol esters inhibit the erythroid differentiation induced by DMSO or butyrate in MEL cells and the spontaneous erythroid differentiation in other lines of Friend erythroleukaemia cells (Rovera, O'Brien and Diamond, 1977; Yamasaki *et al.*, 1977) but, on the contrary, TPA stimulates the erythroid differentiation in two similar cells lines transformed by Rauscher virus (Miao, Fieldsteel and Fodge, 1978). The Rauscher cells deviate from MEL cells, however, by being erythropoietin-sensitive (Sytkowski *et al.*, 1980) and by requiring higher concentrations of TPA (Miao, Fieldsteel and Fodge, 1978).

In two other proerythroid cell lines, HEL and K-562, TPA causes changes into macrophage-like cells (Papayannopoulou et al., 1983) and megakaryocyte-like cells (Tetteroo et al., 1984), respectively (compare Table 23.1). The effect of TPA on HEL cells is similar to that on HL-60 cells which change into macrophage-like cells, similar to but not identical with the monocyte-macrophage phenotype induced by 1,25-$(OH)_2D_3$ (Lotem and Sachs, 1979; Murao et al., 1983; Rovera, O'Brien and Diamond, 1979). In spite of their similar effects, TPA and 1,25-$(OH)_2D_3$ apparently bind to different receptors but, interestingly, a HL-60 variant (R-80) is not induced to differentiation by any of these two drugs (Murao et al., 1983). The TPA-induced differentiation in HL-60 cells can be inhibited by addition of 3-deaza-adenosine which inhibits transmethylation (Feuerstein and Cooper, 1984).

23.3. Cellular responses linked to extraneous signals

23.3.1. Elevation of diacylglycerol and Ca^{2+}

An extracellular signal mediated by a plasma membrane receptor (see Fig.23.3) causes (1) the hydrolysis of phosphatidylinositol (PI) or PI-4-phosphate to diacylglycerol (DG) catalysed by a Ca^{2+}-sensitive phosphodiesterase (phospholipase C) (Downes and Michell, 1982), and (2) elevation of the concentration of intracellular ionized calcium ($[Ca^{2+}]_i$). It is not clear whether the elevation is due to an increased Ca^{2+} influx through the plasma membrane via a receptor-activated Ca^{2+} channel (as suggested in Fig.23.3) or via a Na^+/Ca^{2+} antiport due to inhibition of the Na^+/K^+ pump, or whether Ca^{2+} is mobilized from an intracellular Ca^{2+} pool. In human platelets Rink, Sanchez and Hallam (1983) have shown that addition of DG alone does not cause an elevation of $[Ca^{2+}]_i$.

23.3.2. C-kinase activation

The DG produced enters together with soluble protein kinase C, Ca^{2+}, and phospholipid (phosphatidylserine being most effective) into a quaternary membrane-bound complex, allowing the expression of full catalytic activity of the C-kinase that causes phosphorylation of several proteins (see Fig.23.3), as reviewed in details by Nishizuka (1983, 1984).

Tumour-promoting phorbol diesters, such as TPA (see Fig.23.2), can substitute for DG, apparently without the participation of the signal receptor. Experiments have shown that the C-kinase binds to the membrane within few minutes upon TPA addition, and that TPA strongly increases the apparent affinity of C-kinase for both Ca^{2+} and phospholipid (Kraft and Anderson, 1983; Nishizuka, 1984; Vandenbark, Kuhn and Niedel, 1984). The apparent dissociation constant for binding of a phorbol ester is about 8nM (Nishizuka, 1984).

23.3.3. Ca^{2+}, calmodulin, and cyclic nucleotides

By elevation of $[Ca^{2+}]_i$ the soluble calmodulin binds calcium and changes to a conformation with high affinity for several cellular enzymes and other proteins which are activated (or inactivated) by the binding of Ca^{2+}-calmodulin (reviewed recently by Cheung (1982), Klee and Vanaman (1982) and Means, Tash and Chafouleas, 1982; see also Fig.23.3).

Among the calmodulin-dependent enzymes are adenylate cyclase and cyclic nucleotide phosphodiesterase, catalysing the synthesis of cyclic AMP and breakdown of cyclic AMP and cyclic GMP, respectively. Together with the cyclic nucleotides, Ca^{2+} and calmodulin control a complex pattern of regulatory phosphorylation of enzymes and other functional proteins in cells (Cohen, 1983; Krebs and Beavo, 1979; Rasmussen, 1981). Interestingly, recent observations suggest that the cyclic nucleotides in some cell types inhibit the production of diacyglycerol and thereby counteract the activation of C-kinase (Nishizuka, 1984).

By adding the Ca^{2+} ionophore A23187, $[Ca^{2+}]_i$ can be increased independently of the C-kinase (see Fig.23.3).

23.3.4. Calcium pump

Binding of Ca^{2+}-calmodulin also activates the plasma membrane Ca^{2+} pump and increases the Ca^{2+} sensitivity of the pump. The Ca^{2+} pump is present in most cell types, including erythrocytes, platelets, monocyte/macrophages, neutrophil granulocytes, and lymphocytes (Penniston, 1983; Schatzmann, 1982). The activated, calmodulin-associated Ca^{2+} pump (see Fig.23.3) is capable of reducing $[Ca^{2+}]_i$ to the resting level (about 10^{-7} M or lower) that existed before the stimulation of the cell by the signal.

However, experiments with human erythrocytes suggest that the extrusion of Ca^{2+} from the cell may be delayed compared to the activation of other calmodulin-dependent reactions, primarily due to slow activation of the Ca^{2+} pump at low concentrations of Ca^{2+} or calmodulin, secondarily because the Ca^{2+} pump has to compete with other cellular proteins for calmodulin during activation (Scharff and Foder, 1982; Scharff, Foder and Skibsted, 1983).

23.4. Action of natural signal substances

The capacities of tumour-promoting phorbol diesters like TPA and Ca^{2+} ionophores like A23187 for activating the C-kinase branch and the Ca^{2+}-calmodulin branch, respectively, suggest a mode of action of several extracellular signal substances.

For instance, the synergistic effect of TPA and Ca^{2+} ionophore mimics the effect of thrombin (Yamanishi et al., 1983), ATP (Rink, Sanchez and Hallam, 1983), angiotensin II (Kojima et al., 1983), glucose (Zawalich, Brown and Rasmussen, 1983), and vasopressin (Fain et al., 1984) on secretory exocytosis or enzyme activation in various target cells. This indicates that natural signal substances of widely different composition and molecular sizes (200–34 000 D) can provoke activation of both the C-kinase and the Ca^{2+}-calmodulin system (Fig.23.3).

The mode of action of the natural signal substances effective in the haematopoiesis, that is, erythropoietin, GM-CSF, BPA, and other colony-stimulating factors (see Fig.23.1) is far from understood. By analogy, it is to be expected that these regulatory factors affect cellular phosphorylations, and the need of Ca^{2+} for the action of one of the factors, erythropoietin, has already been clearly demonstrated (see section 23.2.1).

23.5. Possible defects in regulation of cells from bone marrow

The disorders of the haemopoietic regulation characteristic of polycythaemia vera and the various leukaemias may be due to changes in the genetic expression of the involved cells (compare Olsson, 1983). In addition, since the different effects of the C-kinase stimulating phorbol diesters, for example, TPA (compare Table 23.1), indicate that the effect of an induced phosphorylation pattern is strongly dependent on the present developmental phase of the cell, the disorders may be caused by changed cell responses to the signals, leading to, for instance, abnormal protein phosphorylation in the cells. Cell-to-cell interactions may contribute essentially to the disorders, as suggested from the found GM-CSF secretion from human bone marrow cells induced by a phorbol diester (Gerson and Cooper, 1984; Lotem and Sachs, 1979). Besides phosphorylation, other general biochemical reactions such as transmethylation, affecting DNA synthesis, may be changed (compare Feuerstein and Cooper, 1984).

Only little is known about disturbances of the cellular homeostasis of Ca^{2+} and calmodulin in blood diseases originating from the bone marrow. Abnormalities in the calcium content of red cells, for instance, are comparatively unknown (Parker, 1981). Calmodulin is elevated in a number of transformed cells (Means, Tash and Chafouleas, 1982), and also in two human leukaemic cell lines compared to peripheral lymphocytes (Takemoto and Jilka, 1983).

The ATPase activity of the Ca^{2+} pump seems to be reduced by 30% in red cells from polycythaemia vera patients, and the consequences for the Ca^{2+} signals possibly involved in the abnormal cell proliferation have been

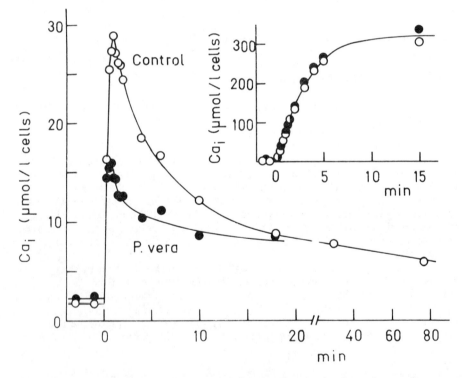

Fig.23.4. A23187-induced change of calcium concentration dependent on time in erythrocytes from polycythaemia vera patient compared with control cells. At zero time A23187 was added to a suspension containing 20% red cells (v/v), 100 μM $CaCl_2$, ^{45}Ca, 1 mM $MgCl_2$, 75 mM KCl, 80 mM NaCl, 6 mM inosine, pH 7.4, 37 °C. The intracellular concentration of total calcium (Ca_i) was measured as ^{45}Ca, as previously (Scharff, Foder and Skibsted, 1983). $[Ca^{2+}]_i$ was about 20% of Ca_i. The transient increase of Ca_i was dependent on the Ca^{2+} pump, hydrolysing ATP. In ATP-depleted cells (inset) Ca_i increased to electrochemical equilibrium with no transient. The Ca^{2+}-ATPase activity of this patient was 80% of control.

discussed previously (Scharff and Foder, 1975). In Fig.23.4 we have investigated 'Ca^{2+} signals' induced by the Ca^{2+} ionophore A23187 in erythrocytes from a polythaemia vera patient compared with a control. According to a mathematical pump-leak model (Scharff, Foder and Skibsted, 1983) a 20–30% reduction of the maximum Ca^{2+} pump activity should imply a 30–40% increase of the amplitude of the Ca^{2+} transient. However, the found amplitude of the Ca^{2+} transient in the polycythaemia vera cells, on the contrary, is less than in the control (Fig.23.4), in spite of identical Ca^{2+} influxes in the two cell speciment (Fig.23.4, inset). Possibly, the Ca^{2+} pump has been activated more rapidly in the polycythaemia vera cells due to, for instance, an increased availability of calmodulin, even though the total cellular calmodulin is found not to be increased (unpublished).

A disorder of the Ca^{2+} homeostasis in polycythaemia vera cells cannot be excluded and may be related to the increased sensitivity to erythropoietin of erythroid progenitors from these patients (compare section 23.1.1).

23.6. Conclusion

In conclusion, the calcium ion participates in the regulation of proliferation and differentiation of bone marrow cells in parallel with a number of regulatory factors and more unspecific natural signal substances. Protein phosphorylations may be important cellular events, accompanying the extraneous signals, but mapping of the specific phosphoproteins and other proteins involved in the network of signal-dependent events, concurrently with the clarification of their functions, seems to be required for a better understanding of the aetiology of polycythaemia vera and the various leukaemias.

References

Adamson, J.W., Fialkow, P.J., Murphy, S., Prchal, J.F. and Steinmann, L. (1976). Polycythemia vera: stem-cell and probable clonal origin of the disease. *N. Engl. J. Med.* **295**, 913–6.

Ash, R.C., Detrick, R.A. and Zanjani, E.D. (1982). *In vitro* studies of human pluripotential hematopoietic progenitors in polycythemia vera. Direct evidence of stem cell involvement. *J. Clin. Invest.* **69**, 1112–8.

Bar-Shavit, Z., Teitelbaum, S.L., Reitsma, P., Hall, A., Pegg, L.E., Trial, J. and Kahn, A.J. (1983). Induction of monocytic differentiation and bone resorption by 1,25-dihydroxyvitamin D$_3$. *Proc. Natl. Acad. Sci. USA* **80**, 5907–11.

Breitman, T.R., Selonick, S.E. and Collins, S.J. (1980). Induction of differentiation of the human promyelocytic leukemia cell line (HL-60) by retinoic acid. *Proc. Natl. Acad. Sci. USA* **77**, 2936–40.

Bridges, K., Levenson, R., Housman, D. and Cantley, L. (1981). Calcium regulates the commitment of murine erythroleukemia cells to terminal erythroid differentiation. *J. Cell Biol.* **90**, 542–4.

Casadevall, N., Vainchenker, W., Lacombe, C., Vinci, G., Chapman, J., Breton-Gorius, J. and Varet, B. (1982). Erythroid progenitors in polycythemia vera: demonstration of their hypersensitivity to erythropoietin using serum free cultures. *Blood* **59**, 447–51.

Chapman, L.F. (1980). Effect of calcium on differentiation of Friend leukemia cells. *Devel. Biol.* **79**, 243–6.

Cheung, W.Y. (1982). Calmodulin: an overview. *Fed. Proc.* **41**, 2253–7.
Cohen, P. (1983). Protein phosphorylation and hormone action - an overview. *Cell. Biol. Int. Rep.* **7**, 479–80.
Collins, S.J., Ruscetti, F.W., Gallagher, R.E. and Gallo, R.C. (1978). Terminal differentiation of human promyelocytic leukemia cells induced by dimethyl sulfoxide and other polar compounds. *Proc. Natl. Acad. Sci. USA* **75**, 2458–62.
Downes, P. and Michell, R.H. (1982). Phosphatidylinositol 4-phosphate and phosphatidylinositol 4,5-bisphosphate: lipids in search of a function. *Cell Calcium* **3**, 467–502.
Eaves, C.J. and Eaves, A.C. (1978). Erythropoietin (Ep) dose-response curves for three classes of erythroid progenitors in normal human marrow and in patients with polycythemia vera. *Blood* **52**, 1196–210.
Fain, J.N., Li, S., Litosch, I. and Wallace, M. (1984). Synergistic activation of rat hepatocyte glycogen phosphorylase by A23187 and phorbol ester. *Biochem. Biophys. Res. Commun.* **119**, 88–94.
Fauser, A.A. and Messner, H.A. (1981). Pluripotent hemopoietic progenitors (CFU-GEMM) in polycythemia vera: analysis of erythropoietin requirement and proliferative activity. *Blood* **58**, 1224–7.
Feuerstein, N. and Cooper, H.L. (1984). Studies of the differentiation of promyelocytic cells by phorbol ester. II. A methylation inhibitor, 3-deazaadenosine, inhibits the induction of specific differentiation proteins. Lack of effect on early and late phosphorylation events. *Biochim. Biophys. Acta* **781**, 247–56.
Gallien-Lartigue, O. (1976). Calcium and ionophore A-23187 as initators of DNA replication in the pluripotent haemopoietic stem cell. *Cell Tissue Kinet.* **9**, 533–40.
Gerson, S.L. and Cooper, R.A. (1984). Release of granulocyte-specific colony-stimulating activity by human bone marrow exposed to phorbol esters. *Blood* **63**, 878–85.
Goldwasser, E. (1981). Erythropoietin and red cell differentiation. *Prog. Clin. Biol. Res.* **66A**, 487–94.
Iscove, N.N., Roitsch, C.A., Williams, N. and Guilbert, L.J. (1982). Molecules stimulating early red cell, granulocyte, macrophage, and megakaryocyte precursors in culture: similarity in size, hydrophobicity, and charge. *J. Cell. Physiol.* **Suppl.1**, 65–78.
Klee, C.B. and Vanaman, T.C. (1982). Calmodulin. In *Advances in Protein Chemistry, vol. 35,* Anfinsen, Edsall, Richards (eds.), pp. 213–321. Academic Press, New York
Kojima, I., Lippes, H., Kojima, K. and Rasmussen, H. (1983). Aldosterone secretion: effect of phorbol ester and A23187. *Biochem. Biophys. Res. Commun.* **116**, 555–62.
Koury, M.J., Bondurant, M.C., Duncan, D.T., Krantz, S.B. and Hankins, W.D. (1982). Specific differentiation events induced by erythropoictin in cells infected *in vitro* with the anemia strain of Friend virus. *Proc. Natl. Acad. Sci. USA* **79**, 635–9.
Kraft, A.S. and Anderson, W.B. (1983). Phorbol esters increase the amount of Ca^{2+}, phospholipid-dependent protein kinase associated with the plasma membrane. *Nature (Lond.)* **301**, 621–3.
Krebs, E.G. and Beavo, J.A. (1979). Phosphorylation-dephosphorylation of enzymes. *Ann. Rev. Biochem.* **48**, 923–59.
Lotem, J. and Sachs, L. (1979). Regulation of normal differentiation in mouse and human myeloid leukemic cells by phorbol esters and the mechanism of tumor promotion. *Proc. Natl. Acad. Sci. USA* **76**, 5158–62.
Mager, D. and Bernstein, A. (1978). Early transport changes during erythroid

differentiation of Friend leukemic cells. *J. Cell Physiol.* **94**, 275–86.
Marks, P.A. and Rifkind, R.A. (1978). Erythroleukemic differentiation. *Ann. Rev. Biochem.* **47**, 419–48.
McCarthy, D.M., San Miguel, J.F., Freake, H.C., Green, P.M., Zola, H., Catovsky, D. and Goldman, J.M. (1983). 1,25-dihydroxyvitamin D_3 inhibits proliferation of human promyelocytic leukemia (HL-60) cells and induces monocyte–macrophage differentiation in HL-60 and normal human bone marrow cells. *Leukemia Res.* **7**, 51–5.
McCulloch, E.A. (1983). Stem cells in normal and leukemic hemopoiesis (Henry Stratton Lecture, 1982). *Blood* **62**, 1–13.
Means, A.R., Tash, J.S. and Chafouleas, J.G. (1982). Physiological implications of the presence, distribution, and regulation of calmodulin in eukaryotic cells. *Physiol. Rev.* **62**, 1–39.
Metcalf, D. (1981). Control of hemopoietic cell proliferation and differentiation. *Prog. Clin. Biol. Res.* **66A**, 473–86.
Miao, R.M., Fieldsteel, A.H. and Fodge, D.W. (1978). Opposing effects of tumour promoters on erythroid differentiation. *Nature (Lond.)* **274**, 271–2.
Misiti, J., Spivak, J.L. (1979). Erythropoiesis *in vitro*. Role of calcium. *J. Clin. Invest.* **64**, 1573–9.
Murao, S., Gemmeli, M.A., Callaham, M.F., Anderson, N.L. and Huberman, E. (1983). Control of macrophage cell differentiation in human promyelocytic HL-60 leukemia cells by 1,25-dihydroxyvitamin D_3 and phorbol-12-myristate-13-acetate. *Cancer Res.* **43**, 4989–96.
Nishizuka, Y. (1983). Calcium, phospholipid turnover and transmembrane signalling. *Phil. Trans. R. Soc. Lond. B* **302**, 101–12.
Nishizuka, Y. (1984). The role of protein kinase C in cell surface signal transduction and tumour promotion. *Nature (Lond.)* **308**, 693–8.
Olsson, I. (1983). Is the maturation arrest in myeloid leukemia reversible?. *Acta Med. Scand.* **214**, 261–72.
Papayannopoulou, T., Nakamoto, B., Yokochi, T., Chait, A. and Kannagi, R. (1983). Human erythroleukemia cell line (HEL) undergoes a drastic macrophage-like shift with TPA. *Blood* **62**, 832–45.
Parker, J.C. (1981). Effects of drugs on calcium-related phenomena in red blood cells. *Fed. Proc.* **40**, 2872–6.
Penniston, J.T. (1983). Plasma membrane Ca^{2+}-ATPases as active Ca^{2+} pumps. In *Calcium and Cell Function, vol.4*. W.Y. Cheung (ed.), pp. 99–149. Academic Press, New York.
Rasmussen, H. (1981). *Calcium and cAMP as Synarchic Messengers*. John Wiley and Sons, New York.
Rimmer, E.F. and Horton, M.A. (1984). Expression of myeloid-specific antigens on two human erythroleukemia cell lines, HEL and K562. *Leukemia Res.* **8**, 207–11.
Rink, T.J., Sanchez, A. and Hallam, T.J. (1983). Diacylglycerol and phorbol ester stimulate secretion without raising cytoplasmic free calcium in human platelets. *Nature (Lond.)* **305**, 317–9.
Rovera, G., O'Brien, T.G. and Diamond, L. (1977). Tumor promoters inhibit spontaneous differentiation of Friend erythroleukemia cells in culture. *Proc. Natl. Acad. Sci. USA* **74**, 2894–8.
Rovera, G., O'Brien, T.G. and Diamond, L. (1979). Induction of differentiation in human promyelocytic leukemia cells by tumor promoters. *Science* **204**, 868–70.
Sawyer, S.T. and Krantz, S.B. (1984). Erythropoietin stimulates $^{45}Ca^{2+}$ uptake in Friend virus-infected erythroid cells. *J. Biol. Chem.* **259**, 2769–74.
Scharff, O. and Foder, B. (1975). Decreased ($Ca^{2+} + Mg^{2+}$)-stimulated ATPase

activity in erythrocyte membranes from polycythemia vera patients. *Scand. J. Clin. Lab. Invest.* **35**, 583–9.

Scharff, O. and Foder, B. (1982). Rate constant for calmodulin binding to Ca^{2+}-ATPase in erythrocyte membranes. *Biochim. Biophys. Acta* **691**, 133–43.

Scharff, O., Foder, B. and Skibsted, U. (1983). Hysteretic activation of the Ca^{2+} pump revealed by calcium transients in human red cells. *Biochim. Biophys. Acta* **730**, 295–305.

Schatzmann, H.J. (1982). The plasma membrane calcium pump of erythrocytes and other animal cells. In *Membrane Transport of Calcium*, E. Carafoli (ed.), pp.41–108. Academic Press, New York.

Scher, W., Tsuei, D., Sassa, S., Price, P., Gabelman, N. and Friend, C. (1978). Inhibition of dimethyl sulfoxide-stimulated Friend cell erythrodifferentiation by hydrocortisone and other steroids. *Proc. Natl. Acad. Sci. USA* **75**, 3851–5.

Sieber, F., Stuart, R.K. and Spivak, J.L. (1981). Tumor-promoting phorbol esters stimulate myelopoiesis and suppress erythropoiesis in cultures of mouse bone marrow cells. *Proc. Natl. Acad. Sci. USA* **78**, 4402–6.

Smith, R.L., Macara, I.G., Levenson, R., Housman, D. and Cantley, L. (1982). Evidence that a Na^+/Ca^{2+} antiport system regulates murine erythroleukemia cell differentiation. *J. Biol. Chem.* **257**, 773–80.

Spivak, J.L. and Graber, S.E. (1980). Erythropoietin and the regulation of erythropoiesis. *Johns Hopkins Med. J.* **146**, 311–20.

Suda, S., Enomoto, S., Abe, E. and Suda, T. (1984). Inhibition by $1\alpha,25$-dihydroxyvitamin D_3 of dimethyl sulfoxide-induced differentiation of Friend erythroleukemia cells. *Biochem. Biophys. Res. Commun.* **119**, 807–13.

Sytkowski, A.J., Salvado, A.J., Smith, G.M., McIntyre, C.J. and DeBoth, N.J. (1980). Erythroid differentiation of clonal Rauscher erythroleukemia cells in response to erythropoietin or dimethyl sulfoxide. *Science* **210**, 74–6.

Takemoto, D. and Jilka, C. (1983). Increased content of calmodulin in human leukemia cells. *Leukemia Res.* **7**, 97–100.

Tetteroo, P.A.T., Massaro, F., Mulder, A., Schreuder-Van Gelder, R. and Borne, A.E.G.K. (1984). Megakaryoblastic differentiation of proerythroblastic K562 cell-line cells. *Leukemia Res.* **8**, 197–206.

Till, J.E. (1982). Stem cells in differentiation and neoplasia. *J. Cell Physiol.* **Suppl. 1**, 3–11.

Till, J.E. and McCulloch, E.A. (1961). A direct measure of the radiation sensitivity of normal mouse bone marrow cells. *Radiat. Res.* **14**, 213–22.

Vandenbark, G.R., Kuhn, L.J. and Niedel, J.E. (1984). Possible mechanism of phorbol diester-induced maturation of human promyelocytic leukemia cells. Activation of protein kinase C. *J. Clin. Invest.* **73**, 448–57.

Wasserman, R.H. and Fullmer, C.S. (1983). Calcium transport proteins, calcium absorption, and vitamin D. *Ann. Rev. Physiol.* **45**, 375–90.

Weinstein, I. (1973). Clinical manifestations. In *Polycythemia. Theory and Management*, H. Klein (ed.), pp. 90–111. C.C. Thomas, Springfield, Illinois.

Whitfield, J.F., Boynton, A.L., MacManus, J.P., Rixon, R.H., Sikorska, M., Tsang, B. and Walker, P.R. (1980). The roles of calcium and cyclic AMP in cell proliferation. *Ann. N.Y. Acad. Sci.* **339**, 216–40.

Yamanishi, J., Takai, Y., Kaibuchi, K., Sano, K., Castagna, M. and Nishizuka, Y. (1983). Synergistic functions of phorbol ester and calcium in serotonin release from human platelets. *Biochem. Biophys. Res. Commun.* **112**, 778–86.

Yamasaki, H., Fibach, E., Nudel, U., Weinstein, I.B., Rifkind, R.A. and Marks, P.A. (1977). Tumor promoters inhibit spontaneous and induced differentiation of murine erythroleukemia cells in culture. *Proc. Natl. Acad. Sci. USA* **74**, 3451–5.

Zawalich, W., Brown, C. and Rasmussen, H. (1983). Insulin secretion: combined effects of phorbol ester and A23187. *Biochem. Biophys. Res. Commun.* **117**, 448–55.

THE STATE OF CALCIUM IN SICKLE CELL ANAEMIA RED CELLS

Robert M. Bookchin, Olga E. Ortiz, Austin Hockaday, Maria Isabel Sepulveda and Virgilio L. Lew.

Albert Einstein College of Medicine, Bronx, New York, USA

and

The Physiological Laboratory, University of Cambridge, UK

Abstract

A critical review of the available evidence suggests that the elevated Ca in sickle cell anaemia (SS) red cells cannot acount for the cell dehydration, rigidity, and altered Na and K contents which characterize the irreversibly sickled forms. Nor is it possible to explain the high cell Ca content by increased cytoplasmic Ca buffering, or by any of the mild Ca pump abnormalities reported, since the reserve Ca extrusion capacity of the pump can easily balance the largest measured sickling-induced Ca leaks. To explain this paradox, we proposed that Ca is sequestered in endocytic inside-out vesicles. In serial electronmicroscopic sections of SS cells we find many vesicles associated with, and containing, electron-dense material. The combined volume of these vesicles requires that the Ca within be concentrated 10–100 times relative to plasma if it is to account for the observed Ca content of these cells.

In 1973 Eaton *et al.* and Palek each reported that red cells from persons with sickle cell anaemia (SS cells) have a high Ca content and accumulate Ca when the cells are sickled *in vitro* by anaerobic incubation. This immediately seemed important because it provided a possible mechanism by which the membrane abnormalities secondary to haemoglobin S polymerization could be linked to the red cell alterations thought to be crucial in the pathophysiology of the disease.

The main clinical features of sickle cell anaemia, chronic haemolysis and microvascular occlusion, can be attributed most directly to the mechanical fragility and cell rigidity which accompany intracellular polymerization of deoxyhaemoglobin S. Nevertheless, even fully oxygenated SS cells which contain no haemoglobin polymers exhibit rheological abnormalities (Chien, Usami and Bertles, 1970); these are most pronounced with the 'irreversibly sickled cells' (ISCs), a subpopulation of SS cells whose elongated shape deformities persist not only in the oxygenated cells, but also in the membranes after lysis (Jensen, Bromberg and Barefield, 1969). and even in their cytoskeletons following extraction of the ghosts with the detergent Triton X100 (Lux, John and Karnovsky, 1976). The main cause of cell rigidity of the oxygenated ISCs was shown to be their high internal viscosity, due to dehydration (Clark, Mohandas and Shohet, 1980). This is reflected in a high intracellular Hb concentration (Bertles and Milner, 1968) and in a very low K^+ with only slightly increased Na^+ content (Clark, Unger and Shohet, 1978; Glader and Nathan, 1978). ISCs were also found to have the highest Ca content among the SS cell population (Bookchin, Raventos and Lew, 1981; Eaton et al., 1973; Palek, 1977).

When Eaton presented his findings of the high Ca in SS cells at a meeting, a listener quipped, 'Are you telling us that the irreversibly sickled cell is the analogue of the avian egg?' Although the remark was made in jest, the notion that the high Ca content is directly or indirectly responsible for SS cell and/or membrane rigidity arose when the high levels were first reported, and is still commonly held. Several possible mechanisms have been proposed for such a cause–effect relationship. Weed, LaCelle and Merrill (1969) had suggested earlier that the Ca accumulating in ATP-depleted red cells induced a sol-to-gel transition at the cytosol–membrane interface which was responsible for an observed loss in cell filterability, and this was the first mechanism advanced for a 'cell-fixing' effect of the high SS cell Ca (Eaton et al., 1973). More recently, it was shown that neither ATP depletion (Feo and Mohandas, 1977; Meiselman, Evans and Hochmuth, 1978). nor Ca loading of red cells in the range of Ca found in SS cells, using the ionophore A23187 (Clark et al., 1981) produced any substantial changes in their membrane elasticity. Both manoeuvres can transform red cells from biconcave discs to spheroechinocytes, whose decreased surface-to-volume ratio result in lowered cell deformability (Clark et al., 1981). The minimum red cell Ca content required for this transformation is not established, however, and in any case, very few SS cells appear echinocytic.

Another mechanism by which a high red cell Ca content might result in membrane rigidity was described by Lorand et al. (1976) who showed that increased levels of red cell Ca^{2+} can activate a transglutaminase enzyme, leading to γ-glutamyl-ϵ-lysine crosslinks between membrane proteins, with the appearance of high molecular weight aggregates. Other investigators reported similar findings (Anderson, Davis and Carraway, 1977; Coetzer and Zail, 1979) but all have not agreed about the amount of cytoplasmic Ca^{2+} (between 50 and 500 μM) needed to activate this mechanism. Palek and Liu (1979) found no high molecular weight aggregates in SDS gels of membranes from freshly separated ISCs, and there is as yet no clear evidence that significant crosslinking of membrane proteins occurs in these or other abnormal red cells by this process.

The proposed mechanism for dehydration of SS cells which has received most attention is the Ca-sensitive K permeability or 'Gardos channel' (Gardos, 1966). Although Tosteson and his coworkers (Tosteson, 1955; Tosteson, Carlson and Dunham, 1955) had shown earlier that deoxygenation-induced sickling was accompanied by a reversible increase in both Na and K permeabilities of the SS cells, there was no clear indication that the resulting net fluxes could lead to the ionic composition found in ISCs. A selective increase in K permeability induced by the increased Ca in SS cells, however, could provide a ready explanation for their marked K depletion with net cation and water loss. Eaton et al. (1978) observed that SS cells exposed to Ca and ionophore acquired many features common to ISCs, while Glader and Nathan (1978) concluded that ATP depletion of the red cells in the presence of Ca, with activation of the 'Gardos effect', were the conditions required for SS cell dehydration and ISC generation *in vitro*, and represented the mechanism operating *in vivo*. Although the shrunken ISCs have reduced ATP per unit cell volume (Glader *et al.*, 1978), the ATP concentration per unit cell water is not reduced (Clark, Unger and Shohet, 1978).

We have confirmed the high Ca content in SS cells (Bookchin and Lew, 1980) but all of our attempts to document ongoing Ca-dependent K fluxes in SS cells have produced negative results: a measurable quinine-inhibitable K-flux was seen only transiently upon mobilization of endogenous SS cell Ca with ionophore and EGTA (Lew and Bookchin, 1980), and even with density-fractionated ISCs, which have the highest Ca content, neither quinine, ATP depletion (in the absence of external Ca) nor extraction of nearly all the cell Ca with ionophore and EGTA had any effect on passive ^{86}Rb influx (Bookchin, Ortiz and Lew, 1984).

As we discussed recently (Bookchin, Ortiz and Lew, 1984) the high Ca in SS cells seems to sustain none of the responses observed when red cells are loaded with similar levels of Ca using the ionophore; besides failure to activate the K permeability, the Ca pump is not activated to extrude either the endogenous Ca or most of the additional Ca accumulated during a period of sickling *in vitro* (Bookchin and Lew, 1980); the SS cell Ca exchanges with external ^{45}Ca at a very slow rate (Bookchin and Lew, 1980); and while it seemed reasonable that the high Ca in ISCs might be responsible for the Na pump inhibition observed in the intact cells, despite normal Na–K ATPase activity in their membranes (Bookchin, Ortiz and Lew, 1984; Clark, Morrison and Shohet, 1978), we found that a reduced V_{max} of the ouabain-sensitive ^{86}Rb influx in ISCs persisted after extraction of nearly all their endogenous Ca in the presence of non-limiting ATP (Bookchin, Ortiz and Lew, 1984). Thus, if Ca were responsible for the Na pump inhibition of ISCs, the inhibition must be totally irreversible despite high ATP levels, an effect not previously described (Brown and Lew, 1983).

The observed findings could be explained if most of the SS cell Ca were membrane bound, as has been suggested (Litosch and Lee, 1980; Palek, 1977); but nearly all the Ca, even in the ISCs, is promptly extracted when the cells are exposed to ionophore and EGTA (Bookchin and Lew, 1980; Bookchin, Ortiz and Lew, 1984). Neither does an unusually high level of cytoplasmic buffering of SS cell Ca account for its 'silence', for in the presence of ionophore, these cells show a low-normal ionized fraction comprising 20–30% of the total cell Ca, over their observed range of endogenous Ca (Bookchin and Lew, 1980). This would indicate mean Ca^{2+} levels of about 8 μM in SS cells, and over 40 μM in some of the ISC fractions. In order for Ca binding to account for Ca silence it would be necessary to postulate, in addition to high Ca-binding affinity and capacity, that the Ca-binding agent in the cell or membrane sharply reduces its Ca affinity in the presence of the ionophore A23187. Such ionophore effects have never been described.

Failure of SS cells to extrude their Ca cannot be ascribed to abnormalities of their Ca^{2+}-Mg^{2+} ATPase *per se*. We have estimated SS cell Ca permeability quantitatively by measuring ^{45}Ca influx after loading the cells non-distruptively with the Ca-chelator Benz 2 to inhibit the Ca pump (Bookchin and Lew, 1983a; Lew *et al.*, 1982), and found it to be near normal in the oxygenated cells, and to rise reversibly to a maximum of about five times normal during deoxygenation-induced sickling. None of the relatively minor reported abnormalities of the powerful Ca pump (Dixon and Winslow,

1981; Gopinath and Vincenzi, 1979; Litosch and Lee, 1980; Niggli et al., 1982) can explain the failure to balance such a small increase in Ca permeability without noticable increase in cytoplasmic Ca^{2+} levels. Observations of Luthra and Sears (1982) suggesting that SS cell Ca^{2+}-Mg^{2+} ATPase activity may be increased above normal makes the explanation of pump-leak imbalance for the net Ca gain of sickled SS cells even less tenable.

The puzzling set of findings led us to postulate (Bookchin and Lew, 1983; Bookchin, Ortiz and Lew, 1983) that most of the Ca in SS cells might be compartmentalized in intracellular endocytic inside-out vesicles (EIOVs) - endocytic vacuoles of the sort described in splenectomized persons (Holroyde and Gardner, 1970) which might persist in sickle cell anaemia patients because of their functional asplenia (Pearson, Spencer and Cornelius, 1969). During periods of increased Ca permeability associated with sickling, such EIOVs might accumulate Ca while the cell membranes also pumped it out. We have described some preliminary experiments with ghosts made from normal and SS cells lysed in cold Mg^{2+}-containing media (to prevent spontaneous endocytosis) in which the finding of ATP-dependent Ca uptake against a large outward Ca^{2+} gradient supported the existence of preformed EIOVs (Bookchin, Ortiz and Lew, 1983).

Basic support for the EIOV hypothesis as an explanation for the functional silence of SS cell Ca outlined above requires demonstration that endocytic vesicles exist within intact SS cells, and that the product of their volume and internal Ca content accounts for most of the cell-associated Ca. Our preliminary electronmicroscopic observations performed by Dr. E.R. Burns (Division of Hematology, Albert Einstein College of Medicine), did confirm the presence of membrane-enclosed vesicles in thin sections of glutaraldehyde-fixed fresh SS cells. Some of these vesicles contained electron-dense material whose origin was obscure. Analysis of vesicles on thin sections is unsatisfactory, however, because their appearance may be indistinguishable from artifacts produced by membrane infoldings. In order to determine that 'vesicles' are fully enclosed within the cells, and to measure their number, size and distribution, it is necessary to follow the three-dimensional structure of a sufficient number of whole cells in serial sections.

Examination of 0.10–0.15 µm serial sections of glutaraldehyde-fixed, fresh, saline-washed SS cells revealed some unexpected features not seen with similar preparations of normal red cells. With the normal cells, serial sections revealed tiny vesicles about 0.12 µm in diameter, usually limited to a single section and with clear contents. Against the dense haemoglobin background of the thick serial sections, it was not possible to identify membrane borders of these vesicles to distinguish them from artifacts. But serial sections of normal ghosts, after lysis of the red cells in 50—100 volumes cold solution containing 1 mM $MgCl_2$ and 2.5 mM K-Hepes, pH 7.5, pelleting at 40 000 x g for 15 min and fixation with 1% osmium, revealed distinct membrane-surrounded vesicles similar in size and frequency (one to three per cell) to those in sections of intact cells. Occasionally, slightly larger (0.25 µm diameter) vesicles were also seen in the normal ghost sections. With the intact SS cells, two distinct types of vesicular compartments were seen, as illustrated in Fig.24.1, each variously located in the centre of the cells or adjacent to the plasma membrane, and showing cross-sectional diameters between 0.05 and 2 µm. One type had sharp circular or oval outlines and clear lumens and since, on successive sections, they always showed openings to the extracellular environment, they cannot be considered genuine intracellular vesicles. In some, however, the opening respresented a very small proportion of the overall surface, as though a nearly complete endocytosis had occurred - but the possibility that such infoldings represent fixation artifacts cannot yet be excluded. The other type of vesicular structures were always wholly intracellular, and consistently contained electron-dense material filling a portion of the lumen; on serial sections, they were generally elongated and fusiform with one or more communicating lobes and often with sections of collapsed lumen. No such convoluted vesicles or inclusions were seen in normal cells or ghosts. The compostion of the electron-dense inclusions is under study. The high density seems inconsistent with fixed, denatured plasma proteins, and it is tempting to speculate that they might represent osmium-stained precipitates of calcium phosphate.

If the vesicle microenvironment favoured an initial precipitation of calcium phosphate, this by itself would tend to stimulate further accumulation of the same material by maintaining a downhill inward gradient for phosphate ions, and a pump-leak balance for Ca^{2+} favouring net vesicular Ca influx.

Intravesicular Ca accumulation turns out to be a critical condition for the viability of the EIOV hypothesis. If the mean vesicular Ca concentration were only equal to that in plasma, as might occur with a steady-state defined by a pump-leak turnover common to both the plasma membrane surrounding the cell and the vesicles, then a total vesicular volume of at least 3% of the total cell volume would be needed to account for a mean SS cell Ca content of about 50 μmol/l cells. This in turn requires an average of one spherical vesicle with nearly 2 μm diameter per cell. Of 75 SS cells we have examined thus far, with up to 36 serial sections 0.10-0.15 μm thick, only 29 of the cells showed vesicles with diameters of over 0.65 μm, and the combined volume of

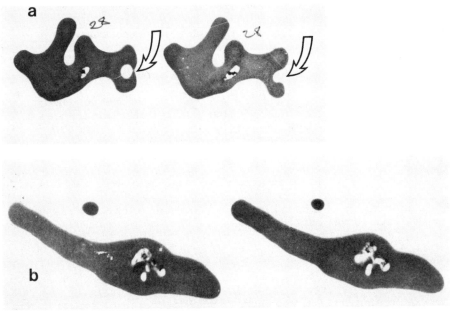

Fig.24.1. Electronmicrographs of two SS cells illustrating the appearance of genuine vesicles and vesicular artifacts. Fresh SS cells were washed in saline and fixed with glutaraldehyde in a cacodilate buffer, and stained with OsO_4. Serial sections were 0.10–0.15 μm thick. x 7000. Two consecutive sections from two different cells. Cells A and B show genuine vesicles, oval in A and multilobar in B, containing ellectron-dense material. The arrows in cell A point to an apparent vesicle with open communication to the extracellular environment.

all the vesicles in any single cell was always less than that of a single 1 μm diameter spherical vesicle. Nearly all the SS cells also contained from one to 20 smaller vesicles with diameters of 0.08–0.17 μm. The SS cells examined thus far permit only an approximate calculation of mean vesicular volume, which is about 0.03–0.3% of the total cell volume. It would then require an intravesicular Ca concentration of 50–100 times that of plasma to account for SS cell Ca. If the observed intravesicular inclusions represent precipitates with such high local Ca concentrations, the Ca should be detectable by X-ray microanalysis. Confirmation or rejection of the EIOV hypothesis may therefore soon be at hand.

Note: Since submission of this symposium chapter we have obtained definitive evidence by electron probe X-ray microanalysis that enclosed vesicles seen in cryosections of SS cells contain very high concentrations of Ca (mean, 35 mmol/kg dry weight) and P, thereby demonstrating that the excess Ca in SS cells is compartmentalized in EIOVs (Lew et al., 1985).

References

Anderson, D.R., Davis, J.L. and Carraway, K.L. (1977). Calcium-promoted changes of the human erythrocyte membrane. *J. Biol. Chem.* 252, 6617–23.

Bertles, J.F. and Milner, P.F.A. (1968). Irreversibly sickled erythrocytes: a consequence of the heterogeneous distribution of hemoglobin types in sickle cell anemia. *J. Clin. Invest.* 47, 1731–41.

Bookchin, R.M. and Lew, V.L. (1980). Progressive inhibition of the Ca-pump and Ca:Ca exchange in sickle red cells. *Nature (Lond.)* 284, 561–2.

Bookchin, R.M. and Lew, V.L. (1983). Calcium transport abnormalities in sickle-cell anemia red cells: Ca-pump failure or concealed IOVs? *Acta Physiol. Lat. Amer.* 32, 64–6.

Bookchin, R.M. and Lew, V.L. (1983). Red cell membrane abnormalities in sickle cell anemia. *Prog. Hematol.* 13, 1–23.

Bookchin, R.M., Raventos, C. and Lew, V.L. (1981). Abnormal vesiculation and calcium transport by 'one-step' inside-out vesicles from sickle cell anemia red cells. Comparisons with transport by intact cells. In *The Red Cell: Fifth International Conference on Red Cell Metabolism and Function:* Brewer (ed.), pp. 163–82. Liss, New York.

Bookchin, R.M., Ortiz, O.E. and Lew, V.L. (1983). Evidence for Ca-accumulating inside-out vesicles (IOVs) in sickle red cells. *Fed. Proc.* 42, 2244.

Bookchin, R.M., Ortiz, O.E. and Lew, V.L. (1984). Silent intracellular calcium in sickle cell anemia red cells. In *The Red Cell: Sixth International Conference on Red Cell Metabolism and Function;* Brewer (ed.). Liss, New York.

Brown, A.M. and Lew, V.L. (1983). The effect of intracellular calcium on the sodium pump of human red cells. *J. Physiol.* 343, 455–93.

Chien, S., Usami, S. and Bertles, J.F. (1970). Abnormal rheology of oxygenated blood in sickle cell anemia. *J. Clin. Invest.* 49, 623–34.

Clark, M.R., Morrison, C.E. and Shohet, S.B. (1978). Monovalent cation transport in irreversibly sickled cells. *J. Clin. Invest.* 62, 329–37.

Clark, M.R., Mohandas, N. and Shohet, S.B. (1980). Deformability of oxygenated irreversibly sickled cells. *J. Clin. Invest.* 65, 189–96.

Clark, M.R., Unger, R.C. and Shohet, S.B. (1978). Monovalent cation composition and ATP and lipid content of irreversibly sickled cells. *Blood* 51, 1169–78.

Clark, M.R., Mohandas, N., Feo, C., Jacobs, M.S. and Shohet, S.B. (1981). Separate mechanisms of deformability loss in ATP-depleted and Ca-loaded erythrocytes. *J. Clin. Invest.* **67**, 531–9.

Coetzer, T.L. and Zail, S.S. (1979). Cross-linking of membrane proteins of metabolically-depleted and calcium-loaded erythrocytes. *Brit. J. Haematol.* **43**, 375–90.

Dixon, E. and Winslow, R.M. (1981). The interaction between (Ca^{2+} + Mg^{2+})-ATPase and the soluble activator (calmodulin) in erythrocytes containing haemoglobin S. *Brit. J. Haematol.* **47**, 391–7.

Eaton, J.W., Skelton, T.D., Swofford, H.S., Kolpin, C.E. and Jacob, H.E. (1973). Elevated erythrocyte calcium in sickle cell disease. *Nature (Lond.)* **246**, 105–5.

Eaton, J.W., Berger, E., White, J.G. and Jacob, H.S. (1978). Calcium-induced damage of haemoglobin SS and normal erythrocytes. *Brit. J. Haematol.* **38**, 57–62.

Feo, C. and Mohandas, N. (1977). Clarification of the role of ATP in red cell morphology and function. *Nature (Lond.)* **265**, 166–8.

Gardos, G. (1966). The mechanism of ion transport in human erythrocytes. *Acta Biochim. Biophys. Acad. Sci. Hung.* **1**, 139–48.

Glader, B.E. and Nathan, D.G. (1978). Cation permeability alterations during sickling: Relation to cation composition and cellular dehydration of irreversibly sickled cells. *Blood* **51**, 983–9.

Glader, B.E., Lux, S.E., Muller-Soyano, A., Platt, O.S., Propper, R.D. and Nathan, D.G. (1978). Energy reserve and cation composition of irreversibly sickled cells in vivo. *Brit. J. Haematol.* **40**, 527–32.

Gopinath, R.M. and Vincenzi, F.F. (1979). (Ca^{2+} + Mg^{2+})-ATPase activity of sickle cell membranes: decreased activation by red blood cell cytoplasmic activator. *Amer. J. Hematol.* **7**, 303–12.

Holroyde, C.P. and Gardner, F.H. (1970). Acquisition of autophagic vacuoles by human erythrocytes. Physiological role of the spleen. *Blood* **36**, 566–75.

Jensen, W.N., Bromberg, P.A. and Barefield, K. (1969). Membrane deformation: A cause of the irreversibly sickled cell (ISC). *Clin. Res.* **17**, 464.

Lew, V.L. and Bookchin, R.M. (1980). A Ca^{2+}-refractory of the Ca-senstive K^+ permeability mechanism in sickle cell anaemia red cells. *Biochim. Biophys. Acta* **602**, 196–200.

Lew, V.L., Hockaday, A., Sepulveda, M.-I., Somlyo, A.P., Somlyo, A.V., Ortiz, O.E. and Bookchin, R.M. (1985). The state of calcium in sickle cell anaemia red cells. *Nature* **315**, 586–9.

Lew, V.L., Tsien, R.Y., Miner, C. and Bookchin, R.M. (1982). Physiological $[Ca^{2+}]_i$ level and pump-leak turnover in intact red cells measured using an incorporated Ca chelator. *Nature (Lond.)* **298**, 478–81.

Litosch, I. and Lee, K.S. (1980). Sickle red cell calcium metabolism: Studies on the Ca^{2+}-Mg^{2+} ATPase and Ca-binding properties of sickle red cell membranes. *Amer. J. Hematol.* **8**, 377–87.

Lorand, L., Weissman, L.B., Epel, D.L. and Bruner-Lorand, J. (1976). Role of the intrinsic transglutaminase in the Ca^{2+}-mediated crosslinking of erythrocyte proteins. *Proc. Natl. Acad. Sci. USA* **73**, 4479–81.

Luthra, M.G. and Sears, D.A. (1982). Increased Ca^{++}, Mg^{++}, and $Na^+ + K^+$ ATPase activities in erythrocytes of sickle cell anemia. *Blood* **60**, 1332–6.

Lux, S.E., John, K.M. and Karnovsky, M.J. (1976). Irreversible deformation of the spectrin-actin lattice in irreversibly sickled cells. *J. Clin. Invest.* **58**, 955–63.

Meiselman, H.J., Evans, E.A. and Hochmuth, R.M. (1978). Membrane mechanical properties of ATP-depleted human erythrocytes. *Blood* **52**,

499—504.

Niggli, V., Adunyah, E.S., Cameron, B.F., Bababunmi, E.A. and Carafoli, E. (1982). The Ca^{2+}-pump of sickle cell plasma membranes. Purification and reconstitution of the ATPase enzyme. *Cell Calcium* **3**, 131—51.

Palek, J. (1973). Calcium accumulation during sickling of hemoglobin S (HbSS) red cells. *Blood* **42**, 988.

Palek, J. (1977). Red cell calcium content and transmembrane calcium movements in sickle cell anemia. *J. Lab. Clin. Med.* **89**, 1365—74.

Palek, J. and Liu, S.C. (1979). Membrane protein organization in ATP-depleted and irreversibly sickled red cells. *J. Supramol. Struct.* **10**, 79—96.

Pearson, H.A., Spencer, R.P. and Cornelius, E.A. (1969). Functional asplenia in sickle cell anemia. *N. Engl. J. Med.* **281**, 923—6.

Tosteson, D.C. (1955). The effects of sickling on ion transport. II. The effect of sickling on sodium and cesium transport. *J. Gen. Physiol.* **39**, 55—67.

Tosteson, D.C., Carlson, E. and Dunham, E.T. (1955). The effects of sickling on ion transport. I. Effect of sickling on potassium transport. *J. Gen. Physiol.* **39**, 31—53.

Weed, R.I., LaCelle, P.L. and Merrill, E.W. (1969). Metabolic dependence of red cell deformability. *J. Clin. Invest.* **48**, 795—809.

MUSCLE CALCIUM ACCUMULATION IN MUSCULAR DYSTROPHY

Genaro M.A. Palmieri and Abbie B. Hinton
Department of Medicine

Tulio E. Bertorini
Departments of Neurology and Pathology

Syamal K. Bhattacharya
Department of Surgery

and

David F. Nutting
Department of Physiology and Biophysics

University of Tennessee Center for the Health Sciences, Memphis, Tennessee 38163, USA

Abstract

Since excessive Ca accumulation plays an important role in muscular degeneration, we determined muscle Ca content in dystrophic hamsters (DH) and patients with Duchenne muscular dystrophy (DMD). Calcium was elevated in heart, diaphragm and rectus femoris of DH by 20-fold, 7-fold and 3-fold, respectively. Similarly, in DMD children and fetuses, muscle Ca was increased 3-fold. Muscle necrosis was common in children but absent in fetuses, suggesting that Ca accumulation precedes necrosis in DMD. We studied the effect on manoeuvres capable of reducing the ingress of Ca into cells: (1) Since parathyroid hormone stimulates Ca influx, parathyroid ablation was performed in DH, resulting in a 52% reduction of Ca in the heart, 25% in the diaphragm and 47% in the rectus femoris, 55 days after surgery. There was a significant reduction in plasma creatine kinase (CK) and some histological improvement. Although hypocalcaemia occurred 6h after surgery, plasma Ca was normal 55 days later. Thus, the beneficial effects of parathyroid ablation in DH were not due to changes in plasma Ca. (2) We tested the Ca antagonist diltiazem (DTZ) in DH. DTZ, 25 mg/kg/day, orally, for 55 days reduced Ca content of heart, diaphragm and rectus femoris by 73%, 61% and 48%, respectively. Plasma CK fell by 37% and there was some histological improvement of skeletal and heart muscle. Parathyroid ablation or DTZ did not affect normal hamsters. The data suggest that manoeuvres capable of reducing cellular Ca uptake may be beneficial in muscular dystrophy.

25.1. Introduction

Although no major endocrine gland dysfunction has been observed in Duchenne muscular dystrophy (DMD), there have been a few attempts to use sex steroids and glucocorticoids in the treatment of this condition (Cohen, Morgan and Shulman, 1972; Drachman, Tokya and Myer, 1974). In spite of the lack of long-term positive results with these agents, the interest for exploring the possible role of hormones in the evolution of the dystrophic process persists. One of the most interesting observations in the area of endocrinology and muscular dystrophy was the finding of Rudman et al. (1972) of a paradoxical negative nitrogen balance following the administration of human growth hormone to children with DMD. That growth hormone may have a deleterious effect on DMD is also supported by the clinical observation of extremely mild manifestations of dystrophy in a 13-year-old boy with DMD and hyposomatotropism; he could walk unassisted and even climb stairs. Other males in his family, who have normal statutory growth and thus presumably normal pituitary function, show the typical manifestations of DMD, including inability to walk by the age of 12 years (Zatz, Betti and Levy, 1981). Similarly, preliminary work of one of us (DFN, unpublished observations) showed some beneficial biochemical and histological effects on dystrophic hamsters after removal of growth hormone by hypophysectomy and adequate replacement of thyroid and adrenal hormones. Therefore, the possibility of an improper response of muscle to endogenous hormones, circulating in plasma in normal concentrations, deserves further investigation.

In our laboratories, we explored this hypothesis further with regard to aspects of the hormonal regulation of Ca metabolism. Increased muscle Ca has been demonstrated in DMD by histological techniques (Oberc and Engel, 1977; Bodensteiner and Engel, 1978), X-ray fluorescence spectroscopy (Maunder-Sewry et al., 1980) and more recently by us, using chemical analysis of tissue extracts (Bertorini et al., 1982). Calcium content is also elevated in the heart of drug-induced (Fleckenstein, 1968) and genetic cardiomyopathic hamsters (Bhattacharya et al., 1982; Wrogeman, Jacobson and Blanchaer, 1973), as well as in skeletal muscles of the latter (Bhattacharya et al., 1982). Parathyroid hormone (PTH), the most important Ca-regulating hormone in mammals, is known to exert its physiological effect in target tissues by stimulating cyclic AMP synthesis and translocation of Ca from the extracellular space to cytosol (Borle, 1968). In supraphysiological concentrations, PTH stimulated the beating rate of heart cells, followed by cell death (Bogin, Massry and Harary, 1981). In hyperparathyroidism an increased accumulation of Ca in brain (Arieff and Massry, 1974) and muscle (Guisado, Arieff and Massry, 1977) have been reported. In order to determine if PTH, albeit in normal concentrations, could have a deleterious effect in muscular dystrophy, we studied the effects of parathyroid ablation in a suitable animal model, the BIO-14.6 hamster. In the present chapter we will also discuss two closely related areas, our experience with the Ca antagonist diltiazem in muscular dystrophy, and secondly, an attempt to determine if muscle Ca accumulation precedes or follows necrosis in DMD. For the latter study, muscle Ca content was determined at various stages of the evolution of the disease, that is, fetal, young ambulatory, and older non-ambulatory boys with DMD.

25.2. Methods

Parathyroid ablation was accomplished by surgical removal of the thyroid and parathyroid glands (TPTX) in normal and BIO-14.6 dystrophic hamsters, aged 35 days. The animals were maintained on physiological replacement of thyroxine and regular laboratory diet for 55 days. The growth rate of operated animals was indistinguishable from that of non-operated controls. At age 90 days, blood was drawn for determination of Ca, Mg and creatine kinase in plasma, and samples of diaphragm, rectus femoris, and heart (ventricles) were obtained for histological and chemical analyses. Ca and Mg were determined in acid extracts of dry, defatted muscles using the sensitive, stoichiometric, nitrous oxide-acetylene flame for atomization in atomic absorption spectrophotometry, which was developed in our laboratories (Bhattacharya, Williams and Palmieri, 1979). Histological evaluation of muscle was carried out using a battery of histochemical stains, including Alizarin Red-S for Ca (Bertorini et al., 1980). All muscle specimens were labelled using a randomized code to ensure blind histological and chemical evaluations. A detailed description of these procedures appeared elsewhere (Palmieri et al., 1981). For studies in patients with DMD or other neuromuscular diseases, and in normal subjects, muscle biopsies were processed using identical techniques.

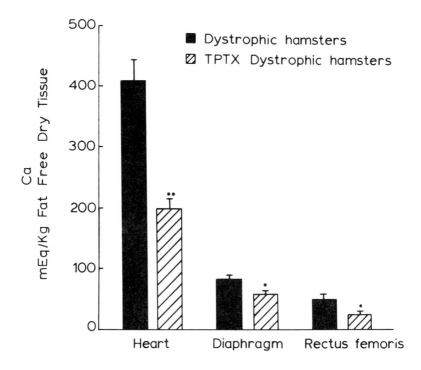

Fig.25.1. Muscle Ca content in nine dystrophic hamsters, 55 days post-TPTX, maintained on physiological throxine replacement, compared with ten intact dystrophic hamsters of the same sex and age. Vertical lines represent SEM. $*p < 0.05$, $**p < 0.001$. In non-dystrophic hamsters the Ca content in the three muscles ranged from 11 to 20 mEq/Kg fat-free dry tissue and was not affected by TPTX. (Replotted from Palmieri et al. (1981) with permission of the publishers).

25.3. Results

25.3.1. Parathyroid ablation

As shown in Fig.25.1, 55 days after parathyroid ablation in dystrophic hamsters, there were reductions of 52%, 25% and 47% in the Ca content of heart, diaphragm and rectus femoris muscles, respectively. The muscle Ca content in non-dystrophic animals was not affected by TPTX.
Semiquantitative histological evaluation of all muscles suggested a more benign degree of dystrophy in hamsters that underwent parathyroid ablation but only the reduction in fibrosis in the heart and muscle atrophy in the rectus femoris reached statistical significance (Palmieri et al., 1981). Plasma creatine kinase also was reduced following parathyroid ablation (1294 ± 416 Sigma U/ml in dystrophic control and 424 ± 67 in TPTX dystrophic animals, $p < 0.02$). As previously observed in other rodents, marked hypocalcaemia occurred 6 h after TPTX, but plasma Ca was normal 8 weeks later. Plasma Mg was unchanged.

25.3.2. Treatment with diltiazem

Similar studies were conducted in dystrophic hamsters treated with diltiazem 25 mg/kg/day orally for 55 days. This Ca antagonist provoked a marked reduction in muscle Ca in dystrophic hamsters: 73% in the heart, 61% in the diaphragm and 48% in the rectus femoris. There was some histological improvement, and a significant drop in plasma creatine kinase was observed (Bhattacharya et al., 1982). No changes were observed in non-dystrophic hamsters treated with diltiazem.

25.3.3. Muscle calcium and magnesium in Duchenne muscular dystrophy

A 3-fold elevation in Ca and a 44% reduction in Mg was observed in skeletal muscle of patients with DMD (Table 25.1). Muscle Ca also rose and Mg fell in most other neuromuscular diseases, but these changes were much less than in DMD ($p < 0.005$). Although non-collagen protein in muscle was decreased in DMD, it did not differ form that in other neuromuscular diseases (Table 25.2). Thus, the mineral changes in DMD cannot be related solely to a reduction in muscle mass.

Muscles from two male fetuses of gestational age 10 and 20 weeks, whose mothers were proven DMD carriers, and from three fetuses of the same gestational age from normal women following voluntary abortions were studied in collaboration with Drs F. Cornelio, I. Dones, F. Dworzak and B. Brambati from Milan, Italy. In addition, muscle from a premature infant (7 months) whose mother was a proven DMD carrier was obtained one week after birth in our medical centre. This boy later showed clinical and laboratory evidence of DMD. A detailed description of the chemical and histological findings on these muscles appears elsewhere (Bertorini et al., 1985). In summary, muscle Ca content was elevated in the fetuses at risk and in the premature infant. In Fig.25.2, the muscle Ca and Mg of these subjects are compared with those obtained in normal subjects and boys with DMD. The latter group was divided according to the severity of the dystrophy in two groups: ambulatory, less than 8 years of age, and non-ambulatory, older than 8 years. Muscle Ca was at least 3-fold higher than the control values in all groups, and no difference was observed between the younger and older DMD groups. There was a mild, non-significant elevation of muscle Mg in fetuses but Mg was markedly reduced in Duchenne boys ($p < 0.0001$; see Table 25.1). Moreover, Mg was significantly lower in non-ambulatory DMD boys than in the ambulatory group ($p < 0.0001$).

Table 25.1. Calcium, magnesium and Mg/Ca ratio in muscle. (reproduced from Bartorini et al (1982), with permission of the publishers.)

Group (n)	Calcium*	Magnesium*	Mg/Ca
A. Normal subjects (Brambati et al., 1980)	8.49 ± 0.36	75.18 ± 3.69	9.06 ± 0.49
B. Duchenne muscular dystrophy (Wrogeman and Rena, 1976)	24.63 ± 0.84	42.36 ± 2.45	1.75 ± 0.11
C. Denervating diseases# (Arieff and Massry, 1974)	16.12 ± 1.07	60.07 ± 3.61	3.98 ± 0.44
D. Polymyositis-dermatomyositis (Bhattacharya et al., 1982)	15.77 ± 1.74	60.80 ± 3.37	4.55 ± 0.71
E. Other neuromuscular diseases** (Bhattacharya et al., 1982)	15.61 ± 1.82	77.26 ± 5.34	5.41 ± 0.51
Significance testing			
A versus B	$P < 0.0001$	$P < 0.0001$	$P < 0.0001$
A versus C	$P < 0.0001$	$P < 0.01$	$P < 0.0001$
A versus D	$P < 0.0001$	$P < 0.02$	$P < 0.0001$
A versus E	$P < 0.0001$	NS	$P < 0.0001$
B versus C,D,E	$P < 0.0001$	$P < 0.005$	$P < 0.001$

* mEq/kg Fat-free dry tissue, mean ± SEM
\# Peripheral neurophathies and spinal muscular atrophy
** Congenital myopathies, limb girdle muscular dystrophy and fascioscapulohumeral muscular dystrophy
NS :not significant

Table 25.2. Non-collagen nitrogen in muscle (reproduced from Bertorini et al. (1982), with permission of the publishers.)

Group (n)	Non-collagen nitrogen[*] ($\mu g/mg$ fresh weight)	P value[#]
Normal subject (Wrogeman, Jacobson and Blanchaer, 1973)	25.53 ± 1.37	------
Duchenne muscular dystrophy (Oldfield, Ellis and Muth, 1958)	13.14 ± 1.22	$P < 0.0001$
Denervating diseases (Fleckenstein, 1968)	14.11 ± 2.04	$P < 0.0001$
Polymyositis-dermatomyositis (Oberc and Engel, 1977)	19.28 ± 3.34	NS
Other neuromuscular diseases (Oberc and Engel, 1977)	17.87 ± 4.99	NS

[*] Mean ± SEM
[#] For comparison with control value; differences between Duchenne muscular dystrophy and all disease control groups were not significant
NS: not significant

The histological examination of muscle from all DMD boys showed opaque fibres, Ca-positive fibres, phagocytosis, fibrosis and necrosis typical of DMD. In fetuses at risk and the premature infant with DMD many Ca-positive fibres and opaque fibres were observed, but not a single necrotic fibre was detected in multiple sections of at least two skeletal muscles. A detailed description of the histological findings appears elsewhere (Bertorini et al., 1982, 1985).

25.4. Discussion

The data presented herein and the results reported by us and others (Bertorini et al., 1982; Bhattacharya et al., 1982; Bodensteiner and Engel, 1978; Maunder-Sewry et al., 1980; Oberc and Engel, 1977) clearly demonstrate that there is an exaggerated accumulation of muscle Ca in both DMD and dystrophic hamsters.

In DMD, Ca accumulation starts very early in life. As early as 10 weeks of gestational age muscle Ca is increased more than 3-fold and this massive Ca overload persists for the rest of life with apparently minor changes. In DMD, muscle necrosis is detected as early as 2 months of age (Hudgson, Pearce and Walton, 1967), but as shown by others (Brambati et al., 1980; Emery and Burt, 1980) and in this communication, no histological signs of necrosis are detected in fetuses at risk of DMD and in a premature infant with DMD. This

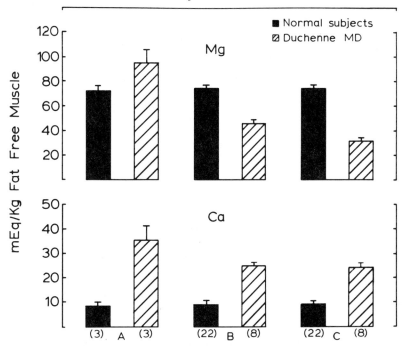

Fig.25.2. Effect of age on Ca and Mg content in skeletal muscle from Duchenne muscular dystrophy and normal subjects. (A) Fetal and premature muscle. Ambulatory patients (B), 8 years old or younger had higher Mg content ($p < 0.0001$) than non-ambulatory children older than 8 years (C). No significant differences in Ca content were observed among groups A, B or C. The bars represent mean, and the vertical lines, SEM.

indicates that Ca accumulation precedes necrosis and, therefore, it may have a role in the pathogenesis of muscle necrosis in muscular dystrophy. Certainly it cannot be only the result of necrosis. That Ca plays a role in the progression of the dystrophic process has been postulated on the basis of a variety of experimental approaches (Ebashi and Sugita, 1979; Godwin, Edwardly and Fuss, 1975; Lossnitzer et al., 1975; Oberc and Engel, 1977; Oldfield, Ellis and Muth, 1958; Wrogeman and Pena, 1976; Wrogeman, Jacobson and Blanchaer, 1973). Three major mechanisms could be involved in the deleterious effect of cellular Ca overload: (a) alterations in energy metabolism (Fleckenstein, 1968; Fleckenstein et al., 1974); (b) activation of specific calcium-dependent proteases (Dayton et al., 1976; Duncan, 1978; Kar and Pearson, 1976; Neerunjun and Dubowitz, 1979); and (c) activation of phospholipases (Epps et al., 1982; Mittnacht, Sherman and Farber, 1979).

Our finding of a beneficial effect of parathyroid ablation in dystrophic hamsters also supports the hypothesis of the important role of cellular Ca accumulation in the progression of the dystrophic process. Parathyroid hormone is known to induce Ca influx into cells of target tissues as part of its physiological cellular action (Rasmussen et al., 1976). In hyperparathyroidism, not only do classical target tissues such as the kidney become overloaded with Ca, but also Ca accumulates in the brain (Arieff and Massry, 1974) and muscle (Guisado, Arieff and Massry, 1977). The myopathy of primary hyperparathyroidism, a condition characterized by hypercalcaemia, is indistinguishable from that of diseases causing secondary hyperparathyroidism, such as osteomalacia, in which plasma Ca is frequently low (Mallette, Patten and Engel, 1975). Thus, high plasma concentration of PTH affects skeletal muscle independently of the levels of plasma Ca. Since the extracellular Ca concentration is approximately four orders of magnitude higher than in the cytosol, the relative small changes of a few mg/dl of plasma Ca observed in human hypercalcaemia and hypocalcaemia accompanying primary or secondary hyperparathyroidism do not appear to be the major cause of muscle damage in any form of hyperparathyroidism. Therefore, excessive plasma PTH concentration is the cause of muscle damage in hyperparathyroidism, possibly by increasing the flux of Ca into cells.

The parathyroid glands of dystrophic hamsters showed a normal histological appearance and the serum Ca was normal. Therefore, there was not evidence of hyperparathyroidism in these animals. Moreover, the plasma concentration of PTH in children with DMD is within normal limits (Palmieri, Bertorini and Hinton, 1985). These data suggest that PTH, albeit in normal concentrations, has a deleterious effect in a disease characterized by defective plasma membranes, such as muscular dystrophy. Parathyroid ablation did not cause chronic hypocalcaemia in dystrophic hamsters. Thus, the beneficial effect of this procedure was not the result of changes in the extracellular Ca concentration, but the elimination of the PTH-induced Ca translocation into dystrophic muscle. It is conceivable that PTH may have also a detrimental effect in other conditions in which genetic or acquired alterations of plasma membrane function may occur, particularly regarding the integrity of Ca pump mechanisms. In further support of this proposal, we observed a doubling of skeletal muscle Ca content in dogs with acute pancreatitis (Palmieri et al., 1983), a condition in which alteration of plasma membranes has been demonstrated (Bockman et al., 1973). Moreover, somatostatin, a naturally occurring Ca antagonist, appears to have a beneficial effect in human and experimental acute pancreatitis (Schwedes et al., 1979; Usadel, Leuschner and Uberla, 1980). The role of intracellular Ca accumulation in the pathogenesis of a variety of diseases is discussed in other chapters throughout this book.

Since PTH stimulates the flux of Ca^{2+} into cells (Borle, 1968), it would be plausible to diminish the secretion of PTH to a minimum in order to reduce the intracellular accumulation of Ca in these conditions. Since the major regulatory determinant of PTH secretion is the plasma concentration of Ca^{2+}, it follows paradoxically that Ca administration could reduce cellular Ca overload by reducing PTH secretion; a small increment in plasma Ca concentration would have a neglible direct effect on cellular Ca (see above) but would be sufficient to diminish the secretion of PTH.

The hypothetical mechanism for a negative Ca balance to increase cell Ca is summarized in Fig. 25.3. Low Ca diets, defective intestinal Ca absorption (vitamin D deficiency or intrinsic intestinal lesions, etc.), or excessive renal loss of Ca that is not compensated by increased intestinal absorption lead to hypocalcaemia that triggers PTH secretion, mobilizing Ca from bone and often nearly normalizing the concentration of Ca^{2+} in the extracellular compartment. A secondary result of this homeostatic regulation of extracellular Ca is that the increased secretion of PTH also increases the flux of Ca into the cells of certain tissues. This is probably more striking if the function of the Ca pump is jeopardized by congenital or acquired defects of plasma membranes. If the alteration of the plasma membrane is severe, as it appears to be in muscular dystrophy, even normal levels of circulatory PTH could enhance Ca influx to muscle, thus aggravating the course of the disease.

Obviously, other hormones that are capable of enhancing cellular Ca influx, such as catecholamines and calcitriol, share with PTH the property of being potentially detrimental in disorders characterized by defective Ca exchange functions of plasma membranes. Moreover, the capacity of different tissues to adjust to Ca overload may differ or may vary with age in the same tissue. In DMD, the fetal muscle contains as much excessive Ca as older DMD patients,

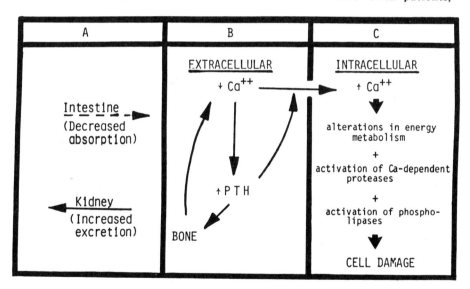

Fig.25.3. Hypothetical role of Ca balance in cellular Ca content. (A) Reduced absorption or increased excretion of Ca (unmatched by increased intestinal absorption) results in (B) hypocalcaemia, that stimulates PTH secretion. PTH mobilizes Ca from bone (and increases renal Ca reabsorption), thus normalizing the extracellular Ca concentration. The elevated PTH also (C) stimulates Ca translocation into cells, particularly in the presence of congenital or acquired defective Ca exchange functions of plasma membranes. The excessive intracellular Ca could cause permanent or reversible cell damage. Thus, suppression of PTH secretion by exogenous Ca might prevent cell damage in certain diseases.

yet no necrotic fibres are present. Perhaps early in life, intracellular homeostatic mechanisms capable of counteracting the Ca challenge are more effective, or less exhausted, by the relatively shorter exposure to high intracellular Ca. The capacity of very young muscle to withstand a Ca challenge is illustrated by the experiments of Chizzonite and Zak (1981) who demonstrated that isolated hearts from normal rats less than 6 days old were unaffected by rapid changes in the Ca concentration of the perfusate, whereas hearts from older animals were rapidly and severely damaged.

Whether the apparently greater resistance to a Ca challenge by fetal or very young muscle is related to the stage of tissue differentiation or to still undetermined protective factors, deserves further investigation. One of these factors could be Mg, which was slightly higher than normal in fetal, non-necrotic DMD muscle but fell to very low levels as muscle necrosis became intense with the progression of the dystrophy. In addition, Mg prevented isoproterenol-induced myocardial damage in rats (Fedelešová et al., 1975), and Ca loaded mitochondria from dystrophic muscle retained many of their functions when Mg depletion was prevented (Wrogeman, Blanchaer and Jacobson, 1980).

The study of intracellular metabolism of divalent cations in muscular dystrophy may provide valuable information for understanding the pathogenesis of this disease and the development of potential remedies.

Acknowledgements

This work was supported in part by a grant from the Muscular Dystrophy Association of America, and by USPHS Grant no. RR002211.

References

Arieff, A.I. and Massry, S.G. (1974). Calcium metabolism of brain in acute renal failure: effects of uremia, hemodialysis and parathyroid hormone. *J. Clin. Invest.* **53**, 387–92.

Bertorini, T.E., Yeh, Y., Trevisan, C., Stadlan, E., Sabesin, S. and DiMauro, S. (1980). Carnitine palmityl transferase deficiency: myoglobinuria and respiratory failure. *Neurology (NY)* **30**, 263–71.

Bertorini, T.E., Bhattacharya, S.K., Palmieri, G.M.A., Chesney, C.M., Pifer, D. and Baker, B. (1982). Muscle calcium and magnesium content in Duchenne muscular dystrophy. *Neurology (NY)* **32**, 1088–92.

Bertorini, T.E., Cornelio, F., Bhattacharya, S.K., Palmieri, G.M.A., Dones, I., Dworzak, F. and Brambati, B. (1984). Calcium and magnesium content in fetuses at risk and pre-nectrotic Duchenne muscular dystrophy. *Neurology (NY)* **34**, 1436–40.

Bhattacharya, S.K., Palmieri, G.M.A., Bertorini, T.E., Nutting, D.F. (1982). Effect of the calcium antagonist diltiazem in dystrophic hamsters. *Muscle Nerve* **5**, 73–78.

Bhattacharya, S.K., Williams, J.C. and Palmieri, G.M.A. (1979). Determination of calcium and magnesium in cardiac and skeletal muscle by atomic absorption spectroscopy using stoichoimetric nitrous oxide-acetylene flame. *Anal. Lett.* **12**, 1451–75.

Bockman, D.E., Schiller, W.R., Suriyapa, C., Mutchler, J.H.W. and Anderson, M.C. (1973). Fine structure of early experimental acute pancreatitis in dogs. *Lab. Invest.* **28**, 584–92.

Bodensteiner, J.B. and Engel, A.G. (1978). Intracellular calcium accumulation in Duchenne dystrophy and other myopathies: a study of 567,000 muscle fibers in 114 biopsies. *Neurology (NY)* **28**, 439–46.
Bogin, E., Massry, S.G. and Harary, I. (1981). Effect of parathyroid hormone on rat heart cells. *J. Clin. Invest.* **67**, 1215–27.
Borle, A.B. (1968). Calcium transport in cell culture and the effects of parathyroid hormone. In *Parathyroid Hormone and Thyrocalcitonin (Calcitonin)*. Talmage and Belanger (eds). ICS No. 159, pp. 258–72. Excerpta Medica, Amsterdam.
Brambati, B., Cornelio, F., Dworzak, F. and Dones, I. (1980). Calcium-positive muscle fibres in fetuses at risk for Duchenne muscular dystrophy. *Lancet* **2**, 969–70.
Chizzonite, R.A. and Zak, R. (1981). Calcium-induced cell death: susceptibility of cardiac myocytes is age-dependent. *Science* **213**, 1508–10.
Cohen, L., Morgan, J. and Shulman, S. (1972). Diethylstilbestrol: Observations on its use in Duchenne muscular dystrophy. *Proc. Soc. Exp. Biol. Med.* **140**, 830–5.
Dayton, W.R., Goll, D.E., Zeece, M.G., Robson, R.M. and Reville, W.J. (1976). A Ca^{++}-activated protease possibly involved in myofibrillar protein turnover. Purification from porcine muscle. *Biochemistry* **15**, 2150–8.
Drachman, D.B., Toyka, K.V. and Myer, E. (1974). Prednisone in Duchenne muscular dystrophy. *Lancet* **2**, 1409–12.
Duncan, C.J. (1978). Role of intracellular calcium in promoting muscle damage: a strategy for controlling the dystrophic condition. *Experientia* **34**, 1531–5.
Ebashi, S. and Sugita, H. (1979). The role of calcium in physiological and pathological processes of skeletal muscle. In *Current Topics in Nerve and Muscle Research*. Aguayo, Karpati (eds). ICS No. 455, pp. 73–84. Excerpta Medica, Amsterdam.
Emery, A.E.H. and Burt, D. (1980). Intracellular calcium and pathogenesis and antenatal diagnosis of Duchenne muscular dystrophy. *Brit. Med. J.* **1**, 355–7.
Epps, D.E., Palmer, J.W., Schmid, H.H.O. and Pfeiffer, D.R. (1982). Inhibition of permeability-dependent Ca^{2+} release from mitochondria by N-acylethanolamines, a class of lipids synthesized in ischemic heart tissues. *J. Biol. Chem.* **257**, 1383–91.
Fleckenstein, A. (1968). Myokardstoffwechsel und Nekrose. In *VI Symposium der Deutsch, Ges. für Fortschritte auf dem Gebiet der Inneren Medizin über 'Herzinfarkt und Schock'*. Heilmeyer and Holtmeier (eds), pp.94–109. Georg Thieme Verlag, Stuttgart.
Fleckenstein, A., Janke, J., Doring, H.J. and Leder, O. (1974). Myocardial fiber necrosis due to intracellular calcium overload; a new principle in cardiac pathophysiology. In *Myocardial Biology. Recent Advances in Studies on Cardiac Structure and Metabolism. No.4*, N.S. Dhalla, pp. 563–80. University Park Press, Baltimore.
Fedelešová, M., Zielgelhöffer, A., Luknárová, O. and Kostolansky, S. (1975). Prevention of K, Mg-aspartate of isoproterenol-induced metabolic changes in the myocardium. In *Recent Advances in Studies on Cardiac Structure and Metabolism. No. 6*. Fleckenstein and Rona (eds), pp. 59–74. University Park Press, Baltimore.
Godwin, K.O., Edwardly, J. and Fuss, C.N. (1975). Retention of ^{45}Ca in rats and lambs associated with the onset of nutritional muscular dystrophy. *Aust. J. Biol. Sci.* **28**, 457–64.
Guisado, R., Arieff, A.I. and Massry, S. (1977). Muscle water and electrolytes in uremia and the effects of hemodialysis. *J. Lab. Clin. Med.* **89**, 322–31.

Hudgson, P., Pearce, G.W. and Walton, J.N. (1967). Pre-clinical muscular dystrophy: histopathological changes observed on muscle biopsy. *Brain* **90**, 565–76.
Kar, N.C. and Pearson, C.M. (1976). A calcium-activated neutral protease in normal and dystrophic human muscle. *Clin. Chim. Acta* **73**, 293–7.
Lossnitzer, K., Janke, J., Hein, B., Stauch, M. and Fleckenstein, A. (1975). Disturbed myocardial calcium metabolism: a possible pathogenetic factor in the hereditary cardiomyopathy of the Syrian hamster. In *Recent Advances in Studies on Cardiac Structure and Metabolism. No. 6.* Fleckenstein and Rona (eds), pp. 207–17. University Park Press, Baltimore.
Mallette, L.E., Patten, B.M. and Engel, W.K. (1975). Neuromuscular diseases in secondary hyperparathyroidism. *Ann. Int. Med.* **82**, 474–83.
Maunder-Sewry, C.A., Gorodetsky, R., Yarom, R. and Dubowitz, V. (1980). Element analysis of skeletal muscle in Duchenne muscular dystrophy using x-ray fluorescence spectrometry. *Muscle Nerve* **3**, 502–8.
Mittnacht, S., Jr., Sherman, C.S. and Farber, J.L. (1979). Reversal of ischemic mitochondrial dysfunction. *J. Biol. Chem.* **254**, 9871–8.
Neerunjun, J.S. and Dubowitz, V. (1979). Increased calcium-activated neutral protease activity in muscles of dystophic hamsters and mice. *J. Neurol. Sci.* **40**, 105–11.
Oberc, M.A. and Engel, W.K. (1977). Ultrastructural localization of calcium in normal and abnormal skeletal muscle. *Lab. Invest.* **36**, 566–77.
Oldfield, J.E., Ellis, W.W. and Muth, O.H. (1958). White muscle disease (myopathy) in lambs and calves. III. Experimental production in calves from cows fed alfalfa hay. *J. Amer. Vet. Med. Assoc.* **132**, 211–4.
Palmieri, G.M.A., Bertorini, T.E. and Hinton, A. (1985). Plasma parathyroid hormone in Duchenne muscular dystrophy. *Neurology (NY)* **35**, Suppl. 1, 164–5.
Palmieri, G.M.A., Nutting, D.F., Bhattacharya, S.K., Bertorini, T.E. and Williams, J.C. (1981). Parathyroid ablation in dystrophic hamsters: effects on calcium content and histology of heart, diaphragm and rectus femoris. *J. Clin. Invest.* **68**, 646–54.
Palmieri, G.M.A., Bertorini, T.E., Nutting, D.F., Bhattacharya, S.K., Luther, R.W. and Pate, J.W. (1983). Muscle calcium accumulation in muscular dystrophy and acute pancreatitis a common pathogenetic mechanism. *Acta Endocrinol. (Suppl)* **256**, 103.
Rasmussen, H., Goodman, D.B.P., Friedmann, N., Allen, J.E. and Kurokawa, K. (1976). Ionic control of metabolism. In *Handbook of Physiology, Section 7 Endocrinology, Vol. VII. Parathyroid Gland,* Greep, Astwood, Aurbach, Geiger (eds), pp.225–64. Physiology Society, Washington, DC.
Rudman, D., Chyatte, S.B., Patterson, J.H., Gerbon, G.G., O'Beirne, I., Barlow, J., Jordan, A. and Shavin, J.S. (1972). Metabolic effects of human growth hormone and of estrogens in boys with Duchenne muscular dystrophy. *J. Clin. Invest.* **51**, 1118–24.
Schwedes, U., Althoff, P.H., Klempa, I., Leuschner, U., Mothes, L., Raptis, S., Wdowinski, J. and Usadel, K.H. (1979). Effect of somatostatin on bile-induced acute hemorrhagic pancreatitis in the dog. *Horm. Metab. Res.* **11**, 655–61.
Usadel, K.H., Leuschner, U. and Uberla, K.K. (1980). Treatment of acute pancreatitis with somatostatin: a multicenter double-blind trial. *N. Engl. J. Med.* **303**, 999.
Wrogemann, K. and Pena, S.D.J. (1976). Mitochondrial calcium overload: A general mechanism for cell-necrosis in muscle diseases. *Lancet* **1**, 672–3.
Wrogemann, K., Blanchaer, M.C. and Jacobson, B.E. (1970). A

magnesium-responsive defect of respiration and oxidative phosphorylation in skeletal muscle mitochondria of dystrophic hamsters. *Can. J. Biochem.* **48**, 1332–8.

Wrogemann, K., Jacobson, B.E. and Blanchaer, M.C. (1973). On the mechanism of a calcium-associated defect of oxidative phosphorylation in progressive muscular dystrophy. *Arch. Biochem. Biophys.* **159**, 267–78.

Zatz, M., Betti, R.T.B. and Levy, J.A. (1981). Benign Duchenne muscular dystrophy in a patient with growth hormone deficiency. *Amer. J. Med. Genet.* **10**, 301–4.

MALIGNANT HYPERTHERMIA

G.A. Gronert

Department of Anesthesiology, Mayo Clinic, Rochester, Minnesota, USA

Abstract

Malignant hyperthermia (MH) is a clinical syndrome that results from abnormal calcium transients in skeletal muscle. Apparently there is a sudden loss of control of intracellular unbound ionized calcium, with compensatory increases in aerobic and anaerobic metabolism to supply energy to drive the calcium pumps. This results, secondarily, in increased heat and acid production. The usual precipitating factors include potent general anaesthetics, such as halothane, and depolarizing muscle relaxants, such as succinylcholine. Once triggered, the course of MH can be fulminant, with body temperature increasing 1°C per 5 min, and blood pH decreasing to less than 7.00. Dantrolene is the specific drug of choice for treatment, because it dramatically and rapidly reverses the signs of increased metabolism, presumably due to its effects in inhibiting SR calcium release without affecting SR calcium uptake. Evaluation of other family members for this inherited myopathy involves, primarily, examination of threshold contracture responses in a viable cut biopsy of skeletal muscle. Susceptible individuals have a lower threshold than normal.

A disorder that might be related to MH is the neuroleptic malignant syndrome, which follows prolonged use of neuroleptic drugs for psychiatric use. The onset is much slower than in MH; the clinical features include stupor, muscle rigidity and high fever.

Malignant hyperthermia is a disorder of wildly uncontrolled metabolism leading to high temperatures and death if not diagnosed early and treated properly (Gronert, 1980). It was first reported by Denborough and Lovell in 1960. For several years thereafter, numerous case reports emphasized the importance of malignant hyperthermia in modern anaesthesia. It was obvious that skeletal muscle was a major factor in the pathogenesis of the acidosis and fever. First, the rapid rise in temperature was frequently accompanied by muscle rigidity. Blood levels of creatine phosphokinase were frequently elevated in affected individuals or their relatives. Viable specimens from muscle biopsies, mounted in temperature-controlled baths, demonstrated abnormal responses to contracture-producing drugs such as caffeine or halothane. Their threshold was less than that of normal muscle. Finally, dantrolene, a skeletal muscle relaxant that lessens calcium release from the sarcoplasmic reticulum, was highly specific in the treatment of malignant hyperthermia, whether porcine or human.

As to aetiology, the data from both clinical and laboratory observations suggest abnormal calcium transients within the muscle fibre (Gronert, 1980; Lopez et al., 1984; Stadhouders et al., 1984). In normal subjects, neuronal depolarization produces acetylcholine release at the neuromuscular junction with a resultant chemical depolarization of the skeletal muscle membrane. This produces an electrical depolarization of the surface muscle membrane. This wave of depolarization is actively propagated down the transverse tubules and in some as yet unexplained manner induces a release of calcium from the sarcoplasmic reticulum. Calcium removes the troponin inhibition of the contractile proteins which proceed into contractile activity with utilization of energy. Further energy is necessary for the calcium pump to transfer this calcium - free, ionized and intracellular - back into the sarcoplasmic reticulum. Thus both contraction and relaxation are energy-requiring conditions (Fig. 26.1). In malignant hyperthermia, this control of calcium by the sarcoplasmic reticulum is apparently lost. As a result, large amounts of ATP are consumed in attempts to pump calcium back into the SR. This leads to dissipation of energy stores within the muscle, accelerated aerobic metabolism and anaerobic glycolysis. Lactate and CO_2 accumulate, leading to an intracellular acidosis. As these diffuse into the surrounding extracellular fluid and venous blood, a whole body metabolic and respiratory acidosis results. This is particularly severe if ventilation cannot

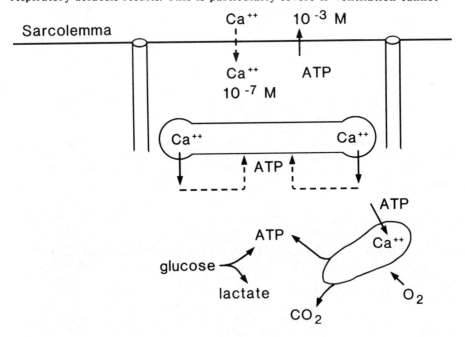

Fig.26.1. Diagram of relaxed skeletal muscle. Calcium fluxes can occur across sarcolemma, SR (between the vertical transverse tubules), and mitochondrion. Calcium transport across the sarcolemma or into the intracellular organelles requires energy (ATP). ATP is produced by glycolysis or by aerobic metabolism. (*Seminars in Anesthesia*, 2:197, 1983, used with permission.)

increase adequately to clear the excess carbon dioxide. Since these reactions are exothermic, excess heat is produced resulting in an increase in temperature of the effluent blood. All of these processes lead to sympathetic stimulation and eventually increased cellular permeability of the muscle cells with a release of potassium, creatine phosphokinase and myoglobin. There is an associated inward movement of water to produce cellular oedema and concentration of intravascular substances such as sodium and haemoglobin. The site of the abnormality in malignant hyperthermia, particularly because of the rapid specific action of dantrolene, is one of or a combination of the following: surface membrane, transverse tubules, the link between the transverse tubules and the sarcoplasmic reticulum and/or the sarcoplasmic reticulum itself. It is theorized that malignant hyperthermia may be a generalized disorder of membranes or calcium-controlling enzyme systems, but direct support is lacking.

The action of dantrolene is rapid and specific and has been documented in both pigs and humans. The effects of dantrolene are remarkably more effective than those of symptomatic treatment alone (Fig. 26.2). Symptomatic treatment is still important, however, and should include cooling by whatever means is available, increased ventilation to remove carbon dioxide produced both by the increased aerobic metabolism and bicarbonate neutralization of lactate, bicarbonate - 2-4 mEq/kg, dantrolene 1-2 mg/kg - as well as

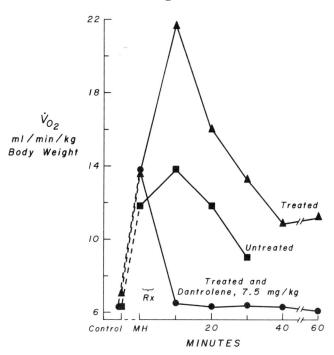

Fig.26.2. Contrasts among groups of swine (five pigs per group) for three treatment options during acute fulminant malignant hyperthermia: none (untreated, ■); treated (symptomatic treatment only, ▲); and treated and dantrolene (symptomatic treatment plus dantrolene, ●). $\dot{V}_{O_2}$ is whole body oxygen consumption. (*Anesthesiology*, 44:488, 1976, used with permission.)

discontinuation of all anaesthetic agents as rapidly as possible. Dantrolene may be repeated each 5–10 min to a total dose of 10 mg/kg. In particular, the potent volatile agents and succinylcholine should be discontinued immediately, and a new machine and tubing used as soon as is feasible. This regimen is associated with good therapeutic results in both pigs and humans.

Anaesthetic triggers include the potent volatile agents such as halothane, enflurane and isoflurane and the depolarizing muscle relaxants such as succinylcholine and decamethonium. These drugs should therefore always be avoided in susceptible individuals. Awake malignant hyperthermia episodes in the absence of exposure to anaesthetic agents are common in susceptible swine and rare in humans. In fact, human awake episodes are not well documented and anecdotal cases need to be reported in considerable detail much as the original malignant hyperthermia case reports were so as to remove all doubt as to their existence. Swine are perhaps more likely to undergo awake triggering because of their 'more pure' inbreeding - in the awake situation they are easily triggered by increased environmental temperature and stress, such as fighting, excitement, shipping, coitus, weaning.

Because of triggering in these situations and because of some problems with affected humans, it has been suggested that sympathetic effects may result in triggering in humans and pigs. This has been discounted in pigs because total spinal blockade with resulting sympathetic denervation does not prevent halothane-induced malignant hyperthermia, and because sympathetic agonists do not trigger malignant hyperthermia under tightly controlled conditions. While this situation is difficult to produce in the intact animals, an isolated, perfused preparation of caudal muscle results in a situation in which there is not a decrease in cardiac output or perfusion and no deterioration of blood pressure or other such necessary factors. Under these conditions alpha-sympathetic and beta-sympathetic agonists do not produce changes consistent with malignant hyperthermia. These changes are, however, produced by increased temperature *per se* or muscle depolarization as produced by carbachol or acetylcholine (Gronert, Milde and Taylor, 1980). Obviously then, individuals susceptible to malignant hyperthermia may, under conditions leading to heat stroke, end up in a situation in which the increase in body temperature *per se* could trigger a hyperthermic response above and beyond that response due to heat stroke of and by itself. However, one need not differentiate among awake high fevers as to the likelihood of malignant hyperthermia. Regardless of aetiology, if control is difficult (as can happen in the very muscular or obese patient), dantrolene should be used as adjunctive therapy to depress muscle metabolism (Jardon, 1982). These febrile states include heat strokes (Jardon, 1982) and certain drug-induced fevers (Smithson, Gronert and Moss, 1983), including the neuroleptic malignant syndrome (Weinberg and Twersky, 1983). Other supportive therapy is still necessary, for example, acid-base, salt, fluid and electrolyte balance, cooling.

Safe anaesthesia for susceptible individuals includes preoperative treatment with dantrolene, barbiturates, nitrous oxide, narcotics, tranquillizers, and non-depolarizing relaxants. Until now, oral dantrolene was recommended as preanaesthetic therapy. However, recent data indicate that intravenous dantrolene is more predictable and therefore preferred (Flewellyn *et al.*, 1983; Gronert, 1983). The dose is 2–3 mg/kg and, if indicated, may be repeated 6–8 hours later. The above management has in general been associated with safe and uncomplicated anaesthesia. There are no situations in which triggering agents can be given to susceptible patients even with the use of dantrolene.

Detection of susceptible patients is still controversial (Gronert, 1983) and involves personal and family history, blood CPK values and muscle biopsy for contracture responses. An elevated blood CPK value in a resting fasting state in a close relative of one known to be susceptible to malignant hyperthermia is virtually diagnostic of susceptibility and no further testing is needed. If the CPK is normal, one has no idea about susceptibility and muscle biopsy must be performed. Contracture responses due to the combination of caffeine and halothane lead to a broad range of responses in both normal and susceptible patients that are very difficult to interpret. As a result, many investigators utilize either caffeine alone or halothane alone in their screening (Rosenberg and Reed, 1983). Overall the responses appear to be consistent in patients surviving malignant hyperthermia episodes and therefore reliable. It is difficult to predict incidence of false-positive reactions because these patients will never be exposed to triggering agents. It is also difficult to recognize false-negative reactions because these patients are frequently treated very cautiously by their anaesthesiologist and hence not exposed to triggering agents.

Other screening tests for susceptibility have not been well established. ATP depletion from platelets, introduced at the University of Colorado, has contradictory evidence concerning its accuracy (Giger and Kaplan, 1983; Gronert, 1983; Verburg et al., 1984). The phosphorylase ratio was introduced by Willner from Columbia University but a followup study by Traynor, Van Dyke and Gronert (1983) fails to confirm it. Calcium uptake from thin strips of muscle likewise is an unconfirmed test (Gronert, 1983) and may have an inordinate number of false positive results (Schwartz, Rockoff and Koka, 1984).

References

Flewellen, E.H., Nelson, T.E., Jones, W.P., Arens, J.F. and Wagner, D.L. (1983). Dantrolene dose response in awake man: implications for management of malignant hyperthermia. *Anesthesiology* 59, 275–80.

Giger, U. and Kaplan, R.F. (1983). Haolthane-induced ATP depletion in platelets from patients susceptible to malignant hyperthermia and from controls. *Anesthesiology* 58, 347–52.

Gronert, G.A. (1980). Malignant hyperthermia. *Anesthesiology* 53, 395–423.

Gronert, G.A. (1983). Controversies in malignant hyperthermia. *Anesthesiology* 59, 273–4.

Gronert, G.A., Milde, J.H. and Taylor, S.R. (1980). Porcine muscle responses to carbachol, α and β-adrenoreceptor agonists, halothane or hyperthermia. *J. Physiol.* 307, 319–33.

Jardon, O.M. (1982). Physiologic stress, heat stroke, malignant hyperthermia - a perspective. *Milit. Med.* 147, 8–14.

Lopez, J.R., Alamo, L., Caputo, C., et al (1984). Measurements of free $[Ca^{2+}]$ in human malignant hyperthermic muscles. *Biophys. J.* 45, 48a

Rosenberg, H. and Reed, S. (1983). *In vitro* contracture tests for susceptibility to malignant hyperthermia. *Anesth. Analg.* 62, 415–20.

Schwartz, L., Rockoff, M.A. and Koka, B.V. (1984). Masseter spasm with anesthesia: incidence and implications. *Anesthesiology* 61, 772–5.

Smithson, W.A., Gronert, G.A. and Moss, K.K. (1983). Dantrolene and potentially fatal hyperthermia secondary to L-asparaginase. *Cancer Tereatment Reports* 67, 318–9.

Stadhouders, A.M., Viering, W.A.L., Verburg, M.P. et al (1984). *In vivo* induced malignant hyperthermia in pigs. III. Localization of calcium in skeletal muscle mitochondria by means of electronmicroscopy and microprobe analysis. *Acta Anaesth. Scand.* **28**, 14.

Traynor, C.A., Van Dyke, R.A. and Gronert, G.A. (1983). Phosphorylase ratio and susceptibility to malignant hyperthermia. *Anesth. Anal.* **62**, 234–6.

Verburg, M.P., Van Bennekom, C.A., Oerlemans, F.T. and De Bruyn, C.H.M.M. (1984). Malignant hyperthermia: adenine incorporation and adenine metabolism in human platelets, influenced by halothane. In *Purine Metabolism in Man IV Part B,* C.H.M.M. De Bruyn, H.A. Simmonds and M.M. Muller (eds), pp. 443. Plenum Publishing Corporation, New York.

Weinberg, S. and Twersky, B.S. (1983). Neuroleptic malignant syndrome. *Anesth. Analg.* **62**, 848–50.

CA^{2+} AND THYROID HORMONE ACTION

Cornelis van Hardeveld

Department of Chemical Pathology, University Hospital, 2333AA Leiden, Holland

and Torben Clausen

Institute of Physiology, University of Aarhus, 8000 Århus C, Denmark

Abstract

In rat skeletal muscle, the effects of Ca^{2+}-mobilizing agents (20 mmol/l K^+_o or 5 mmol/l caffeine) on O_2-consumption, lactate production, glucose uptake and ^{45}Ca-efflux all increase in proportion to the thyroid status. Dantrolene (10^{-5} mol/l), which interferes with the release of Ca^{2+} from the sarcoplasmic reticulum (SR) suppresses these effects. Thyroid hormones increase the amount of SR Ca^{2+}-ATPase, and in keeping with this, the relaxation rate is augmented. In the liver, the stimulating effects of noradrenaline (10^{-6} mol/l) on glucose mobilization and net loss of Ca is considerably reduced by hypothyroidism.

Ca^{2+}-activated enzymes in muscle and liver (Ca^{2+}-ATPase, actomyosin ATPase, phosphorylase kinase, α-glycerophosphate dehydrogenase) show no change in Ca^{2+}-sensitivity (K_A) with the thyroid status, but thyroid hormones induce tissue specific increases in the V_{max} of these enzymes, which may amplify the effects of increased Ca^{2+}-mobilization. The increased Ca cycling induced by thyroid hormones probably contributes to thyroid thermogenesis during muscular work.

Thyroid hormones may control Ca homeostasis by stimulating the proliferation of Ca-storing organelles (SR and mitochondria) and by the induction of changes in membrane composition.

27.1. Introduction

Ca^{2+} is the cytoplasmic mediator of a wide range of biological phenomena involving cell activation. This activation is generally accompanied by an increased energy demand, notably in the skeletal muscles, which constitute a major target for the thermogenic action of the thyroid hormones. In skeletal muscle the clearance of the amount of Ca^{2+} required to activate the contractile filaments demands a substantial fraction of the energy production (Hasselbach and Oetliker, 1983; Rall, 1979). Thyroid hormones have been shown to stimulate the proliferation of sarcoplasmic reticulum and mitochondria in muscle (Gustafsson et al., 1965; Kim, Witzmann and Fitts, 1982; Simonides and van Hardeveld, 1984). Since these organelles represent the major Ca^{2+} clearing systems in the cell, two pertinent questions arise:

(1) Whether thyroid hormones regulate cytoplasmic Ca^{2+} homeostasis.
(2) What is the significance of cytoplasmic Ca^{2+} homeostasis in thyroid hormone induced thermogenesis?

It is the purpose of this short review to present recent information on the effects of the thyroid status on Ca^{2+}-dependent processes and Ca^{2+} kinetics; also the mechanisms by which thyroid hormones could influence intracellular Ca^{2+} movements are considered.

27.2. Metabolic effects of K^+-depolarization and caffeine in muscle

It is an old observation that high extracellular K^+ (20 mmol/l) elicits a pronounced rise (up to 10-fold) in the heat evolution, respiration and glycolysis of isolated muscles (Solandt, 1936). It is supposed that the partial depolarization sets in train the events normally associated with the excitation—contraction coupling process, but without eliciting any contraction. Although the amount of Ca^{2+} released from the SR is insufficient to produce a measurable contraction, several Ca^{2+}-sensitive ATP-requiring processes are activated. These are, notably, the stimulation of actomyosin ATPase, Ca^{2+} transport into the SR, the mitochondria and across the sarcolemma (Fig. 27.1). Dantrolene, an inhibitor of Ca^{2+} release from the SR, prevents the K^+-induced rise in metabolism (van Hardeveld and Kassenaar, 1980). In addition, the presence of Ca^{2+} in the extracellular space is required for a sustained effect of K^+ depolarization (van der Kloot, 1967).

It was of interest to determine whether the metabolic effects of K^+ depolarization in muscle were dependent on the thyroid status. In perfused hind limbs of hypothyroid, euthyroid and hyperthyroid rats high K^+_o (20 mM) stimulated glucose uptake, lactate production and O_2-consumption to levels increasing in proportion to the thyroid status (van Hardeveld and Clausen, 1984; van Hardeveld and Kassenaar, 1980). These effects were abolished by dantrolene (10^{-5} mol/l), in agreement with another study, where dantrolene was found to inhibit the rise in respiration induced by exposing frog sartorius muscle to high K^+_o (Erlij et al., 1982).

Caffeine (5 mmol/l), which induces a release of Ca^{2+} from the SR without depolarizing the plasma membrane, was found to elicit the same pattern of metabolic effects as high K^+_o. This gave further support to the idea that thyroid hormones act on the SR and suggested that Ca^{2+} is mobilized in proportion to the thyroid status. It should be noted that the total calcium contents of hind-limb muscles was unaffected by the thyroid status. The metabolic effects of high K^+_o became transient when extracellular Ca^{2+} was lowered beneath 50 μmol/l, indicating that calcium entry is necessary for the maintenance of the increased substrate utilization (van Hardeveld and Clausen, 1984).

Since thyroid hormones stimulate resting metabolism in skeletal muscle 3-fold, going from the hypothyroid to the hyperthyroid state, it was of interest to assess the effects of dantrolene on resting metabolism. Dantrolene, indeed, inhibited the resting oxygen consumption significantly in perfused skeletal muscle of euthyroid and hyperthyroid rats, the effect being more pronounced in the latter group (van Hardeveld and Kassenaar, 1980). The hypothyroid group showed no response. It has been reported that dantrolene may slightly reduce the resting free Ca^{2+} ion level in bovine skeletal muscle fibres (Jeacocke, 1982).

Experiments with isolated rat soleus muscles (Chinet, Clausen and Girardier, 1977) and perfused rat hind-limbs (van Hardeveld and Kassenaar, 1981) have shown that ouabain induces a slow progressive rise in resting metabolism. In soleus, the exposure to ouabain leads to the development of contractures, indicating that the cytoplasmic Ca^{2+} ion level is augmented. Dantrolene was found to abolish the metabolic effects of ouabain in the perfused rat hind-limb. It should be noted that in hind-limbs obtained from hypothyroid rats, ouabain failed to stimulate oxygen consumption and lactate production.

All the results discussed so far emphasize the importance of dantrolene as a

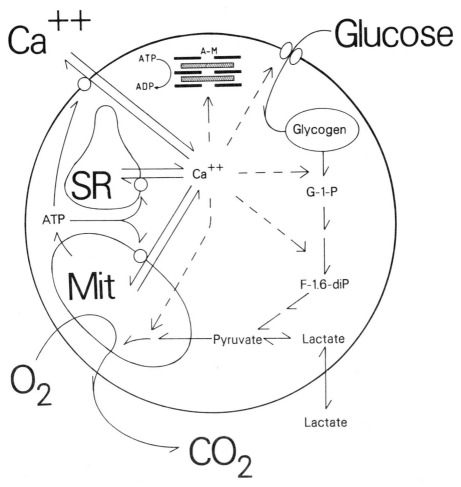

Fig.27.1. Ca distribution and some major transport routes for Ca^{2+} in a typical muscle cell. A rise in cytosolic Ca^{2+} may increase ATP-utilization by stimulating actomyosin ATPase and transport ATPases. Substrate availability may be favoured by activation of the glucose transport system, phosphorylase B kinase, phosphofructokinase, pyruvate dehydrogenase phosphatase, isocitrate dehydrogenase and 2-oxoglutarate dehydrogenase.

tool in identifying the role of cellular calcium pools in the metabolic regulation exerted by thyroid hormones. Obviously precise knowledge about the site of action of this drug could provide a better understanding of thyroid hormone action. However, there is no complete agreement on the site of action of dantrolene. Direct inhibitory effects on isolated SR have been observed (Campbell, Franzini-Armstrong and Shamoo, 1980), but there are also indications that dantrolene interferes with the signal from the T-tubuli to the SR, which normally triggers Ca^{2+} release (Putney and Bianchi, 1974).

The rise in energy consumption elicited by high K^+_o or caffeine requires a supply of metabolizable substrates. These are provided by an increased uptake of glucose and breakdown of glycogen, resulting in a considerable rise in lactate release (van Hardeveld and Clausen, 1984). A coordinated response of increased substrate supply in order to meet the increased energy demand seems possible because glycogenolysis, glycolysis and the oxidation of glycolytic products are all stimulated by a rise in the cytosolic Ca^{2+} concentration (see Fig. 27.1). This occurs by activation of key enzymes in glycogenolysis and glycolysis, notably phosphorylase B kinase (Cohen, 1980) and phosphofructokinase (Patten, Filsell and Clark, 1982), but in addition, three Ca^{2+}-dependent mitochondrial enzymes are stimulated (Denton and McCormack, 1981). Also it has been repeatedly demonstrated that a rise in cytosolic Ca^{2+} is closely correlated with a stimulation of glucose transport (Clausen, 1980). Since glucose uptake and lactate release also increased in proportion to the thyroid status during stimulation with Ca^{2+}-mobilizing agents, taken together the available evidence seemed compatible with the idea that thyroid hormones somehow influence the amount of Ca^{2+} mobilized from the SR. Finally, the stimulation by thyroid hormones of the activities of phosphorylase kinase and α-glycerophosphate dehydrogenase would contribute to the increased substrate supply (Leijendekker, van Hardeveld and Kassenaar, 1984; van Hardeveld, Rusche and Kassenaar, 1976; Werth, Watanabe and Hathaway, 1983). The next step in the analysis, therefore, was to test whether Ca^{2+} ion mobilization could be detected and correlated to the thyroid status.

27.3. Ca^{2+} transport

As the efflux of ^{45}Ca from preloaded tissues is a function of the free cytosolic Ca^{2+} ion level available for transport, the release of Ca^{2+} from cellular pools like the SR may be detected by following the washout of ^{45}Ca from isolated muscles. In both the soleus and extensor digitorum longus muscles, which are representative for the major fibre types in the rat hind-limb, high K^+_o (20 mmol/l) and caffeine (5 mmol/l) stimulated the fractional loss of ^{45}Ca (van Hardeveld and Clausen, 1984). In agreement with earlier studies, dantrolene (10^{-5} mol/l) was found to suppress this response. As shown in Fig.27.2, the stimulation of ^{45}Ca washout was found to increase in proportion to the thyroid status. A similar relationship was observed for soleus muscle fibres obtained from rats of the size used for the hind-limb perfusion experiments. Experiments with mice showed that in soleus muscles obtained from hypothyroid animals the rise in the fractional loss of ^{45}Ca induced by high K^+_o was only 1/10th of that obtained with muscles from euthyroid animals. The stimulating effect of high K^+_o on ^{45}Ca washout from rat soleus was blunted by the omission of Ca from the incubation medium, in keeping with the transient metabolic effects obtained under these conditions. The significance of extracellular Ca was also evident from the observation that high K_o^+ produced a progressive increase in ^{45}Ca uptake which could be suppressed by dantrolene (van Hardeveld and Clausen, 1984).

Taken together, the observations support the idea that thyroid hormones increase the availability of Ca^{2+} in the cytoplasm of muscle cells, primarily by augmenting the mobilization of Ca^{2+} from the SR, but possibly also by favouring Ca^{2+} entry. This could be related to the stimulating effect of

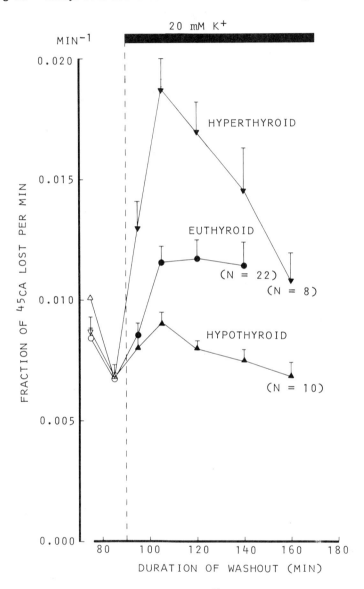

Fig.27.2. The effect of 20 mmol/l K^+ on the washout of ^{45}Ca from soleus muscles obtained from hypothyroid, euthyroid or hyperthyroid rats. Intact muscles prepared from 4-week-old rats were preloaded with ^{45}Ca for 60 min and then washed out in a series of tubes containing non-radioactive Krebs–Ringer bicarbonate buffer (Van Hardeveld and Clausen, 1984). During the interval indicated by the horizontal bar, the K^+ concentration of the washout medium was 20 mmol/l. Each curve represents the mean of 8–22 observations with bars denoting SE.

thyroid hormones on the proliferation of SR (Kim, Witzmann and Fitts, 1982; Simonides and van Hardeveld, 1984) and hence to an increase in the capacity for Ca uptake and release. The thyroid hormone induced increase in SR Ca^{2+}-ATPase activity appears 24–48 hours after the first dose and is most likely the result of *de novo* protein synthesis (Limas, 1978; Simonides and van Hardeveld, 1984).

Thyroid hormones may also influence the energetic efficiency of Ca^{2+} transport, and this may be of importance for the thermogenic effect. Calcium transport in isolated SR vesicles is presumed to be tightly coupled to ATP hydrolysis with a stoichiometry of 2 Ca^{2+}-ions/molecule of ATP hydrolysed, that is, a coupling ratio of 2. However Ca^{2+}/ATP coupling ratios, experimentally determined in native intact SR vesicles from skeletal muscle by kinetic means, yield values in the range 1.5–1.9, down to 1.0 in SR vesicles isolated from heart. Deviations from the expected value of 2.0 have most often been explained on the basis of leaky vesicles. There are indications that hyperthyroidism leads to a decrease in the coupling ratio in the SR of the heart (Suko, 1971), whereas in the SR of hypothyroid skeletal muscle the reverse was observed (Simonides and van Hardeveld, 1984). A thyroid hormone induced increase in the Ca^{2+} permeability of the SR membrane could explain these results. Increased Ca^{2+} permeability has indeed been demonstrated in SR vesicles isolated from the heart and skeletal muscles of hyperthyroid rats (Ash *et al.*, 1972; Limas, 1978). However, in these studies, the Ca^{2+} efflux was considered to represent only a minor percentage of the accumulation rate of Ca^{2+} and could, therefore, hardly influence the coupling ratio. The above-mentioned Ca^{2+}-efflux measurements were made in the absence of ATP. In this respect it is significant that the Ca efflux was much higher in the presence of ATP, which is more like the *in vivo* situation (Gerdes and Møller, 1983). Here, however, it must be emphasized that the Ca efflux was strongly dependent on the Ca-filling level of the SR. An increase in Ca permeability would also be compatible with the presumed proportional Ca^{2+} release during K^+_o or caffeine exposure in different thyroid states, perhaps in addition to the effect of a larger surface area of the SR membrane on Ca^{2+} exchange. Finally, uncoupling of ATPase activity from Ca^{2+} transport could also affect the coupling ratio. This has been shown to occur after the selective modification of SR-membrane lipids (Hidalgo, Petrucci and Vergara, 1982; Navarro Toivio Kinucan and Racker, 1984). For instance, the phosphatidyl-ethanolamine/phosphatidyl-choline ratio are thought to influence the efficiency of Ca transport. In relation to this it is worth noticing that hypothyroidism increases the activation energy of SR Ca^{2+}-ATPase (Simonides and Van Hardeveld, 1984). Changes in the activation energy of membrane-bound enzymes are indicative of a change in phospholipid composition. There is a substantial body of literature describing thyroid hormone effects on membrane phospholipid composition (Hoch, 1982; Hulbert, 1978), and this supports the idea of a possible involvement of SR membrane alterations in the action of thyroid hormones.

27.4. Muscle work

The well-known changes in muscle contractility seen during thyroid dysfunction are obviously an extra indication that Ca homeostasis is controlled by thyroid hormones. The rise in cytosolic Ca^{2+} initiates contraction by binding of Ca^{2+} to the troponin complex and there is a clear relationship between *in vitro* applied Ca^{2+} levels and the height of developed force (Fabiato and Fabiato, 1978). Thyroid hormones shorten contraction time and relaxation time during isometric contractions in heart and skeletal muscle. This suggests that Ca^{2+} is released faster from the SR but is also taken up at a higher rate. Here we encounter a discrepancy. If, as was postulated in the experiments with high K_o^+ and caffeine, Ca^{2+} mobilization is proportional to the thyroid status, one should expect the same for the height of force development. This proves not to be the case (Everts et al., 1981; Skelton et al., 1971). However, the circumstances under which Ca^{2+} cycling takes place during exposure to high K_o^+ and repetitive electrical or neural stimulation are quite different. In the first situation there is continuous Ca^{2+} release and reuptake by the SR at subthreshold levels for contraction. In the second case brief bursts of Ca^{2+} release are followed by longer periods of Ca^{2+} uptake during which continuously changing Ca^{2+} concentrations reach levels which initiate force development. In addition, there is a quite complex temporal relationship between free Ca^{2+} and developed force (Stephenson, 1981). Furthermore, the affinity of Ca^{2+} for the contractile filaments decreases with a lowering of the pH which occurs during contraction (Fabiato and Fabiato, 1978) due to lactate production. Since lactate release increases in the direction hypothyroid-hyperthyroid this could counteract the effect of an increase in Ca^{2+} release on force. Thus, the absence of a proportionality between force and the thyroid status does not exclude that thyroid hormones induce an accelerated Ca^{2+} release during force development. Should this idea be correct, the energy consumption should be proportional to the thyroid status during force deliverance, since increasing amounts of Ca^{2+} would have to be sequestered by the SR in each contraction cycle.

The available data indeed show that thyroid hormones increase the amount of energy which is necessary for delivering a standard amount of force or work (Everts et al., 1981; Leijendekker, Van Hardeveld and Kassenaar, 1983, 1984; Skelton et al., 1971). This was observed under both aerobic and anaerobic conditions. The observed lower energy consumption in contracting hypothyroid muscle could contribute to the development of hypothermia during the stage of shivering thermogenesis in hypothyroid animals exposed to cold. In addition, the administration of dantrolene during contraction seems to decrease energy consumption to a lesser degree in hypothyroid muscle than in euthyroid muscle, whereas the effect on force is the same. It must be emphasized here that in heart and slow skeletal muscle an increase in actomyosin-ATPase activity could contribute considerably to thyroid hormone induced energy—consumption differences, the more so as actomyosin ATPase could account for at least 50% of the energy expended during work. However, in fast skeletal muscle the activation of actomyosin—ATPase only show slight changes with the thyroid status (Nwoye and Mommaerts, 1981), whereas the energy cost for force development is clearly increased by thyroid hormones, strongly suggesting that increased Ca^{2+} cycling is the responsible process. When summarizing the data it seems that thyroid hormones stimulate Ca^{2+} cycling during force deliverance by increasing the rate of uptake and release of Ca^{2+}. This process may contribute significantly to thyroid thermogenesis during work.

27.5. Ca homeostasis and metabolism in the liver

One of the early observations regarding thyroid-hormone regulated Ca homeostasis is the T_4-induced stimulation of Ca transport in the liver

(Wallach et al., 1972. Hyperthyroidism increased Ca influx by 50% and, although irregularly, also Ca efflux by 14–26%. More recent studies have shown that in the liver, α_1-adrenergic agonists and vasoactive peptides appear to cause mobilization of intracellular Ca^{2+}, which through activation of phosphorylase B kinase leads to increased glycogenolysis and glucose output (Fig.27.3). It has been shown that the α_1-agonist phenylephrine causes concentration-dependent activation of glycogen phosphorylase and release of ^{45}Ca from ^{45}Ca-loaded liver cells. The magnitude of both responses to phenylephrine was markedly suppressed after thyroidectomy (Preiksaitis, Kan and Kunos, 1982). In keeping with this is the observation that the ability of noradrenaline (1 μmol/l) to stimulate glycogenolysis is severely impaired in perfused liver obtained from hypothyroid rats. This coincided with diminished phosphorylase A formation and a lower Ca efflux (Storm, van Hardeveld and Kassenaar) (see also Fig.27.3). Also in this case the total calcium content did not differ from euthyroid control values. This indicates that less Ca^{2+} is mobilized from intracellular stores.

Mitochondria and the endoplasmic reticulum constitute important calcium pools in most tissues. Thyroid hormones also seem to stimulate under specific conditions both the energy-requiring Ca influx and the Ca efflux in mitochondria (Evans and Hoch, 1976; Grieff et al., 1982; Harris, Al-Shaikhaly and Baum, 1979; Herd, 1978; Shears and Bronk, 1981). This suggests that Ca cycling in at least liver mitochondria might be increased by thyroid hormones, comparable to the stimulation in SR. Calculations, using in vitro data from liver, indicate that this could account for no more than 3% of the

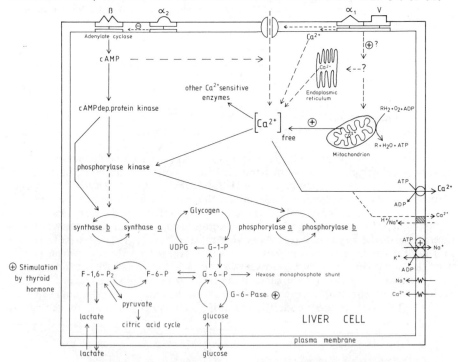

Fig.27.3. Possible mechanisms of action of thyroid hormones on α_1-adrenergic stimulation of glycogenolysis in the liver. The rise in cytosolic Ca^{2+} after α_1-adrenergic stimulation may be potentiated by thyroid hormones through an increase in the number of α_1-adrenoceptors (Preiksaitis, Kan and Kunos, 1982) and/or a facilitatory effect on Ca^{2+} release from intracellular Ca pools.

increased thyroid hormone dependent respiration *in vivo* (Shears and Bronk, 1981). Nevertheless, the changed Ca kinetics in the liver cell could influence steady state cytosolic Ca^{2+} during rest and this aspect awaits further study. Taken together the results implicate that we are dealing here with a more general effect of thyroid hormones on Ca homeostasis, apparently not only restricted to muscle tissue.

References

Ash, A.S.F., Besch, H.R., Harigaya, S. and Zaimis, E. (1972). Changes in the activity of sarcoplasmic reticulum fragments and actomyosin isolated from skeletal muscle of thyroxin-treated cats. *J. Physiol.* **224**, 1–19.

Campbell, K.P., Franzini-Armstrong, C. and Shamoo, A.E. (1980). Further characterization of light and heavy sarcoplasmic reticulum vesicles. *Biochim. Biophys. Acta* **602**, 97–116.

Chinet, A., Clausen, T. and Girardier, L. (1977). Microcalorimetric determination of energy expenditure due to sodium potassium transport in the soleus muscle and brown adipose tissue of the rat. *J. Physiol.* **265**, 43–61.

Clausen, T. (1980). The role of calcium in the activation of the glucose transport system. *Cell Calcium* **1**, 311–25.

Cohen, P. (1980). The role of calmodulin and troponin in the regulation of phosphorylase kinase from mammalian skeletal muscle. In *Calcium and Cell Function. I. Calmodulin*. W.Y. Cheung (ed), pp. 184–98. Academic Press, New York.

Denton, R.M. and McCormack, J.G. (1981). Calcium ions, hormones and mitochondrial metabolism. *Clin. Sci.* **61**, 135–40.

Erlij, D., Shen, W.K., Reinach, P. and Schoen, H. (1982). Effects of dantrolene and D_2O on K^+-stimulated respiration of skeletal muscle. *Amer. J. Physiol.* **243**, C87–95.

Evans, T.C. and Hoch, F.L. (1976). Energy-linked reactions in hypothyroid rat liver submitochondrial vesicles. *Biochem. Biophys. Res. Comm.* **69**, 635–40.

Everts, M.E., van Hardeveld, C., Ter Keurs, H.E.D.J. and Kassenaar, A.A.H. (1981). Force development and metabolism in perfused skeletal muscle of euthyroid and hyperthyroid rats. *Acta Endocrinol.* **97**, 221–5.

Fabiato, A. and Fabiato, F. (1978). Effects of pH on myofilaments and the sarcoplasmic reticulum of skinned cells from cardiac and skeletal muscles. *J. Physiol.* **276**, 233–55.

Gerdes, U. and Møller, J.V. (1983). The Ca^{2+} permeability of sarcoplasmic reticulum vesicles, II. Ca^{2+}-efflux in the energized state of the calcium pump. *Biochim. Biophys. Acta* **734**, 191–200.

Greiff, R.L., Fiskum, G., Sloane, D.A. and Lehninger, A.L. (1982). Influence of thyroid and growth hormone status on the rate of regulated Ca^{2+} efflux from rat liver mitochondria. *Biochem. Biophys. Res. Comm.* **108**, 307–14.

Gustafsson, R., Tata, G.R., Lemborg, O. and Ernster, L. (1965). The relationship between the structure and activity of rat skeletal muscle mitochondria after thyroidectomy and thyroid hormone treatment. *Cell Biol.* **26**, 555–77.

Harris, E.J., Al-Shaikhaly, M. and Baum, H. (1979). Stimulation of mitochondrial calcium ion efflux by thiol-specific reagents and by thyroxin. The relationship to adenosine diphosphate retention and to mitochondrial permeability. *Biochem. J.* **182**, 455–64.

Hasselbach, W. and Oetliker, H. (1983). Energetics and electrogenicity of the sarcoplasmic reticulum calcium pump. *Ann. Rev. Physiol.* **45**, 325–39.

Herd, P.A. (1978). Thyroid hormone-divalent cation interactions. Effect of thyroid hormone on mitochondrial calcium metabolism. *Arch. Biochem. Biophys.* **188**, 220–5.

Hidalgo, C., Petrucci, D.A. and Vergara, C. (1982). Uncoupling of Ca^{2+} transport in sarcoplasmic reticulum as a result of labeling lipid amino groups and inhibition of Ca^{2+}-ATPase activity by modification of lysine residues of the Ca^{2+}-ATPase polypeptide. *J. Biol. Chem.* **257**, 208–16.

Hoch, F.L. (1982). Thyroid control over biomembranes. VII. Heart muscle mitochondria from L-triiodothyronine-injected rats. *J. Molec. Cell. Cardiol.* **14**, 81–90.

Hulbert, A.J. (1978). The thyroid hormones: a thesis concerning their action. *J. Theoret. Biol.* **73**, 81–100.

Jeacocke, R. (1982). Does the sarcoplasmic reticulum achieve chemiosmotic equilibrium in relaxed muscle? *FEBS Lett.* **147**, 225–30.

Kim, D.H., Witzmann, F.A. and Fitts, R.H. (1982). Effect of thyrotoxicosis on sarcoplasmic reticulum in rat skeletal muscle. *Amer. J. Physiol.* **243**, C151–5.

Leijendekker, W.J., van Hardeveld, C. and Kassenaar, A.A.H. (1983). The influence of the thyroid state on energy turnover during tetanic stimulation in the fast-twitch (mixed type) muscle of rats. *Metabolism* **32**, 615–21.

Leijendekker, W.J., van Hardeveld, C., Kassenaar, A.A.H. (1985). Coupled diminished energy turnover and phosphorylase *a* formation in contracting hypothyroid rat muscle. *Metabolism* **34**, 437–41.

Limas, C.J. (1978). Calcium transport ATPase of cardiac sarcoplasmic reticulum in experimental hyperthyroidism. *Amer. J. Physiol.* **235**, H745–51.

Navarro, J., Toivio Kinnucan, M. and Racker, E. (1984). Effect of lipid composition on the calcium/adenosine 5'-triphosphate coupling ratio of the Ca^{2+}-ATPase of sarcoplasmic reticulum. *Biochemistry* **23**, 130–5.

Nwoye, L. and Mommaerts, W.F.H.M. (1981). The effects of thyroid status on some properties of rat fast-twitch muscle. *J. Muscle Res. Cell. Motil.* **2**, 307–20.

Patten, G.S., Filsell, O.H. and Clark, M.G. (1982). Epinephrine regulation of phospho-fructokinase in perfused rat heart. *J. Biol. Chem.* **257**, 9480–6.

Preiksaitis, H.G., Kan, W.H. and Kunos, G. (1982). Decreased α_1-adrenoceptor responsiveness and density in liver cells of thyroidectomised rats. *J. Biol. Chem.* **257**, 4321–7.

Putney, J.W. and Bianchi, C.P. (1974). Site of action of Dantrolene in frog sartorius muscle. *J. Pharmacol. Exp. Ther.* **189**, 202–12.

Rall, J.A. (1979). Effects of temperature on tension, tension-dependent heat, and activation heat in twitches of frog skeletal muscle. *J. Physiol.* **291**, 265–75.

Shears, S.B. and Bronk, J.R. (1981). The effects of thyroxin treatment, in vivo and in vitro, on Ca^{2+} efflux from rat liver mitochondria. *FEBS Lett.* **126**, 9–12.

Simonides, W.S. and van Hardeveld, C. (1985). The effect of hypothyroidism on sarcoplasmic reticulum in fast-twitch muscle of the rat. *Biochim. Biophys. Acta.* **844**, 129–41.

Skelton, C.L., Pool, P.E., Seagren, S.C. and Braunwald, R. (1971). Mechanochemistry of cardiac muscle. V. Influence of thyroid state on energy utilization. *J. Clin. Invest.* **50**, 463–73.

Solandt, D.Y. (1936). The effect of potassium on the excitability and resting metabolism of frog's muscle. *J. Physiol* **86**, 162–70.

Stephenson, E.W. (1981). Activation of fast skeletal muscle: contributions of studies on skinned muscle fibers. *Amer. J. Physiol.* **240**, C1–19.

Storm, H., van Hardeveld, C. and Kassenaar, A.A.H. (1984). The influence of hypothyroidism on the adrenergic stimulation of glycogenolysis in perfused rat liver. *Biochim. Biophys. Acta* **798**, 350–60.

Suko, J. (1971). Alterations of Ca^{2+}-uptake and Ca^{2+} activated ATPase of cardiac sarcoplasmic reticulum in hyper- and hypothyroidism. *Biochim. Biophys. Acta* **252**, 324–7.

van Hardeveld, C. and Clausen, T. (1984). Effect of thyroid status on K^+-stimulated metabolism and ^{45}Ca-exchange in rat skeletal muscle. *Amer. J. Physiol.* **247**, E421–30.

van Hardeveld, C. and Kassenaar, A.A.H. (1980). A possible role for Ca^{2+} in thyroid hormone-dependent oxygen consumption in skeletal muscle of the rat. *FEBS Lett.* **121**, 349–51.

van Hardeveld, C. and Kassenaar, A.A.H. (1981). Evidence that the thyroid state influences Ca^{++}-mediated metabolic processes in perfused skeletal muscle. *Horm. Metab. Res.* **13**, 33–7.

van Hardeveld, C., Rusche, R. and Kassenaar, A.A.H. (1976). Sensitivity of mitochondrial α-glycerophorphate dehydrogenase to thyroid hormone in skeletal muscle of the rat. *Horm. Metab. Res.* **8**, 153–4.

van der Kloot, W. (1967). Potassium-stimulated respiration and intracellular calcium release in frog skeletal muscle. *J. Physiol.* **191**, 141–65.

Wallach, S., Bellavia, J.V., Gomponia, P.J. and Bristrim, P. (1972). Thyroxin-induced stimulation of hepatic cell transport of calcium and magnesium. *J. Clin. Invest.* **51**, 1572–7.

Werth, D., Watanabe, A.M. and Hathaway, D.R. (1983). Mechanisms of enhanced phosphorylase activation in the hyperthyroid rat heart. *J. Molec. Cell. Cardiol.* **15**, 163–71.

CALCIUM REQUIREMENT FOR INSULIN SECRETION

V. Maier

Laboratory of Clinical Chemistry, Centre of Internal Medicine, University of Ulm, West Germany

Abstract

The presence of extracellular calcium is required for glucose or any other secretagogue to stimulate insulin secretion. To substantiate the hypothesis that Ca^{2+} triggers insulin release, uptake of Ca^{2+} by the B-cell of the islets of Langerhans was measured under several experimental conditions. Glucose effects islet Ca^{2+} by

(1) stimulation of Ca^{2+}-influx;
(2) utilization of cellular Ca^{2+};
(3) inhibition of Ca^{2+} efflux by interference with Na^+/Ca^{2+} exchange.

Interactions between cAMP as the second messenger of the action of hormone and Ca^{2+} have been emphasized.

Calmodulin is involved in the stimulus—secretion coupling process for glucose-induced secretion shown by inhibition studies using trifluoperazine. However, is it not clear if the particular Ca^{2+}-dependent protein phosphorylation is directly connected to the stimulation of insulin release.

Clinical observations indicate that serum Ca^{2+} has an important effect on insulin secretion in parathyroid disorders. Hyperparathyroidism worsens coexisting diabetes mellitus. Glucose tolerance values have the tendency to be reduced in idiopathic hypoparathyroidism and pseudohypoparathyroidism.

28.1. Introduction

There is growing evidence that diabetes mellitus is a disease with a genetic predisposition that ultimately results in impaired secretion of insulin from pancreatic islets. However, diabetes is rarely present from birth (Natrass and Santiago, 1983).

Although researchers all over the world are trying to clarify the mechanisms of insulin it is not yet completely understood. Two main factors seem to predominate (Wollheim and Sharp, 1981).

(1) Glucose is the most important secretagogue
(2) There is involvement of calcium ions.

Many substrates and hormones are reaching the islets of Langerhans via the circulation resulting in control of the maintenance of normoglycaemia, modulating the release of insulin from pancreatic islets under direct neural influence via the autonomic nerve fibres.

Alteration in sensitivity of the B-cell to glucose, as well as the dynamics of glucose-stimulated insulin release leads to an impairment of glucose homeostasis ranging from subclinical to overt forms of diabetes.

It is a fact that calcium ions play a regulatory role in many cellular events. An increase in the concentration of ionized calcium in a particular compartment of the cytosol is thought to be responsible for exocytosis of the hormone. It is generally accepted that insulin release evoked by raising the concentration of glucose is only possible at the threshold concentration of 0.05–5 mM Ca^{2+}. When the Ca^{2+} concentration is decreased, the release of insulin also decreases rapidly to basal values.

However, it appears to be extremely difficult to draw any definite conclusion here because Ca^{2+} is required for both ion flux *and* maintenance of membrane integrity.

28.2. Effect of Ca^{2+} on glucose-stimulated biosynthesis and secretion

It is very easy to demonstrate the effect of different calcium concentrations by means of the isolated perfused pancreas (Fussgänger, 1974) or by using isolated perifused islets or islets incubated in an Ca^{2+}-deficient medium (Maier, 1979).

Sustained insulin release is evoked by glucose only in the presence of sufficient amounts of extracellular Ca^{2+}. When the Ca^{2+} concentration of the medium is rapidly decreased to low concentrations, the secretion of insulin induced by glucose rapidly decreases to basal values. Reintroduction of Ca^{2+} again causes stimulation of insulin release. The insulin release from perifused isolated rat islets (Fig.28.1) in response to a sudden rise of glucose in a normal Krebs-Ringer buffer is shown in Fig.28.2. The same experiment was done in a Ca^{2+}-deficient medium.

In contrast to the general biosynthesis, glucose stimulates both the secretion of insulin *and* the process of biosynthesis (Fig.28.3). This is possible by independent mechanisms: low Ca^{2+}-concentrations markedly inhibit insulin

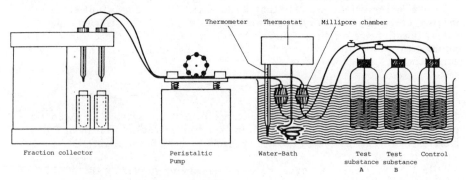

Fig.28.1. Apparatus for the perfusion of pancreatic islets of the rat at 37°C; 80 islets are put into a 'millipore chamber' and perfused continuously with Krebs-Ringer bicarbonate buffer supplemented with 0.2% human albumin. The perfusate is collected automatically.

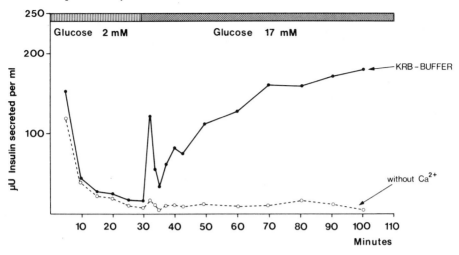

Fig.28.2. Secretion pattern of insulin stimulated by 17 mM glucose in the presence and absence of 2.2 mM calcium; only in the presence of calcium is the hormone released in a biphasic pattern.

release with respect to glucose, whereas the stimulus of insulin biosynthesis in a Ca^{2+}-depleted medium was not impaired. Release of granules appears to be a relatively late consequence of glucose stimulation and is not directly coupled to the biosynthesis.

From these observations it has convincingly emerged that glucose may trigger the release of insulin by provoking intracellular accumulation of calcium. In elderly diabetics glucose as such is no longer recognized. Calcium metabolism does not appear to be disturbed since the sulphonylurea glibenclamide is able to induce a normal secretion of insulin

28.3. Effect of glucose on [^{45}Ca] efflux

Figure 28.4(a) shows that a rise of glucose concentrations (to 16.7 mM) provokes a dual modification of [^{45}Ca] efflux consisting in an initial fall followed by an immediate increase in Ca^{2+} efflux, the secondary rise being concomitant with insulin release. Under these experimental conditions the effluent radioactivity is thought to be exclusively of intracellular origin. These movements are rapid and might therefore be coincidental with the rapid changes in insulin secretion rate in response to glucose.

Malaisse and co-workers produced evidence that the secondary rise in [^{45}Ca] efflux cannot be attributed to insulin release. It reflects the rate of calcium entry into the B-cell through opened voltage-sensitive calcium channels (Herchueltz and Malaisse, 1978; Herchueltz and Malaisse, 1981).

In addition, cobalt and nickel - known as selective blockers of the slow calcium channels - are able to completely abolish the secondary rise in [^{45}Ca] efflux.

In various tissues, including the B-cell the uptake and release of calcium from intracellular organelles may also occur with sodium exchange (Donatsch et al., 1977). However, tetrodotoxin, a specific blocker of sodium channels (Kao, 1966) does not affect glucose-induced [^{45}Ca] efflux.

The two above-mentioned calcium movements are also dependent on the integrity of glucose metabolism. An increase of the islets content of NAD(P)H might also be a link between glucose metabolism and stimulation of calcium influx into the B-cell. By modifying the B-cell redox state glucose might reduce the membrane permeability to potassium, depolarize the membrane and hence activated voltage-sensitive calcium channels.

Therefore, four methods appear to exist solely or in combination with glucose to induce secretion of insulin:

(1) Increase of Ca^{2+} influx from the extracellular influx.
(2) Decreased Ca^{2+} efflux from pancreatic efflux.
(3) Inhibition of intracellular Ca^{2+} sequestration.
(4) Mobilization of calcium from intracellular pools.

The main problem when investigating these mechanisms is that only the *uni*directional influx component of the *bi*directional membrane fluxes can be

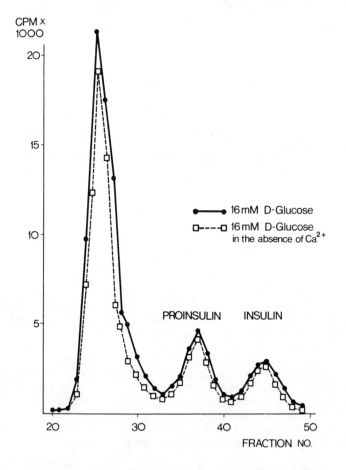

Fig.28.3. Effect of glucose concentration (300 mg%) in the presence and absence of Ca^{2+} on incorporation of [^{3}H]-leucine into the proteins of isolated pancreatic mouse islets. Batches of 25 islets were incubated for 3 h in the presence of [^{3}H]-leucine. After precipitation with trichloroacetic acid, the islet proteins were fractionated on a Sephadex G 50 fine column, using 1 M acetic acid as eluent. The first peak eluting with the void volume represents islet proteins excluded from the gel. Mean curves are shown, derived from six to twelve experiments each.

measured experimentally. In consequence, for example [^{45}Ca], or the accumulation of this compound in pancreatic islets must be linear with time, which can be measured only for a short period.

28.4. Ca-ATPase in pancreatic islets

The nature of calcium movement might be mediated either by a Ca^{2+}

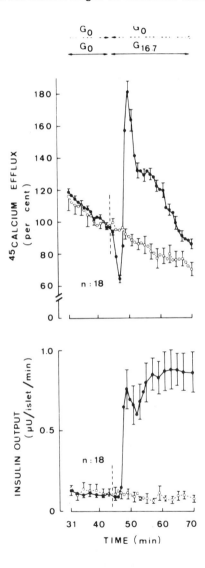

Fig.28.4. Effect of 16.7 mM glucose on [^{45}Ca] efflux and (b) insulin release (●—●) from perifused islets. Control experiments (○—○) were performed in the absence of glucose. Mean values (± SEM) for [^{45}Ca] efflux are expressed as a percentage of the mean value found within the same experiment between the 40th and 44th min of perifusion, and refer to 18 individual experiments. Mean values (± SEM) for insulin output are expressed in µU/islet/min and refer to the same 18 individual experiments. (From Herchueltz and Malaisse, 1981.)

stimulated ATPase, as was demonstrated in erythrocytes and the muscle sarcoplasmic reticulum, or by a Na^+-Ca^{2+} countertransport process which does not appear to involve ATP.

The interpretation is particularly difficult due to the complex compartmentalization of calcium within the cell.

A calcium-stimulated ATPase activity was identified in the plasma membrane-enriched fraction from pancreatic islets. An increase of the intracellular concentration of Ca^{2+} is implicated as a key event of insulin secretion. This Mg-independent Ca-ATPase was found in secretory granules, mitochondria and microsomes from mouse islets (Formby et al., 1976).

As glucose stimulation of insulin secretion is associated with intracellular calcium translocations some metabolite or glucose itself should affect the Ca-ATPase activity. Neither glucose, glucose-6-phosphate, fructose, diphosphate or PEP had any effect on Ca-ATPase activity in the fraction of secretory granules. However, NADH formed during metabolisms of glucose, was found to inhibit Ca—ATPase by 35%. In addition, cAMP and Na^+ also inhibited this enzyme.

28.5. Calmodulin, cAMP, protein phosphorylation and insulin secretion

Calmodulin, the calcium-dependent regulatory protein appears to play a pivotal role in mediating calcium-dependent cellular processes (Tomlinson, Walker and Brown, 1982). Calmodulin is now recognized as a member of the so-called 'troponin-C-superfamily' of calcium-binding proteins which fulfil a fundamental role in muscle concentration and probably also in intestinal calcium transport.

Its structure has been remarkably conserved from earthworm and octopus to pig and ox, and in the plant kingdom too no decisive difference is observed.

A number of neuroleptic drugs have been found to bind to the calmodulin molecule, the most prominent example being trifluoperazine.

Indeed there is evidence that the potency of phenothiazines in inhibiting glucose-stimulating insulin release are parallel to their potency in inhibiting the activation of calmodulin-dependent phosphodiesterase. Therefore, it seems possible that the effects of calmodulin on insulin secretion could be related to some interaction with adenylate cyclase or with the corresponding phosphodiesterase (Fig.28.5).

Calmodulin seems to act at different points in stimulus secretion coupling, influencing cyclic nucleotide metabolisms, protein phosphorylation and exocytosis (Gagliardino et al., 1980).

Many reports have documented the stimulation of insulin release by agents that raise intracellular cAMP levels, for example, glucagon, β-adrenergic agents, inhibitors of phosphodiesterase, cholera toxin and exogenous cAMP. Several reports have documented the glucose increases the cAMP content of islets, but it does not represent a direct stimulator of adenylate cyclase. The demonstration of calmodulin-sensitive adenylate cyclase in the pancreatic islets explains the ability of glucose to increase cAMP levels (Valverde et al., 1979).

There is evidence that both cAMP-dependent and Ca^{2+}-calmodulin-dependent protein kinase can be activated separately. Therefore the possibility remains that cAMP and calmodulin interact either synergistically or consecutively.

At the electron-microscopic level calcium pyroantimonate precipitations were used to localize calcium (Klöppel and Bommer, 1979). Using different

Calcium requirement for insulin secretion 373

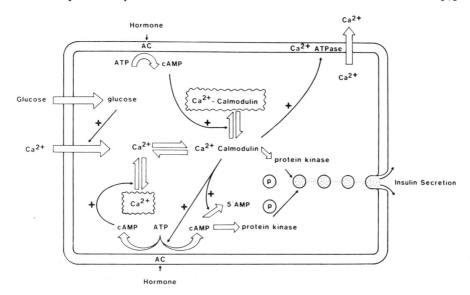

Fig.28.5. Hypothesis of the interrelationships between calmodulin and insulin secretion. The enclosed Ca^{2+} calmodulin and Ca^{2+} represent the binding to membranes. AC: Adenylate cyclase; P: phosphorylation of protein; +: stimulation. (From Tomlinson, Walker and Brown, 1982.)

methods the granules and mitochondria showed the higher concentrations.

28.6. Diabetes and intestinal calcium absorption in the rat

Another observation in the diabetic state is the pardoxical combination of depressed body growth and enhanced intestinal growth. Hence, the effect of diabetes on intestinal transport, such as calcium and fat, is of interest. The decreased body growth in diabetics can be associated with decreased growth of bone and abnormal calcium metabolisms; however, hexose absorption and mucosal disaccharidase activity are increased. The net absorption of calcium in the duodenum is significantly decreased, possibly due to failure to form calcium binding protein. In consequence total and ionized serum calcium are decreased in diabetic rats (Schneider and Schedl, 1972).

28.7. Clinical observations

The range of calcium concentrations in humans is extremely narrow. Therefore only a few reports with respect to pathological calcium values and carbohydrate metabolism exist in the literature.

28.7.1. *Calcitonin*
Calcitonin has many other actions besides its effects on calcium metabolism. Acute calcium administration inhibits glucose-stimulated insulin secretion and impairs glucose tolerance in normal, obese and prediabetic subjects. Calcitonin reduces insulin response to arginine (Ziegler *et al.*, 1972). It has been suggested

that this hormone interferes with the secretion mechanisms by stimulating calcium efflux.

28.7.2. Hypercalcaemia

On the other hand, the influence of calcium excess on insulin release *in vivo* is not clearly defined. Recently, calcium infusion has been proposed as a provocative test for the diagnosis of insulin-secreting tumors. A 2 h infusion of calcium gluconate was performed (4 mg/kg/h) in six patients with islet cell adenoma. As a result, mean calcium plasma levels increased from 9.6 ± 0.4 to 11.6 ± 0.8 mg/100 ml. During calcium infusion, blood glucose and plasma insulin concentrations remained unchanged. These observations suggest that calcium fails to stimulate basal insulin secretion even in cases of organic hyperinsulinism. However, they show that calcium infusion is not helpful as a provocative test in the diagnosis of insulinoma (De Palo et al., 1981).

In another study (Anderson et al., 1982) calcium was infused *intravenously* to fasting diabetic patients and induced a significant decline in blood glucose concentration. This was not the case in healthy volunteers. When glucose was administered *orally* during exogenous hypercalcaemia glucose tolerance decreased significantly in both healthy individuals as well as in diabetics.

Verapamil abolished this hypercalcaemia effect, and even improved the tolerance for oral glucose in patients with non-insulin-dependent diabetes (NIDD), but not in healthy subjects or in patients with chronic hypercalcaemia, or primary hyperparathyroidism. Hypercalcaemia decreases the tolerance for oral glucose in normoglycaemic subjects, and glucose tolerance deteriorates further in patients with already impaired carbohydrate metabolisms.

During hypercalcaemia conflicting results were reported in the literature. Gedik, Akalin and Koray (1980) infused calcium gluconate to non-diabetic, non-obese human volunteers. Serum calcium was increased to 5 mg/100 ml over 4 h. After 1 h an oral glucose tolerance test was started. Insulin release was increased (Gaeke et al., 1975; Kim et al., 1971). The glucose values and glucose areas of the oral glucose tolerance test were normal during the calcium infusion. However, when the serum total calcium level was held above 14 mg/100 ml insulin release was significantly higher than in the normocalcaemic state. On the other hand Ca^{2+} decreases the permeability of cytoplasmatic membranes including glucose; in other words, less glucose is utilized, and in addition the inhibitory effect on the activity of key glycolytic enzymes in peripheral tissues causes hyperglycaemia. In consequence, the insulin secretion is enhanced. In summary, acute hypercalcaemia increases insulin release but does not affect carbohydrate tolerance.

28.7.3. Parathyroid disorders

More complicated appears to be the following situation: concentration of serum calcium varies in patients with parathyroid disorders; that is, patients with primary hyperparathyroidism have high serum calcium, and those with ideopathic hypoparathyroidism and pseudohypoparathyroidism have low serum calcium. In secondary hyperparathyroidism patients have normal serum calcium. To evaluate the role of serum calcium in humans, insulin response after a 100 g oral load was studies (Yasuda et al., 1975).

It was demonstraated that a change in calcium concentration within a pathophysiologically fluctuating range did not affect glucose tolerance, often reported to be abnormal in pseudohypoparathyroidism. The relationship between serum calcium plasma parathyroid hormone level, insulin response to oral glucose load and glucose tolerance in different parathyroid disorders is summarized in Table 28.1.

28.7.4. Coexisting diabetes and hyperparathyroidism

Coexisting diabetes mellitus and hyperparathyroidism is not often recognized (Argun and Ertel, 1978). Removal of parathyroid adenoma may result in frequent hypoglycaemia attacks which leads to a reduction of the administration of insulin. These observations suggest that hyperparathyroidism causes coexisting diabetes mellitus to deteriorate and that one must be aware of increased insulin sensitivity after surgical correction of the hypercalcaemic state.

28.7.5. Hypophosphataemia

One of the most common clinical situations in which hypophosphataemia has been found is poorly controlled diabetes (DeFronzo and Lang, 1980). The negative phosphate balance results from decreased dietary intake, and a large increase in urinary phosphate excretion caused by osmotic diuresis induced by high glucose, metabolic acidaemia and lack of insulin.

Hypophosphataemia may lead to insulin resistance due to an augmented plasma insulin response. This might, in part, explain the abnormality in glucose tolerance observed in patients with hyperparathyroidism.

Since diabetics may also have coexisting hypocalcaemia (because of increased urinary losses, renal insufficiency, or decreaed intake), the plasma level of calcium should be ascertained before phosphate administration, to avoid the clinically serious condition of hypocalcaemia).

If diabetes is under good control there are no disturbances of calcium metabolism.

Table 28.1 Relationship between serum (Ca), serum parathyroid hormone (PTH) and insulin response during oral glucose load and glucose tolerance. Arrows (↓, ↔ and ↑) indicate decreased, normal and increased values respectively. (From Yasuda et al., 1975).

Group	Serum Ca	Parathyroid hormone	Insulin Response	Glucose Tolerance
Primary hyper-parathyroidism	↑	↑	↑	↔
Idiopathic hypo-parathyroidism	↓	↓	↓	↔
Pseudohypopara-thyroidism	↓	↑	↓	↔
Secondary hyper-parathyroidism	↔	↑	↔	↔

Acknowledgment

Grateful thanks to Mrs Merath who typed the manuscript.

References

Akgun, S. and Ertel, N.H. (1978). Hyperparathyroidism and coexisting diabetes mellitus - altered carbohydrate metabolism. *Arch. Intern. Med.* **138**, 1500–2.
Anderson, E.E.H., Röjdmark, S., Hed, R. and Sundblad, L. (1982). Hypercalcemic and calcium antagonistic effects on insulin-release and oral glucose tolerance in man. *Acta Med. Scand.* **211**, 35–43.
DeFronzo, R.A. and Lang, R. (1980). Hypophosphatemia and glucose intolerance: evidence for tissue insenstivity to insulin. *N. Engl. J. Med.* **303**, 1259–63.
De Palo, C., Sicolo, N., Vettor, R. and Federspil, G. (1981). Lack of effect of calcium infusion on blood glucose and plasma insulin levels in patients with insulinoma. *J. Clin. Endocrinol. Metab.* **52**, 804–6.
Donatsch, P., Löwe, D.A., Richardson, B.P. and Taylor, P. (1977). The functional significance of sodium channels in pancreatic B-cell membranes. *J. Physiol. (Lond.)* **267**, 357.
Formby, B., Capito, K., Egeberg, J. and Hedeskov, C.J. (1976). Ca-activated ATPase activity in subcellular fractions of mouse pancreatic islets. *Am. J. Physiol.* **230**, 441–8.
Fussgänger, R. (1974). *Statische und dynamische Aspekte der Insulin-und Glukagonsekretion in vitro unter besonderer Berücksichtigung des isolierten perfundierten Pankreas.* Habilitations-schrift, Universität Ulm.
Gaeke, R.F., Kaplan, E.L., Rubstein, A., Starr, J. and Burke, G. (1975). Insulin and proinsulin release during calcium infusion in a patient with islet-cell tumor. *Metabolism* **24**, 1029–33.
Gagliardino, J.J., Harrison, D.E., Christie, M.R., Gagliardino, E.E. and Ashcroft, S.J.H. (1980). Evidence for the participation of calmodulin in stimulus-secretion coupling in the pancreatic β-cell. *Biochem. J.* **192**, 919–27.
Gedik, O., Akalin, S. and Koray, Z. (1980). Effect of acute hypercalcaemia on glucose tolerance and insulin release in human beings. *Acta Endocrin.* **94**, 196–200.
Henquin, J.C. (1981). Effects of trifluoperazine and pimozide on stimulus secretion coupling in pancreatic β-cells. *Biochem. J.* **196**, 771–80.
Herchueltz, A. and Malaisse, W.J. (1978). Regulation of calcium fluxes in pancreatic islets: dissociation between calcium and insulin release. *J. Physiol. (Lond.)* **283**, 409.
Herchueltz, A. and Malaisse, W.J. (1981). Calcium movement and insulin release in pancreatic islet cells. *Diabète Metab.* **7**, 283–8.
Kao, C.Y. (1966). Tetrodotoxin saxitoxin and their significance in the study of excitation phenomena. *Pharmacol. Rev.* **18**, 997.
Kim, H., Kalkhoff, R.K., Costrini, N.V., Cerletty, J.M. and Jacobson, M. (1971). Plasma insulin disturbances in primary hyperparathyroidism. *J. Clin. Invest.* **50**, 2596–605.
Klöppel, G. and Bommer, G. (1979). Ultracytochemical calcium distribution in B-cells in relation to biphasic glucose stimulated insulin release by the perfused rat pancreas. *Diabetes* **28**, 585–92.
Maier, V. (1979). *Die isolierte Langerhanssche Insel - Biosynthese und Sekretion von Insulin in vitro.* Thieme Verlag, Stuttgart.
Nattrass, M. and Santiago, J.V. (1983). *Recent Advances in Diabetes* 1. Churchill

Livingstone, Harlow.
Schneider, L.E. and Schedl, H.P. (1972). Diabetes and intestinal absorption in the rat. *Am. J. Physiol.* **223**, 1319–23.
Tomlinson, S., Walker, S.W. and Brown, B.L. (1982). Calmodulin and insulin secretion. *Diabetologia* **22**, 1–5.
Valverde, I., Vandermeers, A., Anjaneyulu, R. and Malaisse, W.I. (1979). Calmodulin activation of adenylatecyclase in pancreatic islets. *Science* **206**, 225–7.
Wollheim, C.B. and Sharp, G.W.G. (1981). Regulation of insulin release by calcium. *Physiol. Rev.* **61**, 914–60.
Yasuda, K., Hurukawa, Y., Okuyama, M., Kikuchi, M. and Yoshinaga, K. (1975). Glucose tolerance and insulin secretion in patients with parathyroid disorders - effect of calcium on insulin release. *N. Engl. J. Med.* **292**, 501–4.
Ziegler, R., Bellwinkel, S., Schmidtchen, D. and Minne, H. (1972). Effects of hypercalcemia, hypercalcemia and calcitonin on glucose-stimulated insulin secretion in man. *Horm. Metab. Res.* **4**, 60.

DRUGS AND MUSCLE CONTRACTION

Marcus C. Schaub and Peter G. Waser

Institute of Pharmacology, University of Zürich, CH-8006 Zürich, Switzerland

Abstract

On the molecular level the contractile mechanism is the same for all types of muscles while the intracellular regulation of the contractile process by Ca ions varies from tissue to tissue. In sarcomeric skeletal and heart muscles Ca triggers contraction by binding to troponin on the actin filaments. In addition calmodulin- and cAMP-dependent phosphorylation reactions modulate the contractile process synergistically. Thus in heart muscle drugs which lead to an increase in cAMP such as beta-adrenergic agonists or methylxanthines and drugs such as cardiac glycosides which enhance the availability of cytosolic Ca are both positive inotropic. In contrast, in smooth muscles Ca triggers contraction by calmodulin-dependent phosphorylation of myosin and cAMP-dependent phosphorylation reactions act antagonistically leading to relaxation. The different Ca-regulatory mechanisms and their responses to various classes of drugs, in particular of beta-blockers, cardiac glycosides and Ca-blockers, are discussed.

29.1. Introduction

The aim of pharmacology includes understanding the mechanism underlying successful therapeutic interventions on the molecular level. This opens the rationale for further pharmacological developments. One of the major advances of insight into the regulation of intracellular processes has been achieved in recent years in the field of muscular contraction. A fascinating web of interconnected pathways and feedback loops of regulation has emerged. While the contractile mechanism, on the molecular level involving myosin and actin, is probably the same in all muscles, and even in non-muscle cells the complex regulatory system varies from tissue to tissue (Schaub and Watterson, 1981; Schaub, Watterson and Waser, 1983). This helps explain the flexibility shown by different muscle tissues in their responses not only to the neural and hormonal control but also to pharmacological interventions. A natural division between skeletal muscles on the one hand and heart and smooth muscles on the ohter, is set by their differences in intracellular control. The voluntary skeletal muscles are under strict control of the neuromuscular junction and therefore pharmacological interventions are, in a large part, restricted to this area of neurotransmission. In contrast, the heart as well as smooth muscles are dependent on extracellular Ca ions throughout their activities. In addition, their activities are under significant hormonal influence. This opens up a vast field for pharmacolgical control.

29.2. Calcium as intracellular messenger

For a variety of reasons nature has chosen Ca ions as intracellular messengers to convey the signal received at the surface membrane of the cell to the target site of the physiological response. Beside contraction, Ca activates secretion, mitosis, various metabolic enzyme systems, membrane transport, microtubule assembly—disassembly, phosphorylation reactions and many others. In resting cells the cytosolic-free Ca concentration is below 10^{-7} M. Only against such a low background concentration can Ca ions fulfil their role as information carrier. Ca differs from Mg both in kinetics and thermodynamics; Ca is able to bind to specific protein sites at least 100 times faster than Mg. At such receptor protein sites Ca may bind with a somewhat variable geometry involving six to eight coordination points while Mg is able to interact with proteins specifically in a strict way only, involving six coordination points. Mg could therefore not function in a fast trigger system as is required, for instance, in a fast muscle twitch which lasts for no more than 30—100 ms. Mg is required for many enzyme systems serving as constituent in the nucleotide substrate complex. Since all eukaryotic cells use ATP as an energy source producing inorganic phosphates, a high cytosolic Ca concentration would lead to precipitation of Ca phosphates, thus disrupting cell metabolism altogether. The extracellular Ca concentration is in the millimolar range, so that for a cell at rest an outside-to-inside ratio exceeding 10^4 must be established. The extremely low level of free Ca is maintained by Ca pumps in the cell surface membrane as well as in the intracellular sarcoplasmic reticulum membrane system (SR). Ca is also extruded from the cell by a Na—Ca exchange carrier under the control of the ion gradient established by the Na—K pump of the cell membrane (Fig.29.1 and Fig.29.3). All these pumps require energy to work against the ion gradients. For intracellular activity to begin, Ca must flow from outside through the membrane via potential dependent or receptor-operated Ca-channels and/or the release from the intracellular stores such as SR and plasma membrane-bound Ca. Overflooding of the cytosolic space with Ca, however, leads to cell death. At free Ca concentrations above 10^{-5} M, the mitochondria begin to absorb it and if they take up too much Ca, it leads to disruption of their energy metabolism.

29.3. Intracellular calcium binding proteins

Although regulation of contraction is tissue-specific the first step represents the binding of the released Ca in the cytosol to specific target proteins. Evolution has produced a family of phylogenetically related proteins capable of binding Ca reversibly with high affinity around 10^6 M^{-1} (Kretsinger, 1980; Levine and Dalgarno, 1983). They include calmodulin, troponin-C, myosin regulatory light chain (RLC) and parvalbumin, have molecular weights ranging from 12 000 to 19 000 and are found exclusively within the cell. Of these calmodulin (16 700 daltons) is the ubiquitous intracellular Ca-receptor protein and regulates a variety of enzyme systems including contraction in

Drugs and muscle contraction

smooth muscles and non-muscle cells. Calmodulin is unique for its ability to act as a regulatory subunit in different enzyme complexes such as membrane Ca pumps, protein kinases, adenylate cyclase and cyclic nucleotide phosphodiesterase. By way of the latter two enzymes, calmodulin and therefore Ca itself, are involved in synthesis and degradation of cyclic AMP (cAMP) whose intracellular messenger function has been established. In addition, calmodulin was found to be increased in all transformed tumorous cells. Troponin-C (18 000 daltons) exhibits 50% sequence homology with calmodulin and is directly involved in triggering contraction in sarcomeric muscles (skeletal and heart muscles). Both calmodulin and troponin-C can reversibly bind 4 Ca ions. Thus contraction is always induced by Ca binding to either of these two proteins, so no matter by what kind of stimulus (neural or hormonal), Ca becomes increased in the cytosol. The way these proteins elicit contraction differs, however. A clear picture of how parvalbumin (12 000 daltons) functions has not yet emerged, but it is found in appreciable amounts in fast contracting muscles (Heizmann, Berchtold and Rowlerson, 1982). It may act as a scavenger for the free Ca in the cytosol and so enable cells, which are switched on by a pulse of Ca to a state of high activity, to be rapidly returned to rest. Parvalbumin would thus function as a soluble cytosolic relaxing factor by depleting the trigger sites of troponin-C from Ca

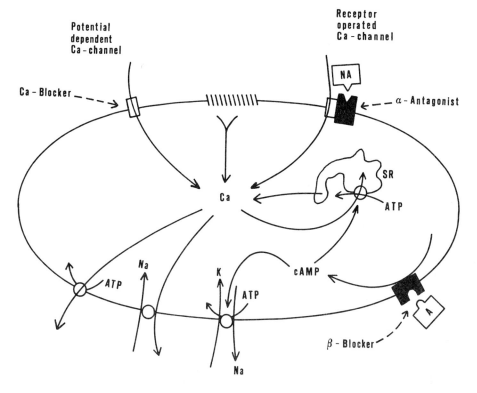

Fig.29.1. Systems regulating intracellular Ca. Smooth muscle cell depicted contains two types of Ca channels (potential dependent and receptor mediated). Intracellular Ca stores are the sarcoplasmic reticulum (SR) and the cell membrane. Energy requiring pumps are the Ca pumps in the SR and cell membrane and the Na-K pump in the cell membrane which latter is electrogenically linked with the Na-Ca-exchange carrier. Pumps are stimulated by cAMP-dependent phosphorylation. (NA) noradrenaline; (A) adrenaline.

faster than the ions are eventually removed from the cytosol by the membrane pumps. The myosin subunit RLC (19 000 daltons) is able to bind both Mg and Ca with high affinity (Watterson, Kohler and Schaub, 1979). During rest Mg is bound to the RLC, and during sustained activity it may become replaced by Ca. Whether this Ca binding to the myosin RLC plays a role in regulation of contraction remains an open question.

29.4. Regulation of skeletal muscle contraction

In principle, contraction is initiated in two ways. In sarcomeric muscles switching on results from a fast direct allosteric protein conformational change induced by Ca ions which bind to troponin-C on the actin filament (Schaub and Watterson, 1981). As a consequence the toponin-tropomyosin complex along the actin filament undergoes a change in its steric disposition yielding the myosin-binding sites on the actin filament acccessible for combination with the myosin crossbridges. Thus the repressed state of the actin filament, which prevents the myosin crossbridges attaching during rest, becomes derepressed and contraction ensues. During prolonged activity or during a tetanus the increased cytosolic Ca leads to a series of phosphorylation reactions stimulating the energy metabolism but also leading to phosphorylation of the myosin RLC. This modulates the contractile properties by affecting the myosin crossbridge movement in such a way that subsequent contractions become more efficient with regard to ATP consumption.

A possible site for efficient pharmacological interference is clearly the neuromuscular junction where the transmission takes place. This is the site of action of the muscle blocking agents used in surgery and treatment of neurological diseases which are accompanied by increased tonus or tetanus. As discussed recently (Schaub, Watterson and Waser, 1983), for clinical use relaxants of the non-depolarizing type with known antagonists and low toxocity are desirable, since they do not produce the damaging side-effects of loss of potassium from the muscle cells. The short-acting vecuronium, a monoquaternary derivative of pancuronium, and the non-steroid bisquaternary atracurium, may be examples of a new generation of such muscle relaxants. Pharmalogical measures which affect presynaptic transmitter metabolism are beyond our scope here. In principle, pharmacological interference at the site of neuromuscular transmission, whether presynaptical or postsynaptical, does not appear to cause malfunction in the ionic pumps, and as a result the intracellular compartmentation of Ca is not disturbed.

There seems to exist only one more recent approach to affect the intracellular handling of the regulatory Ca, that is, in malignant hyperthermia. Here an autosomal genetic disturbance results in faulty membranes, in particular the SR. After administration of anaesthetics and depolarizing muscle relaxants during surgery the SR manifests an impaired capacity to pump Ca and an excess of Ca leaks into the cytosol. Uncontrollable muscle contraction and activation of metabolism ensue, resulting in rising body temperature and a mortality rate of over 60%. This fatal consequence can now be prevented by the hydantoin derivative dantrolene sodium which binds with high affinity only to skeletal muscle SR and blocks leakage of Ca through its membranes. Thereby, skeletal muscle activity is greatly reduced without affecting the neuromuscular transmission and has only a minimal effect on heart and smooth muscles.

29.5. Regulation of smooth muscle contraction

Switching contraction on in smooth muscles and in non-muscle cells represents a slower process operating indirectly via covalent protein modification. Ca binds to calmodulin and Ca-calmodulin then combines with the myosin RLC kinase (130 000 daltons) and finally the myosin RLC becomes phosphorylated. Thus starting contraction depends on phosphorylation of a serine residue in the RLC and relaxation is an even slower process involving dephosphorylation of the RLC again by a protein phosphatase after the Ca level has returned to resting low concentrations. Contractions and relaxations in smooth muscles are therefore running on a different time scale than in fast-twitch skeletal muscles; while in the latter it takes a fraction of a second, in the former these processes take many seconds up to minutes. Beside dephosphorylation of the myosin RLC a second process also leads to relaxation in smooth muscles (Adelstein and Eisenberg, 1980). This second

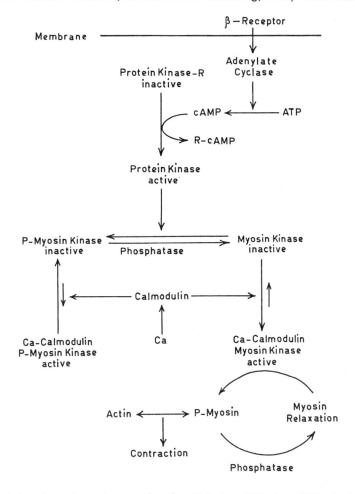

Fig.29.2. Regulation of smooth muscle contraction after Adelstein and Eisenberg (1980), showing the interconnection of the Ca and cAMP pathways. (R), regulatory subunit of cAMP dependent protein kinase. Length of arrows between different forms of myosin RLC kinase indicates shift of equilibrium to either active or inactive state.

mechanism results from beta-adrenergic stimulation which leads to phosphorylation of the myosin RLC kinase by a cAMP-dependent protein kinase (Fig.29.2). When the myosin RLC kinase is phosphorylated it no longer combines readily with Ca-calmodulin and relaxation sets in because of lack of sustained phosphorylation of the RLC despite the Ca which might still be elevated in the cytosol. In addition, however, beta-adrenergic stimulation also leads, via cAMP-dependent protein kinases, to phosphorylation of membrane proteins which are associated with the Ca pump of the SR and the Na—K pump which are thus both enhanced. So in smooth muscle cells beta-adrenergic stimulation achieves a lowering of both the activity of the myosin RLC kinase and the cytosolic-free Ca concentration, both leading to relaxation. Troponin-C, allowing for the fast direct allosteric Ca-switch, does not seem to exist on the actin filaments, although there are Ca-dependent proteins associated with actin (such as caldesmon, and there may be others), which affect the structure of these filaments (Kakiuchi and Sobue, 1983).

In contrast to skeletal muscles, smooth muscles have not developed neuromuscular endplates but possesss a variable population of a number of different receptor types. This variable mosaic of specific receptors on the cell surface ensures that a particular tissue responds in its own characteristic way to the transmitters acetylcholine, catecholamines and various hormones. Some receptors are directly coupled with Ca fluxes, other initiate Ca fluxes through the potential dependent Ca channels as a consequence of depolarization (see Fig.29.1). During contraction in response to alpha-adrenergic, muscarinic cholinergic and H_1 histaminergic agents or 5-hydroxytryptamine, the phosphatidylinositol turnover is activated and appears to be involved in the opening of membranous Ca gates. Consequently, a large number of compounds, both foreign and naturally occurring within the body, leads to contraction of smooth muscles. In all cases, however, activation of the contractile machinery invariably requires an increase in cytosolic Ca. Then in smooth muscles the two intracellular messengers, Ca and cAMP, act antagonistically, as opposed to the situation in heart muscle where they rather act synergistically both producing positive inotropy. Furthermore, in the heart muscle the myosin RLC kinase is much smaller (80 000 daltons) and when phosphorylated itself by the cAMP-dependent protein kinase, it does not stop phosphorylation of the myosin RLC as in smooth muscle. Any pharmacological intervention causing a rise in cAMP therefore favours relaxation in smooth muscles and at the same time increases the rate and force of contraction in the heart muscle. This explains why beta-receptor agonists used against asthma to relax the bronchial smooth muscles, concomitantly stimulate the heart, while beta-blockers employed to calm down heart activity may at the same time elicit an attack of asthma. To develop beta-receptor blockers with selective tissue specificity would be of great help in therapeutics. Whether that will be possible remains to be seen. Whereas the receptor density varies significantly between different tissues, the gross structure of the beta-2 receptor seems to be rather uniform in a variety of tissues, consisting of two subunits (55 000—63 000 daltons) linked together by a disulphide crossbridge (Venter and Fraser, 1983). Beta-1-receptors on the other hand, seem to exist as monomeric protein with a molecular weight around 65 000—70 000 daltons. At present more structural diversity appears between beta-1 and beta-2-receptors within one mammalian species than between beta-2-receptors from different tissues and substantially different species.

29.6. Regulation of heart muscle contraction

The intracellular regulation of contraction in heart muscle is much more complex involving the fast-acting direct Ca switch via troponin-C but also the slower acting phsophorylation reactions which are under control of either Ca-calmodulin or cAMP (Fig.29.3). These intricately intertwined regulatory

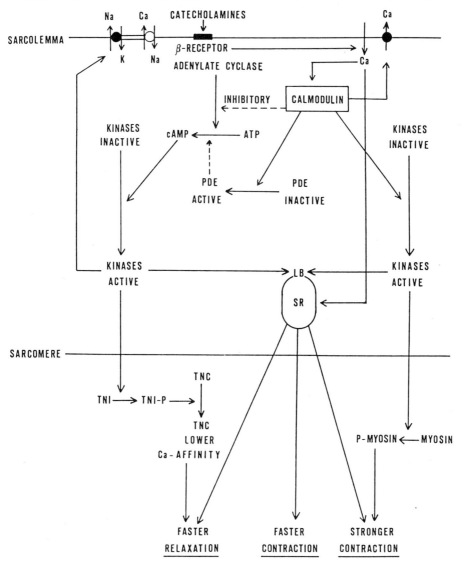

Fig.29.3. Scheme of regulatory pathways in heart muscle cells at the sarcolemmal and sarcomeric levels. Arrows indicate ion fluxes and stimulatory interactions. Dashed arrows represent inhibitory effects. Full circles show energy-requiring pumps in the sarcolemma. Open circle in sarcolemma symbolizes the Na-Ca-exchange carrier electrogenically linked to the Na-K pump. SR, sarcoplasmic reticulum, which also contains an energy-requiring Ca pump; PDE, cyclic nucleotide phosphodiesterase; LB, protein associated with the Ca pump of SR; TNI, troponin-I; TNC, troponin-C.

pathways seem necessary to ensure smooth functioning of the heart pump throughout life. Only part of Ca required for contraction stems from the SR. Half of it or more is handled by the sarcolemmal membrane devices, which comes from the outside of the cell for beat-to-beat activity. As indicated on the right-hand side of Fig.29.3 Ca-calmodulin governs phosphorylation of the myosin RLC and of lamban (LB) of the SR. It further activates the sarcolemmal Ca pump and the cyclic nucleotide phosphodiesterase (PDE), but in heart muscle it seems to inhibit the adenylate cyclase (Potter et al., 1980), thus dampening the production of cAMP. On the left-hand side of the scheme, cAMP governs phosphorylation reactions on all three levels, the Na—K pump of the sarcolemma, the LB of the SR and the tropinin-I of the contractile machine in the sarcomere. Phosphorylation of troponin-I lowers the affinity for Ca binding to troponin-C. In skeletal muscle the troponin-I component cannot be phosphorylated by the cAMP-dependent protein kinase as in its primary structure it is lacking the serine-20, the phosphorylation site. This presents an interesting situation where a tissue specific hormone action is not discriminated by the receptor complex at the cell surface, but the selectivity is comprised in an evolutionary modification of the target protein at the site of the physiological response. A second example is given by the myosin RLC kinase which in the heart has a similar molecular weight as in skeletal muscles (80 000 daltons) and, unlike its larger counterpart in smooth muscles, is not affected in its activity by the cAMP-dependent protein kinase in response to beta-adrenergic stimulation. In the heart both messengers, Ca and cAMP, then act synergistically producing a faster and stronger contraction followed by a faster relaxation.

Figure 29.3 does not include the alpha-1- and alpha-2-receptors known to be present in cardiac muscle beside the more abundant beta-receptors (Exton, 1982). Stimulation of alpha-1-receptors is not associated with changes in cAMP, but results in increased cytosolic-free Ca. Presynaptic alpha-2-receptors mediate the feedback inhibition of noradrenaline release while the postsynaptic receptors lead to a decrease in cAMP synthesis; cAMP seems to be the messenger for all beta-receptor stimulation responses. This fits the observation that carbachol appears to antagonize them by lowering the cAMP levels. It has been reported recently that the human heart contains about equal numbers of both beta-1 and beta-2 receptors (Waelbroeck et al., 1983). Interestingly, the adenylate cyclase activity may be controlled entirely by the beta-2 portion of adrenoceptors. It seems therefore that the beta-2-receptor combines more readily with the stimulatory subunit of the adenylate cyclase complex than do beta-1-receptors. This leaves a question about the function of the beta-1-receptor in this tissue. The smooth muscle cells of both blood vessels and the bronchi, in contrast, contain only beta-2-receptors.

29.7. Cardiac glycosides

On the basis of the scheme shown in Fig.29.3 we can gain some insight into the rationale behind today's usage of drugs, either singly or in combination, in the treatment of heart diseases. For some 200 years the cardiac glycosides (CG) have proved their efficency in the treatment in heart failure. Their action can be differentiated from all other positive inotropic agents such as methylxanthines and their derivatives, catecholamines, glucagon and amrinone, a bipyridine derivative. Under cardiac glycoside treatment the availability of regulatory Ca in the cytosol is increased. Beside its inotropic effect the production of cAMP, if anything, is dampened, so no concomitant undue stimulation of metabolism occurs, as is the case with all other inotropic agents mentioned. In particular this distinction is the major advantage of the pharmacodynamics of cardiac glycosides. The increase in cAMP leads to phosphorylation of the regulatory protein subunit troponin-I which, in turn, lowers the affinity of troponin-C for Ca, so that the Ca can be removed again more readily by the Ca pumps and relaxation sets in more rapidly. There is, of course, still sufficient increase in the cytosolic Ca to produce the effect of positive inotropy, but the shorter duration of the contractions goes together with an elevated demand for oxygen; that is exactly what one would like to avoid. In a heart that works at the brink of insufficiency, about half of the energy is consumed by the various pump systems. Therefore it may already be overstimulated by sympathetic compensatory mechanisms. It is then beneficial to combine the cardiac glycoside treatment with beta-blockers in order to reduce the oxygen requirement of the tissue without affecting the inotropic action of the cardiac glycosides.

The precise mechanism of positive inotropy produced by low doses of cardiac glycosides has still eluded elucidation (Erdmann, 1984). Two types of receptors for cardiac glycosides seem to exist, one with high and a second one with lower affinity. The binding of cardiac glycosides to the high affinity sites is believed to produce the therapeutically desired positive inotropic effect possibly by interaction with the Na-Ca exchange system. On the other hand, all toxic effects are clearly due to inhibition of the Na–K pump. This results in alterations of the ion compartmentation and concomitant disturbance of the electrogenic membrane functions. Most prominent are the loss of intracellular potassium and the accumulation of sodium. It has been shown, however, that under hypoxic conditions the heart cells also loses Mg (Lehr, 1981). In excess, Mg behaves in certain aspects like a Ca antagonist because it competes for binding sites on various regulatory proteins; but when too low, it may enhance undesirable effects due to Ca. Mg can in fact be used therapeutically to relieve symptoms of K deficiency, to lower the sensitivity to dysrhythmias under cardiac glycoside treatment, to reduce vessel spasms and to protect the cell against ischaemic damage. In this context a newer class of drugs needs mentioning, the so-called Ca antagonists (discussed below), which may be used in combination with cardiac glycosides. As soon as the cardiac glycosides begin to show toxic signs use of Ca antagonists may be indicated to counteract the effects due to the increased cytosolic Ca concentration by attenuation Ca influx into the myocardium. In addition, it could be that an endogenous digitalis-like factor could interfere with the pharmacological action of cardiac glycosides. The most prominent problems in cardiac glycoside treatment concern the large variation in response among individuals, a very narrow therapeutic margin indicated by the low therapeutic index of less than two and the risk of occasional fatal dysrhythmias. This has led to the search for new compounds. All cardiac glycosides exhibit qualitatively the same pharmacological effects and distinguish themselves only from one another in their pharmacokinetics. Digitoxin and digoxin remain the standard cardiac glycosides in use, since the introduction of their acetyl-derivatives and methyl-derivatives has neither improved their clinical efficacy nor decreased the risks of toxicity. A new group of non-glycoside, non-catechol cardiotonic agents which can be administered orally are under evaluation at present. Amrinone, a bipyridine derivative, has been introduced recently. It does not inhibit the Na–K pump but seems to increase the Ca influx during the action potential. It also leads to elevation of cAMP. Little is known about toxic effects although thrombocytopenia has been reported and ischaemic side-effects could be potential complications. Most of these new compounds still under evaluation such as sulmazole (a pyridinimodazole derivative), UD-CG 115 (a benzimidazole derivative), RO 13–6438 (a imidazoquinazolinone derivative) or OPC-8212 (a piperazinylquinolinone derivative) are cardiotonic by a different mechanism from cardiac glycosides or catecholamines. They all are accompanied by an increase in cAMP coupled with moderate vasodilatation and in some cases with an increase of the heart rate. In summary, there are still no therapeutic agents which compare favourably with cardiac glycosides.

29.8. Calcium antagonists

In cardiac and smooth muscles excitation–contraction coupling can be disrupted by a number of organic compounds known collectively as Ca antagonists (CA). They also may be referred to as Ca-entry blockers because their main mode of action involves binding to, and thereby, impairing the function of the Ca-channels. Many of them do, however, display additional effects. The negative inotropic, negative chronotropic and coronary vasodilator effects of the calcium antagonists are well established and have a great impact on cardiovascular medicine.

Verapamil, a phenylalkylamine, was first introduced into clinical practice for angina pectoris in 1962 in Germany and was originally considered a beta-blocker. When its antidysrhythmic action became apparent it was thought to have local anaesthetic properties. Fleckenstein (1977) then elucidated its mechanism as inhibitor of the slow inward Ca current in cardiac and smooth muscle cells. Despite of the variety of chemicals affecting the Ca translocation a unitary mechanism for the calcium antagonists has recently been demonstrated (Murphy et al., 1983). At least four main classes of calcium antagonists are distinguished: the dihydropyridines, such as nifedipine; phenylalkylamines, such as verapamil; benzothiazepines, such as diltiazem and the diphenylalkylamines, such as prenylamine. The radioactively labelled dihydropyridines, nitrendipine and nimodipine, bind directly to receptor sites associated with the Ca channel. All other dihydropyridine calcium antagonists inhibit the binding of nitrendipine in a competitive fashion. The other three classes of calcium antagonists bind to a second drug site which is allosterically linked to the dihydropyridine recognition site. The phenylalkylamines and diphenylalkylamines reduce the binding of nitrendipine while the benzothiazepine diltiazem enhances it. Nitrendipine and other dihydropyridines bind with nanomolar affinity to the site that can either block (Cauvin, Loutzenhiser and Van Breemen, 1983) or activate (Schramm et al., 1983) the Ca channels, while the second site binds a diverse group of chemicals with lower affinity which, nevertheless, all affect the Ca translocation. This generalization would allow one to predict that drugs that regulate allosterically nitrendipine binding should behave like calcium antagonists. Ca antagonism was in fact observed on the guinea-pig ileum with several neuroleptics (thioridazine and mesoridazine), antihistamines (dimethindene) and muscarinic anticholinergic agents (biperiden) that mimic the effect of diltiazem on nitrendipine binding. Such binding analyses may help identify clinical effects of drugs related to their calcium antagonist potential. Of the numerous phenothiazines evaluated, thioridazine and mesoridazine were unique in their calcium antagonist properties at therapeutic doses. These two drugs are also eliciting more cardiac side-effects with Q-T interval prolongation than other phenothiazines. Furthermore, these two drugs have been known for a long time to suppress ejaculation, which involves massive contraction of the smooth muscle of the vas deferens.

Employment of radioactively labelled ligands has allowed recently isolation of the dihydropyridine and verapamil/diltiazem recognition sites of the voltage sensitive Ca-channel as an allosterically coupled complex from brian membranes as well as from skeletal muscle transverse tubules (Curtis and Catterall, 1984). In both cases there is a large glycoprotein in the molecular weight range of 210 000 daltons consisting of three subunits of 130 000, 50 000 and 33 000 daltons which are non-covalently linked together. It is worth mentioning also that the voltage-sensitive Na channel is a large glycoprotein of 200 000–300 000 consisting of one large and two smaller subunits. This overall structural similarity may reflect common requirements for rapid voltage gated ion movements across a phospholipid bilayer (Curtis and Catterall, 1984).

Despite the unitary picture which emerges about the main action of the calcium antagonists in blocking the Ca influx into the cell, what are the reasons for the diversity of responses encountered in their practical application? From the vast body of literature certain general aspects may become evident (Cauvin, Loutzenhiser and Van Breeman, 1983; Smith, 1983). The basal tone of vascular smooth muscle is primarily dependent on receptor-mediated intracellular and extracellular Ca pools. The sustained tone of vascular smooth muscle thus represents the resultant from locally released and circulating hormones. This is in contrast to the cardiac tissue where the intracellular Ca metabolism is mainly controlled by Ca channels under the influence of the membrane potential. The calcium antagonists are thought to act primarily on the potential-dependent Ca channels (PDC) which may be activated by mechanical factors such as stretch, autonomous discharges or any agent which leads to depolarization.

The evidence for the existence of a distinct second, receptor-operated type of Ca channel (ROC) stems from the following observations: (1) the tonic noradrenaline contraction is dependent on extracellular Ca, but may occur in the absence of depolarization; (2) Ca influx stimulated by high K depolarization and noradrenaline is additive; (3) calcium antagonists display selective effects on PDC over the ROC. The PDC are always dependent on Ca influx from outside the cell and this influx is attenuated by calcium antagonists. Thus their sensitivity to the calcium antagonists remains rather constant throughout the entire vascular bed. Stimulation of the ROC often is linked to Ca release from intracellular stores such as the SR, and this latter component does not seem to be affected by calcium antagonists measurably at therapeutically active concentrations. In general, at low agonist hormone concentrations, the ROC are more susceptible to calcium antagonists than at high ones. This difference may be attributed to two open states of the ROC, one susceptible to calcium antagonists, the other not. Beside this functional variation in the ROC with regard to calcium antagonist senstivity, there also exist chemical differences. For example the ROC of the cerbral circulation activated by either noradrenaline, thromboxane A_2, serotonine or prostaglandin $F_{2\alpha}$ appear to be more susceptible to inhibition by calcium antagonists. Furthermore, the regional distribution of PDC and ROC varies considerably within the vascular system. In particular, in the venous system contractions induced by agonists which depend on Ca influx are quite resistant to calcium antagonists. Although some calcium antagonists, diphenylalkylamines in particular, bind to calmodulin, there is no correlation between inhibition of calmodulin function and vasodilatation. Tissue selectivity is, of course, also to be expected in view of the diverse chemical nature of calcium antagonists involving differences in physicochemical properties, membrane transport, metabolism and tissue binding.

Variability in sensitivity of various vascular and non-vascular smooth muscle tissues to calcium antagonists appears then to be more complex than that observed in cardiac muscle. Therefore, the therapeutical domain for the calcium antagonists remains, for the time being, where conservation and protection of the heart muscle tissue are required. The lowering of cytosolic Ca levels leads to negative inotropy and chronotropy. The depression in myogenic activity together with lowering of the tone of the coronary vessels, inhibition of platelet aggregation and protection of endothelial integrity - all these effects are beneficial in conditions where myocardial tissue is poorly supplied with blood and the threat of cell damage arises. In cases where anoxic lesions have already occurred, administration of calcium antagonists prevents damage to cells when the tissue becomes reperfused and Ca threatens to flood the intracellular space. Otherwise the mitocondria start taking up Ca ions which leads to interruption of ATP production needed to keep up contractility and the energy-requiring pumping functions. Accordingly, calcium antagonists are clincially in use in tachycardiac dysrhythmias, coronary vasospasm, ischaemic heart disease, intraoperative protection of ischaemic myocardium, acute myocardial infarction, congestive heart failure and hypertrophic cardiomyophathy; they may also be used in combination with cardiac glycosides and even with beta-blockers. Although most calcium antagonists lead to a lowering of the blood pressure in normotensive and hypertensive experimental animals, their role in the chronic management in man has not yet been established (Smith, 1983). It appears that calcium antagonists have no clear advantage over hydralazine as antihypertensive agents. It seems that new drugs from different classes of the calcium antagonists with greater tissue selectivity and better antihypertensive effects must be awaited. Thousands of calcium antagonist derivatives are currently under investigation.

Acknowledgement

Support by the Swiss National Science Foundation no.3. 505.83 is acknowledged.

References

Adelstein, R.S. and Eisenberg, E. (1980). Regulation and kinetics of the actin-myosin-ATP interaction. *Ann. Rev. Biochem.* **49**, 921–56.

Cauvin, C., Loutzenhiser, R. and Van Breemen, C. (1983). Mechanisms of calcium antagonist-induced vasodilation. *Ann. Rev. Pharmacol. Toxicol.* **23**, 373–96.

Curtis, B.M. and Catterall, W.A. (1984). Purification of the calcium antagonist receptor of the voltage-sensitive calcium channel from skeletal muscle transverse tubules. *Biochemistry* **23**, 2113–8.

Erdmann, E. (ed.) (1984). Cardiac glycoside receptors and positive inotropy. Evidence for more than one receptor? Suppl. to *Basic Res. Cardiol.* **79**, 1–162.

Exton, J.H. (1982). Molecular mechanisms involved in alpha-adrenergic responses. *Trends. Pharmacol. Sci.* **3**, 111–5.

Fleckenstein, A. (1977). Specific pharmacology of calcium in myocardium, cardiac pacemakers, and vascular smooth muscle. *Ann. Rev. Pharmacol. Toxicol.* **17**, 149–66.

Heizmann, C.W., Berchtold, M.W. and Rowlerson, A.M. (1982). Correlation of parvalbumin concentration with relaxation speed in mammalian muscles. *Proc. Natl. Acad. Sci. USA* **79**, 7243–7.

Kretsinger, R.H. (1980). Structure and evolution of calcium-modulated proteins. *CRC Crit. Rev. Biochem.* **8**, 119–74.

Kakiuchi, S. and Sobue, K. (1983). Control of the cytoskeleton by calmodulin and calmodulin-binding proteins. *Trends. Biochem. Sci.* **8**, 59–62.

Lehr, D. (1981). Magnesium and cardiac necrosis. *Magnesium Bull.* **3**, 178–91.

Levine, B.A. and Dalgarno, D.C. (1983). The dynamics and function of calcium-binding proteins. *Biochim. Biophys. Acta* **725**, 187–204.

Murphy, K.M.M., Gould, R.J., Largent, B.L. and Snyder, S.H. (1983). A unitary mechanism of calcium antagonist drug action. *Proc. Natl. Acad. Sci. USA* **80**, 860–4.

Potter, J.D., Piascik, M.T., Wisler, P.L., Robertson, S.P. and Johnson, C.L. (1980). Calcium dependent regulation of brain and cardiac muscle adenylate cyclase. *Ann NY Acad. Sci.* **256**, 220–30.

Schaub, M.C. and Watterson, J.G. (1981). Control of the contractile process in muscle. *Trends. Pharmacol. Sci.* **2**, 279–82.

Schaub, M.C., Watterson, J.G. and Waser, P.G. (1983). Pharmacology and muscle contraction. *Trends. Pharmacol. Sci.* **4**, 116–20.

Schramm, M., Thomas, G., Towart, R. and Franckowiak, G. (1983). Novel dihydropyridines with positive inotropic action through activation of Ca^{2+} channels. *Nature (Lond.)* **303**, 535–7.

Smith, R.D. (1983). Calcium entry blockers: key issues. *Fed. Proc.* **42**, 201–6.

Venter, J.C. and Fraser, C.M. (1983). The structure of alpha- and beta-adrenergic receptors. *Trends Pharmacol. Sci.* **4**, 256–8.

Waelbroeck, M., Taton, G., Delhaye, M., Chatelain, P., Camus, J.C., Pochet, R., Leclerc, J.L., De Smet, J.M., Robberecht, P. and Christophe, J. (1983). The human heart beta-adrenergic receptors. *Mol. Pharmacol.* **24**, 174–82.

Watterson, J.G., Kohler, L. and Schaub, M.C. (1979). Evidence for two distinct affinities in the binding of divalent metal ions to myosin. *J. Biol. Chem.* **254**, 6470–7.

CALCIUM ENTRY AND CALCIUM ENTRY BLOCKADE: THERAPEUTIC IMPLICATIONS

T. Godfraind

Laboratoire de Pharmacodynamie Générale et de Pharmacologie, Université Catholique de Louvain, UCL 7350, B-1200 Bruxelles, Belgium

and R.C. Miller

Laboratoire de Pharmacodynamie, Université Louis Pasteur, BP 10, F-67048 Strasbourg Cedex, France

Abstract

Calcium-entry blockers are among calcium antagonists those drugs that specifically interfere with calcium entry through open channels. They belong to four major chemical classes and present specific interactions with calcium channels that can be studied by different methodologies: the measure of the contractile state of the muscles, the measure of calcium fluxes occurring at rest and during excitation, the analysis of competition with specific binding or radioactive ligands such as H3-dihydropyridines, the interaction with the contraction evoked by calcium channel agonists. Such studies permit characterization of the different classes of drugs and show the existence of their tissue selectivity. This pharmacological property is the rationale for the therapeutic indications for calcium entry blockers.

30.1. Introduction

The concept of calcium entry blockade is not very old, but is already well established, and a great quantity of information has been accumulated on a large heterogenous collection of compounds which, as the name implies, are thought to exert their major pharmacological effects by antagonizing the stimulated entry of calcium into cells. Calcium entry blockers are the specific subgroup of a general class of compounds often called calcium antagonists or, perhaps more appropriately, calcium modulators. Because of the diversity of effects of these compounds in various tissues, and because they are at present used mainly to treat hypertension and disorders involving vascular insufficiency, this chapter is limited mainly to calcium entry blockers in vascular smooth muscle.

30.2. Antagonism of depolarization-induced contractions

The idea of calcium antagonism, put forward in 1968 by Godfraind and Polster, is usually demonstrated in smooth muscle *in vitro*, by the antagonism of contractions elicited by the cumulative addition of calcium to preparations depolarized in a calcium-free physiological solution containing 40–100 mM potassium. Exposure to physiological solution enriched in K^+ in the absence of extracellular calcium does not result in a change in the contractile state of the preparation. Depolarization increases membrane permeability to Ca^{2+} by opening specific potential sensitive calcium channels in the plasma membrane (Bolton, 1979) and added Ca^{2+} then enters the cell by flowing along its concentration gradient. Shifts of these Ca^{2+} induced concentration-effect curves to the right can usually be demonstrated after these preparations have been preincubated with a putative calcium entry blocker. The inhibition seems to be of a competitive type in many instances, but in other cases there is an inhibition of maximal contractions. Pseudo-pA2 values are often calculated from these experiments, but no evidence exists for a direct competition between any of these compounds and Ca^{2+}. However, such values do provide a comparative guide to potency differences between compounds which is useful in screening procedures (Spedding and Cavero, 1984).

A rather similar type of experimental protocol consists of contracting tissues in K^+-enriched physiological solution containing a physiological concentration of Ca^{2+}. Using this technique it is possible to study the development of contractions in the presence and absence of putative calcium-entry blockers

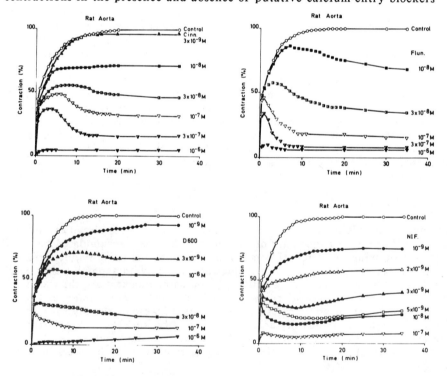

Fig.30.1. Contractions of rat aorta evoked in 100 mM K^+ containing physiological solution in the absence and presence of cinnarizine (Cinn), flunarizine (Flun), D600 and nifedipine (NIF).

(Fig. 30.1). In the rat aorta, depolarization-induced contractions display an initial rapid increase in tension followed by a more slowly developing tonic phase. Contractions elicited after preincubation with calcium-entry blockers such as flunarizine, cinnarizine, D600 and nifedipine are distinguished by an increase in the degree of inhibition of contraction with time in the depolarizing medium. This has been termed 'use dependence' and is almost a general property of calcium-entry blockers. It has been hypothesized that the reason for this use dependence is that the binding site for the compounds becomes more accessible to the blockers upon depolarization, or that there is a change in its affinity for the blockers with changes in membrane polarity. An increase in accessibility might be interpreted as an increase in membrane permeability to these blockers since at least the verapamil group (Hescheler *et al.*, 1982) and probably other compounds have an intracellular site of action associated with inhibition of calcium entry.

It is notable that these depolarization-induced contractions are completely, or nearly completely, abolished in the presence of high concentrations of blockers.

To merely show that a compound inhibits depolarization-induced contractions that are totally dependent on extracellular calcium is not in itself proof that the mechanism of action is by inhibition of calcium entry. For example, inhibition of contractile responses could equally well be due to interference with the activation of the contractile machinery of the cell by preventing the activation of myosin light chain kinase either directly or by interaction with calmodulin, changing its affinity for calcium. Another alternative is an interaction of the compound with actin or myosin, thus preventing development of contraction. More direct evidence for inhibition of calcium entry comes from the demonstration of an inhibition of ^{45}Ca influx induced by depolarization. It has been shown for cinnarizine, flunarizine, nifedipine, nisoldipine and verapamil in vascular tissue that there is a very good correlation between inhibition of contractile responses and of ^{45}Ca entry (Godfraind, 1983; Godfraind and Dieu, 1981; Godfraind and Miller, 1982, 1983; Godfraind, Miller and Socrates Lima, 1982). This inhibition of calcium entry is also associated with a parallel inhibition of calcium efflux. This is to be expected, since any increase in free intracellular calcium above its normal basal level should activate the calcium pump of the plasma membrane and so increase the efflux of calcium from the cell. If the only source of increased free intracellular calcium was calcium entering the cell, then any decrease in entry must be reflected as a decrease in efflux.

However, demonstration of a reduction of calcium influx does not prove that these compounds will always exert calcium entry blocking activity in all situations, since calcium channels are heterogenous (Hagiwara and Byerly, 1981) and may therefore not always be sensitive to the blockers.

Thus it is interesting that flunarizine is about 10-fold more potent as an antagonist of depolarization-induced contractions of rat mesenteric arteries than of rat aorta, while nifedipine is equally potent in both vessels (Godfraind and Miller, 1983). Nifedipine is about 35-fold more potent an antagonist of depolarization-induced contractions of human coronary arteries than of myocardial contractions (Godfraind *et al.*, 1984). Observations of this sort led to the idea of tissue selectivity of calciumn-entry blockers. This concept, if correct, is very important from a clinical point of view as it opens up the possibility of developing calcium-entry blockers that might have very high tissue specificities.

30.3. Antagonism of agonist-induced contractions

Agonists which induce contractions of vascular tissues are known to increase smooth muscle cell membrane permeability to calcium (Bolton, 1979). Agonist-induced contractions are also inhibited by compounds defined as calcium-entry blockers in the type of experiments already discussed (Fig. 30.2). However, higher concentrations of the blockers are generally needed to inhibit this agonist-induced calcium influx than are needed to inhibit calcium entry induced by depolarization in the same tissue (van Breemen et al., 1980; Godfraind and Dieu, 1981). This difference in sensitivity to calcium-entry blockade is consistent with the idea that depolarization and receptor stimulation might open separate populations of calcium channels (Bolton, 1979).

Contractions induced by agonists such as catecholamines differ from those

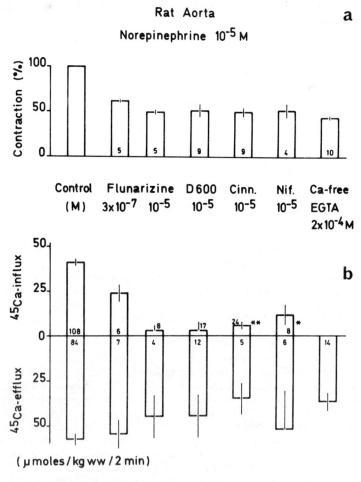

Fig.30.2. Effect of calcium-free solution (containing EGTA 0.2 mM) and the calcium-entry blocking drugs flunarizine, D600, cinnarizine (Cinn), and nifedipine (Nif) at various indicated concentrations, on (a) contractions and (b) ^{45}Ca influx and efflux produced by noradrenaline 10 μM in rat aorta. (Modified from Godfraind and Miller, 1983.)

induced by depolarization in that there is often a component of the contraction independent of extracellular calcium. In rat aorta, maximal noradrenaline-induced contractions are reduced by about 55% in the absence of extracellular calcium (Godfraind, Miller and Socrates Lima, 1982), while contractions of rat mesenteric artery are reduced by about 85–90% (Godfraind and Miller, 1982). Noradrenaline-induced contractions of rabbit aorta are inhibited by about 20% in the absence of calcium (van Breemen et al., 1980). It has also been shown that the part of the contraction inhibited by the blockers is indeed due to an influx of calcium, by direct measurement of agonist-stimulated cellular ^{45}Ca influx and efflux, using the lanthanum method of Godfraind (1976). Lanthanum does not penetrate cell membranes unless they are damaged (dos Remedos, 1981) and displaces extracellular calcium from its binding sites with little or no effect on extracellular calcium (Mayer, van Breemen and Casteels, 1972; van Breemen et al., 1973; Godfraind, 1976). The non-stimulated influx of ^{45}Ca into the La^{3+}-resistant calcium fraction is not much altered by the blockers (Godfraind and Dieu, 1981; Godfraind and Miller, 1982; Church and Zorster, 1980). Noradrenaline-stimulated ^{45}Ca efflux (Fig. 30.3) is also little affected even at elevated (micromolar, μM) concentrations of antagonists (Godfraind and Miller, 1983).

It is evident therefore that the effectiveness of calcium-entry blockers in different tissues, as antagonists of agonist-induced contractions, will depend on the proportion of the contraction dependent on extracellular calcium. This could be interpreted as a type of selectivity of action of the blockers, but in fact there may be no difference at all in the sensitivity of the calcium channels activated, to inhibition by the blocker in question. The apparent selectivity is then due to physiological differences between the tissues.

There are some observations that can be interpreted as illustrating pharmacological selectivity. That is, differences between tissues to the inhibitory effects of calcium-entry blockers at the level of their binding sites, assuming the site to be part of the calcium channel.

Yohimbine is devoid of calcium-entry blocking activity, but two of its stereoisomers, rauwolscine and corynanthine are about equipotent as inhibitors of depolarization-induced contractions and of ^{45}Ca influx in rat aorta. Rauwolscine also inhibits prostaglandin $F_{2\alpha}$-induced contractions in the same tissue in a concentration-dependent manner, although yohimbine and corynanthine do not (Godfraind, Miller and Socrates Lima, 1983). This marked difference in the inhibitory actions of corynanthine could be interpreted as indicating differences between the calcium channels activated in the two cases.

If there is a difference between potential sensitive and agonist-operated calcium channels then the question might also be asked: are calcium channels, activated by stimulation of different receptor types, identical or not? In the case of the rat mesenteric artery both cinnarizine and flunarizine are at least 10-fold more potent as antagonists of noradrenaline than of prostaglandin $F_{2\alpha}$-induced contractions, although nifedipine has a similar potency as an antagonist of both agonists (Godfraind and Miller, 1983).

Binding experiments, using tritiated dihydropyridine derivatives (notably nitrendipine) as ligands, do demonstrate differences between specific binding sites in different tissues taken from the one species (Glossman et al., 1982). This is to be expected if there is a heterogenity of calcium channels between tissues (Hagiwara and Byerly, 1981). However, these studies do not identify multiple binding sites in any one tissue. This might be because dihydropyridines do not distinguish between calcium channel types in the absence of stimulation by either an agonist or by depolarization of the membrane. Indeed, they may not distinguish between calcium channels even when they are activated, as already mentioned. Alternatively, there may be a single calcium channel, which has different 'open state' characteristics in depolarized and non-depolarized (that is, in the presence of agonists) membranes, which manifest as differences in effectiveness of some calcium-entry blockers.

Differences in agonist sensitivity to calcium-entry blockers may exist because of the different effects of agonists on membrane polarity. Agonists such as clonidine have been shown to induce contractions without accompanying changes in cell membrane potential in cat gastric vessels, but phenylephrine decreases the resting membrane potential (Morgan, 1983). Noradrenaline induces contractions of rabbit saphenous arteries without changes in membrane potential, but both contraction and reduction in membrane potential in rabbit ear artery (Holman and Suprenant, 1980). Concentrations of noradrenaline-producing threshold contractions did not

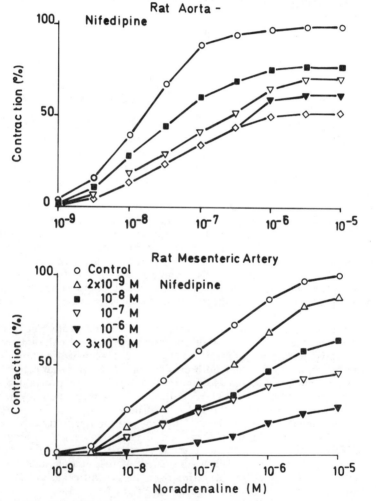

Fig.30.3 Cumulative concentration–effect curves evoked by noradrenaline in rat aorta and mesenteric artery in the absence (o) and presence of nifedipine at concentrations indicated on the graph. (Modified from Godfraind and Miller, 1983.)

Calcium entry and calcium entry blockade

depolarize the membrane in rabbit superior mesenteric artery (Kuriyama and Suzuki, 1978). In other tissues, for example rat small mesenteric arteries, noradrenaline-induced contractile responses have been shown to be correlated with changes in membrane potential (Mulvany, Nilsson and Flatman, 1982). Whether or not agonist-induced contractions are correlated with changes in membrane potential may therefore depend on the agonist, the vessel and the species.

There are thus many possibilities for a type of tissue selectivity, but as already pointed out this apparent selectivity might not be a consequence of a selectivity of interaction of the blockers with calcium channels but a consequence of variations in physiological responses of tissues to stimuli.

30.4. Classifications of Calcium-entry blockers

Compounds generally classified as calcium-entry blockers belong to diverse chemical groups, and Table 30.1 lists a few of them subgrouped on a structural basis. However, in not all cases has it been unequivocally demonstrated that their major mechanism of action is in fact blockade of calcium entry. In view of the diversity of structures, it is not surprising that different compounds exhibit detailed differences in their pharmacological action profiles, and several workers have attempted to classify these compounds on structural, biochemical and pharmacological grounds. On the basis of their selectivity for calcium channels over sodium channels in cardiac tissue Fleckenstein (1983) has divided calcium antagonists into two groups. One group comprises the dihydropyridines, verapamil and diltiazem, and the second the diphenylalkylamines. Binding studies, mainly using [^{3}H] nitrendipine as ligand have shown that the dihydropyridine binding site can be allosterically regulated. Verapamil reduces dihydropyridine binding and diltiazem enhances it. In the light of these observations, Murphy *et al* (1983) have made a division into two groups by setting the dihydropyridines apart from other compounds and Glossman *et al* (1982) distinguished three classes of calcium antagonists. The first group was subdivided into the dihydropyridines and the diphenylpiperazines, the second was exemplified by verapamil and the third by diltiazem. Using pharmacological criteria, Rodenkirchen, Bayer

Table 30.1 Calcium-entry blockers

(1) *Diphenylpiperazines*
cinnarizine, flunarizine, lidoflazine
(2) *Verapamil derivatives*
verapamil, gallopamil, tiapamil
(3) *Dihydropyridines*
nifedipine, nimodopine, nisoldipine, nitrendipine, etc. PY108-068
(4) *Benzothiazepines*
diltiazem, KB-944
(5) *Others*
bencyclane, bepridil, fendiline, perhexiline, prenylamine, yohimbine derivatives, papaverine, amrinone

and Mannhold (1982) and Spedding (1982) have distinguished three different groups. Both laboratories agree that the dihydropyridines are distinguishable from verapamil and diltiazem, which in turn are different from the diphenylpiperazines. There are, however, differences in the subclassifications of verapamil and prenylamine for example (Spedding, 1985). It is as yet too soon to know which, if any, of these classifications will prove most useful from a pharmacological and therapeutic point of view, but to be useful a subclassification must take account of the properties mentioned.

30.5. Influence of endothelium

It has recently become apparent that contractions in vascular smooth muscle can be inhibited by the liberation of a product (or products) from the endothelium (Furchgott, 1983), endothelium-derived relaxing factor (EDRF). There is probably a constant liberation of this factor by the endothelium, but liberation is markedly enhanced by acetylcholine (Holzmann, 1982), and to a lesser extent by alpha-adrenoceptor agonists (Bigaud et al., 1984; Eglème, Godfraind and Miller, 1984; Miller and Stoclet, 1984). The mechanism of action of this factor is unknown, but amongst the many possibilities two obvious ones are that it reduces membrane permeability to calcium and stimulates calcium pumping mechanisms, so reducing free intracellular calcium levels. In this regard it is interesting that in rat aorta, contractions induced by alpha-adrenoreceptor agonists such as clonidine and B-HT 920 are almost completely inhibited in the presence of endothelium, but in the absence of endothelium they elicit contractions of about equal magnitude to those elicited by noradrenaline (Eglème, Godfraind and Miller, 1984; Miller et al., 1984). Agonists such as noradrenaline, phenylephrine and methozamine, that both release intracellular calcium and increase calcium influx, induce concentration-effect curves that are displaced to the right in the presence of endothelium, but maximal responses are not markedly changed (Bigaud et al., 1984; Eglème, Godfraind and Miller, 1984; Godfraind, Eglème and Al-Osachie, 1985). In rat aortic segments the maximal inhibitory effect of flunarizine on noradrenaline, phenylephrine and prostaglandin $F_{2\alpha}$-induced contractions is altered by the presence or absence of the endothelium. The contractions are inhibited to a smaller degree in the absence than in the presence of the endothelium (Godfraind, Eglème and Al-Osachie, 1985; Miller and Stoclet, 1984, 1985). However, when submaximal concentrations of agonists are used to elicit contractions flunarizine has a greater maximal inhibitory effect in the absence than in the presence of the endothelium (Miller and Stoclet, 1984, 1985). The reason for this apparent reversal in efficacicity of the calcium-entry blocker is not as yet clear, but it does seen that EDRF might have more than one action on the vascular smooth muscle. One action is to strongly inhibit contractions dependent on extracellular calcium, and another is to increase the ability of agonists to release intracellular calcium (Miller and Stoclet, 1985). The resultant contraction in the presence of endothelium will therefore depend on the sum of these two opposing effects in the case of strong agonists, that both release intracellular calcium and open calcium channels in the plasmalemma. These observations imply that it is likely that calcium-entry blockers will exhibit marked inhibitory effects on vascular smooth muscle in situations where the vascular endothelium is damaged or changed in some way, so reducing its ability to produce EDRF. Their effects will be much less marked when endothelium is present and fully functional. From these interesting observations it might be predicted that an effective calcium-entry blocker in *in vitro* experiments could display little antihypertensive effect in normal animals or subjects for example, but be an effective antianginal agent if vascular spasm is associated with endothelial changes. This constitutes another type of selectivity.

30.5. Conclusion

It is now evident that the calcium-entry blockers present a new group of compounds with important potential therapeutic applications. An important element in these applications is the apparent tissue selectivity that they exhibit. This selectivity can arise for diverse reasons and is not necessarily a property of an individual compound. Evidently interactions between different cell types in the one tissue can influence calcium handling. Rational use and development of structures with calcium-entry blocking activity as a mechanism of action will depend on an increased understanding of calcium management in normal and pathological situations.

References

Bolton, T.B. (1979). Mechanisms of action of transmitters and other substances on smooth muscle. *Pharmacol. Rev.* **59**, 606–718.

Bigaud, M., Schoeffter, P., Stoclet, J.C. and Miller, R.C. (1984). Dissociation between endothelium-mediated increases in tissue levels and modulation of aortic contractile response. *Naunyn-Schmiedeberg's Arch. Pharmacol.* **328**, 221–3.

Church, J. and Zoster, T.T. (1980). Calcium antagonistic drugs. Mechanism of action. *Can. J. Physiol. Pharmacol.* **58**, 254–64.

dos Remedos, C.G. (1981). Lanthanide ion probes of calcium-binding sites on cellular membranes. *Cell Calcium* **2**, 29–51.

Eglème, C., Godfraind, T. and Miller, R.C. (1984). Enhanced responsiveness of rat isolated aorta to clonidine after removal of the endothelial cells. *Brit. J. Pharmacol.* **81**, 16–8.

Fleckenstein, A. (1983). History of calcium antagonists. *Circ. Res.* **52** (Suppl. 1), 3–16.

Furchgott, R.F. (1983). Role of endothelium in responses of vascular smooth muscle. *Circ. Res.* **53**, 557–73.

Glossmann, H., Ferry, D.R., Lübbecke, F., Mewes, R. and Hoffmann, F. (1982). Calcium channels: direct identification with radioligand binding studies. *Trends Pharmacol. Sci.* **3**, 431–7.

Godfraind, T. (1976). Calcium exchange in vascular smooth muscle, action of noradrenaline and lanthanum. *J. Physiol.* **260**, 21–35.

Godfraind, T. (1983). Actions of nifedipine on calcium fluxes and contraction in isolated rat arteries. *J. Pharmacol. Exp. Ther.* **224**, 443–50.

Godfraind, T. and Dieu, D. (1981). The inhibition by flunarizine of the norepinephrine-evoked contraction and calcium influx in rat aorta and mesenteric artieres. *J. Pharmacol. Exp. Ther.* **217**, 510–15.

Godfraind, T. and Miller, R.C. (1982). Actions of prostaglandin F2alpha and noradrenaline on calcium exchange and contraction in rat mesenteric arteries and their sensitivity to calcium entry blockers. *Brit. J. Pharmacol.* **75**, 229–36.

Godfraind, T. and Miller, R.C. (1983). Specificity of action of Ca^{++} entry blockers. A comparison of their actions in rat arteries and in human coronary arteries. *Circ. Res.* **52 (Suppl. I)**, 81–91.

Godfraind, T. and Polster, P. (1968). Etude comparative de médicaments inhibant la réponse contractile de vaisseaux isolés d'origine humaine ou animale. *Thérapie* **23**, 1209–20.

Godfraind, T., Eglème, C. and Al-Osachie, I. (1984). Role of endothelium in the contractile response of rat aorta to alpha-adrenergic agonists. *Clin. Sci.* **68 (Suppl. 10)**, 655–755.

Godfraind, T., Finet, M., Socrates Lima, J. and Miller, R.C. (1984). The contractile activity of human coronary arteries and human myocardium *in vitro* and their senstivity to Ca entry blockade by nifedipine. *J. Pharmacol. Exp. Ther.* **230**, 514–18.

Godfraind, T., Miller, R.C. and Socrates Lima, J. (1982). Selective alpha 1- and alpha 2-adrenoceptor agonist induced contractions and ^{45}Ca fluxes in the rat isolated arota. *Brit. J. Pharmacol.* **77**, 597–604.

Godfraind, T., Miller, R.C. and Socrates Lima, J. (1983). Effects of yohimbine, rauwolscine and corynanthine on contractions and calcium fluxes induced by depolarization and prostaglandin F2alpha in rat aorta. *Brit. J. Pharmacol.* **80**, 115–21.

Hagiwara, S. and Byerly, L. (1981). Calcium channel. *Ann. Rev. Neurosci.* **4**, 69–125.

Hescheler, J., Pelzer, D., Trube, G. and Trautwein, W. (1982). Does the organic calcium channel blocker D600 act from inside or outside on cardiac cell membrane. *Pflügers Arch.* **393**, 287–91.

Holman, M.E. and Suprenant, A. (1980). An electrophysiological analysis of the effects of noradrenaline and alpha-receptor antagonists on neuromuscular transmission in mammalian muscular arteries. *Brit. J. Pharmacol.* **71**, 651–61.

Holzmann, S. (1982). Endothelium-induced relaxation by acetylcholine associated with larger rises in cyclic GMP in coronary arterial strips. *J. Cyclic Nucleotide Res.* **8**, 409–19.

Kuriyama, H. and Suzuki, H. (1978). The effects of acetylcholine on the membrane and contractile properties of smooth muscle cells of the rabbit superior mesenteric artery. *Brit. J. Pharmacol.* **64**, 493–501.

Mayer, C.J., van Breemen, C. and Casteels, R. (1972). The action of lanthanum and D600 on the calcium exchange in the smooth muscle cells of the guinea-pig taenia coli. *Pflügers Arch.* **337**, 333–50.

Miller, R.C. and Stoclet, J.C. (1984). Influence de l'endothelium sur la sensibilité de l'aorta isolée de rat à la flunarizine. *J. Pharmacol. (Paris)* **15**, 494–5.

Miller, R.C. and Stoclet, J.C. (1985). Modulation by endothelium of contractile responses in rat aorta in absence and presence of flunarizine. *Brit. J. Pharmacol.* In press.

Miller, R.C., Mony, M.C., Schini, V., Schoeffter, P. and Stoclet, J.C. (1984). Endothelial mediated inhibition of contraction and increase in cGMP levels evoked by the alpha-adrenorecptor agonist B-HT 920 in rat isolated aorta. *Brit. J. Pharmacol.* **83**, 902–8.

Morgan, K.C. (1983). Electrophysiological differentiation of alpha-receptors on arteriolar smooth muscle. *Amer. J. Physiol.* **244**, H540–5.

Mulvany, M.J., Nilsson, H. and Flatman, J.A. (9182). Role of membrane potential in the response of rat small mesenteric arteries to exogenous noradrenaline stimulation. *J. Physiol.* **332**, 363–73.

Murphy, K.M.M., Gould, R.J., Largent, B.L. and Snyder, S.H. (1983). A unitary mechanism of calcium antagonist drug action. *Proc. Natl. Acad. Sci. USA* **80**, 860–4.

Rodenkirchen, R., Bayer, R. and Mannhold, R. (1982). Specific and non-specific Ca antagonists. A structure-activity analysis of cardiopressive drugs. *Prog. Pharmacol.* **5**, 9–23.

Spedding, M. (1982). Assessment of 'Ca^{++}-antagonist' effects of drugs in K$^+$-depolarized smooth muscle. Differentiation of antagonist subgroups. *Naunyn-Schmiedeberg's Arch. Pharmakol.* **318**, 234–40.

Spedding, M. (1985). Calcium antagonist subgroups. *Trends Pharmacol. Sci.* **6**, 109–14.

Spedding, M. and Cavero, I. (1984). 'Calcium antagonists': a class of drugs with a bright future. Part II. Determination of basic pharmacological properties. *Life Sci.* **35**, 575–88.

Towart, R. (1981). The selective inhibition of serotonin-induced contractions of rabbit cerebral vascular smooth muscle by calcium-antagoist dihyropyridines. *Circ. Res.* **48**, 650–7.

van Breemen, C., Farinas, P., Casteels, R., Gerba, P., Wuytack, F. and Deth, R. (1973). Factors controlling cytoplasmic Ca^{++} concentrations. *Phil. Trans. R. Soc. B.* **265**, 57–60.

van Breemen, C., Aaronson, P., Loutzenhiser, R. and Meisheri, K. (1980). Ca^{++} movements in smooth muscle. *Chest* **78** (Suppl.) 157–65.

PHARMACOLOGICAL REGULATION OF THE ACTIVITY OF CALMODULIN

Klaus Gietzen

Department of Pharmacology and Toxicology, University of Ulm, Ulm, West Germany

Abstract

Calmodulin (CaM) is probably the major intracellular receptor for Ca^{2+} in eukaryotic cells which confers Ca^{2+} sensitivity on a multitude of enzyme systems and cell functions. CaM has thus become a potential target for pharmacological modification of cellular acitivities. The effects mediated by CaM can be inhibited *in vitro* by a wide range of chemically unrelated substances that vary considerably in their potency and specificity. At present a few highly specific and potent CaM antagonists are available which may be valuable tools for uncovering the significance of CaM in physiological processes. The therapeutic applicability of even specific CaM antagonists is complicated by the fact that blockade of the ubiquitous and multifunctional regulator may lead to numerous undesired side-effects. Antagonism of CaM's activity may become a valuable approach in therapy if sufficiently selective drugs are developed.

31.1. Introduction

Intracellular Ca^{2+} has now been generally accepted as the physiological link in 'excitation–response coupling' (Rasmussen and Waisman, 1983). CaM, a Ca^{2+}-binding protein, mediates many of the intracellular effects of Ca^{2+} by the reversible formation of CaM–Ca^{2+} complex which regulates a multitude of Ca^{2+}-dependent cellular processes and enzyme systems (Cheung, 1980; Klee, Crouch and Richman, 1980). The recognition of CaM's unique role in transforming extracellular information into cellular responses makes pharmacological modification of its effectiveness both a promising and challenging field of research.

Initially, some comments on terminology of Ca^{2+} pharmacology are in order. The term 'Ca^{2+} antagonist' is widely used for compounds like nifedipine or verapamil (Fleckenstein, 1983), which are certainly blockers of Ca^{2+} channels. However, this term implies that these drugs inhibit the effects mediated by Ca^{2+}. Therefore, the author proposes reservation of the term 'Ca^{2+} antagonist' for *all* agents that alter the Ca^{2+} signal or its transformation into a cellular response (Fig.31.1). Ca^{2+} chelators, Ca^{2+}-entry blockers, Ca^{2+}-release inhibitors and CaM antagonists should then be regarded as subgroups of Ca^{2+}

antagonists, all of which are capable of decreasing the effectiveness of CaM.

In this chapter, emphasis is focused on CaM antagonists which are defined as drugs that prevent interaction of CaM with its effector enzymes either by binding to CaM or by binding to effector enzymes (see Fig.31.1). The first part of the chapter deals with the evaluation of the *in vitro* potency and specificity of the currently known putative CaM antagonists to inhibit CaM-dependent functions. In addition a model is discussed describing all activation and inhibition phenomena of CaM-regulated enzymes. The second part of the chapter points out the possible therapeutic applicability of these drugs.

31.2. Biochemical aspects of calmodulin antagonists

The finding of Weiss *et al.* (1974) that trifluoperazine inhibited the activation of phosphodiesterase induced by CaM, and the subsequent demonstration that trifluoperazine and related compounds Ca^{2+}-dependently interact with CaM (Levin and Weiss, 1978) has initiated the search for more effective and specific CaM antagonists. Unfortunately the phenothiazines have often been used uncritically, since their specificity has been considerably overemphasized. Since then, the phenothiazines, and especially trifluoperazine, have been used in numerous investigations as tools to determine whether CaM is involved in cellular functions. Many of these studies are questionable since the phenothiazines and other CaM antagonists are known to produce many other effects at doses similar to those which block the function of CaM.

31.2.1. *Potency of calmodulin antagonists*
Since generally most CaM effector enzymes interact with CaM in a similar

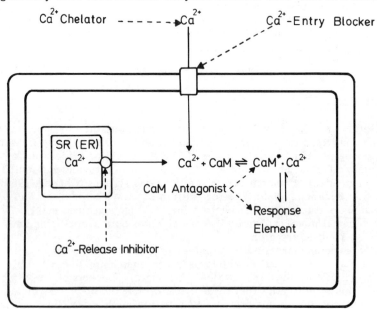

Fig.31.1. Mechanism of action of the different subgroups of Ca^{2+} antagonists.

way, and because of the author's familiarity with the Ca^{2+}-transport ATPase from red blood cells (RBCs), this enzyme is often referred to as an example of a CaM-regulated enzyme. A membrane-bound enzyme (such as Ca^{2+}-transport ATPase) has the advantage over a water soluble enzyme (for example, phosphodiesterase) in that it may allow a more precise statement concerning the specificity of a CaM antagonist. Ca^{2+}-transport ATPase in contrast to phosphodiesterase, may also reflect the unspecific effects of the hydrophobic CaM antagonists caused by perturbation of the lipid matrix in which the enzyme is embedded.

Table 31.1 provides a list (not intended to be complete) of the most important substances that have been shown to inhibit preferentially the CaM-induced stimulation of Ca^{2+}-transport ATPase and phosphodiesterase. It can be seen that a wide range of chemically unrelated substances exhibit CaM antagonistic properties: phenothiazines and butyrophenones, local anaesthetics, vinca-alkaloids, naphthalene sulphonamides, constituents of bee venom, calmidazolium (a complex imidazolium compound) and compound 48/80. A feature common to all CaM antagonists is that they are cationic amphiphiles at physiological pH. It is obvious from Table 31.1 that the potency, expressed as I_{50}, of the various CaM antagonists differs by more than four orders of magnitude. The I_{50} values determined for several CaM antagonists under comparable conditions reveal that calmidazolium, melittin and compound 48/80 are the most potent substances presently available for antagonism of CaM-induced activation of Ca^{2+}-transport ATPase and phosphodiesterase.

The great interlaboratory variability of the I_{50} values for the same substances (for instance, trifluoperazine, chlorpromazine) is striking. However, considering that CaM antagonists are proposed to bind to the hydrophobic $CaM-Ca^{2+}$ complex, and by this action competitively (at least in part) inhibit the interaction with the target enzyme, it becomes evident that several factors may influence the apparent potency of these substances. At least three major parameters were shown to be relevant for the I_{50} of a compound, two of which are due to the (almost) competitive nature of the inhibition:

(1) The concentration of CaM is of crucial importance (Gietzen, Mansard and Bader, 1980; Kobayashi, Tawata and Hidaka, 1979; Raess and Vincenzi, 1980);
(2) The sequence of preincubation of the different components that participate in the inhibitory process (that is, target enzyme, CaM, antagonist) may greatly influence the I_{50} for a compound (Gietzen, Wüthrich and Bader, 1982);
(3) Moreover, the presence of hydrophobic material (such as lipids, proteins, detergents) absorbing the lipophilic antagonist may result in a higher I_{50} value for the same compound (Gietzen, Wüthrich and Bader, 1981).

In the studies from which the I_{50} values of Ca^{2+}-transport ATPase shown in Table 31.1 were obtained, both the CaM concentration (24–180 nM) and the concentration of RBC membrane protein (20–330 µg/ml) varied considerably in the tests.

How incautiously use is made of the apparent potency of CaM antagonists can be seen from many papers in which similarity of the I_{50} values obtained for inhibition of a cellular phenomenon and CaM-sensitive phosphodiesterase is taken as evidence for the participation of CaM in that process. However, by varying the parameters mentioned before (for example, proteins and lipids by using different concentrations of tissue) almost any I_{50} value for a given cellular process may by produced. In the light of the above explanation

Table 31.1. Antagonism of CaM-dependent Ca^{2+}-transport ATPase and phosphodiesterase by various chemical compounds.

Compound	ATPase I_{50} (μM)	ATPase Reference	Phosphodiesterase I_{50} (μM)	Phosphodiesterase Reference
Trifluoperazine	9	(Gietzen, Mansard and Bader, 1980)	10	(Levin and Weiss, 1976)
	18	(Raess and Vincenzi, 1980)	6	(Volpi et al., 1981)
	50	(Levin and Weiss, 1980)		
Chlorpromazine	55	(Kobayashi, Tawata and Hidaka, 1979)	42	(Levin and Weiss, 1976)
	22	(Gietzen, Mansard and Bader, 1980)	47	(Hidaka et al., 1979)
	75	(Raess and Vincenzi, 1980)	6	(Norman, Drummond and Moser, 1979)
	135	(Roufogalis, 1981)		
Fluphenazine	10	(Gietzen, Mansard and Bader, 1980)	7*	
Pimozide	4.5	(Gietzen, Mansard and Bader, 1980)	0.7	(Norman, Drummond and Moser, 1979)
Penfluridol	2.6	(Gietzen, Mansard and Bader, 1980)	2.5	(Levin and Weiss, 1979)
Dibucaine	260	(Volpi et al., 1981)	180	(Volpi et al., 1981)
Vinblastine	35	(Gietzen and Bader, 1980)	16	(Watanabe et al., 1979)
Melittin	0.3*		0.1*	
W-7	100	(Kobayashi, Tawata and Hidaka, 1979)	67	(Hidaka et al., 1980)
W-9	23	(Kobayashi, Tawata and Hidaka, 1979)	45	(Hidaka et al., 1979)
Calmidazolium (= R 24571)	0.35	(Gietzen, Wüthrich and Bader, 1981)	0.2	(Gietzen, Sadorf and Bader, 1982)
	0.4	(Van Belle, 1981)	0.01	(Van Belle, 1981)
Compound 48/80#	0.85	(Gietzen, Sanchez-Delgado and Bader, 1983)	0.3	(Gietzen et al., 1983)

* K. Gietzen, unpublished data
\# µg/ml or µM if assuming a molecular weight of 1000

comparison of the potency of different CaM antagonists is only meaningful if the I_{50} values of the substances were determined under the same experimental conditions with a given enzyme or CaM-dependent activity.

The absolute potency of CaM antagonists seems to depend on several parameters. Norman, Drummond and Moser (1979) could demonstrate that the potency of several compounds to inhibit CaM-sensitive phosphodiesterase correlates with the hydrophobicity of the respective substances. However, the hydrophobicity of a compound does not seem to be the only feature determining the potency of a CaM antagonist since (a) calmidazolium (-log P = 5.32) though less hydrophobic than penfluridol (-log P = 7.6) displays a considerably higher potency than this substance (van Belle, 1984); and (b) compound 48/80, which is rather hydrophilic, approaches the potency of calmidazolium (Gietzen et al., 1983). Most probably electrostatic forces are the second predominant property responsible for the potency of antagonists. Calmidazolium and compound 48/80 may have a much higher affinity for the negatively charged CaM than substances with comparable hydrophobicity because of their distinct ionic properties. Calmidazolium, a quarternary ammonium compound, is positively charged over a wide pH range and compound 48/80 possesses multiple positive charges per molecule at physiological pH. This interpretation is in agreement with the finding of Weiss et al. (1980) who demonstrated that binding of trifluoperazine to CaM is pH-dependent and does not occur below the pI of CaM and above the pKa of the CaM antagonist, indicating the importance of the ionic interaction. Moreover, steric parameters, such as the distance between the hydrophobic region and the positive charge, may influence the potency of CaM antagonists.

31.2.2. *Specificity of calmodulin antagonists*

The wide range of structurally diverse agents acting as antagonists of CaM's function (see Table 31.1) reflects that only minimal structural prerequisites of a compound may be required for interaction with CaM. As mentioned before, CaM antagonists may be regarded as cationic amphiphiles that bind to CaM via a combination of ionic and hydrophobic interactions. Assuming that these general principles are predominantly responsible for the interaction of antagonists with CaM it is most likely that these compounds can interact with additional sites (proteins, lipids) in the cell. There are various criteria available from which the specificity of a putative CaM antagonist may be judged. The following questions should be investigated for a putative CaM antagonist: (a) Does the agent also inhibit proteolysis-activated and anionic amphiphile-activated enzymes? (b) How is the basal (CaM-independent) activity of a CaM-regulated enzyme influenced? (c) Are CaM-independent processes influenced as well? (d) Does the agent interact with other Ca^{2+}-binding proteins?

31.2.2.1. *Proteolysis-activated and anionic amphiphile-activated Ca^{2+}-transport ATPase*.

It has been demonstrated for several enzymes that unsaturated fatty acids, acidic phospholipids, poly-L-aspartic acid and limited tryptic digestion mimic the effect of CaM by activating the enzymes to (almost) the same extent as CaM does, albeit at much higher concentrations (Depaoli-Roach, Gibbs and Roach, 1979; Gietzen, Sadorf and Bader, 1982; Minocherhomjee, Al-Jobore and Roufogalis, 1982; Sarkadi, Enyedi and Gardos, 1980; Taverna and Hanahan, 1980; Wolff and Brostrom, 1976). Surprisingly it could be shown that stimulation of Ca^{2+}-transport ATPase by anionic amphiphiles and limited proteolysis could as well be antagonized by CaM antagonists (Gietzen, Sadorf and Bader, 1982; Gietzen et al., 1983; Vincenzi et al., 1982). In a comparative study (Gietzen, Sadorf and Bader, 1982) it was demonstrated that the phenothiazines, trifluoperazine and penfluridol, antagonize all above-mentioned types of activation of Ca^{2+}-transport ATPase in the same concentration range (Fig.31.2a). Under the same conditions calmidazolium and compound 48/80 exhibited a distinct selectivity for CaM (Gietzen, Sadorf and Bader, 1982; Gietzen, 1983; Gietzen et al., 1983) in that: (a) half-maximal

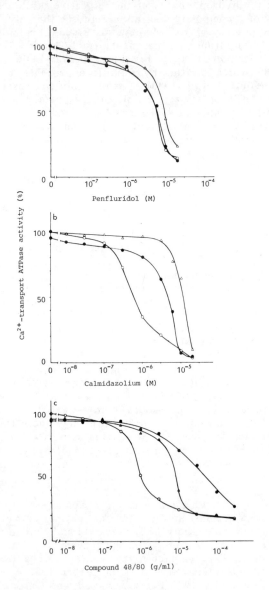

Fig.31.2. Inhibition of activated Ca^{2+}-transport ATPase by penfluridol (a), calmidazolium (b) and compound 48/80 (c). Ca^{2+}-transport ATPase was activated either by CaM (o, 30 nM), oleic acid (Δ, 50 μM) or by limited tryptic digestion (●, for 30 min at 37°C with 0.2 mg of trypsin per mg of RBC membrane protein). Note that the concentrations of compound 48/80 are given as g/ml and those of the other agents as molar concentrations. (Taken from Gietzen, Sadorf and Bader, 1982; and Gietzen et al., 1983.)

inhibition of the stimulation of Ca^{2+}-transport ATPase induced by limited proteolysis or oleic acid required seven and 25 times higher concentrations of calmidazolium respectively (Fig.31.2b); and (b) compound 48/80 antagonized proteolysis-induced or oleic acid-induced activation at 32 and eight times higher concentrations respectively, than were required for CaM antagonism (Fig.31.2c). Antagonism of anionic amphiphile-stimulated Ca^{2+}-transport ATPase by CaM antagonists was taken by Vincenzi et al. (1982) as evidence for a direct effect of these substances on the enzyme itself. However, we have proposed that antagonism of anionic amphiphile-induced activation of CaM-dependent enzymes may occur by complexation of the activator by the cationic amphiphilic inhibitor, via ionic and hydrophobic interactions (Gietzen, Sadorf and Bader, 1982; see also section 31.2.2.3). Inhibition of proteolysis-induced activation by trifluoperazine and penfluridol occurring at similar concentrations as CaM antagonism is much stronger evidence for the assumption that these drugs interact with Ca^{2+}-transport ATPase (Gietzen, Sadorf and Bader, 1982) as well as CaM. The statement seems valid even for compound 48/80, though 32 times higher concentrations of the agent were required for antagonism of proteolysis-induced than for antagonism of CaM-induced stimulation of the enzyme. Inhibition by compound 48/80 of proteolysis-induced activation cannot be attributed to perturbation of the lipid environment of the enzyme since this agent has no inhibitory effect on basal Ca^{2+}-transport ATPase (Gietzen et al., 1983), which in this case should also be inhibited.

31.2.2.2. *Basal (calmodulin-independent) Ca^{2+}-transport ATPase.* Investigation of CaM antagonists on the basal as opposed to the CaM-dependent activity of the same enzyme provides an additional criterion concerning the specificity of a compound. Almost all described inhibitors are more or less unspecific in that they also inhibit the basal activity of CaM-dependent enzymes (Gietzen and Bader, 1980; Gietzen, Mansard and Bader, 1980; Gietzen, Wüthrich and Bader, 1981; Kobayashi, Tawata and Hidaka, 1979; Levin and Weiss, 1976; Luthra, 1982). Table 31.2 summarizes the effect of different CaM antagonists on both basal and CaM-dependent Ca^{2+}-transport ATPase. The ratio of I_{50} value of basal Ca^{2+}-transport ATPase activity to I_{50} value of CaM-dependent fraction of the enzyme's activity is a measure of the specificity of the drug to inhibit the CaM-induced stimulation. Penfluridol and melittin exhibit an extremely low specificity whereas W-7, calmidazolium, and vinblastine are somewhat more specific as indicated by a ratio of 26, 29 and 34, respectively. It has been reported that W-9 when compared with W-7 has even less effect on basal activity and exhibits a 2–3 times higher specificity for CaM as opposed to W-7 (Kobayashi, Tawata and Hidaka, 1979). Compound 48/80, however, is by far the most specific inhibitor of CaM-induced activation of RBC Ca^{2+}-transport ATPase (specificity coefficient: >824) as compared with any of the inhibitors shown in Table 31.2.

31.2.2.3. *Calmodulin-independent processes.* Trifluoperazine and related phenothiazines were shown to interact with and also antagonize numerous sytems not involving CaM, some of them at even lower concentrations than were required for CaM antagonism. For example, trifluoperazine or chlorpromazine inhibited Ca^{2+} uptake into mitochondria (Hirata, Suematsu and Koga, 1982), formation of microtubules by binding to tubulin (Appu Rao and Cann, 1981), neurotransmitter uptake (Seeman, 1980), Ca^{2+}-transport ATPase of sarcoplasmic reticulum (Balzer, Makinose and Hasselbach, 1968) and [$Na^+ + K^+$]-transport ATPase (Green, 1967; Luthra, 1982), to mention a

Table 31.2. Specificity of various compounds for antagonism of CaM-induced stimulation of RBC Ca^{2+} transport ATPase*.

Compound	I_{50} of basal ATPase (µM)	I_{50} of CaM-dependent ATPase (µM)	Coefficient of specificity#
Penfluridol	20	2.6	8
Melittin	3	0.3	10
Trifluoperazine	160	9	18
Fluphenzine	200	10	20
Chlorpromazine	500	22	23
W-7	1100	42	26
Calmidazolium	10	0.35	29
Vinblastine	1200	35	34
Compound 48/80**	>700	0.85	>824

* I_{50} values were determined in our laboratory under the same experimental conditions and were taken in part from Gietzen et al. (1983)

\# The specificity is expressed as the ratio of the I_{50} value of basal Ca^{2+}-transport ATPase activity to the I_{50} value of CaM-dependent fraction of the enzyme's activity

** I_{50} given as µg/ml or µM assuming a molecular weight of 1000

few. In our laboratory we have tested several putative CaM antagonists on different ATPases; two of them were Ca^{2+}-dependent but CaM-independent (Ca^{2+}-transport ATPases from skeletal muscle and cardiac sarcoplasmic reticulum) whereas others were Ca^{2+}/CaM-independent (Mg^{2+}-ATPases of different tissues and [Na^++K^+]-transport ATPase of sarcolemma). In these studies (Gietzen, 1983; Gietzen *et al.*, 1983) compound 48/80 exhibited a remarkably high specificity for CaM-dependent effects since this agent affected the CaM-independent enzymes considerably less, or not at all, compared with any other CaM antagonist tested. Inhibition of CaM-independent enzymes as well as antagonism of basal RBC Ca^{2+}-transport ATPase by putative CaM antagonists might be a consequence of perturbation of the lipid environment in the case of membrane-integral enzymes but could also (or additionally) result from a direct effect on the target enzyme (Au, 1981; Jones, 1980; Luthra, 1982; Raess and Vincenzi, 1980). The remarkably high specificity of compound 48/80 may be determined by the polymeric structure of this agent, as was suggested by Gietzen, Sanchez-Delgado and Bader (1983).

31.2.2.4 *Calcium-binding proteins.* CaM is a member of a family of Ca^{2+}-binding proteins (for example, troponin C, parvalbumin, S-100) that exhibit significant amino-acid sequence homology (Kretsinger, 1980). It has been shown for CaM, troponin C and S-100 that binding of Ca^{2+} to these proteins induces a conformational change exposing hydrophobic regions (Calissano, Alema and Fasella, 1974; LaPorte, Wierman and Storm, 1980; Tanaka and Hidaka, 1980, 1981) which might serve as interfaces for the interaction with the lipophilic antagonists. Indeed, trifluoperazine was found to interact in a Ca^{2+}-dependent manner with both CaM and troponin C, but not with S-100 (Levin and Weiss, 1978). In this study trifluoperazine exhibited a 5-fold higher affinity for CaM compared with troponin C. Tanaka and Hidaka (1981) provided evidence that W-7 also binds to troponin C in addition to CaM, though with two-fold lower affinity. Even compound 48/80, which has been shown to be highly selective regarding various aspects of specificity, also interacts with troponin C. As was demonstrated for CaM (Gietzen *et al.*, 1983) part of the constituents of compound 48/80 bound Ca^{2+}-dependently to a troponin C-sepharose conjugate (K. Gietzen, unpublished results).

32.2.3. *Mechanism of action of calmodulin antagonists*
Based on the results reported by several laboratories, including ours, we have developed a model (Gietzen, Sadorf and Bader, 1982) by which all types of activation and (specific) inhibition phenomena of CaM-regulated enzymes can be described simply by the assumption of hydrophobic and ionic interactions (Fig.31.3).

CaM has two outstanding structural features: acidity (Cheung, 1980) and hydrophobicity, of which the latter was shown to be induced by binding of Ca^{2+} (LaPorte, Wierman and Storm, 1980; Tanaka and Hidaka, 1980). Generally CaM can be considered to be an anionic amphiphile. Substances that can mimic stimulation of enzymes by CaM (such as fatty acids, acidic phospholipids) share with CaM both characteristics. The cationic amphiphilic properties of the CaM receptor site on the enzyme were assumed because of the anionic amphiphilic character of the activators. Binding of CaM and the other activators to an enzyme is thought to suppress an inhibitory peptide sequence which somehow hinders the active centre, resulting in a transformation of the enzyme to a high-activity state (Klee, 1980, Sarkadi,

Enyedi and Gardos, 1980). The more open conformation of the enzyme resulting from interaction of an activator was assumed from a finding reported by Claude Klee (1980). She could show that when treating phosphodiesterase with trypsin in the presence of CaM, cleavage of the inhibitory peptide proceeded faster, which is compatible with the explanation that the enzyme unfolds upon binding of CaM. Activation of CaM-regulated enzymes by limited proteolysis was suggested to be the result of cleavage of the inhibitory peptide sequence (Klee, 1980; Sarkadi, Enyedi and Gardos, 1980). A feature common to all inhibitors of CaM-mediated effects is that they are cationic amphiphiles at physiological pH. Most CaM antagonists were shown to bind Ca^{2+}- dependently to CaM (Levin and Weiss, 1978; Tanaka and Hidaka, 1980). Probably CaM inhibitors act by complexation of the activators because of the above noted complementary structural features via ionic and hydrophobic interactions. As a result the enzyme then flips back into its low-activity state. Our experiments dealing with the inhibition of proteolysis-activated ATPase revealed that in addition a second mode of inhibition is relevant for Ca^{2+}-transport ATPase, that is, direct interaction of CaM antagonists with the enzyme. Attachment of an inhibitor to the ATPase may influence the enzyme by an allosteric effect. We could exclude the possibility that Ca^{2+}-transport ATPase might contain (in addition to that CaM which is Ca^{2+}-dependently bound) CaM as an integral subunit serving as

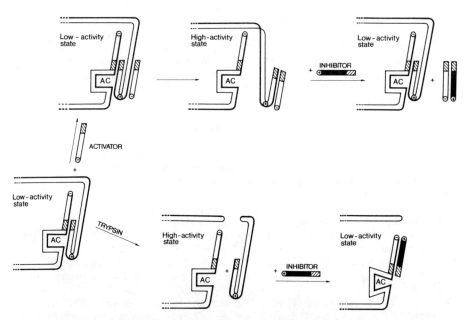

Fig.31.3. Mechanism of stimulation and inhibition of CaM-regulated enzymes. The scheme describes our hypothesis for the mechanism by which CaM-dependent enzymes may be activated and by which the induced activation may be antagonized. The phenomena are described simply on the basis of hydrophobic and ionic interactions. Activators (CaM, oleic acid or phosphatidylserine) are depicted as anionic amphiphiles, whereas CaM inhibitors are depicted as cationic amphiphiles. The hatched areas symbolize hydrophobic regions. An inhibitory peptide sequence of the CaM-regulated enzymes, being positively charged and having a hydrophobic site, is assumed. The inhibitory effect of this polypeptide is suppressed by interaction with anionic activators leading to its displacement or by cleavage with trypsin. The activated enzyme may be inhibited by complexation of the activator by the inhibitor, and for some enzymes in addition by direct interaction of the inhibitor with the enzyme. AC, active centre. Only a section of the CaM-dependent enzyme is shown. (Taken from Gietzen, Sadorf and Bader, 1982.)

binding site for CaM antagonists by demonstrating that Ca^{2+}-transport ATPase purified by CaM affinity chromatography contains only 1 mol CaM per 400 mol of ATPase (Gietzen and Kolandt, 1982).

31.3. Therapeutical aspects of calmodulin antagonists

CaM pharmacology is a promising research field still in its infancy. For biochemical research a few highly specific CaM antagonists are presently available which may be valuable tools for uncovering the significance of CaM in physiological processes. However, the therapeutic applicability of even specific CaM antagonists is complicated by the fact that blockade of the ubiquitous and multifunctional modulator protein may lead to numerous side-effects.

31.3.1. *Pharmacological implications of calmodulin*

The fundamental role of CaM in information transfer from the extracellular space to the cytoplasm resulting in the modulation of vital cellular processes make it a potential target for pharmacological modification. The interest of pharmacologists in CaM was initiated by the important observation of Benjamin Weiss and coworkers (Levin and Weiss, 1976, 1979) that phenothiazines and other antipsychotic drugs interact Ca^{2+}-dependently with CaM. This finding provides an explanation for the observation that these drugs selectively inhibit Ca^{2+}-sensitive forms of cyclic nucleotide phosphodiesterase (Weiss et al., 1974). Because there was fairly good correlation between the *in vitro* potency to inhibit CaM-dependent phosphodiesterase and the clinically observed antipsychotic potency of a variety of neuroleptic drugs it was speculated that binding of these drugs to CaM might be responsible for their therapeutic effects (Levin and Weiss, 1979; Weiss et al., 1980). More detailed studies revealed that this conclusion was probably premature. By investigating additional drugs several other laboratories found no correlation of CaM antagonism and clinical antipsychotic potency (Norman, Drummond and Moser, 1979; Raess and Vincenzi, 1980; Roufogalis, 1981) but good correlation between the hydrophobicity and the potency to inhibit the effect of CaM (Norman, Drummond and Moser, 1979). Moreover, it was demonstrated that antagonism of the effects of CaM by neuroleptics lacks stereospecificity which is in contrast to antipsychotic activity (Norman, Drummond and Moser; Raess and Vincenzi, 1980). Finally, it was shown that several chlorpromazine derivatives which are antipsychotically inactive were equally potent in antagonizing CaM stimulation of RBC Ca^{2+}-transport ATPase (Roufogalis, 1981). Thus CaM antagonism of neuroleptics seems not to be in causal relationship with antipsychotic activity. However, some of the numerous side effects (for example, orthostatic hypotension) may be explained by this biochemical mechanism.

In the following I would like to give some examples of important cellular processes that may be under control of CaM and relate these functions to specific ailments (Table 31.3). Possible treatment of these diseases by CaM antagonists will be evaluated. Some of the above-mentioned ailments are presently treated with drugs (such as vinca alkaloids, tamoxifen, oxatomide, flunarizine, cinnarizine, nimodipine) that were reported to possess CaM antagonistic properties. However, the question is whether or not these drugs exert their therapeutic effect via inhibition of CaM's activity

Table 31.3. Therapeutical implications of CaM.

CaM-dependent cellular process	Related disease	Relevant drugs with CaM antagonistic properties
Cell proliferation	Cancer	Vinca alkaloids
	Psoriasis	Tamoxifen
Smooth muscle relaxation	Sudden deafness	Nimodipine
	Vascular disorders	Flunarizine
	Migraine	Cinnarizine
Histamine release	Anaphylaxis	Oxatomide
	Asthma	
Platelet aggregation	Thrombosis	
	Migraine	
	Metastasis	

31.3.1.1. Cell Proliferation. Knowledge of the role of CaM in cell proliferation might be a new key to a more precise understanding of cancer or psoriasis. It was reported by several laboratories that, compared to normal cells, CaM levels are elevated in exponentially growing tumour cells (LaPorte *et al.*, 1979; Watterson *et al.*, 1976) as well as in cells of psoriatic lesions (Van de Kerkhof and Van Erp, 1983). It was found that CaM is exclusively synthesized during G_1 phase and that in transformed cells the elevated CaM level is due to an increase in the rate of synthesis (Means *et al.*, 1981). It has been speculated that antagonism of CaM by vinblastine (Gietzen and Bader, 1980; Watanabe *et al.*, 1979) and tamoxifen (Lam, 1984) may be related to the antineoplastic activity of these substances. However, it was shown that vinblastine binds to CaM with 10-fold lower affinity as compared with binding to tubulin (Bhattacharyya and Wolff, 1976; Gietzen, Wüthrich and Bader, 1982). Interaction of vinblastine with tubulin was observed at the same concentrations where microtubular depolymerization and antimitotic effects occur (Bhattacharyya and Wolff, 1976). Therefore it seems doubtful that vinca alkaloids exert their microtubule-disrupting effects via antagonism of the activity of CaM (Gietzen, Wüthrich and Bader, 1982).

Though it was shown that CaM antagonists block cell division in cell cultures (Means *et al.*, 1981) the potential use of these substances as antineoplastic drugs seems to be limited. Only drugs that will be preferably accumulated in neoplastic tissue would be of therapeutical value. Otherwise normal cells (with lower CaM levels) would be less resistant to CaM antagonists than transformed cells which accordingly would give raise to severe side-effects. However, it might be that CaM antagonists may become valuable tools in the treatment of psoriasis since topical application of drugs may diminish the risk of systemic side-effects in this case.

31.3.1.2. Smooth muscle relaxation. The key role of CaM in contraction of smooth muscle is well established (Dabrowska *et al.*, 1978; Conti and Adelstein, 1981). Ca^{2+} regulation of smooth muscle contraction is mediated by a Ca^{2+}/CaM-dependent kinase that phosphorylates the light chain of myosin. This stimulates the activity of actomyosin ATPase and hence contraction of the actomyosin system. It has been demonstrated that CaM antagonists, like Ca^{2+}-channel blockers, are capable of inhibiting smooth muscle contraction (Asano, Suzuki and Hidaka, 1982; Fleckenstein *et al.*, 1959) and may therefore be regarded as potential vasodilators. Several clinically approved drugs, like prenylamine, flunarizine, cinnarizine or nimodipine, which are generally classsified as Ca^{2+}-entry blockers (Asano, Suzuki and Hidaka, 1982; Godfraind and Miller, 1982; Towart and Kazda, 1979) possess in addition CaM antagonistic properties (Epstein *et al.*, 1982; Hidaka *et al.*, 1980; Van Belle, unpublished results). The question of whether the therapeutic effect of nimodipine, flunarizine or cinnarizine on vascular disorders (including sudden deafness and migraine) is primarily due to CaM antagonism or blockade of Ca^{2+} channels remains to be determined.

31.3.1.3. Histamine release. There is growing evidence that CaM may be involved in the release process of transmitters, hormones and mediators (DeLorenzo, 1981; Douglas and Nemeth, 1982; Means, Tash and Chafouleas, 1982). Douglas and Nemeth (1982) demonstrated that substances known for their CaM antagonistic properties are capable of antagonizing histamine release induced by any of several secretagogues. In our laboratory we have extended these studies in that various drugs were tested for their ability to inhibit histamine release from mast cells, brain phosphodiesterase and RBC

Ca^{2+}-transport ATPase. The order of potency of the different agents to inhibit CaM-dependent phosphodiesterase and Ca^{2+}-transport ATPase correlates well with their ability to antagonize histamine release (Konstantinova et al., 1984). This implies that the mode of inhibition of histamine release may be related to inhibition of the function of CaM (Konstantinova et al., 1984). Recently it was shown that when rats were pretreated with thioridazine, the anaphylactoid reactions (that is, decrease in blood pressure, heart rate and respiratory rate) induced by a secretagogue were much milder and the recovery was much faster and more complete as compared with those of control rats (Sanchez-Delgado, Grollmuss and Bader, 1984). It was therefore suggested that CaM antagonists may be used as a possible approach to the management of anaphylactoid reactions that are mediated by mast cells or basophils (Sanchez-Delgado, Grollmuss and Bader, 1984). That pretreatment of rats with thioridazine prevents a decrease in blood pressure is indicative of a higher sensitivity of the basophils towards CaM antagonists as compared with vascular smooth muscle. Otherwise antagonism of CaM-dependent myosin light-chain kinase would in contrast results in smooth muscle relaxation and hence in an enhanced decrease in blood pressure. The above-mentioned example clearly shows that if pharmacokinetics of a drug are favourable, CaM antagonists may be used without the risk of severe side-effects.

31.3.1.4. *Platelet aggregation.* Recently it has been shown that platelets contain large quantities of CaM and that various CaM antagonists inhibit platelet aggregation as well as serotonin release from these cells (Nishikawa, Tanaka and Hidaka, 1980; White and Raynor, 1980). Hidaka and coworkers (Nishikawa and Hidaka, 1982; Nishikawa, Tanaka and Hidaka, 1980) provided evidence that Ca^{2+}/CaM-dependent myosin light chain phosphorylation plays an important role in the process of platelet activation. They also demonstrated that the affinity of several naphthalene sulphonamide derivatives correlates with the potency of these substances to antagonize platelet aggregation (Nishikawa and Hidaka, 1982). From these data it has been concluded that CaM might be an important modulator in the phenomenon of platelet activation (Nishikawa and Hidaka, 1982; Nishikawa, Tanaka and Hidaka, 1980; White and Raynor, 1980). Therefore it seems reasonable to speculate whether CaM antagonists may in the future become drugs for the treatment of thrombosis. Since platelet aggregation plays a fundamental role in migraine (Hanington et al., 1981) as well as in the process of metastasis (Mehta, 1984) it is also imaginable that CaM antagonists could be used for these purposes.

31.4 Summary and conclusions

The unique role of CaM in modulating cellular activities initiated research to manipulate its activity by pharmacological modification. In this chapter it has been shown that there is now a wide spectrum of compounds available that differ greatly in their potencies and specificities to block the activity of CaM. None of these presently known antagonists is absolutely specific for CaM. Even compound 48/80, which is in terms of specificity superior to any other CaM antagonist, interacts as well with the homologous Ca^{2+}-binding protein troponin C. Nevertheless, CaM antagonists might be useful biochemical tools for further assessment of CaM's physiological role. However, in order to provide a maximum margin of safety in determining whether CaM is involved in the modulation of enzymes and cellular functions the choice of the appropriate CaM inhibitor is of crucial importance.

From the author's point of view, calmidazolium, W-9 and compound 48/80 are exceptional CaM antagonists in that:

(1) W-9 seems to possess a low membrane perturbating potency (Kobayashi, Tawata and Hidaka, 1979).
(2) Calmidazolium is a potent inhibitor which may allow discrimination between CaM- and anionic amphiphile (for example, fatty acids, acidic phospholipids)-stimulated enzyme activities since this agent was shown to inhibit only weakly anionic amphiphile-induced activation of Ca^{2+}-transport ATPase and phosphodiesterase (Gietzen, Sadorf and Bader, 1982).
(3) Compound 48/80 exhibited high potency and proved to be the most specific CaM antagonist (Gietzen, 1983; Gietzen et al., 1983; see section 31.2).

W-9 has the advantage over calmidazolium and compound 48/80 that it readily passes biological membranes and it is therefore recommended if functions of intact cells are investigated. Compound 48/80 is the agent of choice to determine whether CaM is involved in cell-free systems. However, the disadvantage of calmidazolium and compound 48/80 in being membrane-impermeable may be overcome by entrapment of these agents into liposomes followed by fusion with the target cells or by transient permeabilization of cell membranes.

The few examples of CaM-modulated cellular processes, mentioned in section 3 should demonstrate that the pharmacological implications of CaM are enormous. It has been shown quite convincingly that in vitro inhibition of enzyme activities by clinically approved drugs with CaM antagonistic properties can be explained by their binding to and hence antagonism of CaM. However, the relationship between inhibition of CaM's activity and the therapeutic and/or side-effects of drugs has not yet been established.

CaM pharmacology may be stimulated by retrospective and/or prospective studies to uncover effects of clinically approved drugs with CaM antagonistic properties. For example, it would be interesting to know whether the incidence of asthma or migraine attacks in patients receiving antipsychotics is significantly less as compared with controls. Such a finding could lead to development of drugs turning a side-effect into the major pharmacological property. Moreover, in order to reduce the danger of severe side-effects when aiming at clinical use, CaM antagonists with higher specificity should be developed. This aim may be reached by several approaches: (a) synthesis of agents with fewer effects on CaM-independent processes; (b) development of compounds antagonizing the effect of CaM by interaction with a specific effector enzyme; (c) research for drugs that antagonize CaM in a specific tissue by means of pharmakokinetics. The author is optimistic that with current knowledge we will soon unveil promising applications of CaM antagonists that will finally enable us to selectively modify the fundamental regulatory role of CaM in cellular activity.

Acknowledgements

This work was supported by a grant from the Deutsche Forschungsgemeinschaft. I thank Dr Lin Hymel who critically read the manuscript.

References

Appu Rao, A.G. and Cann, J.R. (1981). A comparative study of the interaction of chlorpromazine, trifluoperazine, and promethazine with mouse brain tubulin. *Mol. Pharmacol.* **12**, 295–301.

Asano, M., Suzuki, Y. and Hidaka, H. (1982). Effects of various calmodulin antagonists on contraction of rabbit aortic strips. *J. Pharmacol. Exp. Ther.* **220**, 191–6.

Au, K.S. (1981). ($Ca^{2+}+Mg^{2+}$)-ATPase of chlorpromazine containing rabbit erythrocyte membrane. *Gen. Pharmacol.* **12**, 285–90.

Balzer, H., Makinose, M. and Hasselbach, W. (1968). The inhibition of the sarcoplasmic calcium pump by prenylamine, reserpine, chlorpromazine and imipramine. *Arch. Pharmacol. Exp. Pathol.* **260**, 444–55.

Bhattacharyya, B. and Wolff, J. (1976). Tubulin aggregation and disaggregation: mediation by two distinct vinblastine-binding sites. *Proc. Natl. Acad. Sci. USA* **73**, 2375–8.

Calissano, P., Alema, S. and Fasella, P. (1974). Interaction of S-100 protein with cations and liposomes. *Biochemistry* **13**, 4535–60.

Cheung, W.Y. (1980). Calmodulin plays a pivotal role in cellular regulation. *Science* **207**, 19–27.

Conti, M.A. and Adelstein, R.S. (1981). The relationship between calmodulin binding and phosphorylation of smooth muscle myosin kinase by the catalytic subunit of 3':5' cAMP-dependent protein kinase. *J. Biol. Chem.* **256**, 3178–81.

Dabrowska, R., Sherry, J.M.F., Aromatorio, D.K. and Hartshorne, D.J. (1978). Modulator protein as a component of the myosin light chain kinase from chicken gizzard. *Biochemistry* **17**, 253–8.

DeLorenzo, R.J. (1981). The calmodulin hypothesis of neurotransmission. *Cell Calcium* **2**, 365–85.

Depaoli-Roach, A.A., Gibbs, J.B. and Roach, P.J. (1979). Calcium and calmodulin activation of muscle phosphorylase kinase: effect of tryptic proteolysis. *FEBS Lett.* **105**, 321–4.

Douglas, W.W. and Nemeth, E.F. (1982). On the calcium receptor activating exocytosis. Inhibitory effects of calmodulin-interacting drugs on rat mast cells. *J. Physiol.* **323**, 229–44.

Epstein, P.M., Fiss, K., Hachisu, R. and Andrenyak, D.M. (1982). Interaction of calcium antagonists with cyclic AMP phosphodiesterases and calmodulin. *Biochem. Biophys. Res. Commun.* **105**, 1142–9.

Fleckenstein, A. (1983). History of calcium antagonists. *Circ. Res.* **52** (Suppl. 1) 3–16.

Fleckenstein, A., Tritthart, H., Fleckenstein, B., Herbst, A. and Grün, G. (1969). Selective inhibition of myocardial contractility by competitive calcium antagonists (iproveratril, D600, prenylamine). *Naunyn-Schmiedeberg's Arch. Pharmakol.* **264**, 227–8.

Gietzen, K. (1983). Comparison of the calmodulin antagonists compound 48/80 and calmidazolium. *Biochem. J.* **216**, 611–6.

Gietzen, K. and Bader, H. (1980). Effects of vinblastine and colchicine on calmodulin-dependent Ca^{2+}-transport ATPase of human erythrocytes. *IRCS Med. Sci.* **8**, 396–7.

Gietzen, K. and Kolandt, J. (1982). Large-scale isolation of human erythrocyte Ca^{2+}-transport ATPase. *Biochem. J.* **207**, 155–9.

Gietzen, K., Mansard, A. and Bader, H. (1980). Inhibition of human erythrocyte Ca^{2+}-transport ATPase by phenothiazines and butyrophenones. *Biochem. Biophys. Res. Commun.* **94**, 674–81.

Gietzen, K., Sadorf, I. and Bader, H. (1982). A model for the regulation of the calmodulin-dependent enzymes erythrocyte Ca^{2+}-transport ATPase and brain phosphodiesterase by activators and inhibitors. *Biochem. J.* 207, 541–8.

Gietzen, K., Sanchez-Delgado, E. and Bader, H. (1983). Compound 48/80: a powerful and specific inhibitor of calmodulin-dependent Ca^{2+}-transport ATPase. *IRCS Med. Sci.* 11, 12–3.

Gietzen, K., Wüthrich, A. and Bader, H. (1981). R 24571: a new powerful inhibitor of red blood cell Ca^{2+}-transport ATPase and of calmodulin-regulated functions. *Biochem. Biophys. Res. Commun.* 101, 418–25.

Gietzen, K., Wüthrich, A. and Bader, H. (1982). Effects of microtubular inhibitors on plasma membrane calmodulin-dependent Ca^{2+}-transport ATPase. *Mol. Pharmacol.* 22, 413–20.

Gietzen, K., Adamczyk-Engelmann, P., Wüthrich, A., Konstantinova, A. and Bader, H. (1983). Compound 48/80 is a selective and powerful inhibitor of calmodulin-regulated functions. *Biochim. Biophys. Acta* 736, 109–18.

Godfraind, T. and Miller, R.C. (1982). Actions of prostaglandin $F_{2\alpha}$ and noradrenaline on calcium exchange and contraction in rat mesenteric arteries and their sensitivity to calcium entry blockers. *Brit. J. Pharmacol.* 75, 229–36.

Green, A.L. (1967). Activity correlations and the mode of action of aminoalkylphenothiazine tranquillizers. *J. Pharm. Pharmacol.* 19, 207–8.

Hanington, E., Jones, R.J., Amess, J.A. and Wachowicz, B. (1981). Migraine: a platelet disorder. *Lancet* 2, 720–3.

Hidaka, H., Yamaki, T., Totsuka, T., and Asano, M. (1979). Selective inhibitors of Ca^{2+}-binding modulator of phosphodiesterase produce vascular relaxation and inhibit actin-myosin interaction. *Mol. Pharmacol.* 15, 49–59.

Hidaka, H., Yamaki, T., Naka, M., Tanaka, T., Hayashi, H. and Kobayashi, R. (1980). Calcium-regulated modulator protein interacting agents inhibit smooth muscle calcium-stimulated protein kinase and ATPase. *Mol. Pharmacol.* 17, 66–72.

Hirata, M., Suematsu, E. and Koga, I. (1982). Calmodulin antagonists inhibit Ca^{2+} uptake of mitochondria of guinea pig peritoneal macrophages. *Biochem. Biophys. Res. Commun.* 105, 1176–81.

Jones, G.L. (1980). Spin label study of phenothiazine interactions with the lipid phase of erythrocyte membranes. *Proc. West. Pharmacol. Soc.* 23, 399–404.

Katoh, N., Raynor, R.L., Wise, B.C., Schatzman, R.C., Turner, R.S., Helfman, D.M., Fain, J.N. and Kuo, J.-F. (1982). Inhibition by melittin of phospholipid-sensitive and calmodulin-sensitive Ca^{2+}-dependent protein kinases. *Biochem. J.* 202, 217–24.

Klee, C.B. (1980). Calmodulin: the coupling factor of the two second messengers Ca^{2+} and cAMP. In *Protein Phosphorylation and Bio-regulation*. Thomas, Podesta, Gordon (eds), pp. 61–9. Karger, Basel.

Klee, C.B., Crouch, T.H. and Richman, P.G. (1980). Calmodulin. *Ann. Rev. Biochem.* 49, 489–515.

Kobayashi, R., Tawata, M. and Hidaka, H. (1979). Ca^{2+} regulated modulator protein interacting agents: inhibition of Ca^{2+}-Mg^{2+}-ATPase of human erythrocyte ghosts. *Biochem. Biophys. Res. Commun.* 88, 1037–45.

Konstantinova, A., Gigl, G., Metz, G., Bader, H. and Gietzen, K. (1984). Effects of cloxacepride and its derivatives on histamine release from mast cells and on calmodulin-dependent enzymes. *Cell Calcium* 5, 302.

Kretsinger, R.H. (1980). Structure and evolution of calcium-modulated proteins. *CRC Crit. Rev. Biochem.* 8, 119–74.

Lam, H.-J.P. (1984). Tamoxifen is a calmodulin antagonist in the activation of

cAMP phosphodiesterase. *Biochem. Biophys. Res. Commun.* **118**, 27–32.
LaPorte, D.C., Wierman, B.M. and Storm, D.R. (1980). Calcium induced exposure of a hydrophobic surface on calmodulin. *Biochemistry* **19**, 3814–9.
LaPorte, D.C., Gidwitz, B.M., Weber, M.J. and Storm, D.R. (1979). Relationship between changes in the calcium dependent regulatory protein and adenylate cyclase during viral transformation. *Biochem. Biophys. Res. Commun.* **86**, 1169–77.
Levin, R.M. and Weiss, B. (1976). Mechanism by which psychotropic drugs inhibit adenosine cyclic 3',5'-monophosphate phosphodiesterase of brain. *Mol. Pharmacol.* **12**, 581–9.
Levin, R.M. and Weiss, B. (1978). Specificity of the binding of trifluoperazine to the calcium-dependent activator of phosphodiesterase and to a series of other calcium binding proteins. *Biochim. Biophys. Acta* **540**, 197–204.
Levin, R.M. and Weiss, B. (1979). Selective binding of antipsychotics and other psychoactive agents to the calcium-dependent activator of cyclic nucletodie phosphodiesterase. *J. Pharmacol. Exp. Ther.* **208**, 454–9.
Levin, R.M. and Weiss, B. (1980). Inhibition by trifluoperazine of calmodulin-induced activation of ATPase activity of rat erythrocyte. *Neuropharmacology* **19**, 169–74.
Luthra, M. (1982). Trifluoperazine inhibition of calmodulin-sensitive Ca^{2+}-ATPase and calmodulin insensitive (Na^++K^+)- and Mg^{2+}-ATPase activities of human and rat red blood cells. *Biochim. Biophys. Acta* **692**, 271–7.
Means, A.R., Tash, J.S. and Chafouleas, J.G. (1982). Physiological implications of the presence, distribution, and regulation of calmodulin in eukaryotic cells. *Physiol. Rev.* **62**, 1–39.
Means, A.R., Chafouleas, J.G., Bolton, W.E., Hidaka, H. and Boyd, A.E. (1981). Calmodulin regulation during the cell cycle of mammalian cells. *Proc. West. Pharmacol. Soc.* **24**, 209–12.
Mehta, P. (1984). Potential role of platelets in the pathogenesis of tumor metastasis. *Blood* **63**, 55–63.
Minocherhomjee, A., Al-Jobore, A. and Roufogalis, B.D. (1982). Modulation of the calcium-transport ATPase in human erythrocytes by anions. *Biochim. Biophys. Acta* **690**, 8–14.
Nishikawa, M. and Hidaka, H. (1982). Role of calmodulin in platelet aggregation. Structure-activity relationship of calmodulin antagonists. *J. Clin. Invest.* **69**, 1348–55.
Nishikawa, M., Tanaka, T. and Hidaka, H. (1980). Ca^{2+}-calmodulin- dependent phosphorylation and platelet secretion. *Nature (Lond.)* **287**, 863–5.
Norman, J.A., Drummond, A.H. and Moser, P. (1979). Inhibition of calcium-dependent regulator-stimulated phosphodiesterase activity by neuroleptic drugs is unrelated to their clinical efficacy. *Mol. Pharmacol.* **16**, 1089–94.
Raess, B.U. and Vincenzi, F.F. (1980). Calmodulin activation of red blood cell $(Ca^{2+}+Mg^{2+})$-ATPase and its antagonism by phenothiazines. *Mol. Pharmacol.* **18**, 253–8.
Rasmussen, H. and Waisman, D.M. (1983). Modulation of cell function in the calcium messenger system. *Rev. Physiol. Biochem. Pharmacol.* **95**, 111–48.
Roufogalis, B.D. (1981). Phenothiazine antagonism of calmodulin: a structurally nonspecific interaction. *Biochem. Biophys. Res. Commun.* **98**, 607–13.
Sanchez-Delgado, E., Grollmuss, O. and Bader, H. (1984). Calmodulin antagonists: a possible approach to the management of mast cell-mediated pathologic reactions. *IRCS Med. Sci.* **12**, 173.

Sarkadi, B., Enyedi, A. and Gardos, G. (1980). Molecular properties of the red cell calcium pump. I. Effects of calmodulin, proteolytic digestion and drugs on the kinetics of active calcium uptake in inside-out red cell membrane vesicles. *Cell Calcium* 1, 287–97.

Seeman, P. (1980). Brain dopamine receptors. *Pharmacol. Rev.* 32, 229–313.

Tanaka, T., Hidaka, H. (1980). Hydrophobic regions function in calmodulin-enzyme(s) interaction. *J. Biol. Chem.* 255, 11078–80.

Tanaka, T. and Hidaka, H. (1981). Hydrophobic regions of calcium-binding proteins exposed by calcium. *Biochem. Int.* 2, 71–5.

Taverna, R.D. and Hanahan, D.J. (1980). Modulation of human erythrocyte Ca^{2+}/Mg^{2+} ATPase activity by Phospholipase A_2 and proteases. A comparison with calmodulin. *Biochem. Biophys. Res. Commun.* 94, 652–9.

Towart, R. and Kazda, S. (1979). The cellular mechanism of action of nimodipine (Bay e 9736), a new calcium antagonist. *Brit. J. Pharmacol.* 67, 409P–10P.

Van Belle, H. (1981). R24571: a potent inhibitor of calmodulin-activated enzymes. *Cell Calcium* 2, 483–94.

Van Belle, H. (1984). The effect of drugs on calmodulin and its interaction with phosphodiesterase. *Adv. Cyclic Nucleotide Res.* 17, 557–67.

Van De Kerkhof, P.C.M. and Van Erp, P.E.J. (1983). Calmodulin levels are grossly elevated in the psoriatic lesion. *Brit. J. Dermatol.* 108, 217–8.

Vincenzi, F.F., Adunyah, E.S., Niggli, V. and Carafoli, E. (1982). Purified red blood cell Ca^{2+}-pump ATPase: evidence for direct inhibition by presumed anti-calmodulin drugs in the absence of calmodulin. *Cell Calcium* 3, 545–59.

Volpi, M., Sha'Afi, R.I., Epstein, P.M., Andrenyak, D.M. and Feinstein, M.B. (1981). Local anesthetics, mepacrine, and propranolol are antagonists of calmodulin. *Proc. Natl. Acad. Sci. USA* 78, 795–9.

Watanabe, K., Williams, E.F., Law, J.S. and West, W.L. (1979). Specific inhibition of a calcium dependent activation of brain cyclic AMP phosphodiesterase activity by vinblastine. *Experientia* 35, 1487–9.

Watterson, D.M., Van Eldik, L.J., Smith, R.E. and Vanaman, T.C. (1976). Calcium-dependent regulatory protein of cyclic nucleotide metabolism in normal and transformed chick embryo fibroblasts. *Proc. Natl. Acad. Sci. USA* 73, 2711–5.

Weiss, B., Fertel, R., Figlin, R. and Uzunov, P. (1974). Selective alteration of the activity of the multiple forms of adenosine 3',5'-monophosphate phosphodiesterase of rat cerebrum. *Mol. Pharmacol.* 10, 615–25.

Weiss, B., Prozialeck, W., Cimino, M., Barnette, M.S. and Wallace, T.L. (1980). Pharmacological regulation of calmodulin. *Ann. NY. Acad. Sci.* 356, 319–45.

White, G.C. and Raynor, S.T. (1980). The effects of trifluoperazine, an inhibitor of calmodulin, on platelet function. *Thromb. Res.* 18, 279–84.

Wolff, D.J. and Brostrom, C.O. (1976). Calcium-dependent cyclic nucleotide phosphodiesterase from brain: identification of phospholipids as calcium-independent activators. *Arch. Biochem. Biophys.* 173, 720–31.

FREE INTRACELLULAR MAGNESIUM IN STRIATED MUSCLE

R. Weingart

Department of Physiology, University of Bern, CH-3012 Bern, Switzerland

Abstract

Magnesium ions are an important constituent of the cytoplasm in striated muscle. As for Ca^{2+}, the total tissue Mg^{2+} due to compartmentalization and binding is much larger than its ionized cytosolic level. The free intracellular Ca^{2+} concentration is in the nM range, and that of Mg^{2+} in the mM range. This means that Mg^{2+} is not passively distributed across the sarcolemma: an outwardly directed transport mechanism is required to maintain the cytosolic $[Mg^{2+}]$. No measurable changes of the cytosolic Mg^{2+} occur during a single twitch, but during a tetanus Ca^{2+} release from the SR seems to be partially compensated for by an uptake of Mg^{2+} by this organelle. While the importance of Mg^{2+} as a regulator of cellular functions and as cofactor of enzyme systems is well recognized, its interactions with the biological processes are poorly understood. In muscle, Mg^{2+} affects a number of events which are directly or indirectly involved in the development of mechanical force. Mg^{2+} depresses the Ca^{2+}-sensitivity of the force-generating mechanism; the capacity for the uptake of Ca^{2+} by the SR increases with increasing $[Mg^{2+}]$, and mitochondrial Ca^{2+} uptake is impaired by this divalent cation.

32.1. Introduction

The cytoplasm of all animal cells contains two divalent cations of general importance, namely Ca^{2+} and Mg^{2+}. A whole spectrum of satisfactory methods are available to quantitatively determine the level of free intracellular Ca^{2+} (Ashley and Campbell, 1979; Tsien, 1983). Therefore, it is not surprising if the knowledge about the functional role of Ca^{2+} is quite advanced for a variety of different tissues (see, for example, Scarpa and Carafoli, 1978). This contrasts sharply with the paucity of information concerning the specific functions of intracellular Mg^{2+}. It is generally accepted that Mg^{2+} participates in a number of intracellular processes (see for example, Wacker, 1969; Polimeni and Page, 1973; Altura, 1981) especially those enzymatic activities involving ATP utiliazation or transfer of phosphate groups. Nevertheless, the regulation of cellular Mg^{2+} homeostasis and its interactions with biological processes are poorly understood. This is partly a consequence of the lack until recently of easy and precise methods for detecting intracellular Mg^{2+}, especially in smaller cells.

Fortunately, the development of ion-selective microelectrodes has enabled the measurement of the free intracellular Mg^{2+} concentration, $[Mg^{2+}]_i$, in relatively small cells (Lanter et al., 1980). This chapter presents some results obtained with these electrodes in different striated muscle preparations (Hess, Metzger and Weingart, 1982). The values determined for $[Mg^{2+}]_i$ are compared with those obtained by other groups using different methods. Furthermore, some possible implications of these measurements are discussed regarding the generation of force in striated muscle.

32.2. Fabrication and properties of Mg^{2+} electrodes

Conventional glass microelectrodes were pulled, bevelled to a tip diameter of approximately 1 μm, siliconized with N-trimethylsilyldimethylamine, and back-filled with Mg^{2+} sensor (kindly provided by Dr D Ammann). The sensor consisted of 20% (w/w) neutral magnesium ligand, ETH 1117 1% sodium tetraphenylborate, and 79% propylene-carbonate (Lanter et al., 1980). The shaft of the electrode was filled with 10 mM $MgCl_2$ dissolved in water. The resin was not sensitive exclusively to Mg^{2+} but also responded to other cations. Interference during intracellular measurement was anticipated for Na^+ and K^+ based on the selectivity coefficients determined by Lanter et al. (1980). Therefore, each electrode was calibrated carefully by means of the constant interference method. This involved the use of solutions which varied in $MgCl_2$ concentration, but contained a constant ionic background simulating the cytoplasmic cation composition. For the studies on mammalian muscle the calibrating solutions were composed as follows (mM): 7 NaCl, 120 KCl, 10 HEPES (pH 7.2), 0.1 EGTA. In case of the amphibian muscle the solutions had the following composition (mM): 12 NaCl, 130 KCl, 10 HEPES (pH 7.0), 0.1 EGTA. In the presence of such background ions, the Mg^{2+} electrode responses usually revealed slopes of 12 mV/decade over the range from 1 to 4 mM $[Mg^{2+}]$, and 16 mV/decade betweeen 4 and 10 mM. The chemical time constant was found to be less than 0.5 s.

32.3. Measurement of intracellular Mg^{2+}

Figure 32.1 illustrates an experiment in which cytoplasmic $[Mg^{2+}]$ was determined in a resting frog skeletal muscle fibre. The upper trace in Figure 32.1A shows the original pen-recorder tracing during impalement and withdrawal of a 3 M KCl reference electrode (V_m). The lower trace shows the corresponding record made with the Mg^{2+} electrode (V_{MgE}). At the time of entry of the reference electrode into the cell, only a minor voltage perturbation was visible on the V_{MgE} trace. Conversely, at the moment of withdrawal of the Mg^{2+} electrode there was almost no disturbance detectable on the V_m trace. This seemed to indicate that the electrode did not produce serious damage to the cell. During the time interval marked with the arrows, V_m was subtracted from V_{MgE} revealing a V_{diff} of 13 mV; V_{diff} was found to be proportional to the free $[Mg^{2+}]_i$. Figure 32.1B shows a plot of the calibration of the Mg^{2+} electrode performed immediately after the intracellular measurement. The curve with the open circles depicts the calibration obtained with the regular background solution used for amphibian muscle (see fabrication and properties of Mg^{2+} electrodes). It ascribed a $[Mg^{2+}]_i$ of 3.4 mM to the measured V_{diff}. The electrode was also calibrated in the presence of a variable $[K^+]$ in order to determine the effect of variations of $[K^+]_i$ on the measurement of $[Mg^{2+}]_i$. The result was that the value obtained for $[Mg^{2+}]_i$ would be overestimated by 13% if $[K^+]_i$ were 150 mM instead of 130 mM, and underestimated by 10% if $[K^+]_i$ were only 100 mM.

Mg^{2+}-selective electrodes with similar characteristics were employed in different types of tissue to compare $[Mg^{2+}]_i$ within sheep Purkinje fibres and ventricular muscle, ferret ventricular muscle, and frog skeletal muscle. The results are collected and summarized in Table 32.1. The tissues under investigation were found to have a mean $[Mg^{2+}]_i$ ranging from 3.1 to 3.5 mM. This seems to imply that no significant difference exists among the tissues studied.

Our Mg^{2+} values, based on direct measurements, are in good agreement with those previously determined by Nanninga (1961; frog skeletal muscle: 3.4 mM; Mg^{2+} binding to nucleotides), Cohen and Burt (1977; frog skeletal muscle: 3.0 mM; NMR), Paradise, Becker and Visscher (1978; rabbit ventricle: 1.5–3.6 mM; $^{28}Mg^{2+}$ flux), Wu et al. (1981; guinea pig ventricle: 2.5 mM; NMR), and Baylor, Chandler and Marshall (1982; frog skeletal muscle: 0.2–6 mM; indicator dyes). The chief disagreeement lies with the value reported by Gupta and Moore (1980; frog skeletal muscle: 0.6 mM; NMR). Taken in sum, the available data suggest a free $[Mg^{2+}]_i$ within the low millimolar range (2–4 mM) for striated muscle.

The Mg^{2+} equilibrium potential, E_{Mg}, given by the Nernst relationship,

$$E_{Mg} = (RT/2F) \ln ([Mg^{2+}]_o/[Mg^{2+}]_i)$$

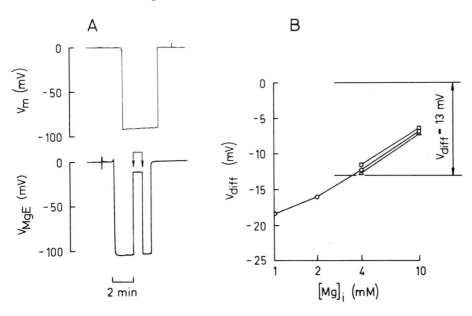

Fig.32.1. Measurement of intracellular free $[Mg^{2+}]$ in a frog skeletal muscle fibre. (A) Top: membrane potential, V_m, recorded with a 3 M KCl microelectrode. (A) Bottom: electromotive force, V_{MgE}, recorded with a Mg^{2+}-selective microelectrode. The arrows (⊓) denote subtraction of V_m (-89 mV) from V_{MgE} (-102 mV), revealing a difference, V_{diff}, of -13 mV. (B) Calibration curve obtained after the intracellular measurement, using calibrating solutions containing different KCl concentrations (100 mM (Δ), 130 mM (○), 150 mM (□)). Assuming a $[K^+]_i$ of 130 mM, the V_{diff} of -13 mV corresponds to a $[Mg^{2+}]_i$ of 3.4 mM. (From Metzger and Weingart, unpublished.)

Table 32.1. Free $[Mg^{2+}]_i$ in unstimulated striated muscle preparations, measured with ion-selective microelectrodes. The values are given as mean ± SE of the mean with the number of observations listed in parentheses. Experimental conditions: $[Mg^{2+}]_o$: 0.5 mM; $[K^+]_o$: 2.5 mM (amphibian tissue) and 4 mM (mammalian tissue), respectively; temperature: 22 °C (From Hess, Metzger and Weingart, 1982.)

Tissue	Membrane potential V_m (mV)	Intracellular Mg concentration $[Mg^{2+}]_i$ (mM)	Equilibrium potential E_{Mg} (mV)
Purkinje fibre, sheep	−70.4 ± 1.5 (17)	3.5 ± 0.3	−24.8
Ventricular muscle, sheep and ferret	−73.3 ± 2.1 (8)	3.1 ± 0.3	−23.1
Skeletal muscle, frog	−83.3 ± 2.0 (10)	3.3 ± 0.4	−24.0

should equal the resting membrane potential, V_m, if Mg^{2+} was distributed according to the electrical gradient across the sarcolemma. However, the equilibrium potential for Mg^{2+} was calculated to be -23 to -25 mV based on a $[Mg^{2+}]_o$ of 0.5 mM and the estimated $[Mg^{2+}]_i$ (Table 32.1, far right-hand column), a value considerably less negative than V_m (-70 to -83 mV; Table 32.1, far left-hand column). Hence, the large inward electrochemical gradient for Mg^{2+} ($V_m - E_{Mg}$) favours passive movement of Mg^{2+} into the cells and requires some energy- dependent process for its cytoplasmic extrusion.

Using sheep Purkinje fibres, we attempted to estimate the Mg^{2+} permeability of the sarcolemma, P_{Mg}. Experimentally, this was achieved by elevating the $[Mg^{2+}]$ from 0.5 to 10 mM for a period up to 10 min, while $[Mg^{2+}]_i$ was monitored. This intervention produced a small and reversible increase in $[Mg^{2+}]_i$. From the initial rate of increase in $[Mg^{2+}]_i$, a maximum rate of gain of $[Mg^{2+}]_i$ was calculated to be 0.42 pmol/cm^{2}s. This flux in turn implied a value of $1.6 \cdot 10^{-8}$ cm/s. Page and Polimeni (1972), studying $^{28}Mg^{2+}$ fluxes in rat ventricle, reported a transmembrane flux of 0.27 pmol/cm^{2}s in the presence of 0.56 mM $[Mg^{2+}]_o$. If one substitutes 10 mM for 0.56 mM of $[Mg^{2+}]_o$, their flux becomes 0.51 pmol/cm^{2}s (Polimeni, personal communication), a value in excellent agreement with the one published by Hess, Metzger and Weingart (1982).

32.4. Implications for the mechanical responses

Skinned muscle fibres have become a very useful preparation for studying the properties and functional interactions of the contractile proteins and sarcoplasmic reticulum (SR). Removal of the sarcolemma either by mechanical microdissection (Natori, 1954) or chemical treatment (Endo and Kitazawa, 1978) leaves the subcellular structures essentially intact, but enables alterations of the ionic environment of the myoplasm. Hence, skinned muscle preparations have been used to investigate the effects of Mg^{2+} on the processes involved in tension development.

With regard to the myofilaments, Mg^{2+} seems to play a dual role. It acts as substrate when complexed to ATP and as regulator when in its ionized form. Brandt, Reuben and Grundfest (1972) examined the effects of MgATP on the tension developed in skinned crayfish muscle fibres. They observed a bell-shaped relationship between tension and pMgATP (negative logarithm of [MgATP] in the presence of physiological $[Ca^{2+}]$. The peak of the curve shifted towards higher [MgATP] upon increasing the $[Ca^{2+}]$ from 10 to 100 or 320 nM. Furthermore, this manoeuvre also increased the maximum developed tension. This observation seems to reflect substrate inhibition of the enzymatic portion of myosin compatible with the concept that Ca^{2+} suppresses the inhibitory interaction between MgATP and the contractile proteins. Assuming physiological concentrations for both [MgATP] and Ca^{2+} (see below), this implies that the muscle remains relaxed until Ca^{2+} is released from the subcellular stores. Similar observations also have been made with skinned cardiac preparations (Fabiato and Fabiato, 1975; see also Best, Donaldson and Kerrick, 1977).

The contractile response in striated muscle not only depends on [MgATP], but also on the level of ionized Mg^{2+}. This was investigated by Fabiato and Fabiato (1975), working with skinned rat ventricular muscle preparations. They studied the effects of Mg^{2+} on tonic tension evoked by changing pCa under conditions where interference from the SR was eliminated by strong buffering with EGTA. Their findings are illustrated in Fig.32.2. This figure shows the normalized developed tension versus pCa of the bathing fluid. In the presence of a constant pMgATP of 2.5 (3.2 mM), elevating the $[Mg^{2+}]$ from 0.032 mM to 0.32 mM shifted the tension-pCa relationship towards larger $[Ca^{2+}]$. A much larger shift in the same direction was produced when $[Mg^{2+}]$ was increased from 0.32 mM to 3.2 mM. This means that a smaller amount of force was developed for a given $[Ca^{2+}]$ after $[Mg^{2+}]$ was increased. However, the maximum tension elicited by a pCa of 5 was not affected by the Mg^{2+} level. Under physiological conditions, [MgATP] and $[Mg^{2+}]$ have been measured to be around 10 mM and 3 mM, respectively (see Polimeni and Page, 1973; Hess, Metzger and Weingart, 1982). Therefore, the curve of tension–pCa obtained in the presence of pMgATP 2.5 and pMg 2.5 may resemble most closely the situation in an intact muscle fibre. In this case, the threshold of pCa for activation of tension must be approximately pCa 6.5 (320 nM), a value corresponding to the free $[Ca^{2+}]_i$ measured with ion-selective microelectrodes (see, for example, Sheu and Fozzard, 1982; Weingart and Hess, 1984). Kerrick and Donaldson (1975) performed similar studies of the influence of Mg^{2+} on Ca^{2+}-activated tension to compare skinned mammalian ventricular muscle and amphibian skeletal muscle. The data obtained from the former tissue agreed with the findings by Fabiato and Fabiato (1975). Most interestingly, the effects of Mg^{2+} on the tonic tension–pCa relationship turned out to be strikingly similar in the two types of muscle. In skeletal muscle the free $[Ca^{2+}]_i$ may be as low as 50 nM (Weingart and Hess, 1984). Assuming a similar [MgATP] and $[Mg^{2+}]$ to cardiac muscle, the foot of the

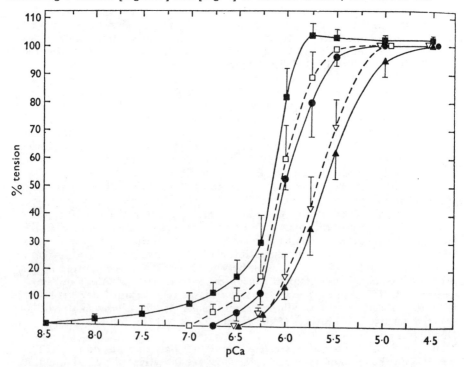

Fig.32.2. Normalized tension developed by skinned rat ventricular muscle cells, plotted as a function of pCa; 100% tension corresponds to the tension measured in the presence of pCa = 5.0, pMg = 3.5, and pMgATP = 2.5. At a constant pMgATP of 2.5, the tension–pCa curve progressively shifts to the right when pMg is decreased from 4.5 (□) to 3.5 (●) and 2.5 (▽). (From Fabiato and Fabiato, 1975.)

tension-pCa relationship under physiological conditions may be almost one order of magnitude above the resting Ca^{2+} level.

32.5. Conclusion

Biological processes investigated in isolated subcellular fractions have yielded quite detailed information concerning the function of such systems. However, precise knowledge of the boundary conditions prevailing within the intracellular ionic milieu is indispensable in order to answer the question of their significance for the intact tissue or animal. Concerning the free $[Mg^{2+}]_i$ in striated muscle, a great number of Mg^{2+}-dependent processes have been studied in subcellular *in vitro* systems in the presence of a submillimolar $[Mg^{2+}]$. Since the $[Mg^{2+}]_i$ appears to be around 2–4 mM, conclusions drawn from such experiments may need critical re-evaluation.

Acknowledgements

Special thanks are due to Dr M.L. Pressler for his comments on the manuscript. This work was supported by the Swiss National Science Foundation (3.565-0.79 and 3.360-0.82).

References

Altura, B.M. (1981). Role of magnesium ions in regulation of muscle contraction (Symposium). *Fed. Proc.* **40**, 2645–79.

Ashley, C.C. and Campbell, A.K. (1979). *Detection and Measurement of Free Ca^{2+} in Cells.* Elsevier/North-Holland, Amsterdam.

Baylor, S.M., Chandler, W.K. and Marshall, M.W. (1982). Optical measurements of intracellular pH and magnesium in frog skeletal muscle fibres. *J. Physiol.* **331**, 105–37.

Best, P.M., Donaldson, S.K.B. and Kerrick, W.G.L. (1977). Tension in mechanically disrupted mammalian cardiac cells: effects of magnesium adenosine triphosphate. *J. Physiol.* **265**, 1–17.

Brandt, P.W., Reuben, J.P. and Grundfest, H. (1972). Regulation of tension in the skinned crayfish muscle fiber. II. Role of calcium. *J. Gen. Physiol.* **59**, 305–17.

Cohen, S.M. and Burt, C.T. (1977). ^{31}P nuclear magnetic relaxation studies of phosphocreatine in intact muscle: determination of intracellular free magnesium. *Proc. Natl. Acad. Sci. USA* **74**, 4271–5.

Endo, M. and Kitazawa, T. (1978). E-C coupling studies on skinned cardiac fibers. In *Biophysical Aspects of Cardiac Muscle.* pp. 307–27. Academic Press, New York.

Fabiato, A. and Fabiato, F. (1975). Effects of magnesium on contractile activation of skinned cardiac cells. *J. Physiol.* **249**, 497–517.

Gupta, R.K. and Moore, R.D. (1980). ^{31}P NMR studies on intracellular free Mg^{2+} in intact frog skeletal muscle. *J. Biol. Chem.* **255**, 3987–93.

Hess, P., Metzger, P. and Weingart, R. (1982). Free magnesium in sheep, ferret and frog striated muscle at rest measured with ion-selective micro-electrodes. *J. Physiol.* **333**, 173–88.

Kerrick, W.G.L. and Donaldson, S.K.B. (1975). The comparative effects of $[Ca^{2+}]$ and $[Mg^{2+}]$ on tension generation in the fibers of skinned frog skeletal muscle and mechanically disrupted rat ventricular cardiac muscle. *Pflügers Arch.* **358**, 195–201.

Lanter, F., Erne, D., Ammann, D. and Simon, W. (1980). Neutral carrier based ion-selective electrode for intracellular magnesium activity studies. *Anal. Chem.* **52**, 2400–2.

Nanninga, L.B. (1961). Calculation of free magnesium, calcium and potassium in muscle. *Biochem. Biophys. Acta* **54**, 338–44.

Natori, R. (1954). The property and contraction process of isolated myofibrils. *Jikei Med. J.* **1**, 119–26.

Page, E. and Polimeni, P.I. (1972). Magnesium exchange in rat ventricle. *J. Physiol.* **224**, 121–39.

Paradise, N.F., Becker, G.W. and Visscher, M.B. (1978). Magnesium net fluxes and distribution in rabbit myocardium in irreversible contracture. *Am. J. Physiol.* **234**, C115–21.

Polimeni, P.I. and Page, E. (1973). Magnesium in heart muscle. *Circ. Res.* **32**, 367–74.

Scarpa, A. and Carafoli, E. (1978). Calcium transport and cell function. *Ann NY Acad. Sci.* **307**, 1–655.

Sheu, S. and Fozzard, H. (1982). Transmembrane Na^+ and Ca^{2+} electrochemical gradients in cardiac muscle and their relationship to force development. *J. Gen. Physiol.* **80**, 325–51.

Tsien, R.Y. (1983). Intracellular measurements of ion activities. *Ann. Rev. Biophys. Bioeng.* **12**, 91–116.

Wacker, W.E.C. (1969). The biochemistry of magnesium. *Ann. NY Acad. Sci.* **162**, 717–26.

Weingart, R. and Hess, P. (1984). Free calcium in sheep cardiac tissue and frog skeletal muscle measured with Ca^{2+}-selective microelectrodes. *Pflügers Arch.* **402**, 1–9.

Wu, S.T., Pieper, G.M., Salhany, J.M. and Eliot, R.S. (1981). Measurement of free magnesium in perfused and ischemic arrested heart muscle. A quantitative phosphorus-31 nuclear magnetic resonance and multi-equilibria analysis. *Biochemistry* **20**, 7399–403.

CARDIAC GLYCOSIDES AND INTRACELLULAR CALCIUM IN MAMMALIAN CARDIAC MUSCLE

J. Daut

Physiologisches Institut der Technischen Universität München, D-8000 München 40

Abstract

Some of the basic mechanisms contributing to the positive inotropic action of cardiac glycosides are briefly characterized in order to point out some specific areas where further experimental research is needed. The interaction of the Na-transporting and Ca-transporting systems is considered in some detail. If there is a Na-Ca-exchange system, either in the sarcolemma or in the mitochondrial membrane, the free intracellular Ca and the free intracellular Na become interdependent. The nature of this interdependence and the possible role of free intracellular Na in mediating slow changes in myocardial contractility are discussed.

33.1. Introduction

The role of calcium in excitation–contraction coupling in cardiac muscle is a very controversial issue. In particular the mechanism of release of Ca from intracellular stores and the redistribution of Ca between various intracellular stores during inotropic interventions are still a matter of conjecture. Not surprisingly, therefore, the mechanism of the positive inotropic action of cardiotonic steroids remains elusive. The only thing that appears certain at present is that the action of cardiac glycosides must finally lead to an increased release of Ca into the cytosol. No consensus has been reached so far about whether there are two different actions of cardiac glycosides, one predominant at low (therapeutic?) concentrations and one predominant at toxic concentrations, or whether all of the effects of cardiotonic steroids are attributable to inhibition of the Na-K pump. Thus this 'overview' cannot provide any clear-cut answers. Instead, an attempt is made to briefly characterize some of the elements that appear to be relevant for the understanding of the action of digitalis, and to outline the possible interaction between the Na-transporting and the Ca-transporting systems in mammalian cardiac muscle.

33.2. Transmembrane calcium fluxes

33.2.1. Second inward current

The existence of a second inward current (I_{si}) in mammalian myocardium was first demonstrated by Beeler and Reuter (1970 a and b) and New and Trautwein (1972 a and b). From these initial measurements it was calculated that the Ca influx during an action potential should be between 5×10^{-7} and 5×10^{-6} M per litre cell volume. In view of the large Ca buffering capacity of the cytoplasm this amount of Ca appeared to be insufficient to appreciably activate the contractile system (Reuter, 1974). Recent measurements in isolated myocardial cells showed that the second inward current is much faster and much larger than the measurements in multicellular preparations had suggested (Isenberg, 1982; Isenberg and Klöckner, 1982; Mitchell et al., 1983). Comparison of these measurements with the relation between free myoplasmic Ca and tension (Solaro et al., 1974; Fabiato, 1981) appears to suggest that enough Ca ions are flowing into the cell during an action potential to account for the contraction (Isenberg, 1982). However, as has been pointed out by many authors, there is no direct correlation between the amplitude of I_{si} and the force of contraction. During a 'positive staircase', for example, I_{si} decreases while the accompanying contraction increases.

Bers (1983) was able to measure a rapid transient depletion of extracellular Ca near the cell membrane during individual action potentials. This transient depletion was suppressed by Ca channel blockers and may be due to I_{si}. However, it increased with the increase in tension during a positive staircase. McDonald, Nawrath and Trautwein (1975) found no effect of ouabain or dihydro-ouabain on I_{si} in cat ventricular muscle whereas Marban and Tsien (1982) found a distinct increase of I_{si} in ferret ventricular muscle and calf Purkinje fibres after application of cardiac glycosides in a wide range of concentrations (down to 20–50 nM strophanthidin). They propose that the increase in I_{si} may be brought about by an increase in intracellular Ca activity.

33.2.2. Electrogenic Na-Ca exchange

A Na-Ca exchange mechanism was first described by Reuter and Seitz (1968). Its properties and possible function have recently been reviewed by Mullins (1981) and by Chapman, Coray and McGuigan (1983b). The changes in intracellular Na activity (a^i_{Na}) and intracellular Ca activity (a^i_{Ca}) during both Na withdrawal contractures and changes in extracellular Na and Ca activities have now been measured in several laboratories (Sheu and Fozzard, 1982; Bers and Ellis, 1982; Chapman, Coray and McGuigan, 1983a; Allen, Eisner and Orchard, 1984). These authors conclude from their studies that the Na-Ca exchanger is probably electrogenic, although no consensus has yet been reached about the number of Na ions transported per Ca ion. Generally the Na-Ca exchange system is thought not to be directly affected by cardiac glycosides. However, Chapman, Coray and McGuigan (1983a) found that Na withdrawal contractures in the presence of strophanthidin were increased even if the initial value of a^i_{Na} was the same as in control preparations.

33.2.3. Intracellular Ca transients

In the cardiac Purkinje fibre the Ca transient measured with aequorin during an action potential shows two distinct peaks labelled L_1 and L_2 (Wier, 1980; Wier and Isenberg, 1982). The first peak is thought to be related to I_{si} whereas the second peak was attributed to release of Ca from intracellular stores. Figure 33.1 shows that both components are increased by application of cardiac glycosides (Blinks *et al.*, 1982). In mammalian ventricular muscle only one peak of the Ca transient can be recorded (Kurihara and Allen, 1982; Morgan and Blinks, 1982) which resembles the slow component of luminescence (L_2) seen in the Purkinje fibre. From these observations it was concluded that in mammalian ventricular muscle most of the Ca which activates contraction comes from intracellular stores. The Ca signal in ventricular muscle, too, is markedly increased by cardiac glycosides (Morgan and Blinks, 1982).

33.2.4. *Isotopic studies of excitation-dependent Ca fluxes*

The classical tracer-flux studies all suggested that the transmembrane Ca flux per beat was small (see review by Lewartowski, 1983). However, these data were obtained by averaging the ^{45}Ca fluxes during many beats. Recently Pytkowski *et al.* (1983) tried to measure the ^{45}Ca influx and efflux during a single contraction in guinea-pig hearts. They found that the uptake of ^{45}Ca during a single beat was in the order of 3×10^{-4} M/kg wet weight, which is two orders of magnitude larger than previously assumed. They also found that most of the Ca that is taken up into this rapidly exchangeable pool (Ca_1) is released from the cells during the same or the next beat. Furthermore they found a more slowly exchanging cellular Ca fraction (Ca_2) which can take up about 1.7 mmol Ca/kg wet weight during continous stimulation. This Ca fraction changes in parallel with the force of contraction and was found to be extruded from the cells during rest within less than 4 min. These results are difficult to reconcile with previous findings. They seem to imply either that only a small fraction of Ca influx is electrogenic or that there may be a direct influx (and efflux) of Ca from the extracellular space into the terminal cisternae of the sarcoplasmic reticulum (see Fabiato and Fabiato, 1979; Isenberg, 1982; Pytowski *et al.*, 1983).

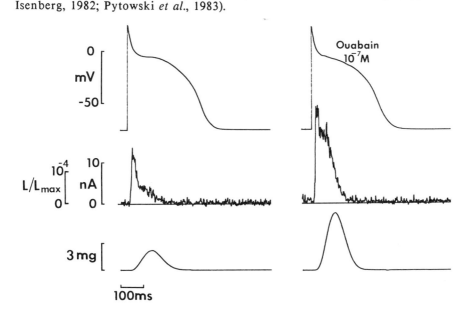

Fig.33.1. Influence of 10^{-7} M ouabain on action potential, aequorin signal, and tension development of a dog Purkinje strand; control records on the left; stimulus intervals 1 s. (From Blinks *et al.*, 1982.)

33.3. The positive inotropic effect of cardiac glycosides

33.3.1. *Dependence on external Ca*

The effect of cardiotonic steroids on amplitude and time-course of isotonic contractions closely resembles the effect of an increase of extracellular Ca (see, for example, Fig.1 of Reiter, 1970). This similarity led Loewi in 1918 to suggest that 'digitalis sensitizes the heart to calcium' (quoted by Reiter, 1970). The changes in intracellular Ca activity produced by changes in external Ca have recently been studied by Eisner, Lederer and Vaughan-Jones (1984). Within the range of $[Ca]_o$ usually applied, the changes in $a^i{}_{Ca}$ parallel the changes in $[Ca]_o$. Ebner and Reiter (1977) carefully studied the interrelations between the positive inotropic effect of dihydro-ouabain (DHO), extracellular Ca concentration and frequency of contraction. Figure 33.2 shows that increasing external Ca shifts the concentration-effect curve of DHO to the left. This means that a given glycoside concentration can have a large or a small positive inotropic effect, depending on external Ca. This implies that comparison of experiments using different preparations can be very misleading if no Ca concentration-effect curve is established.

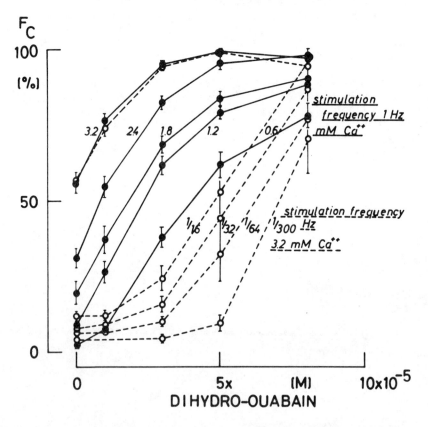

Fig.33.2. Dependence of the positive inotropic effect of dihydro-ouabain on Ca_o^{2+} in comparison to its dependence on stimulation frequency. Ordinate: F_c = peak force of contraction as percent of the maximal glycoside effect at 1 Hz, 3.2 mM Ca. Abscissa: molar concentration of DHO, linear scale. Filled circles: variation of $[Ca]_o$, stimulation frequency 1 Hz; open circles, broken lines: variation of stimulation frequency, $[Ca^{2+}]_o$ 3.2 mM. (From Ebner and Reiter, 1977.)

Cardiac glycosides and intracellular calcium

Much confusion in the literature would have been avoided if more authors had been aware of this. The maximal positive inotropic effect of a cardiac glycoside obtained at a stimulation frequency of 1 Hz at high external Ca has proved to be a good empirical standard to characterize the concentration dependence of a given inotropic intervention. In this context it should be remembered that the amplitude of the twitch obtained under optimum inotropic conditions in cardiac muscle never reaches the maximum tension that can be induced by full activation of the myofilaments after chemical skinning (Fabiato, 1981). The maximum twitch tension seems to be limited by the amount of Ca that can be accumulated in intracellular stores.

Reiter and colleagues have also extensively studied the frequency dependence of the action of cardiac glycosides (see Ebner and Reiter, 1977; Reiter, 1981) and emphasized the similarity between the effects of (a) an increase in frequency, and (b) an increase in the concentration of the cardiotonic steroid on the isometric contraction (see Fig.33.2).

33.3.2. Binding of cardiac glycosides to the Na-K ATPase

The binding of the drug to its receptor is the first step in the positive inotropic action of cardiac glycosides. Recently the binding of dihydro-ouabain (DHO) to intact myocardial cells has been studied with a voltage-clamp technique (Daut and Rüdel, 1982). By measuring the change in electrogenic pump current after rapid application of DHO the binding

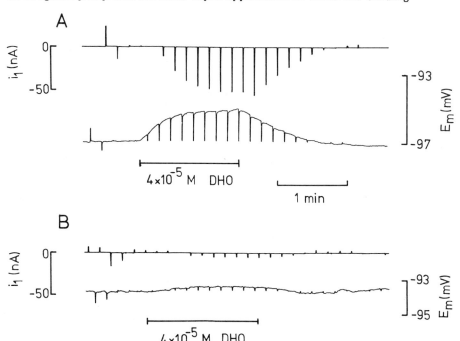

Fig.33.3. Electrophysiological measurement of the binding of DHO to guinea-pig ventricular muscle using the method of Daut and Rüdel (1982). (a) Upper trace: voltage-clamp current; lower trace: membrane potential. The voltage clamp was switched on for 600 ms at regular intervals. The change in voltage-clamp current is proportional to the number of pump sites blocked by DHO. (b) The same experiment repeated 15 min later in the presence of reduced external Na (37 mM instead of 147 mM, substituted by choline).

kinetics and the fractional staturation of the receptor in the steady state could be investigated (Fig.33.3). It was found that the binding affinity of DHO was increased by external Na (Fig.33.3) and decreased by external K (Daut, 1983). This may be the mechanism underlying the well-known Na-dependence and K-dependence of the positive inotropic effect of cardiac glycosides (see, for example, Reiter, 1981).

The rate of binding and unbinding of DHO was then compared to the rate of change of intracellular Na measured under the same conditions with Na-sensitive microelectrodes (Daut, 1982). It was found that the changes in a^i_{Na} induced by short-lasting application of DHO were much slower than the binding of DHO to the Na-K ATPase associated with inhibition of the pump. The increase isometric twitch tension paralleled the change in a^i_{Na} and both decayed with a rate constant of 0.5–1 min^{-1} during washout of DHO. This rate constant of decay of a^i_{Na} (k_{Na}) to a new steady state value agrees well with the rate constant of exchange of a^i_{Na} calculated from the measured density of the electrogenic pump current in guinea-pig ventricular cells (Daut, 1982). The rate constant of the dissociation of DHO from the Na-K pump, however, was found to be faster than that.

These findings rule out any inotropic mechanism in which the binding of the drug to the Na-K ATPase *per se* causes the increase in contractility (Lüllmann and Peters, 1979), at least for the action of DHO on guinea-pig ventricle. With low concentrations of other cardiac glycosides in other cardiac preparations, however, the situation may be more complex (see below).

33.3.3. *Interaction between Na and Ca transporting systems*

It has been postulated for many years that an increase in intracellular Na may be the second step in the sequence of events that eventually leads to the positive inotropic effect of cardiac glycosides (Glynn, 1964; Repke, 1964; Baker *et al.*, 1969), but solid evidence for this has only been obtained after the introduction of intracellular Na-sensitive microelectrodes. Both in the cardiac Purkinje fibre (Ellis, 1977; Deitmer and Ellis, 1978; Lee *et al.*, 1980) and in ventricular muscle (Daut, 1982; Sheu and Fozzard, 1982; Lee *et al.*, 1984) it has now been shown that cardiac glycosides can indeed produce a rise in a^i_{Na}. There is now widespread agreement that this rise in a^i_{Na} is responsible for the increase in contractility seen under these conditions.

Because of the difficulty of measuring the changes in a^i_{Na} associated with positive inotropy it has been suggested that a fast transient rise in a^i_{Na} at the inner side of the cell membrane may be involved (Akera and Brody, 1978). However, calculations of intracellular rates of diffusion showed that the hypothetical intracellular Na gradients should be very shortlived compared to intracellular Ca gradients (see, for example, Fig. 6.2(c) in Chapman, Coray and McGuigan, 1983a). At present it does not seem necessary to invoke such subsarcolemmal Na transients to explain the action of cardiotonic steroids.

On the contrary, the rate constant of exchange of bulk a^i_{Na} seems to be functionally important. It is remarkably similar in different mammalian cardiac preparations. Lado, Sheu and Fozzard (1982) found a value of 0.5–0.75 min^{-1} for k_{Na} in rabbit ventricular muscle. In the sheep Purkinje fibre values in the range 0.5–1.2 min^{-1} have been reported for k_{Na} in 4mM external K or Rb (Eisner and Lederer, 1980; Eisner, Lederer and Vaughan-Jones, 1981). In the canine Purkinje fibre k_{Na} was found to be 0.8–0.9 min^{-1} in 4 mM external K (Gadsby and Cranefield, 1979; Gadsby, 1980).

It should be noted that the rate of exchange of intracellular Na inferred from recent electrophysiological studies is much larger than the estimates of k_{Na} derived from tracer flux studies. Langer and Serena (1970) and Bridge *et al.* (1981) gives values of 0.083 min^{-1} and 0.038 min^{-1}, respectively, for k_{Na} in rabbit myocardium. The reason for this discrepancy may be that the rapid Na efflux in the first few minutes, which had a rate constant of 0.6 min^{-1} (Langer and Serena, 1970), was interpreted as extracellular exchange (Langer, 1974; Bridge *et al.*, 1981). It may well be that this 'phase 1' in fact includes at least part of the cellular Na efflux.

As has been pointed out by Mullins in his stimulating review (1981) the parallel arrangement of ATP-driven ion pump and a Na-Ca exchanger in the cell membrane may be used to extrude both Na and Ca. Either the Na-K pump extrudes Na, and external Na is then used to drive Ca out of the cell via the Na-Ca exchanger, or an ATP-driven Ca pump may extrude Ca and the resulting Ca influx can be used to keep internal Na low via the Na-Ca exchanger running in the opposite direction (for example, in red blood cells). In either case one ATP-driven pump is sufficient to provide the energy of the outward transport of both Na and Ca via such a cascade-like system. If both sarcolemmal Na and Ca pumps are present, the pump with the highest transport rate will determine the direction of the ion movement through the Na-Ca exchanger in the quiescent preparation or during diastole.

One version of the 'classical' model of the action of cardiac glycosides is depicted schematically in Fig.33.4. It is assumed that it is the *diastolic* free Ca that regulates the distribution of Ca between various stores. The continuous arrows indicate the influence of $a^i{}_{Na}$ and $a^i{}_{Ca}$ on the various Na and Ca transport processes during diastole. A rise in $a^i{}_{Na}$ brought about by inhibition of the Na-K pump (or by increase in contraction frequency) will reduce Ca efflux through the Na-Ca exchanger. Thus relatively more Ca will be taken up into the SR by the sarcoreticular Ca pump. It is likely that calmodulin (CaM) also plays a role in mediating this increased uptake of Ca into the SR (see, for example, Johnson, Wittenauer and Nathan, 1983).

A second important assumption is that there are no significant intracellular Na gradients and that intracellular Na is exchanged with a rate constant of

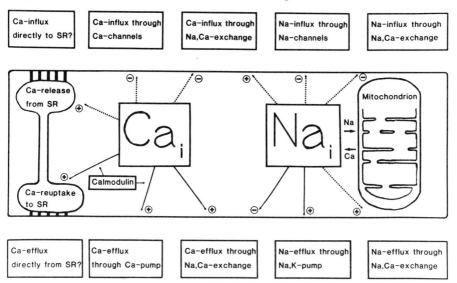

Fig.33.4. Schematic diagram of the influence of free intracellular Na and Ca on the transmembrane movements of Ca and Na during diastole (continuous arrows) and during systole (broken arrows).

$0.5 - 1$ min^{-1}. A further assumption in this system is that diastolic $a^i{}_{Ca}$ will be forced to follow any slow change in $a^i{}_{Na}$ as a result of a Na-Ca exchange mechanism. This follows from the notion that the Na-Ca exchanger can mediate large Ca-fluxes (see, for example, Fig. 4.3 of Mullins, 1981; or Bers and Ellis, 1982) (that is, its 'apparent conductance' is relatively high), which implies that it tends towards equilibrium during rest or during diastole. Now the important question arising at this point is whether the Na-Ca exchanger will move toward equilibrium by changing $a^i{}_{Na}$ or $a^i{}_{Ca}$ or both.

As a first approximation we may regard all the Ca-binding proteins including the SR and the mitochondria as one lumped Ca buffer system. The buffering power of this system in the steady state may then be defined as B (M/pCa-unit). From this definition it follows that the change in $a^i{}_{Na}$ ($\Delta\, a^i{}_{Na}$) associated with an increase of $a^i{}_{Ca}$ (from $a^i{}_{Ca(1)}$ to $a^i{}_{Ca(2)}$) through the Na-Ca exchanger is given by

$$\Delta\, a^i{}_{Na} = n \cdot (f_{Na}/f_{Ca}) \cdot B\, \log_{10}(a^i{}_{Ca(2)}/a^i{}_{Ca(1)}), \tag{1}$$

where n is the number of Na ions exchanged for each Ca ion and f_{Na} and f_{Ca} are the activity coefficients of intracellular Na and Ca. As a first approximation f_{Na} and f_{Ca} may be assumed to be 0.75 and 0.32, respectively, and n may be 3. It follows then from equation (1) that the Na—Ca exchange system will move towards equilibrium by changing $a^i{}_{Ca}$ only, with virtually no change in $a^i{}_{Na}$, if the lumped buffering power is smaller than 1×10^{-4} M.

The Ca-buffering capacity of the axoplasm of squid axon was measured by Brinley et al. (1977) and was found to be even higher than that, about 1 mM per pCa-unit. Inserting this figure in equation (1) shows that a 50% increase of $a^i{}_{Ca}$ by the Na-Ca exchanger would only require a decrease of $a^i{}_{Na}$ of about 1.3 mM. This means that in the steady state the electrochemical gradient of Ca would still follow any imposed change of the electrochemical gradient of Na. This has indeed been found by Sheu and Fozzard (1982; their Fig.13). The relative changes in $a^i{}_{Na}$ and $a^i{}_{Ca}$ produced by the Na-K exchanger will, of course, depend on various conditions such as the initial value of $a^i{}_{Na}$ and the value of B (see equation (1)).

The main conclusion from these considerations is that the filling of an intracellular store (possibly store Ca_2 of Pytkowski et al., 1983) may depend on the value of diastolic free Ca and thus on $a^i{}_{Na}$. This store may then influence the force of the subsequent contraction.

However, the sarcolemmal Na-Ca exchanger is only one possible mechanism by which slow change in $a^i{}_{Na}$ may influence diastolic $a^i{}_{Ca}$. As has been pointed out by Carafoli and co-workers (Carafoli, E. 1982), the surface area of motochondria is about 100 times larger than the surface area of the sarcolemma, and there is a powerful Na-Ca exchange system at the mitochondrial membrane as well. The Na and Ca affinities of this system are of the right order of magnitude to provide for a Na-Ca exchange between mitochondria and cytoplasm which could also force $a^i{}_{Ca}$ to follow slow changes in $a^i{}_{Na}$. Further studies are needed to clarify the relative importance of these two Na-Ca exchange systems.

During contraction Ca buffering is much more complex because the time-dependence of the binding of released Ca to various Ca-binding proteins such as calmodulin, myosin, troponin etc. has to be considered (see Robertson, Johnson and Potter, 1981). The influence of intracellular Na and Ca on the transmembrane Na and Ca fluxes during contraction is schematically indicated by the dotted lines in Fig.33.4.

33.3.4. Effects of stimulation frequency

An increase in stimulation frequency also produces a rise in $a^i{}_{Na}$ and $a^i{}_{Ca}$. After the end of rapid stimulation both $a^i{}_{Na}$ and $a^i{}_{Ca}$ were found to decay in parallel with a rate constant of 0.3–0.5 min^{-1} in the Purkinje fibre (Cohen, Fozzard and Sheu, 1982; Lado, Sheu and Fozzard, 1982; Lederer and Sheu, 1983). Thus in this case, too, $a^i{}_{Ca}$ appears to follow the change in $a^i{}_{Na}$. Interestingly the 'positive inotropic effect of activation' (PIEA; Blinks and Koch-Weser, 1961) was also found to decay with a rate constant of 0.5–1 min^{-1} in a wide variety of species (see Koch-Weser and Blinks, 1963).

It is tempting to speculate that the *slow* changes in contractility produced by a change in stimulation frequency, that is the 'positive inotropic effect of activation' (Blinks and Koch-Weser, 1961), may be due to the same mechanism postulated above for the action of cardiotonic steroids. A similar hypothesis has been proposed before by Langer and Serena (1970) and Reiter (1981). Thus, despite our ignorance about the actual mechanism of excitation–contraction coupling, it appears possible to identify $a^i{}_{Na}$ as the intracellular messenger that mediates changes in myocardial contractility occurring within 1-2 min. The *fast* frequency-dependent changes in contractility which occur within a few seconds seem to be mediated by different mechanisms; they may be related to the filling of a more rapidly exchanging intracellular Ca store (Ca$_1$ of Pytkowski *et al.*, 1983).

33.3.5. Different effects of low concentrations of cardiac glycosides

Ever since the inhibition of the Na-K ATPase by cardiac glycosides has been measured there has been a controversy about a stimulatory effect of low concentrations of cardiac glycosides. In isolated Na-K ATPase preparations the general pattern seems to emerge that with rather crude enzyme preparations stimulation of the Na-K pump by low drug concentrations can be observed, whereas with highly purified Na-K ATPase normally only inhibition is seen. In 1976 Cohen, Daut and Noble drew attention to the fact that in the sheep Purkinje fibre low concentrations of ouabain usually shift the steady-state current-voltage relation in the direction opposite to that observed with high concentrations, and in 1978 Deitmer and Ellis (1978) reported that low concentrations of cardiotonic steroids may produce a decrease of intracellular Na. These findings have since been confirmed, but their interpretation and functional significance is still controversial (Noble, 1980). Stimulation of the Na-K pump by low concentrations of ouabain was offered as one possible interpretation amongst others (Cohen, Daut and Noble, 1976), but unambiguous evidence for this is still lacking.

T.W. Smith and co-workers (Hougen, Spicer and Smith, 1981; Lechat, Malloy and Smith, 1983) showed that the effects of low concentrations of ouabain on guinea-pig atrium may be at least partly mediated through release of catecholamines by the cardiac glycoside. A negative inotropic effect of low concentrations of ouabain has also been found in the sheep Purkinje fibre (Hart, Noble and Shimoni, 1983) and in guinea-pig atrium (Grupp *et al.*, 1983). However, this effect in guinea-pig atrium disappeared when stimulation with needle electrodes instead of field stimulation was applied.

In view of these findings the usual word of caution will not be omitted here; in order to demonstrate stimulation of the Na-K pump in vivo both a rise in intracellular Na and neural effects (cholinergic and adrenergic) would have to be ruled out in the same experiment. This has not been done yet.

In rat ventricular muscle the dose–response curve of the effects of ouabain on contractility clearly shows a 'hump' at low concentrations. Ouabain-binding studies in cultured heart cells from neonatal rats (Werdan *et* al., 1984) and chicken embryos (Kazazoglou et al., 1983) confirmed the existence of a low and a high affinity binding site for ouabain. In these studies it was found that the positive inotropic effect of ouabain and the inhibition of the Na-K pump were correlated with binding of the drug to the low affinity site. However, recently Grupp et al. (1985) found that both the 'low-dose' and the high-dose effect of ouabain on rat ventricular muscle were correlated with an increase in a^i_{Na} and with an increase in twitch tension. It is not yet clear whether the high affinity binding site is of any functional significance in vivo and whether it exists in mammalian cardiac preparations other than rat heart.

The search for an endogeneous digitalis-like substance ('endodigin') has been intensified in recent years (Gruber, Whitaker and Buckalew, 1980; Godfraind et al., 1982). Studies on the nature of the interaction of this substance with the Na-K ATPase will no doubt help to clarify our ideas about the clinical significance of the 'low-dose effects' of cardiotonic steroids. If the only effect of such an endogeneous substance is inhibition of the Na-K pump it is difficult to see how digitalis therapy can be improved by introducing new cardiotonic steroids. On the other hand if the high affinity sites prove to be functionally important it may be possible to find cardiotonic steroids that bind more specifically to these sites.

References

Akera, T. and Brody, T.M. (1978). The role of Na^+, K^+-ATPase in the inotropic action of digitalis. Pharmacol. Rev. 29, 187–220.

Allen, D.G., Eisner, D.A. and Orchard, C.H. (1984). Factors influencing free intercellular calcium concentration in quiescent ferret ventricular muscle. J. Physiol. 350, 615–30.

Baker, P.F., Blaustein, M.P., Hodgkin, A.L. and Steinhardt, R.A. (1969). The influence of calcium on sodium efflux in squid axons. J. Physiol. 200, 431–58.

Beeler, G.W.Jr. and Reuter, H. (1970a). Voltage clamp experiments on ventricular myocardium fibres. J. Physiol. 207, 165–90.

Beeler, G.W.Jr. and Reuter, H. (1970b). Membrane calcium current in ventricular myocardium fibres. J. Physiol. 207, 191–209.

Bers, D.M. (1983). Early transcient depletion of extracellular Ca during individual cardiac muscle contractions. Amer. J. Physiol. 244, H462–8.

Bers, D.M. and Ellis, D. (1982). Intracellular calcium and sodium activity in sheep heart Purkinje fibres. Effect of changes of external sodium and intracellular pH. Pflügers Arch. 393, 171–8.

Blinks, J.H. and Koch-Weser, J. (1961). Analysis of the effects of changes in rate and rhythm upon myocardial contractility. J. Pharmacol. Exp. Ther. 134, 373–89.

Blinks, J.R., Wier, W.G., Morgan, J.P. and Hess, P. (1982). Regulation of intracellular (Ca^{++}) by cardiotonic drugs. In Advances in Pharmacology and Therapeutics II, vol. 3, H. Yoshida, Y. Hagihara and S. Ebashi (eds.), pp. 205–16. Pergamon Press, Oxford.

Bridge, J.H.B., Cabeen, W.R.Jr., Langer, G.A. and Reeder, S. (1981). Sodium efflux in rabbit myocardium: relationship to sodium-calcium exchange. J. Physiol. 316, 555–74.

Brinley, F.J., Tiffert, T., Scarpa, A. and Mullins, L.J. (1977). Intracellular calcium buffering capacity in isolated squid axons. J. Gen. Physiol. 70, 355–84.

Carafoli, E. (1982). The transport of calcium across the inner membrane of mitochondria. In *Membrane Transport of Calcium*. E. Carafoli (ed.), pp. 109–35. Academic Press, New York.

Chapman, R.A., Coray, A. and McGuigan, J.A.S. (1983a). Sodium/calcium exchange in mammalian ventricular muscle: a study with sodium-sensitive microelectrodes. *J. Physiol.* **343**, 253–76.

Chapman, R.A., Coray, A. and McGuigan, J.A.S. (1983b). Sodium–calcium exchange in mammalian heart: the maintenance of low intracellular calcium concentration. In *Cardiac Metabolism*. A.J. Drake-Holland and M.I.M. Noble (eds), pp. 117–49. John Wiley and Sons Ltd, Chichester.

Cohen, C.D., Fozzard, H.A. and Sheu, S. (1982). Increase in intracellular sodium ion activity during stimulation in mammalian cardiac muscle. *Circ. Res.* **50**, 651–62.

Cohen, I., Daut, J. and Noble, D. (1976). An analysis of the action of low concentrations of ouabain on membrane currents in Purkinje fibres. *J. Physiol.* **260**, 75–103.

Daut, J. (1982). The role of intracellular sodium ions in the regulation of cardiac contractility. *J. Molec. Cell. Cardiol.* **14**, 189–92.

Daut, J. (1983). Inhibition of the sodium pump in guinea-pig ventricular muscle by dihydro-ouabain: effects of external potassium and sodium. *J. Physiol.* **339**, 643–62.

Daut, J. and Rüdel, R. (1982). The electrogenic sodium pump in guinea-pig ventricular muscle: inhibition of pump current by cardiac glycosides. *J. Physiol.* **330**, 243–64.

Deitmer, J.W. and Ellis, D. (1978). The intracellular sodium activity of cardiac Purkinje fibres during inhibition and re-activation of the Na-K pump. *J. Physiol.* **284**, 241–59.

Ebner, F. and Reiter, M. (1977). The dependence on contraction frequency of the positive inotropic effect of dihydro-ouabain. *Naunyn Schmiedeberg's Arch. Pharmacol.* **300**, 1–9.

Eisner, D.A. and Lederer, W.J. (1980). Characterization of the electrogenic sodium pump in cardiac Purkinje fibres. *J. Physiol.* **303**, 441–74.

Eisner, D.A., Lederer, W.J. and Vaughan-Jones, R.D. (1981). The dependence of sodium pumping and tension on intracellular sodium ion activity in voltage-clamped sheep Purkinje fibres. *J. Physiol.* **317**, 163–87.

Ellis, D. (1977). The effect of external cations and ouabain on the intracellular sodium activity in sheep heart Purkinje fibres. *J. Physiol.* **273**, 211–40.

Fabiato, A. (1981). Myoplasmic free calcium reached during the twitch of an intact isolated cardiac cell and during calcium-induced release of calcium from the sarcoplasmic reticulum of a skinned cardiac cell from the adult rat or rabbit ventricle. *J. Gen. Physiol.* **78**, 457–97.

Fabiato, A. and Fabiato, F. (1979). Calcium and cardiac exitation–contraction coupling. *Ann. Rev. Physiol.* **41**, 473–84.

Gadsby, D.C. (1980). Activation of electrogenic Na^+/K^+ exchange by extracellular K in canine cardiac Purkinje fibres. *Proc. Natl. Acad. Sci. USA* **77**, 4035–9.

Gadsby, D.C. and Cranefield, P.F. (1979). Electrogenic sodium extrusion in cardiac Purkinje fibres. *J. Gen. Physiol.* **73**, 819–37.

Glynn, I.M. (1964). The action of cardiac glycosides on ion movements. *Pharmacol. Rev.* **16**, 381–407.

Godfraind, T., De Pover, A., Castaneda Hernandez, G. and Fagoo, M. (1982). Cardiodigin: endogenous Digitalis-like material from mammalian heart. *Arch. Int. Pharmacodyn. Ther.* **258**, 165–7.

Gruber, K.A., Whitaker, J.M.C. and Buckalew, V.M.Jr. (1980). Endogenous digitalis-like substance in the plasma of volume expanded dog. *Nature (Lond.)* **287**, 743–5.

Grupp, G., Grupp, I., Ghysel-Burton, J., Godfraind, T., De Pover, A. and Schwartz, A. (1983). Contractile force effects of low concentrations of ouabain in isolated guinea-pig, rabbit, cat, and rat atria and ventricles. In *Current Topics in Membranes and Transport, Vol. 19*. K. Bronner and A. Kleinzeller, pp. 897–906. Academic Press.

Grupp, I., Im, W.-B., Lee, C.O., Lee, S.-W., Pecker, M.S. and Schwartz, A. (1985). Relation of sodium pump inhibition to positive inotropy at low concentrations of ouabain in rat heart muscle. *J. Physiol.* **360**, 149–60.

Hart, G., Noble, D. and Shimoni, Y. (1983). The effects of low concentrations of cardiotonic steroids on membrane currents and tension in sheep Purkinje fibres. *J. Physiol.* **334**, 103–31.

Hougen, T.J., Spicer, N. and Smith, T.W. (1981). Stimulation of monovalent cation active transport by low concentrations of cardiac glycosides. Role of catecholamines. *J. Clin. Invest.* **68**, 1207–14.

Isenberg, G. (1982). Ca entry and contraction as studied in isolated bovine ventricular myocytes. *Z. Naturforsch.* **37c**, 502–12.

Isenberg, G. and Klöckner, U. (1982). Calcium currents of isolated bovine ventricular myocytes are fast and of large amplitude. *Pflügers Arch.* **395**, 30–41.

Johnson, J.D., Wittenauer, L.A. and Nathan, R.D. (1983). Calmodulin, Ca^{2+}-antagonists and Ca^{2+} transporters in nerve and muscle. *J. Neural Transmission*, **Suppl. 18**, 97–111.

Kazazoglou, T., Renaud, J.F., Rossi, B. and Lazdunski, M. (1983). Two classes of ouabain receptors in chick ventricular cardiac cells and their relation to (Na^+, K^+)-ATPase inhibition, intracellular Na accumulation, Ca^{2+} influx and cardiotonic effect. *J. Biol. Chem.* **258**, 12163–70.

Koch-Weser, J. and Blinks, J.R. (1963). The influence of the interval between beats on myocardial contractility. *Pharmacol. Rev.* **15**, 601–52.

Kurihara, S. and Allen, D.G. (1982). Intracellular Ca^{++} transients and relaxation in mammalian cardiac muscle. *Jap. Circ. J.* **46**, 39–43.

Lado, M.G., Sheu, S.-S. and Fozzard, H.A. (1982). Changes in intracellular Ca^{2+} activity with stimulation in sheep cardiac Purkinje fibres. *Amer. J. Physiol.* **243**, H133–7.

Langer, G.A. (1974). Ionic movements and the control of contraction. In *The Mammalian Myocardium*. G.A. Langer and A.J. Brady (eds), pp. 193–217. Wiley, New York.

Langer, G.A. and Serena, S.D. (1970). Effects of strophanthidin upon contraction and ionic exchange in rabbit ventricular myocardium. *J. Molec. Cell. Cardiol.* **1**, 65–90.

Lechat, P., Malloy, C.R. and Smith, T.W. (1983). Active transport and inotropic state in guinea-pig left atrium. *Circ. Res.* **52**, 411–22.

Lederer, W.J. and Sheu, S.-S. (1983). Heart rate dependent changes in intracellular sodium activity and twitch tension in sheep Purkinje fibres. *J. Physiol.* **345**, 44P.

Lee, C.O., Kang, D.H., Sokol, J.H. and Lee, K.S. (1980). Relation between intracellular Na ion activity and tension of sheep cardiac Purkinje fibres exposed to dihydro-ouabain. *Biophys. J.* **29**, 315–30.

Lewartowski, B. (1983). Calcium exchange. In *Cardiac Metabolism*. A.J. Drake-Holland and M.I.M. Noble (eds), pp. 101–6. John Wiley and Sons, New York.

Lüllmann, H. and Peters, J. (1979). Action of cardiac glycosides on the excitation-contraction coupling in heart muscle. *Prog. Pharmacol.* 2, 1–57.

Marban, E. and Tsien, R.W. (1982). Enhancement of calcium current during digitalis inotropy in mammalian heart: positive feed-back regulation by intracellular calcium? *J. Physiol.* 329, 589–614.

McDonald, T.F., Nawrath, H. and Trautwein, W. (1975). Membrane currents and tension in cat ventricular muscle treated with cardiac glycosides. *Circ. Res.* 37, 674–82.

Mitchell, M.R., Powell, T., Terrar, D.A. and Twist, V.W. (1983). Characteristics of the second inward current in cells isolated from rat ventricular muscle. *Proc. Roy. Soc. Lond. B* 219, 447–69.

Morgan, J.P. and Blinks, J.R. (1982). Intracellular Ca^{2+} transients in the cat papillary muscle. *Can. J. Physiol. Pharmacol.* 60, 524–8.

Mullins, L.J. (1981). *Ion Transport in Heart*. Raven Press, New York.

New, W. and Trautwein, W. (1972a). Inward membrane currents in mammalian myocardium. *Pflügers Arch.* 334, 1–23.

New, W., Trautwein, W. (1972b). The ionic nature of slow current and its relation to contraction. *Pflügers Arch.* 334, 24–38.

Noble, D. (1980). Mechanism of action of therapeutic levels of cardiac glycosides. *Cardiovasc. Res.* 14, 495–514.

Pytkowski, B., Lewartowski, B., Prokopczuk, A., Zdanowski, K. and Lewandowska, K. (1983). Excitation- and rest-dependent shifts of Ca in guinea-pig ventricular myocardium. *Pflügers Arch.* 398, 103–13.

Reiter, M. (1970). Cardioactive steroids with special reference to calcium. In *A Symposium on Calcium and Cellular Function.* A.W. Cuthbert (ed.), pp. 270–9. MacMillan (1970)

Reiter, M. (1981). The positive inotropic action of cardioactive steroids on cardiac ventricular muscle. In *Handbook of Experimental Pharmacology, vol. 56,* K. Greef (ed.), pp. 187–219. Springer, Heidelberg.

Repke, K. (1964). Über den Wirkungsmodus von Digitalis. *Klin. Wochenschr.* 42, 157–65.

Reuter, H. (1974). Exchange of calcium ions in the mammalion myocardium. *Circ. Res.* 34, 599–605.

Reuter, H. and Seitz, N. (1968). The dependence of calcium efflux from cardiac muscle on temperature and external ion composition. *J. Physiol.* 195, 451–70.

Robertson, S.P., Johnson, J.D. and Potter, J.D. (1981). The time-course of Ca^{2+} exchange with calmodulin, troponin, parvalbumin and myosin in response to transient increases in Ca^{2+}. *Biophys. J.* 34, 559–69.

Sheu, S.-S. and Fozzard, H.A. (1982). Transmembrane Na^+ and Ca^{2+} electro-chemical gradients in cardiac muscle and their relationship to force development. *J. Gen. Physiol.* 80, 325–51.

Solaro, R.J., Wise, R.M., Shiner, J.S. and Briggs, F.N. (1974). Calcium requirements for cardiac myofibrillar activation. *Circ. Res.* 34, 525–30.

Werdan, K., Wegenknecht, B., Zwissler, B., Brown, L., Krawietz, W. and Erdmann, E. (1984). Cardiac glycosides receptors in cultured heart cells II. Characterization of a high affinity and a low affinity binding site in heart muscle cells from neonatal rats. *Biochem. Pharmacol.* 33, 1873–86.

Wier, G. (1980). Calcium transients during excitation–contraction coupling in mammalian heart: aequorin signals of canine Purkinje fibres. *Science* 207, 1085–7.

Wier, G. and Isenberg, G. (1982). Intracellular (Ca^{2+}) transients in voltage-clamped cardiac Purkinje fibres. *Pflügers Arch.* 392, 284–90.

MECHANISM OF THE CARDIOTONIC AND SMOOTH MUSCLE-RELAXING EFFECTS OF THE METHYLXANTHINES

Hasso Scholz, Michael Böhm, Wilfried Meyer and Wilhelm Schmitz

Abteilung Allgemeine Pharmakologie, Universitäts-Krankenhaus Eppendorf, D-2000 Hamburg 20, West Germany

Abstract

The positive inotropic effect (PIE) of the methylxanthines, notably of theophylline, is similar to the PIE of β-adrenoceptor stimulating agents in that is is very rapid in onset and inhibited by slow channel- blockers (such as manganese ions or verapamil). At concentrations up to 1 mM it is likely to be due to an increase in slow calcium inward current during the cardiac action potential which, in turn, is probably the result of an inhibition of phosphodiesterase activity with a subsequent increase in myocardial cAMP levels. The prolongation of the myocardial contraction (this effect is opposite to the β-adrenergic effect) observed at concentrations higher than 1 mM does not necessarily contradict this view. It is probably due to an additional cAMP-independent inhibition of calcium uptake by the sarcoplasmic reticulum and is not shared by more potent phosphodiesterase inhibitors such as 1-methyl-3-isobutylxanthine which shortens the isometric contraction as do β-adrenoceptor stimulating agents. At high concentrations a sensitization of the contractile proteins to Ca^{2+} may also contribute to the positive inotropic effect of the methylxanthines. The increase in cAMP levels is probably also responsible for the relaxant effects of the methylxanthines in smooth muscle. Finally, the mechanical effects of these drugs in heart and smooth muscle are unlikely to result from an adenosine-antagonistic action. In the heart, for example, adenosine produces a negative inotropic effect in atria but a positive inotropic effect, if any, in ventricular myocardium, whereas the positive inotropic effect of theophylline is similar in both tissues.

34.1. Introduction

The methylxanthines produce a great variety of pharmacological effects: they stimulate the central nervous system as they do in heart and skeletal muscle; they relax smooth muscle, produce diuresis, stimulate gastric and catecholamine secretion, inhibit secretion by mast cells, and increase glycogenolysis and lipolysis (Rall, 1980). The present report is only concerned with the cellular mechanisms of the positive inotropic effect of these drugs, notably of theophylline, in cardiac muscle and with a brief treatment of the relaxing effects of the methylxanthines in smooth muscle.

34.2. Characteristics of the positive inotropic effects of theophylline in cardiac muscle

The positive inotropic effect of theophylline, the concentration-dependency of which is shown in Fig.34.1, resembles that of β-adrenergic stimulation in that it develops very rapidly, is independent of the extracellular Na^+ and K^+ concentrations, is impaired by slow channel-blockers such as manganese or verapamil and stimulates Ca^{2+} movements through the sarcolemma (Tsien, 1977, Scholz, 1980, 1983, Farah, Alousi and Schwarz, 1984). Some effects produced with high millimolar concentrations of the classical methylxanthines are, however, opposite to those of the β-adrenergic agonists (Scholz, 1980). Most importantly, they prolong rather than reduce the duration of the

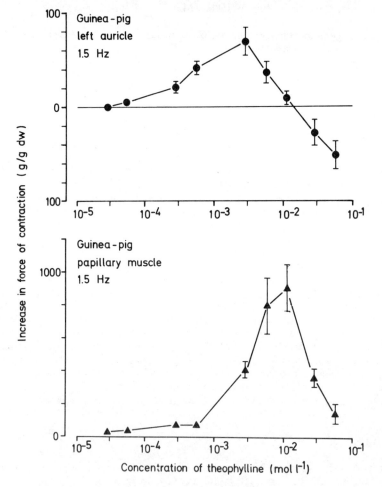

Fig.34.1. Concentration–response curves for the effects of theophylline ($2.78 \times 10^{-5} - 5.5 \times 10^{-2}$ M, incubation time for each concentration 5 min) on force of contraction in guinea-pig electrically driven left auricles (upper panel) and right ventricular papillary muscles (lower panel). Stimulation frequency 1.5 Hz. Ordinate: increase in force of contraction over predrug values in g/g dry weight. $n = 5-13$. Predrug values (g/g dry weight): 86.6 ± 6.19 (●, $n = 60$) and 420 ± 42.9 (▲, $n = 96$). At concentrations of 5.55×10^{-4} M and higher, only one concentration was studied in one preparation; lower concentrations were administered cumulatively. (Modified from Bechthold, 1971.)

isometric contraction at concentrations higher than 1 mM. Moreover, their positive inotropic effect does not remain stable but declines during prolonged exposure (Fig.34.2) and they increase rather than decrease high potassium-induced contractures.

34.3 Possible mechanisms of the positive inotropic effect of theophylline

It is generally accepted that the cardiotonic effect of any drug ultimately results from an increase in the concentration of intracellular free Ca^{2+} $[Ca^{2+}]_i$ to interact with the contractile proteins or to an increased sensitivity of the myofilaments for Ca^{2+}, or both. The positive inotropic effect of the methylxanthines may be attributed to:

(1) A direct effect on intracellular Ca^{2+} stores, namely an inhibition of Ca^{2+} uptake by the sarcoplasmic reticulum;
(2) an increase in slow Ca^{2+} inward current I_{si} resulting from an inhibition of phosphodiesterase activity and a subsequent increase in cellular cAMP content;
(3) a sensitization of the contractile proteins to Ca^{2+}; and
(4) a blockade of receptors for endogenous adenosine.

In the following paragraphs it is discussed briefly how these steps might be affected during the cardiotonic action of theophylline. It is important to note in this context that the maximal theophylline plasma concentrations reached during therapy are not higher than 50—100 μM (9—18 mg/l; Rall, 1980).

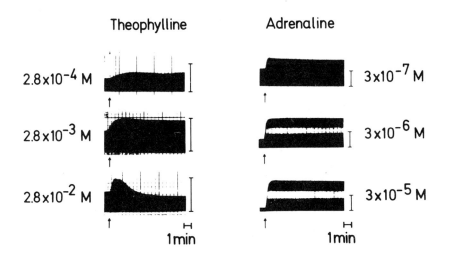

Fig.34.2. Effect of theophylline (2.8 x 10^{-4} M — 2.8 x 10^{-2} M) and adrenaline (3 x 10^{-7} M — 3 x 10^{-5} M) on force of contraction in guinea-pig electrically driven left auricles. Stimulation frequency 1.5 Hz. The time of drug addition is marked with arrows (↑). Vertical bars denote 1 g. (Modified from Bechthold, 1971.)

34.4. Inhibition of Ca^{2+} uptake by the sarcoplasmic reticulum

At concentrations of 1 mM or more, that is, at concentrations which are beyond those reached during therapy, theophylline and other methylxanthines have been shown to release Ca^{2+} from, and to impair the sequestration of Ca^{2+} by, the sarcoplasmic reticulum in a direct cAMP-independent manner (Chapman and Léoty, 1976; Endo, 1977; Scholz ,1980). The inhibitory effect on Ca^{2+} uptake has been postulated to be one of the mechanisms of the cardiotonic action of these drugs (Blinks et al., 1972, Fabiato and Fabiato, 1973). However, this effect is conceivably related to the positive inotropic action of theophylline only at millimolar concentrations. It seems more likely that it rather serves to decrease the filling of intracellular stores with Ca^{2+} that can be released during depolarization. This effect may thus counteract the theophylline-induced increase in slow Ca^{2+} inward current discussed below, and may therefore account for the decline of the positive inotropic

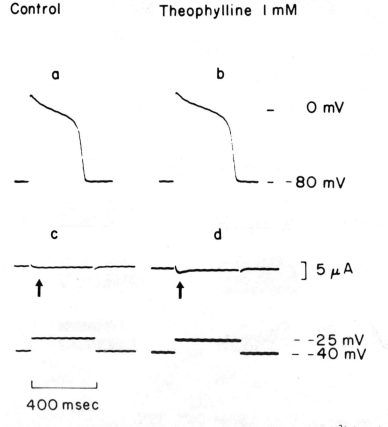

Fig.34.3. Voltage-clamp experiment to demonstrate the effect of theophylline on slow Ca^{2+} inward current in a cow right ventricular trabecula. Action potential (upper panel), membrane potential (lower panel) and membrane current (middle panel) under control conditions and in the presence of 1 mM theophylline. The holding potential was kept constant at -40 mV in order to inactivate the rapid Na^+ inward current. For details of method see Reuter and Scholz (1977). Frequency of stimulation 0.3 Hz, extracellular Ca^{2+} concentration 0.45 mM. It is evident that theophylline increases the slow Ca^{2+} inward current (downward deflection marked with arrows in the middle panel). Unpublished experiment of H. Reuter and H. Scholz.

effect of the drug at high concentrations and for the prolongation of the isometric contraction mentioned above. It may also account for the increase in high potassium-induced contractures.

34.5. Effects of theophylline on Ca^{2+} movements and cAMP system of the heart

Tracer and electrophysiological studies have provided evidence that the positive inotropic action of theophylline is closely related to an increased entry of Ca^{2+} into the myocardial cell during the excitation process (Scholz, 1980). Theophylline accelerates the uptake of $^{45}Ca^{2+}$ in beating but not in resting preparations and enhances the slow Ca^{2+} inward current. This has been shown in voltage-clamp experiments (Scholz and Reuter, 1976) such as that illustrated in Fig.34.3 and by measuring the rate of rise of Ca^{2+}-dependent slow action potentials. It has also been shown that the effects of theophylline on slow inward current and force of contraction have similar time and concentration dependencies. In all these respects, the effects of theophylline resemble those observed with β-adrenoceptor stimulating agents although they are not due to a release of endogenous catecholamines, at least in isolated organs, or to a direct stimulation of β-adrenoceptors.

The similarity between the positive inotropic effects of the β-adrenergic catecholamines and theophylline suggests that an increase in myocardial cAMP levels might be involved not only in the former but also in the latter. Figure 34.4 shows that theophylline inhibits phosphodiesterase activity,

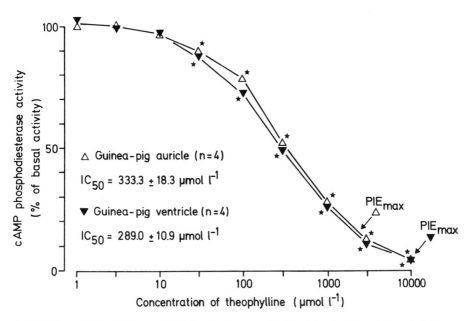

Fig.34.4. Effect of theophylline on cAMP phosphodiesterase (PDE) activity in left auricles and right ventricles from guinea pigs. PDE activity was determined as described by Schmitz et al. (1980). Substrate concentration 1 μM. The inhibition of PDE activity became significant ($p < 0.05$) at 30 μM theophylline (marked with asterisks). Each point represents the mean ± SEM of four experiments. PIE_{max} denotes the concentrations at which theophylline produces maximal positive inotropic effects as shown in Fig.34.1.

likewise in atrial and ventricular cardiac preparations, in a concentration-dependent manner with an IC_{50} of about 300 μM. It should be noted that phosphodiesterase (PDE) activity was inhibited by about 25% at the 'therapeutic' concentration of 100 μM and that about 75% phosphodiesterase inhibition was achieved at 1 mM theophylline.

Figure 34.5 shows that the inhibition in phosphodiesterase activity actually leads to an increase in myocardial cAMP content, and that this and the PDE inhibiting as well as the positive inotropic effect are similarly concentration-dependent. This supports the view of a causal relationship between these events, which is not necessarily contradicted by the finding that the curves diverge at the high concentration of 5 mM theophylline. Here, cAMP continued to increase whereas the positive inotropic effect of the drug was not enhanced further. This effect, however, can probably be explained by a cAMP-independent effect of theophylline to decrease the Ca^{2+} concentration of the sarcoplasmic reticulum, as was discussed above. Thus, it is reasonable to conclude that the increase in cAMP in the case of the methylxanthines is due to an inhibition of the degradation of cAMP, while the β-adrenergic positive inotropic response is due to an increase in cAMP formation. In both cases, the increase in cAMP probably leads to the increase

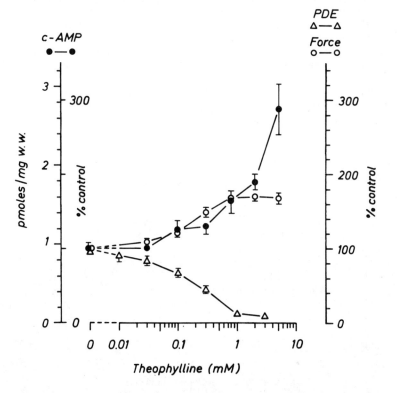

Fig.34.5. Effect of theophylline (0.01–5 mM) on cAMP and force of contraction in guinea-pig electrically driven (frequency 3 Hz) left auricles. Pretreatment with reserpine. Incubation time 60 s. From Dönges et al. (1977). Also shown is the phosphodiesterase (PDE) activity of guinea-pig left auricles as determined according to Schmitz et al. (1980) (n = 5). It is evident that cAMP content and positive inotropic effect on the one hand and cAMP content and inhibition of phosphodiesterase activity on the other are similarly concentration-dependent up to 2 mM. ● = cAMP; Δ = phosphodiesterase; ○ = force of contraction.

in slow Ca^{2+} inward current which, in turn, finally produces the cardiotonic effect.

This explanation has been questioned because of the fact noted above that the classical methylxanthines, caffeine and theophylline, especially at high concentrations, have some effects that are opposite to those of the β-adrenergic agonists (see, for instance, Blinks et al., 1972; Sobel and Mayer, 1973). However, these effects - prolongation of the contraction, increase in high-potassium contractures, and instability of the cardiotonic effect - can readily be explained by the direct cAMP-unrelated effects of the drugs on the sarcoplasmic reticulum which are not shared by β-adrenergic agonists. These effects thus do not exclude the possibility that cAMP serves as a mediator of the positive inotropic effect of the methylxanthines. Instead, they merely suggest that the classical methylxanthines have cAMP-independent 'side-effects'. In accord with this explanation, the more potent phosphodiesterase inhibitor, 1-methyl-3-isobutylxanthine, which has no direct action on the sarcoplasmic reticulum, increases the force of contraction and decreases the duration of contraction in exactly the same way as do β-adrenergic agonists (Korth, 1978; Scholz, 1984).

In summary, biochemical data are in accord with the view that an inhibition in phosphodiesterase activity with a subsequent increase in cAMP is involved in the positive inotropic effect of theophylline. It is important to note that the inhibition in phosphodiesterase activity actually occurs already at so-called therapeutic concentrations. The present hypothesis is not necessarily contradicted by the fact that theophylline evokes effects opposite to those of β-adrenoceptor stimulating agents because these are readily explained by cAMP- independent supplementary actions of the drug on the sarcoplasmic reticulum.

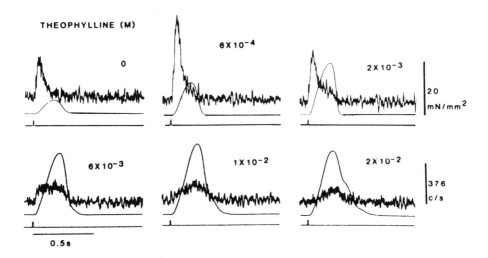

Fig.34.6. Influence of theophylline on aequorin signals (noisy traces) and tension development of a cat papillary muscle. Numbers in individual panels indicate molar concentrations of theophylline. Stimulus interval 4 s. (From Morgan and Blinks, 1982.)

34.6. Sensitization of the contractile proteins to Ca^{2+}

Fabiato and Fabiato (1973, 1976; Fabiato, 1981) reported that the Ca^{2+} sensitivity of skinned single cardiac cells was increased by caffeine (10–20 mM), theophylline (10 mM) and 3-isobutyl-1-methylxanthine (0.1 mM). The potential importance of this effect has also been considered by Morgan and Blinks (1982) who observed that theophylline at concentrations of 2–20 mM in cat papillary muscles produced large positive inotropic effects without consistently increasing the amplitude of the signal of the Ca^{2+}-sensitive bioluminescent protein aequorin (Fig.34.6). The fact that the aequorin signal did not rise was attributed to a decreased release of Ca^{2+} from the sarcoplasmic reticulum, and the positive inotropic effect was explained, at least in part, by an increase in Ca^{2+} sensitivity of the contractile apparatus. Similar effects, that is, an increase in tension with a decrease in the accompanying light transient, were obtained by Allen and Kurihara (1980) with 2–15 mM caffeine in rat and cat papillary muscles, by Allen et al. (1983) with 10 mM caffeine in ferret papillary muscles, and by Hess and Wier (1984) with 2–10 mM caffeine in dog cardiac Purkinje fibres. However, in contrast to these results, Bers and Ellis (1982) reported that 5–10 mM caffeine in sheep heart Purkinje fibres increased the magnitude of both the rise in tension (contracture) and intracellular Ca^{2+} (measured with Ca^{2+}-sensitive microelectrodes) induced by lowering the extracellular Na^+ concentration. Moreover, Herzig, Feile and Rüegg (1981) found that theophylline or caffeine in concentrations up to 10 mM had no effect on the Ca^{2+} sensitiviy of porcine contractile proteins. Finally, in human atrial cardiac muscle theophylline (0.6–10 mM) produced a similarly pronounced increase in the amplitudes of the aequorin and tension signals, where the aequorin signal was prolonged at theophylline concentrations of 6 and 10 mM but remained unchanged or was even shortened at lower concentrations. In summary, this matter remains controversial. Notwithstanding that different results have been observed with different experimental approaches, an increase in Ca^{2+} sensitivity of the contractile proteins consistently has been observed only at theophylline or caffeine concentrations of 2 mM or more. Thus, this effect may conceivably contribute to the cardiotonic action of the methylxanthines, but only at millimolar concentrations.

34.7. Blockade of receptors for endogenous adenosine

It has been suggested that some effects of the methylxanthines might be due to an adenosine-antagonistic action, that is to a blockade of receptors for endogenous adenosine (for example, Fredholm, 1980a, 1980b, Rall, 1980, 1982; Osswald, 1982; Persson, Erjefält and Andersson, 1983). According to this view, theophylline also increases the cellular cAMP content which, however, is not the result of decreased phosphodiesterase activity. Instead, cAMP rises because theophylline antagonizes the adenosine-produced decrease in cAMP. This mechanism is probably of great importance in other systems (in the central nervous system, for example), but it is rather unlikely as a mechanism underlying the positive inotropic effect of theophylline in the heart, as has also been pointed out by Collis, Keddie, and Torr (1984). It is true that adenosine is capable of inhibiting adenylate cyclase activity in homogenates (Brückner et al., 1984), but there is no inhibitory effect at all of adenosine on the cAMP content of *intact* cardiac muscle preparations (Fig.34.7). Adenosine is also ineffective in decreasing myocardial cAMP levels if these have previously been raised by the PDE inhibitor IBMX (Fig.34.8). Finally, the main argument in this context is that the inotropic effects of theophylline and adenosine are opposite only in atria. The cardiotonic effect of theophylline is similar in atrial and ventricular preparations (see Fig.34.1). As shown in Fig.34.9, however, adenosine is negatively inotropic only in the

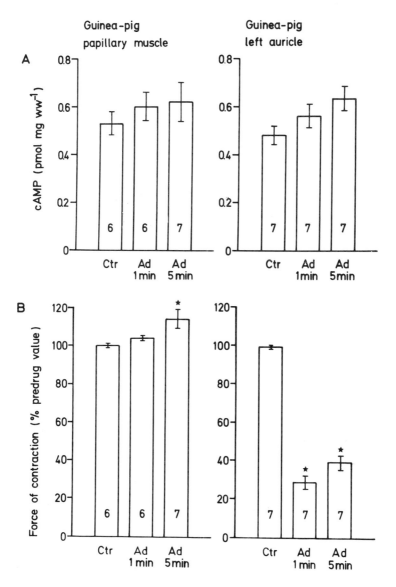

Fig.34.7. Effects of 100 μM adenosine or drug-free bathing solution (Ctr = control) on cAMP content (A) and force of contraction (B) in guinea-pig isolated right ventricular papillary muscles (left) and left auricles (right). The incubation time for adenosine was 1 min and 5 min respectively. The preparations were electrically driven at 1 Hz (papillary muscles) or 3 Hz (auricles). The numbers in the columns give the numbers of experiments. The predrug values for force of contraction were 0.224 ± 0.029 mN mg ww^{-1} (papillary muscles; n = 19) and 0.134 ± 0.009 mN mg ww^{-1} (auricles; n = 21) respectively. Significant differences from controls are marked with asterisks. (Modified from Fenner and Nobis, 1981 and Brückner et al., 1984.)

atrium, but not in ventricular cardiac muscle. In the latter, adenosine does not change force of contraction at concentrations up to 10 µM, and higher concentrations of adenosine even produce a slight positive inotropic effect. It should also be noted that cardiostimulant and adenosine-antagonistic effects of the xanthine derivatives can be separated from each other. Enprofylline (3-propylxanthine), for example, exerts cardiotonic effects at concentrations devoid of an adenosine-antagonistic action (Persson, Erjefält and Andersson, 1983) whereas the opposite (adenosine-antagonistic without cardiostimulant effects) is found with 8-phenyltheophylline (Collis, Keddie and Torr, 1984).

34.8. Relaxing effect of theophylline in smooth muscle

It is widely accepted that the relaxing effects of theophylline and other phosphodiesterase inhibitors on coronary arteries and other smooth muscles are closely related to their cAMP-increasing action. Thus, the theophylline-produced decrease in phosphodiesterase activity and the subsequent increase in cAMP is probably also responsible, at least in part, for the relaxant effects of theophylline in this and other smooth muscle preparations, as has been summarized by Kukovetz, Pöch and Holzmann (1981; see also Polson, Krzanowski and Szentivanyi, 1982). It is in accordance with this view that in smooth muscle, too, the inhibitory effects of the methylxanthines closely resemble those of the β-adrenergic agonists (see Bolton, 1979; Bülbring, Ohashi and Tomita, 1981). The effect occurs without changes of the resting potential or of action potentials which are elicited by electrical stimulation. The relaxant effect is probably ultimately due to a cAMP-produced increase in the sequestration of Ca^{2+} by intracellular stores (especially the sarcoplasmic reticulum) and hence to a decrease in the intracellular free Ca^{2+} concentration. In smooth muscle, too, an adenosine-antagonistic action does not appear to play a major role in the mechanical effects of theophylline (Karlsson, Kjellin and Persson, 1982; Persson, Karlsson and Erjefält, 1982).

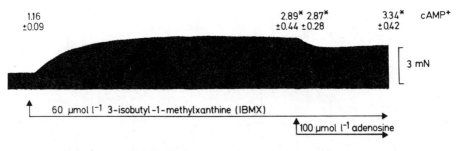

Fig.34.8. Force of contraction (original recording) and cAMP content (pmol mg ww^{-1}; n = number of preparations) in guinea-pig electrically driven (frequency 1 Hz) right ventricular papillary muscles in the presence of 60 µmol l^{-1} 3-isobutyl-1-methylxanthine (IBMX; 0–20 min) and 100 µmol l^{-1} adenosine (15–20 min). IMBX increases force of contraction and cAMP content. Adenosine antagonizes the positive inotropic but not the cAMP-increasing effect of IBMX. (Unpublished experiments of J. Neumann, W. Schmitz and H. Scholz, 1984.)

34.9. Conclusion

The positive inotropic effect of theophylline and the effect of this drug to increase the slow inward current I_{si} have similar time-dependencies and concentration-dependencies. This indicates that both effects are causally related to each other. It appears also likely that these actions result from an inhibition of phosphodiesterase activity with a subsequent increase in myocardial cAMP levels. One current hypothesis to relate the changes in both the Ca^{2+} and the cAMP system to each other suggests that the increase in cAMP may lead to a cAMP-dependent phosphorylation of sarcolemmal proteins and, as a result, to an increase in the number of functional Ca^{2+} channels which, in turn, is due to an increase in the probability that the channels are in an open state when the cell membrane is depolarized (see Reuter, 1983). At concentrations higher than 1 mM, a sensitization of the contractile proteins to Ca^{2+} and a direct, cAMP-independent inhibitory effect on the Ca^{2+} uptake by the sarcoplasmic reticulum may also contribute to the cardiotonic effect of theophylline. In contrast, an adenosine-antagonistic action of theophylline is not likely to play a major role in the positive inotropic effect of this drug.

The theophylline-induced prolongation of the myocardial contraction and

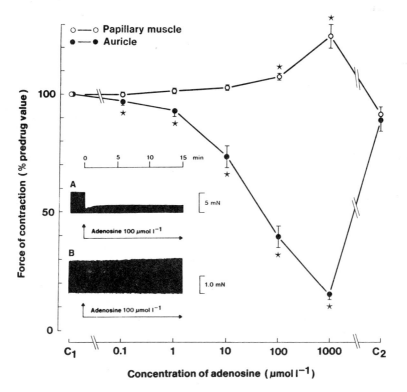

Fig.34.9. Cumulative concentration–response curves for the effects of adenosine (incubation time 5 min for each concentration) on force of contraction in guinea-pig electrically driven left auricles (●; frequency of stimulation 3 Hz) and right ventricular papillary muscles (○; frequency of stimulation 1 Hz). C_1 = predrug control. C_2 = force of contraction after 10 min wash in drug-free buffer. n = 7–8. *$p < 0.05$ versus C_1. (Inset) Original recordings illustrating the inotropic effects of 100 µmol l^{-1} adenosine in a guinea-pig left auricle (upper panel) and a right ventricular papillary muscle (lower panel).

the poor persistence of the cardiotonic effects of theophylline (these effects are opposite to those of the β-adrenergic agonists) do not necessarily contradict the view that cAMP is involved in the positive inotropic action of the drug. They are probably also due to the additional cAMP—independent inhibition of Ca^{2+} uptake by the sarcoplasmic reticulum.

The increase in cAMP levels is probably also responsible for the methylxanthine-produced relaxant effects in smooth muscle.

Acknowledgement

This work was supported by the Deutsche Forschungsgemeinschaft.

References

Allen, D.G., Eisner, D.A., Lab, M.J. and Orchard, C.H. (1983). The effects of low sodium solutions on intracellular calcium concentration and tension in ferret ventricular muscle. *J. Physiol.* **345**, 391—407.

Allen, D.G. and Kurihara, S. (1980). Calcium transients in mammalian ventricular muscle. *Eur. Heart J.* **1 (Suppl. A)**, 5—15.

Bechthold, H. (1971). *Uber die positiv inotrope Wirkung von Theophyllin und Adrenalin an Vorhöfen und Papillarmuskeln des Meerschweinchenherzens bei Normaltemperatur und Hypothermie.* Inaugural-Dissertation, Universität Mainz.

Bers, D.M. and Ellis, D. (1982). Intracellular calcium and sodium activity in sheep heart Purkinje fibres. Effect of changes of external sodium and intracellular pH. *Pflügers Arch.* **393**, 171—8.

Blinks, J.R., Olson, C.B., Jewell, B.R. and Braveny, P. (1972). Influence of caffeine and other methylxanthines on mechanical properties of isolated mammalian heart muscle. Evidence for a dual mechanism of action. *Circ. Res.* **30**, 367—92.

Bolton, T.B. (1979). Mechanisms of action of transmitters and other substances on smooth muscle. *Physiol. Rev.* **59**, 606—718.

Brückner, R., Fenner, A., Meyer, W., Nobis, T.-M., Schmitz, W. and Scholz, H. (1984). Cardiac effects of adenosine and adenosine analogs in guinea-pig atrial and ventricular preparations. Evidence against a role of cAMP and cGMP. *J. Pharmacol. Exp. Ther.* In Press.

Bülbring, E., Ohashi, H. and Tomita, T. (1981). Adrenergic mechanisms. In *Smooth Muscle: an Assessment of Current Knowledge.* E. Bülbring, A.F. Brading, A.W. Jones, T. Tomita (eds.), pp. 219—48. Arnold, London.

Chapman, R.A. and Léoty, C. (1976). The time-dependent and dose-dependent effects of caffeine on the contraction of the ferret heart. *J. Physiol.* **256**, 287—314.

Collis, M.G., Keddie, J.R. and Torr, S.R. (1984). Evidence that the positive inotropic effects of the alkylxanthines are not due to adenosine receptor blockade. *Brit. J. Pharmacol.* **81**, 401—7.

Dönges, C., Heitmann, M., Jungbluth, H., Meinertz, T., Schmelzle, B. and Scholz, H. (1977). Effectiveness of theophylline to increase cyclic AMP levels and force of contraction in electrically paced guinea-pig auricles. Comparison with isoprenaline, calcium and ouabain. *Naunyn-Schmiedeberg's Arch. Pharmacol.* **301**, 87—97.

Endo, M. (1977). Calcium release from the sarcoplasmic reticulum. *Physiol. Rev.* **57**, 71—108.

Fabiato, A. (1981). Effects of cyclic AMP and phosphodiesterase inhibitors on the contractile activation and the Ca^{2+} transient detected with aequorin in skinned cardiac cells from rat and rabbit ventricles. *J. Gen. Physiol.* **78**, 15a–16a.

Fabiato, A. and Fabiato, F. (1973). Activation of skinned cardiac cells. Subcellular effects of cardioactive drugs. *Eur. J. Cardiol.* **1/2**, 143–55.

Fabiato, A. and Fabiato, F. (1976). Techniques of skinned cardiac cells and of isolated cardiac fibers with disrupted sarcolemmas. With reference to the effects of catecholamines and of caffeine. In *Recent Advances in Studies on Cardiac Structure and Metabolism, Vol. 9, The Sarcolemma.* P.-E. Roy and N.S. Dhalla (eds), pp. 71–94. University Park Press, Baltimore.

Farah, A.E., Alousi, A.A. and Schwarz, R.P. (1984). Positive inotropic agents. *Ann. Rev. Pharmacol. Toxicol.* **24**, 275–328.

Fenner, A. and Nobis, T.-M. (1981). *Uber die Wirkung von Adenosin, N^6-Phenylisopropyl-adenosin und 2',5'-Dideoxy-adenosin auf die Kontraktionskraft und den cAMP- und cGMP-Gehalt von Meerschweinchenherzen.* Inaugural-dissertation, Medizinische Hochschule, Hannover.

Fredholm, B.B. (1980a). Are methylxanthine effects due to antagonism of endogenous adenosine? *Trends in Pharmacol. Sci.* **1**, 129–32.

Fredholm, B.B. (1980b). Theophylline actions on adenosine receptors. *Eur. J. Respir. Dis.* **61 (Suppl. 109)**, 29–36.

Herzig, J.W., Feile, K. and Rüegg, J.C. (1981). Activating effects of AR-L 115 BS on the Ca^{2+} sensitive force, stiffness and unloaded shortening velocity (V_{max}) in isolated contractile structures from mammalian cardiac muscle. *Arzneim. Forsch. (Drug Res.)* **31**, 188–91.

Hess, P. and Wier, W.G. (1984). Excitation–contraction coupling in cardiac Purkinje fibers. Effects of caffeine on the intracellular (Ca^{2+}) transient, membrane currents, and contraction. *J. Gen. Physiol.* **83**, 417–33.

Karlsson, J.-A., Kjellin, G. and Persson, C.G.A. (1982). Effects on tracheal smooth muscle of adenosine and methylxanthines, and their interaction. *J. Pharm. Pharmacol.* **34**, 788–93.

Korth, M. (1978). Effects of several phosphodiesterase-inhibitors on guinea-pig myocardium. *Naunyn-Schmiedeberg's Arch. Pharmacol.* **302**, 77–86.

Kukovetz, W.R., Pöch, G. and Holzmann, S. (1981). Cyclic nucleotides and relaxation of vascular smooth muscle. In *Vasodilatation.* P.M. Vanhoutte and I. Leusen (eds), pp. 339–53. Raven, New York.

Morgan, J.P. and Blinks, J.R. (1982). Intracellular Ca^{2+} transients in the cat papillary muscle. *Can. J. Physiol. Pharmacol.* **60**, 524–8.

Morgan, J.P., Chesebro, J.H., Pluth, J.R., Puga, F.J. and Schaff, H.V. (1984). Intracellular calcium transients in human working myocardium as detected with aequorin. *J. Am. Coll. Cardiol.* **3**, 410–8.

Osswald, H. (1982). Die Theophyllinwirkung in therapeutischen Dosen als Antagonismus zum endogenen Adenosin. In *Theophylline and Other Methylxanthines.* N. Rietbrock, B.G. Woodcock and A.H. Staib (eds), pp. 107–15. Vieweg, Braunschweig and Wiesbaden.

Persson, C.G.A., Karlsson, J.-A. and Erjefält, I. (1982). Differentiation between bronchodilatation and universal adenosine antagonism among xanthine derivatives. *Life Sci.* **30**, 2181–9.

Persson, C.G.A., Erjefält, I. and Andersson, K.E. (1983). Positive inotropic and chronotropic effects and coronary vasodilation *in vitro* by two antiasthmatic xanthines with different abilities to antagonize adenosine. *J. Cardiovasc. Pharmacol.* **5**, 778–85.

Polson, J.B., Krzanowski, J.J. and Szentivanyi, A. (1982). Inhibition of a high affinity cyclic AMP phosphodiesterase and relaxation of canine tracheal smooth muscle. *Biochem. Pharmacol.* **31**, 3403–6.

Rall, T.W. (1980). Central nervous system stimulants. The xanthines. In *Goodman and Gilman's The Pharmacological Basis of Therapeutics*. A.G. Gilman, L.S. Goodman and A. Gilman (eds), pp. 592–607. MacMillan, New York.

Rall, T.W. (1982). Evolution of the mechanism of action of methylxanthines: from calcium mobilizers to antagonists of adenosine receptors. *Pharmacologist* **24**, 277–87.

Reuter, H. (1983). Calcium channel modulation by neurotransmitters, enzymes and drugs. *Nature* **301**, 569–74.

Reuter, H. and Scholz, H. (1977). A study of the ion selectivity and the kinetic properties of the calcium dependent slow inward current in mammalian cardiac muscle. *J. Physiol. (Lond.)* **264**, 17–47

Schmitz, W., Hackbarth, I., Scholz, H. and Wetzel, E. (1980). Effects of vanadate on the cAMP system of the heart. *Basic Res. Cardiol.* **75**, 438–43.

Scholz, H. (1980). Effects of beta- and alpha-adrenoceptor activators and adrenergic transmitter releasing agents on the mechanical activity of the heart. In *Handbook of Experimental Pharmacology, Vol. 54/I, Adrenergic Activators and Inhibitors, Part I*. L. Szekeres (ed.), pp. 651–733. Springer, Berlin, Heidelberg and New York.

Scholz, H. (1983). Pharmacological actions of various inotropic agents. *Eur. Heart J.* **3 (Suppl. A)**, 161–72.

Scholz, H. (1984). Inotropic drugs and their mechanisms of action. *J. Am. Coll. Cardiol.* **4**, 389–97.

Scholz, H. and Reuter, H. (1976). Effect of theophylline on membrane currents in mammalian cardiac muscle. *Naunyn-Schmiedeberg's Arch. Pharmacol.* **293 (Suppl.)**, R 19.

Sobel, B.E. and Mayer, S.E. (1973). Cyclic adenosine monophosphate and cardiac contractility. *Circ. Res.* **32**, 407–14.

Tsien, R.W. (1977). Cyclic AMP and contractile activity in heart. *Adv. Cycl. Nucleotide Res.* **8**, 363–420.

MOLECULAR MECHANISM OF ANTIARRHYTHMIC ACTION OF DIPHENYLHYDANTOIN (DPH): ITS EFFECT ON Ca^{2+}-ATPASE ACTIVITY

H.M. Rhee

Department of Pharmacology, Oral Roberts University, School of Medicine, Tulsa, Oklahoma 74171, USA

and S. Dutta

Department of Pharmacology, Wayne State University, School of Medicine, Detroit, Michigan, 48201, USA

Abstract

Calcium ion is essential for cardiac biochemical and electrophysiological functions. Especially, slow inward current plays an important role in the sinoatrial node under certain pathological conditions such as cardiac arrhythmias. To understand the molecular mechanism of several typical antiarrhythmic agents, DPH (8 mg/kg), DMP (4 mg/kg) and KCl (0.6 mEq/kg.h) including quinidine were used to counteract ouabain-induced cardiac arrhythmias with varying degree of success. Upon confirming the effectiveness of the antiarrhythmic agents, cardiac membrane vesicles were prepared. $^{45}Ca^{2+}$ transport and Ca^{2+}-dependent ATPase activity were determined *in vitro* to correlate their antiarrhythmic efficacy observed *in vivo*. Most drugs decreased significantly (up to 40%) both Ca^{2+} transport and Ca^{2+}-ATPase activities, although it required relatively large dose *in vitro*. These drugs also inhibited active uptake of monovalent cation, as demonstrated using $^{86}Rb^{+}$ tracer technique. The inhibitory effect of the drugs on active uptake of monovalent cation was more pronounced in the sinoatrial node or Purkinje fibre than the contracting ventricular muscle. We concluded that the effectiveness of the antiarrhythmic agents is related to their ability to modulate Ca^{2+} flux directly and indirectly via their effect on the monovalent cation movement in specialized cardiac tissues.

35.1. Introduction

Diphenylhydantoin (DPH) is a very effective antiarrhythmic agent agiainst certain types of cardiac arrhythmias such as digitalis-induced arrhythmias (Bigger, Bassett and Hoffman, 1968; Chai, Lee and Wang, 1976; Damato, 1969; Helfant, Scherlag and Damato, 1967). This drug decreases membrane excitability and reduces Na$^+$ influx during the phase four diastolic period, which may be the electrophysiological antiarrhythmic mechanism of the drug. DPH also increases membrane responsiveness and conduction velocity under certain experimental conditions (Hoffman, 1966; Rosati et al., 1967; Smith and Haber, 1971; Sasyniuk and Ogilvie, 1975), which may be beneficial if the conduction velocity is reduced for some reasons. The biochemical basis of the antiarrhythmic action of DPH is poorly understood, although both electrophysiological and ionic aspects of DPH action are well documented. Therefore, the primary objectives of this research was to examine the effect of DPH on Ca^{2+}-ATPase activity and Ca^{2+}-transport in cardiac membrane vesicles.

35.2. Materials and methods

Intact, open-chest dog hearts were prepared and physiological parameters were recorded as described (Rhee, Dutta and Marks, 1976; Rhee, Huang and Askari, 1981). Cardiac arrhythmias were induced by an injection of ouabain (40 μg/kg, i.v.) and the arrhythmias were maintained by a continuous infusion of ouabain at a rate of 72 ng/kg/min. In another group of dogs a treatment with DPH (8 mg/kg, i.v.) was initated after a confirmation of the persistent cardiac arrhythmias. Antiarrhythmic action of DPH was maintained by a multiple injection of the drug (1 mg/kg, i.v.) every 10 min.

For the study of ouabain binding, ^{3}H-ouabain was treated as described (Rhee, Dutta and Marks, 1976) and determined ^{3}H-ouabain binding in either cardiac homogenate or in the partially purified Na$^+$, K$^+$-ATPase (Rhee, Dutta and Marks, 1976). The plasma concentration of ouabain was also determined at 30 min intervals. Protein was determined by the procedure of Lowry et al. (1951).

Na$^+$, K$^+$-ATPase was partially purified as described (Rhee, Dutta and Marks, 1976) and determined in control, ouabain-treated and ouabain-plus DPH-treated groups. Inorganic phosphate released was determined according to Fiske and Subbarow (1925). Cardiac membrane vesicles were also prepared according to the method of Jones, Phan and Besch (1978), and Ca^{2+}-ATPase activity and Ca^{2+}-uptake in the vesicles were compared in both ouabain-treated and ouabain-plus DPH treated groups (Jones, Phan and Besch, 1978).

35.3. Results and discussion

Figure 35.1 illustrates a typical electrocardiographic effect of ouabain and ouabain-plus DPH treatment. Ouabain was injected intravenously at time O after recording of other cardiac parameters, such as blood pressure, intraventricular pressure and its first derivative including end-diastolic pressure (data not shown). In this typical experiment ouabain induced ventricular tachycardia in 37 min after ouabain treatment. The persistency of the arrhythmias was observed several hours when the treatment with DPH was initated at 180 min. Within 6 the repetitive ventricular arrhythmias were converted into normal sinus rhythm. Multiple injection of DPH effectively eliminated the cardiac arrhythmias even though ouabain was continuously infused for a total of 5h. At this time, plasma concentration of oubain was no different in the two groups of dog, as summarized in Table 35.1.

In an attempt to understand the mechanism of the antiarrhythmic action of

DPH potential differential binding of ouabain to cardiac homogenate and Na$^+$, K$^+$-ATPase was determined in the two groups. In five dog experiments ouabain content in cardiac homogenate in the ouabain-treated group was no different from the ouabain-plus DPH-treated group (3.6 ± 0.4 versus 3.6 ± 0.1 pmol/mg protein). Na$^+$, K$^+$-ATPase (E.C. 3.6.1.3) have been considered as putative receptors for cardiac glyosides (Lee and Klaus, 1971; Schwartz, Lindenmayer and Allen, 1975). Cardiac glycoside such as ouabain-induced toxicity may be indeed due to the inhibition of the enzyme, although the mechanism of positive inotropic action of cardiac steroids is still controversial (Rhee, Dutta and Marks, 1976; Rhee, Huang and Askari, 1981; Okita, Richardson and Roth-Schecter, 1973; Noble, 1980). Ouabain inhibited the enzyme activity from the control 9.6 ± 0.3 µmol Pi released/mg protein/h to 6.0 ± 0.6. However, DPH treatment did not increase the inhibited Na$^+$, K$^+$-ATPase activity (see Table 35.1), although it restored the cardiac arrhythmias to the normal sinus rhythm (see Fig.35.1). This result basically confirms our previous finding (Rhee, 1983a), which is internally consistent with the above data that DPH was not able to displace ouabain from either the homogenate or Na$^+$, K$^+$-ATPase preparation. This indicates quite clearly the antiarrhythmic action of DPH is not mediated by the reactivation of Na$^+$, K$^+$-ATPase, which was inhibited by a high dose of ouabain (Rhee, 1983a).

Rhythmic cardiac excitation–contraction coupling depends not only on Ca^{2+}, but also K$^+$ and Na$^+$. Ca^{2+}, in particularly, is essential for cardiac

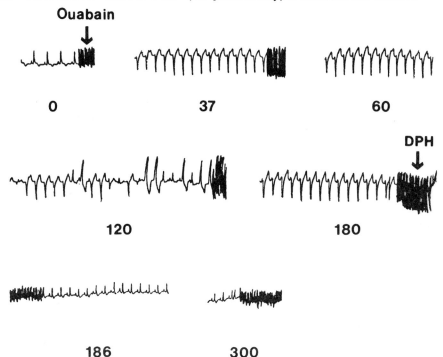

Fig.35.1. Electrocardiogram of intact, open-chest dog hearts after ouabain alone or after ouabain plus diphenylhydantoin (DPH) treatment. Ouabain (40 µg/kg, i.v.) was given at time 0, when ouabain infusion (72 ng/kg/min) was also begun to maintain the plasma level of ouabain. After an induction of cardiac arrhythmias DPH (8 mg/kg, i.v.) was given at time 180 min, which restored the normal sinus rhythm within 6 min. Numbers below each tracing represent time in minutes after an initial ouabain treatment.

Table 35.1. Comparison of several parameters in ouabain-treated dogs and dogs treated with ouabain plus diphenylhydantoin (DPH). See the method for the dose of ouabain and DPH. For each group at least five dogs were used; each value represents mean ± SE.

Group	Plasma ouabain (pmol/ml)	Ouabain binding to homogenate (pmol/mg protein)	Ouabain binding to Na^+,K^+-ATPase (pmol/mg protein)	Na^+,K^+-ATPase activity (μmol/mg/h)
Control	--	--	--	9.6 ± 0.3
Ouabain	28 ± 5.1	3.6 ± 0.4	13.6 ± 1.0	6.0 ± 0.6
Ouabain +DPH	23 ± 1.7	36 ± 0.1	12.4 ± 1.8	7.2 ± 0.5

contraction and this ion plays an important role in depolarization of certain tissues in pathological condition (Cranefield, 1975). Thus, the effect of DPH on Ca^{2+}-ATPase activity and cardiac Ca^{2+}-uptake was investigated in cardiac membrane vesicles of dog heart. The membrane vesicles represent primary sarcoplasmic reticulum (Jones, Phan and Besch, 1978). However, when they are exposed to detergent such as sodium dodecylsulphate, the latent Na^+, K^+-ATPase activity is manifested, which has been considered as sarcolemmal membrane (Jones, Phan and Besch, 1978). In the preparation DPH had no effect on Mg^{2+}-dependent ATP hydrolysis (Fig.35.2), but it inhibited Ca^{2+}-dependent ATPase activity in the DPH dose dependently. The inhibition was significant with a high dose of DPH such as 5×10^{-4} M. DPH also inhibited Ca^{2+} transport in the membrane vesicles (data not shown), which may be consistent with the inhibition of Ca^{2+}-ATPase.

Many Ca^{2+}-channel blockers such as verapamil and diltizem have potent antiarrhythmic properties, although the exact molecular mechanism of action may not be clear. However, if the agents block or reduce the so-called slow calcium channel, then there may be a net decrease in cytosolic Ca^{2+} level. In this study DPH reduced Ca^{2+} as reported (Rhee, 1983b), which indicates that there may be an actual increase in cytosolic calcium level. This suggests that a specific local pool of Ca^{2+} in certain cardiac tissues may be important for the genesis of cardiac arrhythmias and for the effectiveness of antiarrhythmic agents such as DPH.

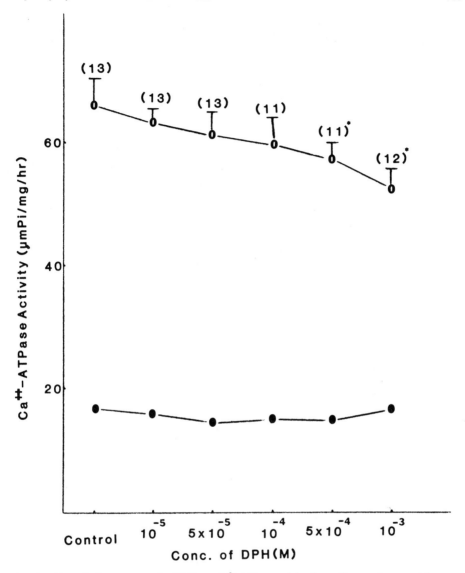

Fig.35.2. Effect of diphenylhydantoin (DPH) on Ca^{2+}-ATPase activity in cardiac membrane vesicles prepared from dog hearts. Canine left ventricle was homogenized and centrifuged to prepare the membrane vesicles according to Jones, Phan and Besch (1967). ATP hydrolysis by the vesicles was carried out without DPH (control) or in the presence of the indicated dose of DPH for 10 min at 37°C. ATPase activity assayed in the presence of Mg^{2+} alone (●--●) or in the presence of 50 μm Ca^{2+} (o--o) was determined. Numbers in parentheses indicate number of assays performed in at least five animals and vertical bars indicate standard error.

References

Bigger, J.T. Jr., Bassett, A.L. and Hoffman, B.F. (1968). Electrophysiological effect of diphenylhydantoin on canine Purkinje fibers. *Circ. Res.* 21, 221–36.

Chai, C.Y., Lee, T.M. and Wang, S.C. (1976). Effects of diphenylhydantoin on cardiac arrhythmias induced by carotid occlusion in the cat. *Arch. Int.*

Pharmacodyn. **219**, 180–92.
Cranefield, P.F. (1975). The conduction of cardiac impulse: the slow response end cardiac arrhythmias. Futura Publishing Co., Mount Kisco, New York.
Damato, A.N. (1969). Diphenylhydantoin: pharmacological and clinical use. *Prog. Cardiovasc. Dis.* **12**, 1–15.
Fiske, C.H. and Subba Row, Y. (1925). The colorimetric determination of phosphorus. *J. Biol. Chem.* **66**, 375–400.
Helfant, R.H., Scherlag, B.J. and Damato, A.N. (1967). The electrophysiological properties of diphenylhydantoin sodium as compared to procaine amide in the normal and digitalis-intoxicated heart. *Circulation* **36**, 108–18.
Hoffman, B.F. (1966). The genesis of cardiac arrhythmias. *Prog. Cardiovasc. Dis.* **8**, 319–29.
Jones, L.R., Phan, S.H. and Besch, Jr., H.R. (1978). Gel electrophoretic and density gradient analysis of the (K^++Ca^{2+})-ATPase and the (Na^++K^+)-ATPase activites of cardiac membrane vesicles. *Biochem. Biophys. Acta* **514**, 294–309.
Lee, K.S. and Klaus, W. (1971). The subcellular basis for the mechanism of inotropic action of cardiac glycosides. *Pharmacol. Rev.* **23**, 193–261.
Lowry, O.H., Rosebrough, N.J., Farr, A.L. and Randall, R.J. (1951). Protein measurement with the Folin phenol reagent. *J. Biol. Chem.* **193**, 265–75.
Noble, D. (1980). Review: mechanism of action of therapeutic levels of cardiac glycosides. *Cardiovasc. Res.* **14**, 495–514.
Okita, G.T., Richardson, F. and Roth-Schecter, B.F. (1973). Dissociation of the positive inotropic action of digitalis from inhibition of sodium- and potassium-activated adenosine triphosphatase. *J. Pharmacol. Exp. Ther.* **185**, 1–11.
Rhee, H.M., Dutta, S. and Marks, B.H. (1976). Cardiac Na^+, K^+-ATPase activity during positive inotropic and toxic actions of ouabain. *Eur. J. Pharmacol.* **37**, 141–53.
Rhee, H.M., Huang, W. and Askari, A. (1981). Relationship between the positive inotropic effect of ouabain and its inhibitory effects on Na^+, K^+-ATPase and active transport of Rb^+ in the dog heart. *Eur. J. Pharmacol.* **70**, 273–8.
Rhee, H.M. (1983a). Evidence against an involvement of Na^+, K^+-ATPase in antiarrhythmic mechanism of phenytoin. *Naunyn-Schmiedeberg's Arch. Pharmacol.* **332**, 78–84.
Rhee, H.M. (1983b). Effects of some antiarrhythmic agents on $^{45}Ca^{++}$ transport in dog heart membrane vesicles and $^{86}Rb^+$ transport in specialized cardiac tissue. In *Calcium-binding Protein*. Bernard et al. (ed), pp. 293–5. Elsevier Science Publishers.
Rosati, R.A., Alexander, J.A., Schaal, S.F. and Wallace, A.G. (1967). Influence of diphenylhydantoin on electrophysiological properties of the canine heart. *Circ. Res.* **21**, 757–65.
Sasyniuk, B.I. and Ogilvie, R.I. (1975). Antiarrhythmic drugs: electrophysiological and pharmacokinetic considerations. *Ann. Rev. Pharmacol.* **15**, 131–55.
Schwartz, A., Lindenmayer, G.E. and Allen, J.C. (1975). The sodium and potassium adenosine triphosphatase: pharmacological, physiological and biochemical aspects. *Pharmacol. Rev.* **27**, 3–134.
Smith, T.W. and Haber, E. (1971). The clinical value of serum glycoside concentration in the evaluation of drug toxicity. *Ann. NY. Acad. Sci.* **179**, 322–37.

Index

A23187 (calcium ionophore), 72, 139, 143–5, 155, 182, 247, 312, 314, 315, 319, 328
acetylcholine, 350, 352
acetylstrophanthidin, 235, 243–5
acidosis, 349, 350
aconitine, 291–3
actin, 213, 214
actin–myosin complex, 221
action potential, 235–44
  effect of nickel on, 238
actomyosin, 180, 417
  ATPase, 355, 356, 361, 417
adenosine receptors, blockade by methylxanthines, 454–6
adenylate cyclase, 128, 129, 145
α-adrenergic agents, 67, 75
β-adrenergic agents, 30, 31, 67, 74, 75
aequorin, 7, 8, 58, 214, 218–21, 435
  derivatives of, 8
aerobic metabolism, 349, 350
agonist-induced contractions, antagonism of, 396–9
aldosterone secretion, 143, 146
alkali cations, 100
amiloride, 36
amrinone, 386
anaemia
  anaplastic, 312
  sickle cell, 327–32
    cell dehydration in, 327
    membrane rigidity of red cells in, 327
    state of calcium in, 327–32
anaerobic
  glycolysis, 350
  metabolism, 349
angiotensin II, 143, 146, 319
anticalcium drugs, effect of
  on prevention of plaque formation, 257–73
  on reversal of plaques, 273–8
antipyrylazo III, 3, 4, 10, 235, 236, 239–41
arachidonic acid, 143, 165–7, 171–3
arrhythmias, cardiac
  and calcium channel blockers, 304–6, 464
  effect of diphenylhydantoin on, 461–3
  and intracellular calcium, 291–306
  ouabain-induced, 461–3
  pathophysiological mechanisms of, 291, 295–304
arsenazo III, 3, 4, 10
atherogenesis

cellular events of, 256
extracellular events of, 256
primary stimuli of, 256
role of calcium in, 255–7
atherosclerosis
effect of calcium antagonists, 253–78
auxin, 248, 251
awake episodes, in malignant hyperthermia, 352

BAPTA, 8–10
  derivatives of, 8–10
barbiturates, 352
barium (Ba^{2+}), binding of, 100, 101
BAY K 8644, 31
benz-2, 329
benzothiazepines, 388
bepridil, 399
bidentate chelator, 98
bioluminescence, 6–8
biperiden, 388
blood
  diseases, 311–21
  pH, 349
bone marrow cell differentiation
  effect of
    butyrate on, 315, 316
    calcium on, 315
    dimethylsulphoxide on, 315–17
    erythropoietin on, 314, 316
    phorbol esters on, 316–18
    retinoic acid on, 316, 317
    vitamin D$_3$ on, 316, 317
bone marrow cells
  calcium regulation of, 311–21
  defects in regulation of, 319–21
  burst-forming unit, 312, 313, 317
  burst-promoting activity, 312
butyrate, effect of, on cell differentiation, 315, 316

cadmium, (Cd^{2+}), binding of, 100, 101
caffeine, 84–6, 88, 90, 301, 349, 353, 355, 360, 454
depolarization, metabolic effects of, 356–8
calcification, 253, 256, 273, 278
calcineurin, 122–4, 127–31
calciotriol, 343
calcitonin, effect of
  on calcium efflux, 374
  on insulin secretion, 373, 374
calcium (Ca^{2+})
  binding of, 95–103

to calmodulin, 125–8
to troponin complex, 360
–calmodulin complex, 125, 318, 405–7
–chlortetracycline complex, 6, 59
-dependent
  enzymes, 380
  proteases, 342
determination of intracellular, 1–11
electrodes selective for, 1, 2, 10, 239
effect of
  on calmodulin, 125–8
  on cell differentiation, 315
  on chromosome replication, 179–98
  on conformation of membranes, 105–19
  on fusion of membranes, 105, 111–17
  in G_1 phase, 184–91
  on insulin biosynthesis, 368, 369
  on insulin secretion, 368, 369
  on lateral phase separation, 105, 107–11, 113
  on mitosis, 57–61
  on phase transition, 105–7
  in pre-G_1 stages, 191, 192
  on structure of biological membranes, 105, 117–19
extracellular, and cardiac glycosides, 436, 437
in G_1 phase, 184–91
intracellular
  and cardiac arrhythmias, 291–306
  effect of cardiac glycosides on, 433–42
photoprotein activated by, 1, 6–8
in plants, 247–51
in pre-G_1 stages, 191, 192
role of, in hypertension, 283–90
-sensitive potassium permeability, 328, 329
in sickle cell anaemia red cells, 327–32
slow inward current of, 434
  effect of methylxanthines on, 451–3, 457
and thyroid hormone action, 355–63
tolerant cells, 17
calcium absorption, intestinal, effect of diabetes on, 373
calcium accumulation, muscle, in muscular dystrophy, 335–44
calcium activation, in skeletal muscle, 213–23
  effect of
    fibre length, 214–17
    multiple cross-bridge states, 222, 223
calcium antagonists, 229, 235–9, 247, 250, 251, 302, 306
classification of, 388, 399, 400
definition of, 405, 406
effect of
  on atherosclerosis, 253–78
  on formation of plaques, 257–73
  on muscle contraction, 388–90
  on muscular dystrophy, 335–8, 342
  on reversal of plaques, 273–8
indications for, 390
photoinactivation of, 237–9
therapeutic implications of, 393–401
calcium ATPase, 95, 100, 101, 355, 407–15, 418, 419 (see also calcium pump)
anionic amphiphile-activated, 409–11
basal, 411, 412
calmodulin-dependent, 121, 125, 128, 129, 231, 232
effect of
  calmodulin antagonists on, 407–15, 418, 419
  diphenylhydantoin on, 461–5
  lanthanum on, 232
of endoplasmic reticulum, 60–2, 154–62, 227, 231, 232
in pancreatic islets, 371, 372
in plants, 248–51
of plasma membrane, 58, 140, 227–32
in polycythaemia vera erythrocytes, 320, 321
proteolysis-activated, 409–11
of sarcoplasmic reticulum, 236, 355, 360
in sickled cells, 329, 330
in vascular smooth muscle, 289
calcium binding
  to calmodulin, 125–8
  effect of fibre length change, 218–21
  to troponin, 218
calcium-binding proteins, 121–31, 380–2, 405
  homology of, 122–4
  interaction of
    calmodulin antagonists, 413
    with phenothiazines, 126, 127
  sequence of, 122, 124
  vitamin D-dependent, 123, 127
calcium–calcium exchange, 39
calcium channel, 35, 140, 235–9, 244, 247, 250, 251, 315, 388–90
activation
  time course of, 18–21
  kinetics of, 32–5
agonist, 31

Index

photoinactivation of, 244
in cardiac cells, 15–31, 291–3
conductance properties, 23–5
effect of La^{3+} on, 228
gating
  kinetics of, 21–3
  voltage dependence of, 16
inactivation kinetics of, 25–9
phosphorylation, 30
  by β-adrenergic agents, 30, 31
in plasma membrane, 229
receptor-operated, 145, 180, 227, 229, 257, 318, 389, 397, 398
  antagonism of, 387, 398
voltage-operated, 180, 227, 229, 286, 292, 295, 306, 369, 370, 394–7
  antagonism of, 394, 395
calcium channel blocker, 30, 31, 244, 288, 304–6, 405, 417
  antiarrhythmic effects of, 464
  classification of, 388, 399, 400
  therapeutic implications of, 393–401
calcium chelator, 291, 329, 405, 406
  fluorescent, 4–6, 10
  nuclear magnetic resonance by, 1, 8–10
calcium concentration
  in acinar cells, 154–62
  intracellular free, 1–11, 15, 47–9, 140–6, 151
    regulation of, 57, 58, 151–62
  in mitochondria, 67–76
calcium content
  muscle, in muscular dystrophy, 335–41
    effect of age on, 336, 338, 341, 344
  sickled cell, 327–32
calcium current, 16–31
  in heart muscle, 235–45, 434
  effect of methylxanthines on, 451–3, 457
calcium cycling
  induced by thyroid hormones, 355
  Na^+-dependent, 68–72, 76
    control of, 74–6
  Na^+-independent, 68–71
calcium electrode, 1, 2, 10, 152, 239
calcium efflux, 360
  effect of
    calcitonin on, 374
    glucose on, 369–71
    thyroid hormone on, 362
  from mitochondrial matrix, 67, 69, 362
  Na_o-dependent, 37–9
  Na^+-independent, 67
calcium entry, 207–9, 288, 289, 304, 359, 393–401
calcium entry blocker, 253–78, 388–90, 405, 406
  classification of, 388, 399, 400
  effect of endothelium on, 400, 401
  therapeutic implications of, 393–401
calcium fluxes
  excitation-dependent, 435
  transmembrane, 434, 435
calcium homeostasis, 253
  control of, by thyroid hormones, 355–63
  in polycythaemia vera, 311
calcium indicator, 125, 209, 235
  fluorescent, 1, 4–6, 10
  metallochromic dye as, 1–4, 10, 236, 239
calcium influx
  agonist-induced, 396–9
  calciotriol-stimulated, 343
  into cardiac cells, 15, 16, 239–44, 292, 295
  catecholamine-stimulated, 343
  effect of thyroid hormone on, 362
  into mitochondrial matrix, 67–75, 362
  Na_i-dependent, 37–9
  parathyroid hormone-stimulated, 335, 336, 342–4
calcium leak, sickling induced, 327
calcium messenger system, 121, 128, 139–46, 248, 251, 380
  gain control in, 139, 141–5
calcium metabolism, regulation of, 151–62, 336
calcium mobilization, 203–9
  effect of thyroid hormone on, 355, 361
calcium mobilizing agents, 355, 358
calcium overload, cellular, 342–4
calcium permeability, SR membrane induced by thyroid hormone, 360
calcium pump, 35, 179, 180, 292, 294, 319, 342, 343, 350 see also Calcium ATPase
  of erythrocytes, 49–52
  biochemistry of, 52
  effect of calmodulin on, 51
  function of, 49–51
  lipid requirement of, 51
  of plasma membrane, 47–53
  effect of calmodulin on, 141, 144, 145, 319
  in polycythaemia vera erythrocytes, 320, 321
  reaction cycle of, 48–51
  of sarcoplasmic reticulum (SR), 47–51
  in sickled cells, 329

calcium regulation
　of bone marrow cells, 311–21
　by calcium cycling, 71–4
　in mitochondria, 67–76
calcium release, 216, 236, 302
　agonist-dependent, 230
　calcium-induced, 84–6, 88, 301
　depolarization-induced, 79–84
　in diseased muscle, 88–90
　drug-induced, 84–6
　effect of
　　magnesium on, 425
　　thyroid hormone on, 355–61
　from endoplasmic reticulum, 203, 208, 227–30
　by inositol triphosphate, 143–5, 151, 157–62, 177, 203, 208
　from intracellular stores, 15
　from mitochondrial matrix, 68, 294, 362
　molecular components involved in, 86–8
　Na^+-independent, 68, 70
　from sarcoplasmic reticulum, 79–90, 219, 294, 349–53, 355–61
calcium-release inhibitors, 79, 88, 405, 406
calcium requirement for insulin secretion, 367–75
calcium sequestration
　by endoplasmic reticulum (ER), 57–62
calcium signal, 240
calcium store, 292, 294
　agonist-sensitive, 228–30
calcium transients, 1–11, 139–46, 219, 220
　cardiac, 434, 435
　in malignant hyperthermia, 349–53
　in polycythaemia vera erythrocytes, 311, 321
calcium transport, 350
　effect of
　　diphenylhydantoin on, 461, 465
　　thyroid hormone on, 355–63
　efficiency of, 360
　by endoplasmic reticulum (ER), 57–62
　mechanisms in heart muscle, 340–45
　mechanisms in smooth muscle, 230–2
　by mitochondria, 48, 67–76
　in plants, 248–51
calcium uniporter, 67–75
　activation by α-adrenergic agonists, 67, 75
calcium uptake
　by acinar cells, 152–6
　effect of methylxanthines on SR, 450, 451
caldesmon, 129, 384
caligulin, 123
calmidazolium (R24571), 419
　effect of
　　on calcium ATPase, 407–12
　　on cell proliferation, 186
　　on phosphodiesterase, 407, 408
　hydrophobicity of, 409
　specificity of, 410–12, 419
calmodulin, 103, 121–31, 209, 318, 380–6
　basic protein binding to, 127
　calcium binding to, 125–8
　-dependent enzymes, 128–31, 318, 319
　hysteresis in, 141, 142
　modulation of, 141–44
　in plants, 249, 250
　drug interaction, 126, 389, 405–19
　effect of
　　on chromosome replication, 179–98
　　in G_1 phase, 184–91
　　on insulin secretion, 367, 372, 373
　　in pre-G_1 stages, 191, 192
　-independent processes, 411–13
　in leukaemic cells, 320
　pharmacological implications of, 415–19
　pharmacological regulation of the activity of, 405–19
　in plants, 247–51
　in transformed cells, 320
calmodulin antagonists (see also specific compounds)
　biochemical aspects of, 406–15
　coefficient of specificity, 411, 412
　effect of
　　on anaphylactoid reactions, 418
　　on calcium ATPase, 407–15, 418, 419
　　on calcium-binding proteins, 413
　　on calcium uptake into mitochondria, 411
　　on calmodulin-independent processes, 411–13
　　on cell proliferation, 417
　　on histamine release, 417, 418
　　on magnesium ATPase, 413
　　on myosin light chain phosphorylation, 418
　　on phosphodiesterase, 406–19
　　on platelet function, 418
　　on smooth muscle function, 417
　　on sodium–potassium ATPase, 411–13
　hydrophobicity of, 407–9
　mechanism of action of, 413–15
　pharmacokinetics of, 418, 419

potency of, 406–9
specificity of, 409–13
therapeutical aspects of, 415–19
calmodulin-binding proteins, 128–31
calmodulin blockers, 184–6
calsequestrin, 121
carbachol, 352
carboxyl groups, 95
cardiac glycoside, 244, 298, 302–3
  and intracellular calcium, 433–42
  effect of, on muscle contraction, 386–8
  effects of low concentrations, 441, 442
  positive inotropy of, 387, 436–42
    and binding to sodium–potassium ATPase, 437, 438
    dependence on extracellular calcium, 436, 437
    and sodium–calcium exchange, 438–40
cardiac muscle, effect of cardiac glycosides on, 433–42
cardiomyopathia, 336
cation channel, calcium-sensitive, 295, 297, 299
CBP-18, 123
cell deformability, 328
cell dehydration, 327–9
cell differentiation, bone marrow
  effect of
    butyrate on, 315, 316
    calcium on, 315
    dimethylsulphoxide on, 315–17
    erythropoietin on, 314, 316
    phorbol esters on, 316–18
    retinoic acid on, 316, 317
    vitamin D_3 on, 316, 317
cell division
  effect of
    calcium on, 179–92
    calmodulin on, 179–91
    cAMP on, 192–8
    cAMP-dependent protein kinase, 192–8
cell filtrability, 328
cell proliferation, 416, 417
chelator, 95, 98, 99
chlorpromazine
  effect of
    on calcium ATPase, 407, 408, 411, 412
    on phosphodiesterase, 407, 408
  specificity of, 412
chlortetracycline, 1, 6
  complex with calcium, 6, 59
chromosome replication
  control of, 179–98

cinnarizine, 395, 397, 399, 415–17
circus movement (re-entry), 291, 302–4
clonidine, 400
C-kinase, 139–46, 203, 208, 209 (see also protein kinase C)
  modulation of, 144, 145
  activation of, 208, 209, 318
    by angiotensin II, 319
    by phorbol esters, 318, 319
    by thrombin, 319
    by vasopressin, 319
cobalt (Co^{2+}), binding of, 100, 101
colony-forming unit, 312, 313, 317
colony-stimulating factor, 312
  granulocyte/macrophage, 312, 313
compartmentalization
  of calcium, 425
  of magnesium, 425
compound 48/80
  binding of calcium binding proteins, 413
  effect of
    on calcium ATPase, 407–12
    on magnesium ATPase, 413
    on phosphodiesterase, 407, 408
    on sodium–potassium ATPase, 413
  hydrophobicity of, 409
  specificity of, 409–13, 418, 419
concanavalin A, 182
conformation
  of head group, 105, 117–19
  of membranes, effect of calcium on, 105–19
conformational changes, 95
contractile proteins, effect of methylxanthines on, 454
contraction
  agonist-induced, antagonism of, 396–9
  depolarization-induced, antagonism of, 394, 395
contraction time, effect of thyroid hormone on, 361
contracture-producing drugs, 349
co–ordinating groups, basic, 96
copper (Cu^{2+}), binding of, 100, 101
corynanthine, 397
creatine kinase, 335–8
creatine phosphokinase, 349–53
  value, 352, 353
cross-bridge
  attachment, 213, 214
  detachment, 214
  model, 222
crosslinking, 328
cryptate, 102

cyclic AMP (cAMP), 318
  effect of
    on chromosome replication, 179–98
    in G_1 phase, 192–7
    on insulin secretion, 372, 373
    in pre-G_1 stages, 197, 198
    on protein kinase, 30, 31, 127–30, 248
  messenger system, 121, 128, 140, 248
  synthesis of, 336

D-600, 30, 250, 395
dantrolene, 79, 85–8, 349–53, 355–9
decamethonium, 352
decremental conduction, 291
  and circus movement, 291, 302–4
depolarization-induced contractions, antagonism of, 394, 395
depolarizing afterpotentials, 291, 298–302
determination of intracellular calcium, 1–11
dexamethasone, 317
diabetes
  effect of, on intestinal calcium absorption, 373
  and hyperparathyroidism, 375
  mellitus, 367, 373, 375
diacylglycerol, 142–5, 177, 191, 203, 204, 208, 209, 318
  kinase, 204
dibucaine, 408
dichlorophosphonazo III, 3, 4
digitonin, 232
1,4-dihydropyridines, 30, 31, 388
diltiazem, 464
  effect of,
    calcium channels, 291, 304–6, 388, 389, 399
    in muscular dystrophy, 335–8
    on plaque formation, 269
dimethylsulphoxide, effect of
  on cell differentiation, 315–17
diphenylalkylamines, 388
diphenylhydantoin
  antiarrhythmic actions of, 461–3
  effect of
    on calcium ATPase, 461–5
    on calcium transport, 461, 464
    on ouabain binding to sodium–potassium ATPase, 462–4
    on sodium influx, 462
    on sodium–potassium ATPase activity, 463, 464
diphosphonate derivatives, effect of
  on calcium channels, 253

  on plaque formation, 257–73
  on plaque reversal, 273–8
divalent metal ions, 95
  in muscular dystrophy, 335–44
DNA
  synthesis, 313, 320
  replication, 179, 184–98
docosahexaenoic acid, 170–3
docosatetraenoic acid, 170–3
doxorubicin, 36
Duchenne muscular dystrophy, 335–44

EF hand, 122
elastolysis, 255, 256
endocytic inside-out vesicles, 327, 330–2
  hypothesis of, 330–2
endocytosis, 331
endodigin, 442
endoplasmic reticulum (ER), 159
  calcium release from, 151, 158, 203, 207–9
  calcium sequestration by, 57–62
  calcium uptake by, 151
β-endorphin, 127
endothelium
  -derived relaxing factor, 400
  influence of
    on calcium entry blocker, 400, 401
    on contractions, 400, 401
energy metabolism, effect of calcium on, 342
enflurane, 352
epidermal growth factor, 183
erythropoiesis, 311, 317
erythropoietin, 311–17
  binding to progenitor cells, 313
  effect of, on cell differentiation, 314, 316
ethylenediamine, binding of metal ions by, 102
ethylenediaminetetraacetate, binding of metal ions by, 99
  effect of
    on calcium ATPase, 461–5
    on calcium transport, 461, 464
    on ouabain binding to sodium–potassium ATPase, 462–4
    on sodium influx, 462
    on sodium–potassium ATPase activity, 463, 464
excitation-contraction coupling, 356, 388
  role of calcium
    in heart muscle, 235

Index

in skeletal muscle, 213–23
in smooth muscle, 227–32

fever, drug-induced, 352
fibrosis, 338, 341
flexibility of proteins, 102
flunarizine, 395, 397, 399, 415–17
fluphenazine, 408, 412
fodrin, 129, 130
forskolin, 145
fusion
  effect of calcium on, 105, 111–17
  of membranes, 105, 111–17

G_1 phase
  effect of
    calcium in, 184–91
    calmodulin in, 184–91
    cAMP in, 192–7
    protein kinase in cAMP-dependent, 192–7
G_1 stages, pre-
  effect of
    calcium in, 184–91
    calmodulin in, 184–91
    cAMP in, 192–7
    protein kinase in cAMP-dependent, 192–7
Gardos channel, 328
Gardos effect, 329
gelsolin, 123
glucose
  effect of, on calcium efflux, 369–71
  tolerance, 367, 373–5
α-glycerophosphate dehydrogenase, 355, 358
glycogen synthase kinase, 129, 130
glycolysis, anaerobic, 350
granulocyte/macrophage colony
  stimulating factor, 312
  secretion of, 320
granulopoiesis, 311, 317
growth hormone, 336

haematopoiesis, 311–13
haemoglobin S polymerization, 327, 328
halothane, 84, 88–90, 349–53
head group
  conformation of, 105, 117–19
  of lipids, 106, 107
  of proteins, 105, 117–19
heart muscle contraction
  effect of
    calcium on, 235–45

calcium antagonists on, 388–90
calmodulin-dependent
  phosphorylation on, 385, 386
cardiac glycosides on, 386–8
cyclic AMP-dependent
  phosphorylation on, 385, 386
methylxanthines on, 447–58
regulation of, 385, 386
heat stroke, 352
histamine release, 416–18
hydrocortisone, 317
hypercalcaemia, 342
  effect of, on insulin secretion, 374
hyperparathyroidism, 336, 342
  and diabetes, 375, 376
hypertension
  environmental factor of, 284
  genesis of, 290
  genetic factor of, 284–6
  humoral factor of, 284
  and intracellular electrolytes, 283–90
  and renin level, 285
  role of calcium in, 283–90
  volume-expansion, 284
hypocalcaemia, 335, 338, 342
hypoparathyroidism
  effect of, on insulin secretion, 367, 374, 375
  idiopathic, 367, 374
  pseudo, 367, 374
hyposomatotropism, 336
hypothyroidism, 355

ion-selective electrodes, 1, 2, 10, 426, 428
inositol-1,4,5-triphosphate, 143–5, 151–62, 177, 203–9
  metabolism of, 158
  receptor, 208
inotropy, positive
  of cardiac glycosides, 387, 436–42
    and binding to sodium–potassium
      ATPase, 437, 438
    and extracellular calcium, 436, 437
    and sodium–calcium exchange, 438–40
    and stimulation frequency, 441
  of methylxanthines, 447, 448
    mechanisms of, 449, 457, 458
inside-out vesicles, endocytic, 327, 330–2
insulin biosynthesis
  effect of calcium on, 368, 369
  glucose-stimulated, 368, 369
insulin secretion, 145
  calcium requirement of, 367–75
  effect of

calcitonin on, 373, 374
calmodulin on, 372
cyclic AMP on, 372
hypercalcaemia on, 374
parathyroid disorders on, 367, 374, 375
protein phosphorylation on, 372
glucose-stimulated, 367–9
intestinal calcium absorption, effect of diabetes on, 373
iron (Fe^{2+}), binding of, 100, 101
islets of Langerhans, 367.
isocitrate dehydrogenase, NAD-linked, 71–4
isoflurane, 352

lanthanum (La^{3+}), effect of
  on calcium ATPase, 232
  on calcium channels, 228, 257
  on plaque formation, 257–73
  on plaque reversal, 273–8
lead (Pb^{2+}), binding of, 100, 101
length–tension relation, in skeletal muscle, 215–17, 221
leucotriene, 255, 256
leukaemia, 311, 312
  myeloblastic, 313
lidocaine, 295
linoleic acid, 170–3
linolenic acid, 170–3
lipid bilayer, destabilisation, 113–15
lipid dehydration, 113
lipid-lowering drugs, 255
lipid phase transition, 105–7
lipid-protein interaction, 117–19
lipoprotein
  low density, 255–7
  very low density, 257
lymphocyte
  B, 311
  T, 311
lymphopoiesis, 311

magnesium (Mg^{2+})
  binding of, 100–2
  compartmentalization of, 425
  content in muscle, 337–9, 341, 344
  effect of
    on calcium uptake capacity of SR, 425
    on mitochondrial calcium uptake, 425
    on tension development, 429–31
  intracellular free concentration
    determination of, 426–9
    in striated muscle, 425–31
magnesium ATP, 429, 430

magnesium ATPase, effect of calmodulin antagonists on, 413
magnesium electrodes, 426–9
  fabrication of, 426
  properties of, 426
magnesium equilibrium potential, 428, 429
magnesium homeostasis, cellular, 425
magnesium ligand (ETH 1117), 426
magnesium permeability, sarcolemma, 429
magnesium transport mechanism, 425
  energy-dependent, 429
magnesium uptake, sarcoplasmic reticulum, 425
malignant hyperthermia, 89, 90, 349–53
  awake episodes of, 352
  calcium transients in, 349, 350
  metabolism in, 349, 350
  muscle rigidity in, 349
  release of
    creatine phosphokinase in, 349, 351
    myoglobin in, 351
    potassium in, 351
  screening tests for, 352, 353
  symptomatic treatment of, 351, 352
malonic acid as metal chelator, 98, 99
manganese (Mn^{2+}), binding of, 100, 101
MAP-2, 129, 130
megakaryocytopoiesis, 311
melittin, 127
  effect of
    on calcium ATPase, 407, 408, 411, 412
    on phosphodiesterase, 407, 408
  specificity of, 411, 412
membrane
  effect of calcium on, 105–19
  potential, 243
  protein, 105, 117–19
mercury (Hg^{2+}), binding of, 100, 101
mesoridazine, 388
metabolic acidosis, 350
metabolic effect
  of caffeine-depolarization, 356–8
  of potassium-depolarization, 356–8
metabolism
  aerobic, 349, 350
  anaerobic, 349, 350
  effect of thyroid hormones on, 361–3
  in liver, 361–3
metal-chelating properties of organic compounds, 95–103
metal ions, binding of, by proteins, 95
  selectivity of, 95

Index 475

metallochromic dye
  as calcium indicator, 1–4, 10
1-methyl-3-isobutylxanthine, 447, 454
methylxanthines, 447–58
  blockade of adenosine receptors by, 454–6
  cardiotonic effect of, 447–58
  effect of
    on contractile proteins, 454
    on phosphodiesterase, 451–3, 457
    on slow calcium inward current, 451–3, 457
    on SR calcium uptake, 450, 451
  pharmacological properties of, 447
  smooth muscle relaxation by, 456
mitochondria
  proliferation of, 356
  regulation of calcium in, 67–76
    by calcium cycling, 71–4
murexide, 3, 4
muscle atrophy, 338
muscle, calcium accumulation in, 335–44
muscle calcium content, 335–41, 344
muscle contraction
  and β-blockers, 379
  and calcium antagonists, 388–90
  and cardiac glycosides, 386–8
  and drugs, 379–390
  regulation of
    heart, 385, 386
    skeletal, 382, 425–31
    smooth, 303, 384
muscular dystrophy, 335–44
  and magnesium, 337–9, 341, 344
muscle fibre, skinned, 429
muscle magnesium content, 337–9
muscle necrosis, 341
  pathogenesis of, 342
muscle relaxant, 350
  non-depolarizing, 352
muscle rigidity, 349
muscle work, control of, by thyroid hormones, 360, 361
myeloblastic leukaemia, 313
myelopoiesis, 311
myelopoietic stem cells, 311
myocardial damage, 344
myoglobin, release of, 351
myopathy, 349
myosin, 213, 214
myosin light chain, 123, 380–4
  phosphorylation, 143, 383, 384
myosin light chain kinase, 125, 129, 383, 384, 395

Na^+–Ca^{2+} carrier, see Sodium–calcium carrier
Na^+–Ca^{2+} exchange, see Sodium–calcium exchange
NAD kinase, 247–50
Na^+–H^+ antiporter, see Sodium–proton antiporter
Na^+–K^+ cotransport, see Sodium–potassium contransport
Na^+–K^+ pump, see Sodium–potassium pump
Na^+–Li^+ countertransport, see Sodium–lithium countertransport
Na^+–Na^+ exchange, see Sodium–sodium exchange
naphthalene sulphonamides, 184, 207, 418
narcotics, 352
natriuretic hormone, 285, 290
  and sodium transport, 285, 286
neuroleptic malignant sydrome, 352
nicardipine, 253
nickel (Ni^{2+}), 238, 239
  binding of, 100, 101
nifedipine, 235, 388, 395–9
  effect of
    on atherosclerosis, 269
    on calcium channels, 30, 236–9, 253
    on calmodulin, 415–17
  photoinactivation of, 236–9
nimodipine, 388, 399
  effect of
    on calcium channels, 30, 417
    on calmodulin, 415–17
nisoldipine, 395, 399
  photoinactivation of, 236–9
nitrendipine, 388, 399
  effect of, on calcium channels, 30
nitrogen as electron donor, 95
nitrogen balance, 336, 340
nitrous oxide, 352
non-depolarizing muscle relaxant, 352
norepinephrine, 289
nuclear magnetic resonance (NMR) of ^{19}F-labelled calcium chelators, 1, 8–10

obelin, 7, 8
octahedral symmetry, 102
oleic acid, 170–3
oncomodulin, 123
organic compounds, metal-chelating properties of, 95–103
osteomalacia, 342

ouabain, 288, 299, 435
  binding to sodium–potassium ATPase, 461–3
  dihydro-, 436
oxatomide, 415, 416
oxidative metabolism
  calcium-control of, 74–6
  and heart contractility, 74
α-oxoglutarate dehydrogenase, 71–4
oxygen as electron donor, 95

pacemaker, ectopic
  automaticity of, 291, 295–8
palmitoleic acid, 170–3
pancreatic
  acinar cells, 151–62
  islets, 367
    calcium ATPase in, 371, 372
parathyroid
  ablation, 335–8, 342
  disorders, effect of, on insulin secretion, 367, 374, 375
parathyroid hormone
  effect of
    on calcium influx, 335–8, 342–4
    on cAMP synthesis, 336
    on muscular dystrophy, 335–8, 342, 343
  secretion of, 343
parvalbumin, 103, 122, 123, 127, 380–6, 413
patch-clamp, 15, 16, 229, 299
penfluridol
  effect of
    on calcium ATPase, 408–12
    on phosphodiesterase, 408
  hydrophobicity of, 409
  specificity of, 409–12
penicillamine as metal chelator, 99
perhexiline, 399
phagocytosis, 341
phase separation, lateral, 105, 107–11
phase transition, of lipids, 105–7
phenothiazines, 406–8, 415
phenothiazine-sepharose, 127
phenoxybenzamine, 188
phentolamine, 188
phenylalkylamines, 188
phorbol ester, 139, 142–5, 203, 208, 320
  effect of, on cell differentiation, 316–18
phosphatidic acid, 203, 207, 209
phosphatidylinositol, 140–4, 176, 203
  hydrolysis of, 318
phosphatidylinositol-4-4-biphosphate, 143, 151, 158, 203–5
phosphatidylserine, 177
phosphodiesterase, calcium-sensitive, 204, 318
phosphodiesterase, cyclic nucleotide, 125, 127, 179, 192
  calmodulin-dependent, 128–31
  calmodulin-independent, 128, 130
  effect of
    calmodulin antagonists on, 406–19
    methylxanthines on, 451–3, 457
phosphofructokinase, 129, 358
phosphoinositide
  breakdown, 203–9
  in calcium mobilization, 203–9
  cycle, 204–6
phosphokinase, creatine, 349–53
phospholamban, 129, 130
phospholipase, activation of, 342
phospholipase A_2, 165–77, 179, 191, 192, 197
  effect of
    calcium, 175–7
    neutral lipids, 169–72
    unsaturated fatty acids, 172–7
  inhibition of, 167–74
  purification of, 167, 168, 174
phospholipase C, 176, 191, 204, 205, 318
phospholipid transfer protein, 204
phosphorylase
  A, 362
  kinase, 125–30, 355, 358, 362
  ratio, 353
photoinactivation, 235–45
photoprotein
  calcium-activated, 6–8, 10
  luminescence, 6–8
pimozide, 408
plants
  role of
    calcium in, 247–51
    calmodulin in, 247–51
plaque, fibrous
  prevention of formation of, 257–73
    biochemical findings on, 263–6
    effect of anticalcium drugs on, 257–73
    mechanism of, 270–3
    morphological findings on, 266–8
  reversal of, 273–8
    biochemical findings on, 273–7
    effect of anticalcium drugs on, 273–7
    mechanism of, 278
    morphological findings on, 277, 278
plasma membrane, alteration of, in

Index

muscular dystrophy, 343
platelet
 activation, 146
 aggregation, 416, 418
 -derived growth factor, 181, 192, 255
 phospholipase A_2, 165–77
 secretion, 143, 416, 418
pluripotent stem cell, 311, 312
polycythaemia vera, 311–13, 319–21
polyphosphoinositides, 203
 breakdown, 203–209
potassium (K^+)
 binding of, 100
 depolarization, metabolic effects of, 356–8
 release of, 351
prenylamine, 388, 399
progenitor cell, 311–13
prolactin secretion, 145
proliferation
 of mitochondria, 356
 of sarcoplasmic reticulum, 355
prostaglandin, 191, 192
 biosynthesis of, 165, 166, 176
protease, calcium-dependent, 342
protein kinase, 247
 calcium-phospholipid-dependent, 140, 187, 208
 calmodulin-dependent, 249, 250, 417
 cAMP-dependent, 30, 31, 127–30, 179, 181, 248
 in G_1 phase, 192–7
 in pre-G_1 stages, 197, 198
protein kinase C, 121, 177, 179, 188, 191, 315, 318 *see also* C-kinase
 blocker, 188
protein phosphorylation, 379, 382–90
 abnormal, 320
 calcium-dependent, 367
 effect of, on insulin secretion, 372, 373
pump-leak balance, 331
pyruvate dehydrogenase phosphatase, 71–6

QT-syndrome, long, 302
quadridentate chelator, 98
quercetin, 84, 85
quin-2, 1, 4–6, 10, 209
quinate: NAD^+ oxidoreductase, 247–50
quinquedentate chelator, 98

R24571, *see* Calmidazolium
rauwolscine, 397
relaxation time, effect of thyroid hormone on, 361
release
 of creatine phosphokinase, 351
 of myoglobin, 351
 of potassium, 351
respiratory chain, 68, 69, 74–6
retinoic acid, effect of, on cell differentiation, 316, 317
RNA synthesis, 313
ruthenium red, 79, 87, 88
ryanodine, 301

S-100, 123, 127, 413
salicylic acid as metal chelator, 99
sarcolemma, removal of, 429
sarcoplasmic reticulum (SR)
 calcium ATPase, 355
 calcium release from, 79–90, 349–53, 355
 proliferation of, 355
 storage of calcium in, 289
screening tests for malignant hyperthermia, 352, 353
secretin, 127
selectivity in binding of metal ions, 95
serotonin secretion, 143, 418
sexadentate chelator, 98, 99
sickle cell anaemia, state of calcium in, 327–32
sickled cells,
 dehydration of, 327–9
 irreversibly, 327
 membrane rigidity of, 327, 328
 potassium content of, 327–9
 sodium content of, 327–9
skeletal muscle contraction
 effect of
  calmodulin-dependent phosphorylation on, 379, 382
  cyclic AMP-dependent phosphorylation on, 379, 382
 regulation of, 382
skeletal muscle, activation by calcium, 213–23
 effect of
  fibre length, 214–17
  multiple cross-bridge states, 222, 223
  magnesium, 425–31
skinned muscle fibre, 429
sliding filament model, 214
slow inward current, 434
smooth muscle,
 excitation-contraction coupling in, 227–32

relaxation by methylxanthines, 456
role of calcium in, 227–32
vascular, 227, 253, 283–90
tone of, 289
smooth muscle contraction
effect of
calcium antagonists on, 388–90
calmodulin antagonists, 416, 417
calmodulin-dependent
phosphorylation on, 379, 383, 384
cyclic AMP-dependent
phosphorylation on, 379, 383, 384
regulation of, 383, 384
sodium (Na^+), binding of, 100
sodium–calcium carrier, 67–72
activation of
by β-adrenergic agonists, 75
by glucagon, 75
inhibition of
by calcium, 67, 74, 75
sodium–calcium exchange, 35–44, 47, 52, 53, 69, 162, 230, 231, 236, 240–4, 283–99, 315, 318, 434
effect of cardiac glycosides on, 438–40
partial reactions of, 35–9
in plasma membrane vesicles, 41–3
physiological role of, 43, 44
stoichiometry of, 40, 41, 288
sodium–lithium countertransport, 283, 284
sodium–potassium ATPase, 231, 232, 294, 329
binding of ouabain to, 462–4
effect of
calmodulin antagonists on, 411–13
cardiac glycosides on, 437, 438, 441, 442
diphenylhydantoin on, 463, 464
sodium–potassium cotransport, 283, 284
sodium–potassium pump, 47, 52, 283–90, 291, 294, 300, 318, 433, 437–42, see also sodium–potassium ATPase
endogenous inhibitor of, 284, 442
sodium–proton antiporter, 68, 69
sodium pump, see sodium–potassium pump
sodium retention, 285
sodium-sensitive electrode, 438
sodium–sodium exchange, 283, 284
sodium transport, 283–90
natriuretic hormone, effect of, 285, 286
somatostatin, 342
spheroechinocyte, 328

stimulus–secretion coupling, 160, 367
strontium (Sr^{2+}), binding of, 100, 101
strophanthidin, 434
succinylcholine, 349, 352
sulphur as electron donor, 95
sustained cellular response, 139–45

tamoxifen, 415–17
tau factor, 129, 130
TCBP-10, 123
tetracaine, 36, 79, 88
theophylline, 447
blockade of adenosine receptors by, 454–6
effect of
on contractile proteins, 454
on phosphodiesterase, 451–3, 457
on slow calcium inward current, 451–3, 457
on SR calcium uptake, 450, 451
positive inotropy of, 448, 449
smooth muscle relaxation by, 456
thioridazine, 388, 418
thrombin, 319
thyrocalcitonin, 253, 270
thyroid hormone
effect of
on calcium cycling, 355
on calcium homeostasis, 355, 356, 361–3
on calcium transport, 358–60
on liver, 361–3
on metabolism, 361–3
on muscle work, 360, 361
on proliferation, 355, 356
on thermogenesis, 355, 356
thyroid status
effect of
calcium-dependent processes, 356
thyroid thermogenesis, 355
tolbutamide, 145
TPA (phorbol ester), 182, 188–91
tranquillizers, 352
transglutaminase, 328
transition metal ions, 100
transmethylation, effect of, on DNA synthesis, 320
transverse tubule, 79–85, 217, 350, 351
tridentate chelator, 98
trifluoperazine
binding to calcium-binding proteins, 413
effect of
on calcium ATPase, 407–12

Index

on cell proliferation, 186
on phosphodiesterase, 407, 408
specificity of, 410–13
trimazosin, effect of
on phosphodiesterase, 253
on plaque formation, 257–73
troponin, 103, 350
binding of calcium to, 218, 379, 382
complex, 360
troponin C, 122–4, 127, 380–6
binding of calmodulin antagonists, 413, 418
tubulin, 180
tumour promoter, 188, 208

unidentate chelator, 98

vanadate, 156
vasoactive intestinal peptide (VIP), 127
vasopressin, 319
verapamil, 247, 250, 251, 289, 295, 388, 405, 464
effect of
on calcium channels, 253, 288, 388, 395, 399
on plaque formation, 269
veratrine, 291–3
vesicles, endocytic inside-out, 327, 330–2
voltage clamp, 236–8
villin, 123
vinblastine
binding to
calmodulin, 417
tubulin, 417
effect of
on calcium ATPase, 407, 408, 411, 412
on phosphodiesterase, 407, 408
vinca alkaloids, 415–17
vitamin D_3, 311
deficiency, 343
effect of, on cell differentiation, 316, 317

W-7
binding to calcium-binding proteins, 413
effect of
on calcium ATPase, 408, 411, 412
on cell proliferation, 186
on phosphodiesterase, 408
specificity of, 411, 412
W-9, 411
specificity of, 419

W-13, effect of
on cell proliferation, 184

Yohimbine, 397

Zeatin, 251
Zinc (Zn^{2+}), binding of, 100, 101